continued on back

Statistical Survey Techniques

Statistical Survey Techniques

RAYMOND J. JESSEN

Professor of Business Statistics
School of Management
and
Professor of Biostatistics
School of Public Health

University of California
Los Angeles

John Wiley & Sons,
New York · Chichester · Brisbane · Toronto

Library of Congress Cataloging in Publication Data:

Jessen, Raymond James, 1910–
    Statistical survey techniques.

    (Wiley series in probability and mathematical statistics) (A Wiley publication in applied statistics)
    Includes bibliographies.
    1. Sampling (Statistics) 2. Statistics. I. Title.

HA31.2.J48   519.5   77–21476
ISBN 0–471–44260–7

Printed in the United States of America

10 9 8 7 6 5 4 3 2 1

Preface

This book is based on my research and teaching in the field of statistical surveys over a period of years. The research began in 1938 when I, a graduate student then, was put onto a newly established research project, rather confidently entitled, "Sample Census Methodology," sponsored by the U.S. Department of Agriculture with the support and accommodation of the Statistical Laboratory of Iowa State University. A course of instruction on sampling techniques was established at the Laboratory a few years later by Professor William G. Cochran and I believe has been a part of its teaching curriculum to the present time.

The statistical survey, as a field of inquiry, has undergone a remarkable growth since then, thanks to the innovative and experimental leadership of P. C. Mahalanobis of India, Morris H. Hansen and William G. Hurwitz of the U.S. Bureau of the Census, Frank Yates of England, and many others. Moreover important advances in sampling and other survey methodology have taken place in such diverse fields as public opinion measurement, sociology, political science, economics, business, various governmental agencies, biology (e.g., wildlife and fisheries), engineering (e.g., traffic studies), urban planning and management, ecological studies, and many, many others. An attempt is made in this book to describe techniques that seem to be basic or could be basic to the great bulk of these fields and to provide a means for evaluating their properties. Since data quality is often as important to a survey as the validity and efficiency of its sampling plan, one chapter (13) is devoted to the more common problems found here.

Although I have used much of the material in the book for courses of instruction for many years, I have varied the topics selected, the emphasis, and the order of presentation, partly to accommodate the particular interests of the students present at the time and partly just to experiment. I have no general recommendations to make on the matter. The variety and abundance of material (for a one- or even a two-quarter course) gives the instructor ample opportunity to design the course that suits him best. The references at the end of each chapter provide a guide to the more pioneering works rather than the later and perhaps tidier papers. Some may help the more curious student to seek out the more advanced or exotic ideas and techniques.

Although some of the examples and exercises consist of synthetic data in order to help the student fix certain concepts and to get a feel for the computations in a simple setting, many of them are based on real data from actual studies. Perhaps the student may have no particular interest in rats, corn yields, political attitudes, and the like, yet real data should give him a feel for the distribution of variation in the real world and the remarkable degree of carry-over of natural patterns from one field of inquiry to another. I would have liked to have more in the nonagricultural fields but I failed to keep good notes when the opportunities were there.

My debt to colleagues for helping me become acquainted with statistical surveys is great indeed. I was very fortunate to be exposed to Professors Cochran's and Snedecor's statistical philosophy when I was still impressionable. For a number of years Messrs. Morris Hansen and the late William Hurwitz were very influential on my thinking and were excellent critics in the full and proper meaning of the word. However, despite these associations, whatever errors appear in this book are mine and mine alone.

RAYMOND J. JESSEN

Los Angeles, California
January 1978

Contents

CHAPTER 1

The Statistical Survey

1.1. Introduction

A simple example of a statistical survey and one with which most of us in the U.S. are familiar is the census of population. This survey, required by the Constitution, has been made every 10 years since 1790. Its original purpose was to determine the number and location of inhabitants in the United States in order to provide a basis for determining the number of Congressional representatives to be allocated for each state. During more recent times the census of population has been used for a number of other purposes and has

1

become an important source of information for research and administration in business, labor, government, and other fields of activity. The population censuses of today, in addition to enumerating or counting persons, obtain a variety of characteristics about each individual person· (such as age, education, marital status, and income), characteristics of the household, and other data that help illuminate the nature of the nation and its people.

Another statistical survey with which most of us are familiar is the "opinion poll". This type of investigation, undertaken to determine the opinions and attitudes of a "public" on certain issues of interest, usually involves the interrogation of a number of people with a set of prepared questions. It is usually carried out on only a small fraction of the total persons making up the public, the pooled opinion of which is being determined.

Surveys (we shall now drop "statistical", since this book will be concerned only with surveys for statistical or quantitative information) that are based on samples rather than on complete coverage may for convenience be termed *sample* surveys, as distinguished from *census* surveys. It will be seen later that the sample survey is far more common in practice. From the point of view of principles, the census survey may be regarded as a special case of the sample survey—that is, one based on a 100% sample.

Before we attempt to define a survey more precisely, it may be well to point out a number of other examples of surveys in order to indicate their varied nature and the variety of fields in which we find them. Listed below are a few examples.

Physical sciences and engineering. Geologic surveys to detect mineral deposits and other resources; watershed surveys to determine requirements for flood control; soil surveys to determine and map kinds of soils by type; establishment of stations to record weather data; quality control systems for manufactured products; cosmic and other ray-counting investigations; highway traffic flow surveys; acceptance sampling.

Biological sciences and agriculture, fisheries, forestry, medicine. Wildlife surveys to determine number of game and fish; plant disease and pest surveys; animal disease and pest surveys; ecological surveys; pollen density surveys; epidemiologic surveys; crop area and yield surveys for estimates and forecasts; surveys for farm animal numbers and production; oceanic surveys of plankton.

Social sciences and business, education, communications, public health. Survey of production and consumption of goods; marketing studies of consumer wants; public health surveys; political polls; measurement of public attitude and opinion; studies of marketing channels, costs; leadership, listenership and viewership surveys; dietary surveys; educational surveys; housing surveys.

1.2. Types of Investigations

Since statistical investigations differ considerably in nature as well as in complexity, it is usually helpful to classify them into groups. First, we may distinguish between two broad classes of investigations primarily on the role the investigator plays. If he observes what is going on without disturbing it, the investigation is generally called a *survey* (from the Latin work *supervidere*, to oversee). If, on the other hand, because of curiosity about "what does what and how much", he does something to the set of elements or to the system, then we are likely to say he is carrying out an *experiment* (from the Latin work *experiri*, to try). A second criterion for classifying is that of an investigation's purpose or objective. Is it to be primarily *descriptive*, or is it to be *analytical*, looking for possible relationships? The analytical type is concerned with possible associations between certain factors and usually with causative mechanisms that may "explain" what is observed. Third, we may ask whether we are observing a simple set of elements with fixed values, or a multifactor system in which there are one or more cause-and-effect mechanisms. Distinguishing on the number of factors involved in investigations essentially distinguishes their degrees of complexity.

The classification scheme just outlined is summarized in Table 1.1, where

Table 1.1. Classifications of investigations

Type of investigation	Investigator's role in application of "treatment"	Objective of investigation			
		Descriptive (to measure and describe)		Analytical (to detect and/or ascribe relationships)	
		Nature of phenomena		Nature of phenomena	
		One-factor	Multifactor	One-factor	Multifactor
Survey	Passive role a "natural process" "Uncontrolled" "Endogenous"	*Simple survey* (e.g., an opinion poll)	*Comparative survey* (e.g., U.S. Census of Population)		*Survey research* (e.g., factors affecting educational achievement)
Experiment	Active role a "planned process" "Controlled" "Exogenous"	*Simple experiment* (e.g., determining the speed of light)	*Comparative experiment* (e.g., agricultural field trials)		*Experimental research* "Factorial" experiment (e.g., genetic research)

alternative labels and expressions are shown. This book will be concerned primarily with the simple survey and somewhat less with the comparative survey. Occasionally comments will be made on related types of investigations. *Survey research*, a term commonly used in the social sciences, generally refers to a research undertaking in which many of the data will be supplied by simple and comparative surveys, but which may involve rather difficult problems of detecting and attributing cause-effect relationships. If such inquiries could be carried out as properly designed experiments, they would be much easier to deal with statistically and inferentially, but this is usually impossible. *Econometric research*, like survey research, is also highly dependent on data from surveys, usually a series of simple and comparative surveys taken over time.

1.3.   Survey Method

Usually a survey can be carried out in a number of different ways. If, for example, we wish to find the average height of students at a college, we can measure every student for height and calculate the required average. If carried out properly, this procedure should provide the data required for an exact answer. Alternatively, we could send each student a letter requesting him to obtain the measurement in an approved manner and return us the information. In each of these cases we obtain a measurement from each and every student. The presumption is that an exact answer is required and a complete census (that is, a 100% sample) is therefore necessary.

Let us suppose that an exact answer is not required and that a reasonably close estimate would be satisfactory for our purposes. With this new objective we have a number of possible ways of carrying out the survey. For example, we might station ourselves at the doorway of a selected classroom and, as the students pass through, pick out 10 students who we think are about representative of all students on our campus and measure their heights. Or, rather than use our judgment, we could get a list of *all* the names of the students in the college, write them on cards, put the cards in some kind of bowl, mix them up, draw out 10 at random, and make our measurements on the 10 students. Or we might write to the 10 students asking their height and then analyze whatever data were returned by those interested enough to comply with the request. We might even consider a scheme whereby, on a sample of students, we ask each to give his height as best he can remember it or guess it.

It is quite reasonable to suspect that the "estimates" resulting from these schemes might differ considerably, and one is tempted to ask, "Which scheme gives the most *accurate* result and how *accurate* is it?" If all were equally accurate, we might look for a further criterion for discriminating among them, such as, "Which scheme is least costly and how much does it cost?" These questions suggest the two main classes of problems in the design of surveys: (i) the determination of the accuracy of alternative schemes of investigation and (ii) the determination of the costs of each.

Types of Problems in Investigations Based on Surveys. The kinds of problems that arise when a survey is planned and executed are not the same from survey to survey. However, a number of problems occur frequently enough to be regarded as typical of surveys.

GROUP 1. Determining the objectives of the proposed investigation and putting them in specific terms:
1. The universe to be dealt with.
2. Concepts, characteristics, etc., to be counted or measured.
3. Averages (or totals) to be determined or structure of process to be examined.

GROUP 2. General considerations of survey design:
1. Choice of the unit (e.g., family or person; farm or area, individual or group) on which determinations (measurements, observations) are to be made.
2. Whether determinations will be confined to a sample or a complete census should be taken.
3. How the determinations are to be made (interview, mail, machine, human observation, etc.).
4. Funds and other resources available.
5. Estimation of costs of each operation.
6. Estimation of variances and biases.
7. Degree of accuracy required; amount and kind of risk of error that can be taken.

GROUP 3. Sampling design (if sampling is to be done):
1. Stratification (if so, what kind).
2. Choice of sampling unit (size and nature).
3. Number and type of stages.
4. Number and type of phases.
5. Number of sampling units to be taken at each stage and at each phase.
6. Cost of each operation.
7. Method of selecting sampling units (probability or judgment?).
8. Method of dealing with "hard-to-get" determinations through sampling.
9. Method of making estimates from the sample data.

GROUP 4. Making the determinations or observations:
1. Direct versus indirect methods.
2. Construction of the recording form (the questionnaire, the schedule, the ballot, the diary, recording tape, etc.).

3. Method of selecting, training, and controlling the investigator (the interviewer, the observer, etc.).
4. Method other than sampling in dealing with "hard-to-get" determinations.

GROUP 5. Processing the data:
1. Method of treating missing data.
2. Coding.
3. Tabulation.

GROUP 6. Estimates and their reliability:
1. Choice of estimator.
2. Exact and approximate methods of estimating confidence limits.

GROUP 7. Examination for quality (how "good" are the results?):
1. Checks: internal and external.
2. Reasonableness.

GROUP 8. Presentation of the results:
1. How much detail.
2. Frankness.

Survey Method and Scientific Method. Regardless of the nature of an investigation and the particular procedure chosen for carrying it out, a body of general principles and logic should always be borne in mind by every investigator. This body of principles and logic is the *scientific method*, the general nature of which is well stated by Karl Pearson (1900, pp. 12-13):

Now this is the peculiarity of scientific method, that when once it has become a habit of mind, that mind converts *all* facts whatsoever into science. The field of science is unlimited; its material is endless, every group of natural phenomena, every phase of social life, every stage of past or present development is material for science. *The unity of all science consists alone in its method, not in its material.* The man who classifies facts of any kind whatever, who sees their mutual relation and describes their sequences, is applying the scientific method and is a man of science. The facts may belong to the past history of mankind, to the social statistics of our great cities, to the atmosphere of the most distant stars, to the digestive organs of a worm, or to the life of a scarcely visible bacillus. It is not the facts themselves which form science, but the method in which they are dealt with. The material of science is coextensive with the whole physical universe, not only that universe as it now exists, but with its past history and the past history of all life therein. When every fact, every present or past life therein, has been examined, classified, and co-ordinated with the rest, then the mission of science will be completed. What is this but saying that the task of science can never end till man ceases to be, till history is no longer made, and development itself ceases?

Edgar G. Miller, Jr., of the Department of Biochemistry, Columbia University, also is worth quoting on an aspect of this subject (1949):

In a discussion of "Science and Security" (*Science*, June 25, 1948, p. 659) E. U. Condon states that "there is still a great reluctance . . . to accept and extend" the use of the scientific method "in the fields of sociology, economics, and politics," and he hopes for "the full acceptance and use of the scientific method in the social sciences."

In this there is the implication that the methods of these sciences are not scientific now—a view that has appeared repeatedly in notes and comments on the needs of social research, and one that is not particularly helpful to those needs. R. O. Bender, in a recent communication (*Science*, December 10, 1948, p. 665), points out some of the discouraging aspects, although he apparently does not altogether accept as final the assumed definition of scientific method. The definition on which such comments are based, as shown in Dr. Condon's discussion, is an equating of science with controlled experimentation. This description of science and scientific method is certainly inadequate.

Astronomy has done very well without subjecting the behavior of stars and planets to experimental controls. Geology has not used this method to any notable extent. Most of Darwin's contribution to biology did not come from such an experimental basis, either in its conception or verification. A problem need not be inaccessible to science if controlled experiments are impossible . . . .

It would be more useful if the physical and natural scientists, instead of preaching a particular procedural definition of scientific method and commending this to their social science colleagues, would recognize that the division of science into the accepted disciplines is merely a convenience arising from the varieties of methods and interests most useful in investigating different aspects of experience and expectation. *Science is not any particular method or set of techniques. It is a way of reasoning. The standards are intellectual rather than procedural. The method of observation, formalization, and testing must vary with the nature of the problem.* [Italics added.]

There is no unique kind of difficulty facing the social sciences that is not met, at least to some degree, in the natural sciences. All sciences must, in some phase and field, and to some extent, try to cope with such restrictive troubles as a high degree of indeterminacy, a highly heterogeneous universe for sampling, the impossibility of controlling variables, the difficulty of translating a factor or concept into an operationally definable identification or measurement, the need for including the inaccessible past history of a datum in an observation, the effect of observer on observed, and a host of other puzzles which challenge the scientific ingenuity and skill of the investigator. Most of these loom much larger in the social sciences, which have in general a more difficult task than the natural sciences. The need for studying and developing the nonexperimental approaches of scientific method is great, since the applicability of controlled experimentation is so limited.

The survey method is not a substitute for the scientific method—it is one aspect of it and therefore must meet its overall requirements.

Quantitative Inference, Sampling, and Statistics.   Nearly every scientific investigation is confined to a sample of some sort. Those who determine a crop's response to fertilizer treatment, for example, conduct their investigation on a small number of plots—comprising perhaps not more than an acre of land. Inferences may then be made for an area covering hundreds of thousands of acres. The accuracy of these inferences or generalizations can be evaluated exactly, provided certain procedures are followed. This subject is an important branch of statistics known as *statistical inference.*

Another type of inference problem arising in many scientific investigations is concerned with the interpretation of the relationship between certain factors and an observed result. For example, it is observed that a certain field of corn yielded 75 bushels per acre, and we have reason to believe that a number of factors "affected" this yield—the amount of rainfall during the season, other climatic conditions, the kind of seed, the kind and amount of tillage, fertilizers, and perhaps other factors. By proper experiment, those individual "effects" can sometimes be conveniently determined; otherwise, they might be *inferred* from numerous observations. Here we have observed values only on the "net" effects, which we believe are made up of individual effects ascribable to individual factors or perhaps to certain combinations of them. We cannot observe the individual effects—we can only observe the aggregate and the amount of each factor present. We must somehow try to infer values for the individual effects. For this problem of inference, statistics may be of great help.

Since the survey type of investigation is so frequently confronted with these problems of inductive inference (for descriptive surveys, those arising from sampling; for analytical surveys, those arising from both sampling and structural interpretation), we shall give considerable weight to these problems in considering appropriate designs.

1.4. Planning a Survey

Before undertaking an investigation, whether a survey or experiment, it is necessary, or certainly highly advisable, for the investigator to do some thinking about the kinds of things he would like to know. He may be interested in simple measurement facts, such as the total number of persons in the United States hospitalized in 1960, or the fraction of adult persons in New Jersey possessing a certain attitude about some issue. On the other hand, he may not be concerned with such facts of measurement but may wish to obtain *ideas* (such as the techniques farmers use in growing a certain crop; what housewives do with flour sacks; suggestions for improving a TV show). In either case he should determine what his ultimate objective is and be prepared to state it clearly as his research goal.

The Research Goal.   This is the goal beyond the immediate aims of a study—

one that we would like to reach but do not necessarily hope to with the investigation at hand. Examples of such goals might be: "What do people think of the government?" "How can market information help improve the marketing of a product?" "What kind of a refrigerator do consumers want?" "What effect does amount of income have on a family's consumption of milk?" "How will agricultural production be affected by certain acreage or production control schemes?" "What is happening to the California sardine?" "How effective is radio advertising?"

It may be noticed that these examples of "research goals" look very much like practical problems that a research man may undertake as a *study* or a *project*. They are *broad* enough to have some general application or interest when achieved, yet *narrow* enough to have meaning and a reasonable chance for some progress within, say, a few years' time. They are like goals in a game in the sense they seem to be achievable, yet may require some effort. The speed and cost of reaching the goal depends on a number of things, including the skill of the players and the obstacles met (many of which are unforeseen).

The research goal may arise quite naturally, or it may be the creation of the research director or even the research worker. But, whether or not the individual research worker decided on the goal, he may well find it worth thinking about, especially from time to time as he prepares the details of his plans.

The research goal determines our strategy. Our next step deals with the tactics.

The Working Objectives.   Now we decide how our big problem can be broken into a number of smaller ones, the solutions to which will help us solve the big problem. A good research tactician readily recognizes these smaller problems and the importance of their solution to the solution of the overall problem. He can also recognize those that may be quite interesting but irrelevant to his research goal and therefore unacceptable (for *this* investigation, anyway).

Consider the goal implied by the question: "How effective is radio advertising?" Suppose a seller of a commodity wants to know whether his expenditure of funds for radio advertising is "worth its cost"—whether more should be spent or less. In this case some possible *working objectives* might be:

1. How many radios are tuned to the *program*?
2. How many persons listen?
3. What are their characteristics? Are they potential buyers of the product?).
4. What do they think of the program?
5. Can they identify the advertiser?

Notice that an answer to any one of these objectives will not in itself completely answer our goal question: "How effective is radio advertising?"

Each deals with some aspect that might be of some help. These objectives, however, seem much simpler than our goal, and it appears that we have a more reasonable chance of "getting somewhere" with them.

The General Method of Investigation. At this step we consider the way to accomplish the working objectives. Continuing with our radio advertising problem, assume that the program in which we are interested is presented over six different broadcasting stations once each week during the year. Such problems as these might be considered:

1. Looking at working objective 1, which deals with the number of radios tuned to the program, we could raise several questions of procedure or method. For example, if we like rather complete information before we feel satisfied, we may want to take a complete census of all households within reach of the radio signals of the six stations and find out how many of them have radios, and how many of those radios were tuned to each of the 52 programs. This, however, would be rather expensive.

2. Several methods of taking the census may be considered. For example, we might have investigators call on each house personally and obtain the required information by interview. Or a list of names of all householders in the region could be obtained and required information from these sought through a mailed appeal. Perhaps telephones could be used.

3. But all this might be rather costly, if carried out carefully and completely—perhaps too costly, considering the possible value of the results. We may also feel that complete accuracy is not vital—that approximations to the facts may be adequate. In this case, we may consider confining our investigations to only a portion of all the households, selecting that portion in such a way that it will "represent" the remaining ones. We may also consider a sample of weeks from the 52 and perhaps even a sample of the six stations.

At this point we may even raise a new question, such as: "Would not a comparison of the product's sales before and after the radio advertising be a good way to assess the 'effect' of advertising?" Such a scheme was overlooked in our partial list of working objectives. As a matter of fact, if we are convinced that the results from this scheme satisfy our research goal, we should completely revise our list of working objectives. Revisions of this sort may arise when we consider our methods more fully, and they are to be expected. Rather than trying to confine our considerations of method to the working objectives exclusively, we should permit completely new ideas or schemes to be brought forth for examination.

How the Measurements, or Counts, Are to Be Made. Suppose we are still pursuing the working objectives (1) to determine the number of radios tuned to the program and (2) to determine the number of persons listening to the program. We now face the question: "What is to be the nature of the basic data?" Sometimes the answer is obvious, but not always. For example, it would seem that a radio set would be a suitable unit on which to obtain such information as:

1. Is it, or is it not, within reach of our six stations?
2. Is it, or is it not, tuned to our program during its broadcast?
3. If the radio is tuned to the program, how many persons are listening?

In this case we regard the radio set as our *unit of observation* or element and all the radio sets in the region covered by the six broadcasting stations as our *universe*.

On the other hand, we might consider the person as a unit on which to obtain our data. For example, we find out directly for each person in our region whether or not he listens to our program. In this case our *unit of observation* is a person and all persons in this region our *universe*.

A decision on which unit, the radio or the person, is to be the subject of our measurements cannot always be made until we consider the problems, both logical and practical, that each may involve. One of these is the *manner* in which the measurements are to be made.

Several ways of obtaining the measurement data are immediately apparent. An investigator may be employed to interrogate a person (by personal interview, telephone, or mail) and record the answers he obtains. Or he may observe a radio, or a person, over a period of time and note the relevant facts required (this may be hard to do, but we are not trying to be entirely realistic at the moment). There is a possibility of having a machine attached to the radio that tape-records such data as when the radio is turned on and to what station (in terms of the frequency of the broadcast signal). It cannot determine whether anyone is listening, however. (A gadget of this sort actually is used by a commercial organization for radio and television audience measurement.)

These considerations illustrate that alternatives exist as to what sort of unit may be used and what methods may be employed to obtain the required measurements or counts. Decisions are required on these matters to specify the plan of the investigation.

Statistical Design; Sampling Plan.   Suppose we have arrived at tentative decision for each of the foregoing steps, and we have decided that the cost of obtaining measurements on all radios or all persons in our radio listening region is prohibitive and that a sampling scheme of some sort will be suitable. We now ask: how many units should we measure, and in what sequence or arrangement? Should we make a large number of measurements on one of the 52 weeks and none on the remaining 51, or should we make a small number on each of the 52 weeks? Suppose we would like to compare the listening characteristics of the persons living in cities with those living in the suburbs and rural areas, and suppose that four out of five radio listeners live in cities. Should we take equal numbers from both groups, or should we take one-fifth in the rural and suburban areas and four-fifths in cities, or should we use some other proportion? Should we pick certain listeners for our sample because we "think" they are "representative", or should we use some randomization scheme? How important is it, when we get through, to have some measure of the reliability of our estimates?

These and similar considerations involve the *statistical design* of the study. They are concerned with making our investigation efficient and logical for the objectives we have set forth. This area of investigational research procedure is the primary subject of the present book.

Plan of Tabulation and Analysis. An aspect of investigation that is easily overlooked in planning is that of tabulation and analysis of the data. It is at this stage that research workers frequently find serious deficiencies in their planning—embarrassing deficiencies that very likely could have been avoided if some thought had been given at the planning stage.

A useful way to do this planning is to draw up blank tables for the arguments of the study, or at least for the most important ones. These should be complete in every detail, and the exact meanings of the concepts involved should be examined closely, particularly if comparisons are to be made with other sources of information. Attention should also be given to the kind of statistical treatment, if any, that is to be given the data. If analysis of variance is to be used, write out the summary for the analysis of variance and examine it for adequacy. Are the degrees of freedom sufficient? Similar examination should be made for the number of observations expected for each class or subclass in the classification tables. If weaknesses develop, consider modification elsewhere that may correct or avoid them.

Possible Results to Possible Conclusions. Finally, no planning can be complete without a consideration of the possible results. For each possible result, determine all the possible conclusions that could follow. This is usually a very interesting part of the planning operation. Investigations may actually be discarded at this point, because it cannot be seen how any possible result could lead to a conclusion of sufficient value to justify the undertaking. Much effort, time, and money (not to mention later disappointment) has been saved by careful thinking at this point. This examination may also reveal serious flaws in the plan that call for drastic revision.

The steps outlined here need not be taken in sequence. Nor should the reader think that a plan can always be set forth clearly and completely in one smooth sweep. Most often it takes form gradually with frequent modification, sometimes mild, sometimes drastic. But for those interested in sound conclusions, it is hardly necessary to overemphasize the need for sound planning. The results of an investigation can never be any better than the *design* of that investigation.

1.5.   The Sample—Preliminary Concepts

Suppose that the universe with which we plan to deal has been clearly defined. Sometimes at this point it is clear that the investigation must be carried out by obtaining an observation on each and every element of that universe. An accountant, for example, obtains data on the transactions of a business firm to determine its financial standing. In this case transactions (sales and purchases) can be regarded as the elements of some universe specified for some period of time. Suppose the objective is to determine the exact numerical difference between the aggregate dollar value of those transactions that are sales and the aggregate dollar value of those that are purchases. If this difference is to be determined to the exact penny, then every transaction made by the firm during the period under study must be examined. In this case the financial survey requires a 100% sample (or a census) of the universe of transactions.

In many investigations such complete accuracy is not required; some inaccuracy is tolerable, *particularly* if the amount of inaccuracy, or some measure of it, is known. In other cases, even though a complete coverage would be desirable, for one reason or another it may be impossible or infeasible. In these cases it will be necessary to get along somehow with only partial coverage of the elements in the universe. An example would be an investigation of ammunition to determine the fraction of all shells that do not fire when placed in a gun. To obtain the required observation on a shell it must be fired. Only by destroying each element in the universe can we obtain full knowledge on the proportion of its elements that are, or were, duds.

If we confine our observations to anything fewer than all the elements of the universe in which we are interested, we shall say we are using a *sample*. Complete coverage (a census) may be regarded as a sample of 100% of the elements. It is, therefore, convenient to regard sampling as the general case, and a census as a special case. If less than a 100% sampling is used, the incompleteness of the resulting information about the universe of inquiry makes it necessary to speculate on what the results would have been if complete coverage had been obtained.

Before we pursue this matter further, let us define concepts.

Some Preliminary Definitions.   The following definitions are to be regarded as first approximations. Definitions of a more technical nature will be developed later.

1. *Universe of inquiry (or investigation).*   That set of elements, the

characteristics of which we wish to study or investigate, will be called the universe of inquiry, or more briefly the universe. If only a sample of elements can be examined, the universe is the set of elements about which we wish to generalize from the sample.

For example, the persons whose names appear in a telephone directory might be a universe of inquiry. We might wish to know the ages of all such persons, how many are women, how many have names with one, two, three, four, or more syllables, and so on. Note that the person is the object under observation. The characteristics of persons considered are: (i) age, (ii) sex, (iii) name possessing one, two, three, four, or more syllables.

2. *Population.* This word will usually be used to describe a set of characteristics of a universe. Thus, in the above example the set of ages for the persons whose names appear in the telephone directory constitutes a *population of ages.* Likewise, the set of data on the sex of each person constitutes a population of sex membership.

3. *Observation unit or element.* This is the unit on which the observations are obtained. In the case of the universe of persons whose names appear in a telephone directory, the individual person is the observation unit. Ordinarily the observation unit is an obvious element of the universe, but not always. (See Chapters 4 and 12.)

4. *Sample.* In the broad sense, a sample is any fraction of all the elements in a universe. Ordinarily it is a fraction taken in a manner such that it will "represent" the universe.

5. *Judgment selection.* This is a method of selecting the elements of a sample by directly judging which elements are appropriate and which are not. Criteria for appropriateness may vary according to the judge, so long as the selection of the sample does not follow some chance system (such as random).

6. *Probability selection.* This is a procedure whereby the ultimate selection of an individual element of a sample is left to probability. For example, in order to achieve a random selection, we must insure that each element has an equal and independent probability to be selected.

7. *Estimate.* An estimate, in the broad sense, is a number or value constructed by some rule from the sample results that stands for some true population number or value, which is usually unknown.

8. *Estimator.* A calculating scheme or formula for obtaining an estimate from a set of observations is referred to as the estimator.

In order to help describe the concepts to follow, it is sometimes helpful to look upon a given sample as one of many possible such samples that *could have been selected* in the same manner. This is not difficult to see where a randomized selection is taken, because it was *pure chance* that this particular set of elements was chosen for the sample. Sampling of this sort can be regarded as a repetitive process, like the tossing of a coin to determine the fraction of times a head turns up. It is possible to regard the judgment type of selection as being a repetitive process, in which the finally selected elements comprising the sample may vary from trial to trial with the whim of the selector. However, even if the constitution of the judgment sample cannot be regarded as varying from trial to trial, this does not invalidate the definitions to follow—in fact it may make them clearer.

9. *Expected value.* The expected value of an estimator is the *average* value of all possible estimates from repeated trials. It is not necessarily a value that can actually occur. (For example, the expected value of the number of dots on the face turned "up" in the toss of a true die is 3.5, because this is the average that would be obtained from repeated tosses; but on any single toss it is impossible to get 3.5 dots.)

10. *Bias.* If, on repeated trials, the average value of estimates differs from the quantity being estimated, the estimator is regarded as biased and the difference is the magnitude of the bias. In other words, an estimator is regarded as biased if its *expected value* differs from the true value.

11. *Reliability (or precision).* In considering reliability we shall be referring to a measure of the closeness of each observation to its own average over repeated trials.

12. *Accuracy (or validity).* Accuracy refers to *a measure* of closeness to the targeted value. For example, an estimator may have high reliability but low accuracy.

13. *Size of sample.* The number of sampling units contained in a sample will be its size. In the case of randomized samples with no restrictions, it is the number of individually randomized units contained in the sample. If the randomized units are *groups* of observation units, the size of the sample is the number of groups not the number of individual observation units.

Precision (Reliability) versus Accuracy; Bias. Suppose we are evaluating the shooting performance of four guns, A, B, C, and D. Each gun is fired at a target a number of times, and the record of hits, shown in Fig. 1.1, is examined. Note that gun D is right on target, whereas the others are off but differ somewhat in the nature of their "offness". Gun C has a pattern similar to D except that the hits are clustered to the lower right corner of the target. This hit pattern is described as precise because of the compactness of the hits, but biased because they center on a point off target. Hit patterns of this sort may be corrected by adjusting the gun's sights or modifying the aiming procedure. Gun B's hits being widely scattered, it is regarded as having low precision. Since the pattern gives no evidence that the gun is shooting in any direction other than the bull's-eye, this indicates that it is unbiased. Adjusting the sights will not help here. Gun A has the low-precision pattern of B but has

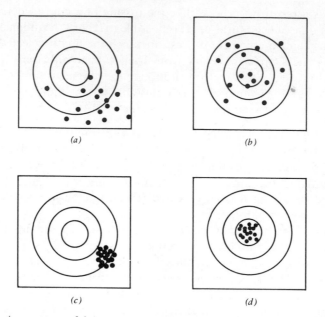

Fig. 1.1.   Various patterns of shots at a target. (*a*) Low precision, high bias, low accuracy. (*b*) Low precision, low bias, low accuracy. (*c*) High precision, high bias, low accuracy. (*d*) High precision, low bias, high accuracy.

a tendency to favor the lower right corner of the target. It has not only low precision but also a bias. If the bias in A were removed, such as by proper sight adjustment, it would behave like B.

Although gun C is precise, its precision may not be very useful, since its shots always fail to hit the target. The imprecise guns A and B hit the bull's-eye at least on occasion. Hence either precision or reliability alone may be a useless property of samples, since each may give us consistently wrong results.

Bias may exist in statistical estimates also, and it is sometimes more difficult to detect and perhaps more difficult to remove than in the case of our guns. Sometimes we find that high-precision but biased statistical estimates are more accurate than those of low precision but unbiased, and therefore the former may be preferred. Therefore, bias in itself is not to be regarded as sufficient reason to reject a statistical estimating scheme.

1.6.   Basic Types of Samples

Two factors seem to enter into the consideration of all sample selection schemes: (i) the presence or absence of a chance mechanism and (ii) the presence or absence of objectivity on the part of the sampler. An objective role is simply one that is clear and unambiguous; hence it is one that, if followed by every sampler, would produce the same sample (or one with the same properties). A subjective role is one that permits the sampler to exercise his judgment or feelings as to what is a "good" sample. Considering the two factors, both at the two levels each, we can distinguish four types of samples, as shown in the accompanying table.

Role played by sampler	Selection procedure	
	Probability	Nonprobability
Objective	Random or probability samples	Purposive samples
Subjective	Quasi-random samples	Judgment samples

Under probability-based samples we have two types: those that are selected on objective rules for random selection and those that are not. The former are the genuine random samples, the latter only quasi-random. In the latter the probabilities are imputed to have been present because the person considering the sample feels it is reasonable to do so. For example, the instructor may regard the persons in his classroom as a quasi-random sample of all students in his school. His class is not a random sample (unless students in the school actually were assigned to his class by lot), but he may feel that circumstances are such that he would like to regard it as a random one and so characterize its properties. The soundness of this decision depends purely on circumstances, hence generalizations are difficult.

In the case of nonprobability-based samples there are also two types: purposive and judgment, depending on whether the rules of selection were objective or not.

All four types of samples are used in practice. Statisticians generally prefer random or probability samples, because a well-established theory is available to aid them in understanding the behavior of such samples. The properties of the others are less well known, but some examples are given below on judgment and purpose samples. Random sampling will receive a great deal of attention in Chapters 2 and 3. A more general type of sample, called a probability sample, will be presented in Chapter 8.

Judgment Samples. The direct use of judgment is probably the most common method of deliberately selecting samples. A simple example of the method and its weakness is given by Yates (1960) in selecting samples of wheat plants (shoots) to determine their average height in small fields (plots). The investigator selected two shoots from a quarter-meter row of shoots. It was found that these shoots averaged $+3.3$ cm too high when sampled on May 31 and -2.7 cm too low when sampled on June 28. The difference in the direction of the bias was attributed to the difficulty of observing the shoot heights; when plants are half-grown, only top leaves are apparent, but when they are full-grown, all shoots of all lengths can be seen. However, in both cases the investigators tended to choose samples that were not free of bias. In this case, if reliance were placed on the two judgment samples to estimate growth rate, a 10% underestimate would have occurred.

An Experiment in Selecting Samples of Stones.   In order to see more clearly the characteristics of judgment versus random methods of selection, an experiment was planned and carried out on a universe that could be fully viewed and simply dealt with. This universe was a collection of 126 stones of various sizes and shapes displayed on a table. Each member of a class was instructed to look at the display and then pick out a sample that, in his judgment, accurately represented the average weight of the whole lot. In each case the selector was requested to pick three replicates of each sample size, where the specified sample sizes were 1, 2, 5, 10, and 20 (only two replicates for size 20 were required). A replicate in this case is merely a different selection of the same number of stones. The stones were numbered, so that after a selector decided upon a certain set of stones for his sample he merely recorded the identification numbers. Subsequently, the weights of the indicated samples were determined and the results summarized. The averages (in grams) for the 16 selectors were as follows:

Size of sample:	1	2	5	10	20
Mean of 16 selectors:	103.1	105.1	91.6	99.1	95.6

Since the true mean was 98.4 grams, no great average bias is apparent, but biases on the part of the individual selectors are apparent.

The comparison of the results of the 16 selectors with those obtained when the stones are drawn at random (by the use of random numbers) is interesting. (The results for the random samples are based on *126* of size 1, *30* of size 2, *90* of size 5, *60* of size 10, and *10* of size 20.) As a measure of accuracy the deviations of each sample of a given size from the true (i.e., population) value were averaged (ignoring sign). The results are shown below, in grams:

Size of sample:	1	2	5	10	20
Mean deviation of judgment samples:	40.0	44.9	35.3	38.5	31.0
Mean deviation of random samples:	80.6	71.4	41.3	34.1	26.2

These mean deviations were then plotted for each sample size (see Fig. 1.2). It can be seen that the judgment sample was more accurate on the average for the

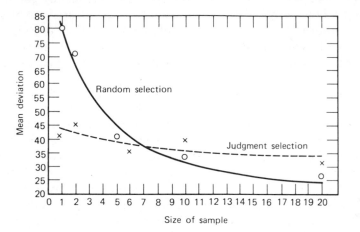

Fig. 1.2.   Plot of mean deviations of sample means from universe mean. Comparison of random and judgment methods of selecting samples of different sizes. Universe of 126 stones.

smaller sample sizes, but less accurate for the larger sizes. The critical size of sample in this case seemed to be about 8, where both methods provided about the same accuracy.

Other examples have appeared in the literature, and there may be many that have not. The evidence seems to be that in most, if not, all cases of judgment selection a bias will occur, and in many cases the direction of the bias can be anticipated. The bias can be small or large, depending on the selector and the material he is working with. The reliability of such samples seems to be higher than that of random samples of comparable size. Accuracy, however, the combined sampling error and bias, is difficult to predict and may be treacherous—particularly in the hands of the innocent.

Judgment versus Random Selection.   Although the results shown in the foregoing experiment constitute a very limited case, they do seem to agree quite well with general observation and other experiences. Judgment selection has its strongest case where (i) the sample is small, (ii) the universe is fairly small and visible or known to the selector, (iii) the elements in the universe vary considerably in the character under investigation, and (iv) the selector has great and proven skill in this art.

Perhaps one of the strongest points for the chance method is one that has not been mentioned yet. By randomizing the selection, we can obtain a measure of a sample's accuracy from the sample itself (for example the standard error of the estimate). Ordinarily this measure of accuracy is not available from samples obtained by judgment selection, and therefore the value of such samples is very much diminished.

Purposive Samples.   Samples of this type are selected by following an objective procedure but not using probability. A great variety of procedures can be envisioned, and some have been proposed and tried. Two classes have become prominent. The first utilizes correlation, the second the principle of persistence, as selection criteria. Each will be considered separately.

Correlation Criterion.   Intuitively, if a sample can be picked such that it has the same mean as the population on some known characteristic, x, then it should have a mean on the unknown but desired-to-be-known characteristic, y, either exactly or very nearly equal to the true population mean. This being the general idea, procedures have been developed to exploit it. A famous example of this type was presented by Gini and Galvani (1929) in proposing a sample of the Italian 1921 census of population for archival storage. The method was somewhat popular among statisticians, at least in the 1920s and 1930s, but lost much of its appeal after a criticism by Neyman (1934), who argued the advantages of an alternative procedure, called stratified-random selection.

Persistence Criterion.   If a particular community over a series of elections in a state always votes with the winner, then chances are good, if it is the only one or one of a few that has done so, that it will continue to vote with winners, so it becomes known as a "bellwether" or "typical" community and is sought out for special attention for "in-depth" studies by opinion gatherers and others. It becomes a favorite sample choice because it has a unique statistical track record for "representativeness" of the state. Hence it is expected that it will continue to behave similarly in the future.

Strand and Jessen (1943) studied a number of applications of this idea to the problem of finding suitable small samples of large areas to represent counties. (In this case townships are about one-sixteenth the size of counties.) They tried a number of methods, mostly based on the performance of individual townships over a period of nine years (1927–1935) in representing their county with regard to eight characteristics, such as acres in farm, acres in oats, corn production, and number of sows bred for spring farrow. They concluded that "purposive selection does not provide samples of greater accuracy than stratified random selection."

General Remarks on the Four Types.   Although the findings so far indicate that purposive samples are somewhat disappointing in their performance against randomized selections, the studies are not very exhaustive— particularly in dealing with very small samples. In the case of judgment sampling, there is evidence that in very small samples it can be superior to simple random sampling. Perhaps the same is true for purposive samples under certain conditions. More study of this matter would be welcome.

The "quasi-random" samples cannot be discussed and compared with the other three types in any simple and satisfactory manner. Each quasi-random case appears to be unique. The only generalization we can make at this point seems to be that such samples should be used very carefully.

◇   ◇   ◇

1.7. Brief History of the Statistical Survey

The theory of sampling is probably one of the oldest branches of statistical theory. However, only during the past 75 years has there been much progress in applying that theory to, and developing new theory for, statistical surveys. Kiaer (1895) of Norway and Bowley (1906, 1913) of England were the leading early users and advocates of sampling for obtaining social and economic data.

Renewed interest in sampling in the 1920s gave rise to discussions of methodology by Bowley (1926) and Jensen (1926a, b), who were concerned mainly with the problem of how samples should be selected—that is, whether purposive (see Section 1.6) or random selection was to be preferred. The idea of stratification was developed and the advantages of stratified-random over simple random sampling became known.

During the 1930s a great demand for social and economic data dereloped in the United States and Europe. Sample censuses were taken in Poland, Bulgaria, and Sweden. Sampling methodology was being developed in India to estimate crop production, in the U.S. Bureau of the Census to deal with social data, and in the U.S. Bureau of Agricultural Economics and Iowa State University to estimate farm characteristics and inventories and production of crops and livestock. From these centers came the area as a sampling unit and its wide application to farms, households, and field crops. At the same time American pollsters and social scientists were experimenting with sampling methods and question-asking techniques, and the method known as *quota sampling* was developed and popularized. During the 1930s more sophisticated methods of estimation and the theory of subsampling and double sampling were devised by workers associated with the Rothamsted Agricultural Experimental Station in England.

In the 1940s came the development of unequal probability sampling theory in the U.S. Bureau of the Census, and so-called area-probability sampling methods became widely used and quite sophisticated in both theory and practice. Interest in nonsampling errors and the means to isolate and measure them became quite strong. The 1945 census of agriculture in the United States was collected in part on a sample, and a well-designed evaluation of the quality of that census was made—also on a sample. The 1950 Census of population made extensive use of built-in samples to increase its accuracy and reduce costs.

In the 1950s the statistical survey became an important tool of several agencies of the United Nations for encouraging countries to improve their fact-gathering quantity and quality. Methodology continued to be developed and applications extended to newer areas (e.g., fisheries, wild populations, soil resources). In the 1960s and 1970s the trends continued—new problems arising and new solutions being tried. Theory and practice continue to become more useful, extensive, and sophisticated.

1.8. Summary

A statistical survey is an investigation of some aggregate or universe of physical things, elements, events, or the like to determine its quantitative characteristics, such as the fraction of elements possessing some quality, the average magnitude of some measurement, or the frequency of some event. These characteristics, or their parameters, are of some interest to us—usually as the basis for understanding the value or power of that aggregate so that decisions about it can be made more intelligently. Sometimes we may survey a process or a complex of interrelations in order to better understand some underlying mechanism or the nature of interrelationships. The statistical survey can be an orderly, reliable, and efficient way to go about making such investigations.

A statistical survey undertaken to measure difficult concepts in a large universe can call for considerable planning if it is to measure what it is intended to measure and to do it efficiently. The general strategy employed, commonly called *design*, involves not only statistical considerations but also a somewhat more than casual knowledge of the general setting of the problem, costs of various operations, their probable effectiveness, and the usual problems of monitoring quality and being ready with corrective measures. And this is to be done with constraints on budget, time, and reliability of personnel.

Four general classes or types of samples have been described. They differ in regard to the presence or absence of probability and the presence or absence of objectivity in their selection. Only one type, random, has a theory developed to describe its properties. Two others, purposive and judgment, have to be tested empirically; because so many factors can influence outcome, it will take some time to come to useful simple conclusions. However, for very small samples, where random samples may perform erratically, the purposive and judgment methods have strong intuitive appeal. Little can be said about quasi-random samples.

In the chapters to follow, simple random samples will not be relied upon very extensively, but random samples with one or more restrictions placed on them will be. A solution to the sampling problem has only begun.

1.9. Review Illustrations

1.   Classify each of the statistical investigations listed below, as to whether it is essentially (i) a survey or experiment, (ii) descriptive or analytical.

(*a*)   Salescorp is a large organization with dealers scattered over the entire United States. In order to strengthen its dealer organization it undertook a study to determine the relationship between sales and the size of the territory (in inhabitants, say) served by each dealer. From this information it is expected the optimum size of dealer territory can be determined.

(*b*)   A bag of seeds is inspected for the fraction of impurities it contains. A sample is drawn out and examined. It is concluded the bag contains 10% impurities by weight.

(*c*)   A bag of seeds is inspected for the fraction of seeds not viable. A sample is drawn out and put into a chamber where proper moisture, heat, and so on are provided for germination. After suitable time, the fraction of nonviable seeds is determined.

(*d*)   Same as (*c*) except that the purpose is now to study the effects of different temperature and moisture levels on viability.

(*e*)   The public's view on a certain proposal, not yet announced, is investigated by means of interviewing a representative sample of adults to obtain their views.

(*f*)   Same as (*e*) except that a number of variations of the proposal are put to the respondents to study their reactions.

(*g*)   A study of two methods of carrying out farm advisory work by the Iowa Agricultural Extension Service consisted (in essence) of the random selection of 32 of the state's (say) 1500 communities. In 16 of these, chosen at random, method A was used and in the remaining 16 method B. Farmers in these communities were interviewed for relevant characteristics just before the programs were put into effect and again at the end of five years.

Solutions

(*a*)   (i) Survey—because the data were obtained from the universe of interest as it existed, with salesman and assigned territories presumably determined for reasons other than this study. This could be regarded as an "uncontrolled experiment", where territory size is the "treatment" of interest and sales the response. (ii) Analytical—because the relationship between sizes of territories and sales is being sought.

(*b*)   (i) Survey—because the impurity content is to be determined for the bag "as is". No treatment is involved. (ii) Descriptive—because no relationship of any sort is being sought.

(c) (i) Survey—because the purpose is to determine the fraction of nonviable seeds where conditions for viability are held at a standard. (ii) Descriptive—because no relationship (such as response to different treatments) is being sought.

(d) (i) Experiment—because the response (germination or not) to a given treatment (temperature, moisture, etc.) is being determined. (ii) Analytical—because relationships are now being sought.

(e) (i) Survey—because we do not generally regard the interview as a "treatment", sometimes to our regret. (ii) Descriptive—if a single treatment (e.g., one line of questioning) is given.

(f) (i) Survey—for the same reason as (e)(i). (ii) Descriptive—presumably no relationships are under study. This case differs from (e)(ii) only in that a number of different inquiries are being made on the same investigational units (persons). Could be regarded as a comparative survey.

(g) (i) Experiment—because the treatment is method of advisory work and the experimental unit is the community. (ii) Descriptive—because only a comparison of effectiveness is sought.

It will be noted that surveys and experiments can often be difficult to distinguish, as can their descriptive or analytical nature. However, arbitrary as they may be in some cases, these distinctions still serve a useful purpose in communicating.

2. The following data on estimated and actual yields of sugar beets in the State of Colorado (in tons per acre) are taken from the *1950 Proceedings of the American Society of Sugar Beet Technologists*, p. 626.

Year	Manager's judgment estimate	Estimate from random sample	Actual harvest
1941	14.8	14.4	14.8
1942	13.9	12.3	12.0
1943	13.9	12.9	12.8
1944	12.6	12.4	12.6
1945	13.3	12.2	12.2
1946	12.6	13.0	13.0
1947	13.7	14.5	15.3
1948	13.1	13.6	13.7
1949	14.7	16.0	16.5
Total	122.6	121.3	122.9

Regard the production of sugar beets in the state as under the surveillance and management of representatives of the sugar processing plants. Each year, just prior to harvest, each manager is asked to supply an estimate of what he judges the acreage under his jurisdiction will produce. The average of these estimates for each year appears in column 1. The estimate from random sampling is based on digging and weighing beets in small plots selected at random in the beet growing area. Actual harvest figures are obtained at the close of the season from the records of all processing plants.

(*a*) Describe the surveys implied here in regard to (i) working objective, (ii) general method of investigation, (iii) nature of measurements obtained, (iv) whether limited to a sample or not, (v) type of sample (if used).

(*b*) Compare the two methods on accuracy of performance; examine for possible explanation of performance differences.

Solution

(*a*) (i) In both cases the working objective is the determination of the tonnage of sugar beets in the fields just prior to harvest; (ii) one involves a survey of managers reporting for their jurisdictions, the other a survey of plots; (iii) one obtains a person's presumed expert judgment on the tonnage of beets in an area, the other carries out an actual harvest of plots; (iv) one is a census (or 100% sample), the other a sample, presumably a small fraction of the total acreage; (v) a random sample.

(*b*) The errors (differences between estimates and actual harvest values) are:

Year:	1	2	3	4	5	6	7	8	9
Judge:	0	1.9	1.1	0	1.1	$-.4$	-1.6	$-.6$	-1.8
Random:	$-.4$	.3	.1	$-.2$	0	0	$-.8$	$-.1$	$-.5$

and the mean square errors are $12.35/8 = 1.54$ and $1.20/8 = .15$ for judgment and random, respectively. Hence the random is about 10 times ($= 1.54/.15$) as accurate as the judgment estimates. Both schemes seem to provide estimates without overall bias. This is surprising for the judgment estimates! (but see later).

However, if we look into the matter further and regard the yield each year as a moving target, we see that yields plotted on time appear to have no pronounced trend or cyclic pattern over the nine years. Look at the accompanying plot of the data. The random estimates appear to follow the actual yields fairly closely, but the judgment estimates seem to go along somewhat independently.

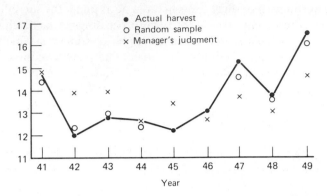

A plot of the judgment estimates against the actual yields shows a strong tendency for the managers to overestimate low yields and underestimate high yields, which is not so apparent in the random estimates. Hence judgment estimates here have a bias dependent on yield and just happen to average out over this nine-year period.

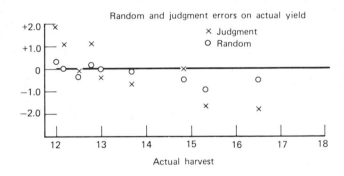

In conclusion, the survey of field plots to estimate new harvest yield is far more accurate than that based on managers' judgments—even though the latter is a complete census. The judgments are inaccurate and appear to contain little useful information on yield changes from year to year.

3.   The following is an excerpt from an article appearing in *Harper's*, May 1973, dealing with a poll of political opinion on likely issues in the 1976 elections.

Seeking answers, *Harper's* engaged pollster Oliver Quayle to survey the voters of Illinois, as he did for us last year with fascinating results. A paradise for psephologists, Illinois is the nation's leading "swing state", mainly because its socioeconomic makeup is a microcosm of the nation's. Illinois has not backed a Presidential loser since 1916; in the past seven Presidential elections, its average deviation from the national popular vote has been less than 1 percent.

The Quayle surveyors assembled a panel of 455 Illinois voters of both parties, who represent the state's makeup in key urban, suburban, and rural areas. In mid-February, 263 of the panelists (a precise cross-section of the state's 1972 Presidential voters) agreed to lengthy telephone interviews in which they answered thirty questions about their racial attitudes and Kennedy's chances in 1976.

(a) Considering only the problem of selecting one or more states to represent the 50, what type of sample can this be regarded as?

(b) Critically evaluate the suitability of this sample (of states) for the purposes stated and/or implied. Ignore the matter of selecting the sample of voters within the state.

Solution

(a) The selection procedure appears to be a purposive one, where the voting records of the 50 states were examined and Illinois was found to have the "best statistical record" of all. Presumably some judgment was also used—that is, some states may have been omitted from consideration because of possible credibility problems, imagined or real. For example, a state such as Alaska or Hawaii might be excluded even if it had a better record, simply because readers would doubt its representativeness because of its geographic uniqueness.

(b) If the pollster is confined to a sample of one, the procedure is probably a good one. However, claims of accuracy should be kept very, very moderate. It is usually better to spread observations over more than one state if the United States is to be "accurately" represented.

1.10. References

Bowley, A. L.
1906

Address to the Economic Science and Statistics Section of the British Association for the Advancement of Science. *Journal of the Royal Statistical Society*, **69:** 540–558. (An early appeal for statistical sampling.)

Bowley, A. L.
1913

Working Class Households in Reading. *Journal of the Royal Statistical Society*, **76:** 672–701 (June 1913). (A pioneer household survey using primitive statistical sampling.)

Bowley, A. L.
1926

Measurement of the Precision Attained in Sampling. *Bulletin of the International Statistical Institute*, **22:** 1–62 (l'ere Livre). (Early paper on statistical sampling.)

Cochran, W. G.
Watson, D. J.
1936

An Experiment on the Observers' Bias in the Selection of Shoot-Heights. *Empire Journal of Experimental Agriculture*, **4:** 69–76. (For examples of selection bias in judgment samples.)

Cohen, M. R.
Nagel, E.
1934

An Introduction to Logic and Scientific Method. (467 pp.) New York: Harcourt, Brace and Co. (Especially Chapter 20, pp. 391–403, an early presentation of what may be regarded as the probabilistic nature of science.)

Deming, W. E.
1950
Some Theory of Sampling. New York: John Wiley & Sons, Inc. (Chapter 1, pp. 1–21. A good discussion of problems arising in planning a survey.)

Gallup, G. H.
Rae, S. F.
1940
The Pulse of Democracy; the Public-Opinion Poll and How It Works. New York: Simon & Schuster. (Statement of procedures used in the early Gallup Poll.)

Gini, C.
Galvani, L.
1929
Di Una Applicazione del Metodo Rappresentativo All "ultimo Censimento Italiano della Popolanizianione", (1 ˜diciembre 1921). *Annali de Statistica,* Series VI, 4. [A presentation of a purposive sampling procedure that led to Neyman's (1934) critical discussion of it.]

Jensen, A.
1926a
Report on the Representative Method in Statistics. *Bulletin de l'Institute International de Statistique,* **22:** 356–376. (Early discussion of stratified sampling.)

Jensen, A.
1926b
Report on the Representative Method in Practice. *Bulletin de l'Institute International de Statistique,* **22:** 377–436. (Early discussion of stratified sampling.)

Jessen, R. J.
Kempthorne, O.
Deming, W. E.
1949
Observations on the 1946 Elections in Greece. *American Sociological Review,* **14:** 11–16. (An example of planning a survey for determining the fairness of an election.)

Kameda, T.
1931
Application of the Method of Sampling to the First Population Census. *Bulletin of the Institute of International Statistics,* **25:** 121–132. (An early and interesting example of how reliance was placed on a sample because of an unusual event and how subsequent findings proved its value.)

Kiaer, A. N.
1895
Observations et Experiences Concernant les Dénombrements Répresentatifs. *Bulletin de l'Institute International de Statistique,* **9:** 176–183. (One of the earliest reports on use of sampling in social surveys.)

Marks, E. S.
Mauldin, W. P.
Nisselson, H.
1953
The Post-enumeration Survey of the 1950 Census: A Case History in Survey Design. *Journal of the American Statistical Association,* **48:** 220–243. (A good discussion of problems arising in planning a survey, and their resolution for a survey. A good case study.)

Miller, E. G., Jr.
1949
Scientific Method and Social Problems. *Science,* **109:** 290–291 (March).

Neyman, J.
1934
On the Two Different Aspects of the Representative Method: The Method of Stratified Sampling and the Method of Purposive Selection. *Journal of the Royal Statistical Society,* **97:** 558–606. (A classic paper on statistical sampling.)

Pearson, K.
1900
The Grammar of Science. (2nd ed., 548 pp.) London: Adam and Charles Black. (Chapter 1. An old but still excellent discussion of the nature of science.)

Snedecor, G. W.
1939
Design of Sampling Experiments in the Social Sciences. *Journal of Farm Economics,* **21:** 846–855. (Excellent historical statement on statistical sampling for surveys.)

Snedecor, G. W.
1950
The Statistical Part of the Scientific Method. *Annals of the New York Academy of Science,* **52:** 792–799. (A clear statement on the nature of statistics.)

Stephan, F. F.
1948
History of the Uses of Modern Sampling Procedures. *Journal of the American Statistical Association,* **43:** 12–39. (Excellent historical statement on statistical sampling for surveys.)

Stephan, F. F.
McCarthy, P. J.
1958
Sampling Opinions: An Analysis of Survey Procedure. New York: John Wiley & Sons, Inc. (Chapter 1. A good statement on the problems of planning a survey. See Chapters 14, 15, and 16. Also, a good detailed description and discussion of various sampling methods used by practitioners. Chapters 2–8.)

Strand, N. W.
Jessen, R. J.
1943
Some Investigations on the Suitability of the Township as a Unit for Sampling Iowa Agriculture. *Iowa Agricultural Experiment Station Bulletin.* **315**: 615–650. (A study of some purposive samples in agriculture.)

Yates, F.
1949
1960
Sampling Methods for Censuses and Surveys. London: Charles Griffin & Co. (An excellent discussion of practical problems arising in the planning of a survey.)

1.11. Mathematical Notes

Accuracy is sometimes measured by *mean square error*, which is defined as:

$$\text{MSE } (y) = E[y - \mu]^2,$$

which then can be decomposed into the two components, variance and bias, as follows:

$$\begin{aligned}
\text{MSE } (y) &= E[y - E(y) + E(y) - \mu]^2 \\
&= E[y - E(y)]^2 + [E(y) - \mu]^2 \\
&= \text{Var } (y) + (\text{bias})^2,
\end{aligned}$$

where y = an observation of (or shot directly at) μ (the bull's-eye),

E = the "mathematical expectation" or average of (see Chapters 2 and 3 for further explanation).

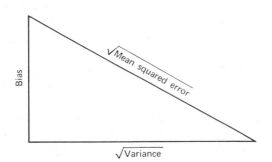

Represented diagrammatically, mean square error or inverse accuracy is the hypotenuse of a right triangle with legs comprising variance and bias squared.

1.12. Exercises

1. Below are listed brief descriptions of various investigations, some actual, some artificial. Examine each and decide whether it is (i) an *experiment* or a

survey and (ii) *descriptive* or *analytical*, and explain why you have classified it as such.

(*a*) The producer of El Weedo Cigarettes wanted to know what type of package would maximize its sales. Preliminary study of various package designs reduced the number of candidates to two, A and B. A number of "trial" stores were selected in which El Weedos were offered for sale in both package types. After a period of time, type B was selected because its sales were greater.

(*b*) A manufacturer is planning a product that fits onto dogs. In order to minimize the number of different sizes required, he must know certain dimensions of the nation's dog population. The data are obtained through the help of a sample of veterinarians.

(*c*) A number of years ago a considerable sum of money was spent to determine more precisely the speed of light. In the investigation, the light under measurement was flashed through a mile-long tube at near vacuum.

(*d*) In order to determine the shooting quality of a shipment of cartridges, a random sample of them are fired from a test stand.

(*e*) Galton, the "discoverer" of regression, found by examining data on a number of fathers and their sons that sons of tall fathers tended to be shorter than their fathers and that sons of short fathers tended to be taller.

2. A study of about 100 farms in Greece revealed that grapes under irrigation produced very little more fruit than those not being irrigated. There was some lack of unanimity on the conclusion that irrigating grapes in this region was a waste of time, water, and money. The farms were a random sample of all farms; some farmers irrigated their grapes and others didn't. The region generally has limited rainfall. What do you conclude?

3. A critic, perhaps a high-level official, who is very familiar with the universe being sampled has an opportunity to examine a random sample of 100 countries in the United States. The critic states that several of the counties selected contain very little of the item of interest, or that for one reason or another many of the counties selected simply are not representative. He feels that money spent on a survey in these counties would be wasted. How would you respond?

4. Following are excerpts from a news article that appeared in the Los Angeles *Times*, November 3, 1968.

MODOC COUNTY, CALIFORNIA
HAS 2,280 DEMOCRATS,

1,541 REPUBLICANS,
2 PROHIBITIONISTS,
79 NON-PARTISANS,
AND AN UNCANNY KNACK
FOR VOTING WITH THE
STATEWISE WINNERS IN
THE LAST 20 GENERAL ELECTIONS.
MODOC COUNTY IS NOW READY TO TRY FOR 21

Since 1928 Modoc has voted with the state in every election for President, Governor, and Senator—and no other of the 58 counties has done this.

This curious and suggestive statistic was discovered by Eugene C. Lee, of the Institute of Governmental Studies at Berkely, who is not inclined to take too ponderous a view of what he describes as a "mathematical accident" and has no ready explanation for why Modoc, with its 8,000 ranchers, farmers, lumbermen and millworkers, should be such a reliable indicator of the way California votes. Yet, accident or not, the fact exists, and its existence was what lured me up to Modoc, where I spent several days recently testing the political climate with an eye on Election Day.

(*a*)   How do you regard Modoc County for basing a study of political opinion in California? Explain why you take the position you do.

(*b*)   Can you help Professor Lee explain why Modoc has performed as it has? If so, elaborate.

5.   An article in *Time* magazine, June 2, 1958, p. 70, described a study on abortion by the so-called Kinsey group of Indiana University and another by a practicing Baltimore doctor, Dr. Timanus. The Kinsey study was based on a sample of 5293 women (let us assume a random one from the United States) reporting 1044 abortions. Dr. Timanus based his study on 5210 abortions he performed on patients sent to him by 353 physicians during 31 years who presumably would not perform the illegal operation themselves. In quoting Timanus' results, *Time* states: "Since Dr. Timanus performed almost five times as many abortions as turned up in the Kinsey study, his analysis may have wider application." Critically evaluate this statement in view of facts stated and implied.

CHAPTER 2

Elements of Random Sampling

2.1. Introduction

One of the most important arguments for random selection over judgment or purposive selection is that random samples behave in a manner that can be easily and logically expressed, whereas judgment or purposive samples generally do not. Random samples follow certain predictable behavior patterns that can be simply expressed mathematically. A random sample is the simplest type of objective sample and is the basis for many more complex types, some of which will be discussed in this book. We shall now define and examine some elementary concepts in order to see the principles involved.

2.2. The Finite Universe and Its Population

Suppose we have a set of four stones and would like to know their average weight. It is convenient to distinguish the following concepts: (1) the set of physical elements as the *universe*, (2) the set of measured values on those elements (in this case, weights) as the *population* of interest, and (3) the population mean as a *parameter* of the population. A given universe may have a number of populations of interest. In the case of the set of stones, we may also be interested in the mean volume of the stones, or the fraction of the stones that are black. Hence the set of volumes and the set of designated blackness are populations of the same universe of stones, and the mean and proportion are their parameters of interest, respectively.

A distinction is being made here between universe and population. The word unverse will be used to denote the *physical* set of elements of objects to be dealt with, whereas population pertains to a set of measurements or observations taken on some characteristic of those objects. Hence for a single universe we may be interested in examining one or more populations, and·for each population we may be interested in one or more of its parameters. In the case of the stones we have designated three populations of interest; two are *quantitative* (based on measurements) and one is *qualitative*.

Let us designate these concepts mathematically. Let the number of elements in the universe by N, and let U_i be any element i. Then

$$U_i: \ i = 1, 2, 3, \ldots, N$$

is the universe of elements and

$$\left. \begin{array}{c} X_i: \\ Y_i: \\ Z_i: \end{array} \right\} \ i = 1, 2, 3, \ldots, N$$

are the populations of volumes, weights, and blackness, respectively. For parameters, let us designate

(2.1)
$$Y = \sum_{i=1}^{N} Y_i, \quad \text{the total of the } y\text{-population,}$$

(2.2)
$$\bar{Y} = Y/N, \quad \text{the mean,}$$

(2.3)
$$S^2 = \sum_{i=1}^{N} (Y_i - \bar{Y})^2/(N-1), \quad \text{the variance}$$

of that population, and likewise for the x- and z-populations. In the case of a population of zeros and ones, such as the z-population, where $Z_i = 1$ if the quality is possessed by element i (e.g., stone i is black) and 0 otherwise, we may designate $\bar{Z}$ by P, the proportion of elements that are black.

Example 2.1. Suppose we have the following values for the volume, weight, and blackness populations for the four-stone universe. Calculate the parameters of possible interest. We have the observations:

(element)	i:	1	2	3	4
(volumes)	X_i:	6	8	11	15
(weights)	Y_i:	2	5	7	14
(blackness)	Z_i:	0	1	1	1

$$Y = \sum_{i=1}^{N} Y_i = 28, \qquad X = \sum_{i=1}^{N} X_i = 40, \qquad Z = \sum_{i=1}^{N} Z_i = 3,$$

$$\bar{Y} = 28/4 = 7.0, \qquad \bar{X} = 40/4 = 10, \qquad \bar{Z} = 3/4 = 0.75.$$

$$S_y^2 = \frac{\sum_{i=1}^{N}(Y_i - \bar{Y})^2}{N-1} = \frac{(-5)^2 + (-2)^2 + (0)^2 + (7)^2}{4-1} = \frac{78}{3} = 26,$$

$$S_x^2 = \frac{\sum_{i=1}^{N}(X_i - \bar{X})^2}{N-1} = \frac{(-4)^2 + (-2)^2 + (1)^2 + (5)^2}{4-1} = \frac{46}{3} = 15\tfrac{1}{3},$$

$$S_z^2 = \frac{\sum_{i=1}^{N}(Z_i - \bar{Z})^2}{N-1} = \frac{(-3/4)^2 + (1/4)^2 + (1/4)^2 + (1/4)^2}{4-1} = \frac{12/16}{3} = \frac{1}{4}.$$

A more convenient form of Eq. 2.3 for computation is given by a little algebraic manipulation, giving the variance of a population,

(2.4)
$$S^2 = \frac{\sum X_i^2 - (\sum X_i)^2/N}{N-1}.$$

2.3. Possible Samples of a Finite Universe

Suppose we regard a *sample* as some *fraction* of a total. If a universe consists of N elements, any subset of these N elements constitutes a sample of that universe. Since there may be a great many different subsets of elements into which a universe may be divided, there may therefore be a great many samples available to choose from. Let us consider the matter of counting and constructing possible samples from a finite universe.

Example 2.2. We have a universe of four stones. How many different samples of size two are there, and what elements do they include? The possible samples are:

$$A,B; \quad A,C; \quad A,D; \quad B,C; \quad B,D; \quad C,D;$$

where A,B denotes a sample consisting of elements A and B, and so on. A total of six samples are possible.

When samples are allowed to contain only unduplicated elements, regardless of arrangements, they are called *nonreplacement* samples. If they are allowed to include duplicated elements, they are called *replacement* samples. The terms are naturally applicable when samples are drawn physically, such as balls from an urn in the classical examples. In our example of the four stones, the total number of possible replacement samples would be $16 (= 4^2)$, since the arrangement or sequence of the draws distinguishes different replacement samples. The possible replacement samples include the 12 possible nonreplacement samples plus the four in which elements are duplicated: A,A; B,B; C,C; and D,D.

In the case of nonreplacement samples, for any size sample, n, taken from a universe of size N, the total number of *possible* samples, T, is given by the number of combinations of N things taken n at a time, or

$$T = \frac{N(N-1)\,(N-2)\cdots(N-n+1)}{n!}$$

(2.5)
$$= \frac{N!}{(N-n)!n!},$$

which is a well-known formula from the theory of combinations.

Hence we have two ways to define possible samples, either replacement or nonreplacement. Having chosen one of these definitions, we still have the problem of choosing a *particular* sample from the set of T possible samples. Since both types of samples have certain advantages in certain circumstances, the properties of both will be examined.

This matter will be pursued in Section 2.5. Section 2.4 will be a brief presentation of a mathematical method of averaging, known as mathematical expectation, a technique to be applied in evaluating random samples.

2.4. Mathematical Expectation

In drawing inferences about the population from simple random samples, we shall be generally concerned with the mean and variance of the sample itself—that is, with $\bar{y}$ and s^2. Since we are interested in how accurately our sample can reflect the characteristics of the population, we may like to understand the manner in which $\bar{y}$ and s^2 of the sample are related to $\bar{Y}$ and S^2 of the population. To do this rather easily, a technique of averaging called *mathematical expectation* is useful.

Suppose a random variable x_i must assume the values $x_1, x_2, \ldots, x_N$, and the probabilities associated with those values are $P\{x_1\}, P\{x_2\}, \ldots, P\{x_N\}$, where $\sum_{i=1}^{N} P\{x_i\} = 1$. *Then the expected value* of this random variable is defined to be the quantity

$$(2.6) \qquad E[x] = \sum_{i=1}^{N} x_i P\{x_i\}.$$

Suppose we select an element from a universe of N elements at random. What is the expected value of x? In this case our random variable $x_i = X_i$, $i = 1$, $2, \ldots, N$, and the probability of observing an X_i is $P\{U_i\} = 1/N$,

$$E[x] = \sum_{i=1}^{N} X_i P\{U_i\} = \sum_{i=1}^{N} X_i \frac{1}{N} = \bar{X},$$

from Eq. 2.2.

Suppose instead of x_i we consider some function of x_i, say $g(x_i)$, instead of x_i when x_i is observed. For example, $g(x)$ could be the square of x_i, say $g(x) = x^2$. The expected value of $g(x)$ in general is given by

$$(2.7) \qquad E[g(x)] = \sum_{i=1}^{N} g(x_i) P\{x_i\}.$$

Referring again to the case of randomly selecting a single element from a universe of N, what is the expected value of x^2? Now $g(x) = x^2$ and $x_i = X_i$; then, using Eq. 2.6, we obtain

$$E[x^2] = \sum_{i=1}^{N} X_i^2 \frac{1}{N},$$

which, using Eq. 2.4, becomes

$$(2.8) \qquad E[x^2] = \left(\frac{N-1}{N}\right) S^2 + \bar{X}^2.$$

From the definition, Eq. 2.7, it follows that the expected-value operator E possesses the following properties in regard to adding a constant and multiplying by a constant, respectively:

$$(2.9) \qquad E[g(x)+c] = E[g(x)]+c,$$

$$(2.10) \qquad E[cg(x)] = cE[g(x)].$$

With a little more background in probability, it can be shown that

$$(2.11) \qquad E[g(x)+h(y)] = E[g(x)]+E[h(y)];$$

hence in general the expected value of the sum (or the difference) of two random variables is the sum (or the difference) of the expected values of each random variable.

2.5.   RR and RNR Sampling; Expectation of $\bar{y}$ and s^2

Section 2.3 presented the concept of *possible* samples for each of the two basic types of samples, replacement and nonreplacement. The matter of how to choose among the possible samples of either type was not considered. The simplest statistical method of choosing is to give all T samples an equal chance of being selected. Let us first consider samples of size $n = 1$, since in this case both types are identical.

In this case we have $T = N$ possible samples, so there is a one-to-one correspondence between samples and elements. To obtain the expected value of the sample mean, $\bar{y}$, we can use Eq. 2.6, where P_i, the probability that element i appears in a sample, is $1/N$; thus

$$E[\bar{y}] = \sum_{i=1}^{N} Y_i P_i \{Y_i\}$$

(2.12)
$$= \sum_{i=1}^{N} Y_i \frac{1}{N} = \bar{Y}.$$

Since $E[\bar{y}] = \bar{Y}$, we regard the means of random samples of $n = 1$ as providing unbiased estimates of the population mean.

For samples of size $n = 2$ or more we must distinguish between the two types. For the nonreplacement case one can generate and list all possible samples, T, number them from 1 to T, and pick one at random by the use of random numbers. This would be very cumbersome if T were large. Simple and just as effective is to choose the sample in sequence, first select an element at random from the original N, second one at random from the remaining $N - 1$, and so on, until n elements are selected. It can be shown that in this case each of the T possible samples, although not explicitly generated, is given an equal chance of being selected.

A great variety of procedures can be used to select random samples correctly. A method is valid if it meets the following two conditions:

1.   Each element must have an equal chance of appearing in the sample of n; that is, $P_i = n/N$.
2.   If $n \geq 2$, each pair of elements must have an equal chance of appearing in the sample; that is, $P_{ij} = n(n-1)/N(N-1)$.

If it can be shown that both these conditions have been met in the selection procedure, we have sufficient evidence that simple random samples are being generated.

One procedure for selecting RNR (random nonreplacement) samples is the classical ball-and-urn case. An urn contains N balls, and n are selected by thoroughly mixing the balls each time a ball is to be selected without replacement. If this is carried out under ideal circumstances, it can be shown that the probability that a given sample of n will contain a given ball i is $P_i = n/N$. With this property it can be shown that the mean, $\bar{y}$, of any set of measurements taken on the sample elements will have an expectation identical to the population mean, $\bar{Y}$. That is,

(2.13) $$E[\bar{y}] = \bar{Y}$$

for all RNR samples.

Also it can be shown that the variance, s^2, of the sample will have an expectation equal to the population variance, namely,

(2.14) $$E[s^2] = S^2.$$

Example 2.3.   To demonstrate these properties empirically, let us consider the universe of four stones presented in Section 2.2. The stone weights, Y_1, were as follows:

i:	1	2	3	4
Y_i:	2	5	7	14

For samples of size two, the six possible samples, their Y_i values, the computed $\bar{y}$s, and the s^2s, where $\bar{y} = \sum_{i=1}^{n} Y_i/n$ and $s^2 = \sum (Y_i - \bar{y})^2/(n-1)$, are given in Table 2.1.

Table 2.1   Sample values for a universe of four stones

Possible sample (t)	Sample elements, U_i	Sample Y_i	$\bar{y}_t$	s_t^2
1	1,2	2,5	3.5	4.5
2	1,3	2,7	4.5	12.5
3	1,4	2,14	8.0	72.0
4	2,3	5,7	6.0	2.0
5	2,4	5,14	9.6	40.5
6	3,4	7,14	10.5	24.5
$\sum$			42.0	156.0
Average			7.0	26.0

Since $\bar{Y} = 7.0$ and $S^2 = 26.0$, it will be noted that under RNR sampling, wherein each possible sample, t, has $P\{t\} = 1/T = 1/6$ of being selected,

$$E[\bar{y}] = \sum_{t=1}^{T} \bar{y}_t P\{t\} = \frac{42.0}{6} = 7.0 = \bar{Y};$$

$$E[s^2] = \sum_{t=1}^{T} s_t^2 P\{t\} = \frac{156.0}{6} = 26.0 = S^2.$$

Hence, in this case both $\bar{y}$ and s^2 are unbiased estimators of $\bar{Y}$ and S^2.

For RR Samples.　If the sample balls in the balls-and-urn case are returned each time before the next selection, we will generate a random replacement sample of n' "balls," which are really *draws*. In this case, the sample mean, $\bar{y}$, defined as

(2.15)
$$\bar{y} = \frac{1}{n'} \sum_{i=1}^{n'} Y_i,$$

has an expectation $E[\bar{y}] = \bar{Y}$ and is therefore an unbiased estimator of the population mean. The sample variance for this case, which is defined as s_0^2 to distinguish it from the RNR case, is calculated by

(2.16)
$$s_0^2 = \frac{1}{n'-1} \sum_{i=1}^{n'} (Y_i - \bar{y})^2$$

and has the expectation

(2.17)
$$E[s_0^2] = \left(\frac{N-1}{N}\right) S^2$$

—slightly different than the s^2 of RNR sampling.

2.6.　The Variance of $\bar{y}$ and Its Estimated Variance

To resume the case of samples of $n = 1$ (either RR or RNR), let us now define the *variance* of a random variable, x, as

$$\text{var}(x) = E[(x - E[x])^2].$$

Then, since $y = Y_i$, the variance of y for samples of $n = 1$ is given by

$$\begin{aligned}
\text{var}(y) &= E[(Y_i - \bar{Y})^2] \\
&= E[Y_i^2 - 2Y_i\bar{Y} + \bar{Y}^2] \\
&= E[Y_i^2] - \bar{Y}^2
\end{aligned}$$

(2.18)
$$= \left(\frac{N-1}{N}\right) S^2$$

by using Eqs. 2.8 and 2.11. Likewise the variance of $\bar{y}$ for RNR sampling for any sample size n can be shown to be

$$\begin{aligned}
\text{var}(\bar{y}) &= E[(\bar{y} - E[\bar{y}])^2] \\
&= E[(\bar{y} - \bar{Y})^2]
\end{aligned}$$

(2.19)
$$= \left(\frac{N-n}{N}\right)\left(\frac{S^2}{n}\right).$$

The first term in parentheses is called "the finite population correction" or FPC. To confirm this result, we can apply it to the data in Table 2.1. Working out the computations, we obtain:

Sample t:	1	2	3	4	5	6
$\bar{y} - Y$:	-3.5	-2.5	$+1.0$	-1.0	$+2.5$	$+3.5$
$(\bar{y} - \bar{Y})^2$:	12.25	6.25	1.00	1.00	6.25	12.25

$\sum(\bar{y} - Y)^2 = 39.00$, hence $E[(\bar{y} - \bar{Y})^2] = 39.00/6 = 6.50$. And from Eq. 2.19 we have

$$\text{var}(\bar{y}) = \left(\frac{N-n}{N}\right)\frac{S^2}{n}$$

$$= \left(\frac{4-2}{4}\right)\frac{26.0}{2} = 6.50;$$

hence, we have confirmation. It may be noted that if $n = 1$, then Eq. 2.19 reduces to Eq. 2.18. Note also that when sample size is increased to include the whole universe, then the variance of $\bar{y}$ becomes zero, a comforting result.

Estimated Variance of $\bar{y}$. Normally we do not know S^2 for the population being estimated. In either RNR or RR sampling, S^2 can be estimated unbiasedly by the sample variance, by s^2 in the case of RNR sampling and by $[N/(N-1)] s_0^2$ in the case of RR sampling. (See the preceding section.) To estimate var $(\bar{y})$, we shall replace S^2 by s^2 and place a "hat" over var. Thus

$$\widehat{\text{var}}(\bar{y}) = \left(\frac{N-n}{N}\right)\frac{s^2}{n}$$

will denote the value of Eq. 2.19 as estimated from a sample.

For RR Samples. In the case of RR sampling where repeated draws are taken, we can express the variance of the mean over n' draws as an average of n replications of the case where $n = 1$. Thus, we have

(2.20) $$\text{var}(\bar{y}_{RR}) = \left(\frac{N-1}{N}\right)\frac{S^2}{n'}$$

which may be compared to Eq. 2.19, the RNR case.

2.7. The Expectation of, and Variance of, p

The properties of samples presented so far in this chapter apply to any kind of numerical data. In the case where $Y_i = 0$ or 1, indicating the presence or absence of some quality possessed by an element i, sometimes called a *binomial type population*, the formulas can be simplified. For example, for Eq. 2.2, where $\bar{Y} = Y/N$, if we let $\sum Y_i = N_1$, the number of elements with $Y_i = 1$, then $\bar{Y}$ becomes the proportion of elements possessing the property, or

(2.21) $$\bar{Y} = P = \frac{N_1}{N}$$

and Eq. 2.3, the variance of Y_i, becones

(2.22)
$$S^2 = \frac{NPQ}{N-1},$$

where $Q = 1 - P$.

Likewise for an RNR sample, p or $\bar{y}$ is the proportion of 1s; hence

(2.23)
$$\bar{y} = p = \frac{n_1}{n}, \qquad s^2 = \frac{npq}{n-1},$$

and

(2.24)
$$\text{var}\,(p_{\text{RNR}}) = \left(\frac{N-n}{N}\right)\frac{NPQ}{(N-1)n}.$$

For RR Samples.   As in the RNR case, p is the proportion of 1s; the sample variance

$$s_0^2 = \frac{n'pq}{n'-1}$$

and

(2.25)
$$\text{var}\,(p_{\text{RR}}) = \left(\frac{N-1}{N}\right)\frac{NPQ}{(N-1)n'} = \frac{PQ}{n'}.$$

2.8. Measures of Sampling Performance; Precision and Efficiency

In order to compare different sampling procedures, we need some measures of performance. A simple and effective measure is based on a comparison of variances of the sample means that each scheme possesses. Let us define *precision* of a sample estimate, $\bar{y}$, as

(2.26)
$$P(\bar{y}) = \frac{1}{\text{var}\,(\bar{y})}.$$

A way to compare the precision of two schemes, A and B, for estimating $\bar{Y}$ is given by the ratio *relative precision*, where the precision of estimator $\bar{y}_A$ relative to that of estimator $\bar{y}_B$ is given by

$$\text{RP}\,(A/B) = \frac{P(\bar{y}_A)}{P(\bar{y}_B)} = \frac{1/\text{var}\,(\bar{y}_A)}{1/\text{var}\,(\bar{y}_B)}$$

(2.27)
$$= \frac{\text{var}\,(\bar{y}_B)}{\text{var}\,(\bar{y}_A)}.$$

If we let RNR and RR sampling be schemes A and B, respectively, then from

Eqs. 2.19 and 2.20 we can express Eq. 2.27 as

$$(2.28) \qquad \text{RP (RNR/RR)} = \frac{[(N-1)/N]\,(S^2/n')}{[(N-n)/N]\,(S^2/n)} = \left(\frac{N-1}{N-n}\right)\frac{n}{n'},$$

which indicates that a comparison of precision is heavily dependent on the sizes used in the schemes under comparison. By holding sample sizes constant, we can compare precision free of this complication. Let us call this comparison *relative efficiency*, where

$$(2.29) \qquad \text{RE } (A/B) = \text{RP } (A/B), \qquad \text{where } n_A = n_B.$$

Applied to our comparison of RNR with RR sampling, we have

$$(2.30) \qquad \text{RE (RNR/RR)} = \left(\frac{N-1}{N-n}\right)\frac{n}{n'} = \frac{N-1}{N-n}.$$

and since and since $(N-1)/(N-n) \geq 1$ for all sample sizes, it can be said that RNR sampling is always more efficient than RR sampling except in the trivial case when $n = 1$ or when N is very large, in which cases both are equivalent.

Hence by sample design, in this case the choice of nonreplacement rather than replacement sampling, more *efficient samples*—that is, *better samples*—are selected for estimating the mean. Much of this book will be devoted to the consideration of *design* techniques that will provide greater efficiency in estimating population parameters holding either sample size constant or costs constant.

2.9. Problems of Randomization; Sampling Frames

A common method of selecting samples is through the use of random digits, or "numbers," from tables of random digits or "numbers." (See Table A in the Appendix.) Suppose we have 768 employees and wish to draw a random sample of 20 of them. First let us number them serially from 1 to 768. Next we turn to Table A and arbitrarily drop our finger on the page; say (for convenience) we land on the upper left-hand corner. The digits in the first line of this box of 5×5 digits are

03991

and we arbitrarily decide to take columns 2, 3, and 4—that is, digits 399—and proceed down these columns. [*Note:* Random digit tables are truly international: one may prefer to read from left to right as in English, or from right to left as in Hebrew, or from top to bottom as in Chinese!] Since 399 is within our range of interest—that is, the numbers 001 to 768—we accept it and designate employee number 399 as a member of our sample. Proceeding down, we come to which we reject because it exceeds our range of interest. Next, we come to

$$754$$
$$264$$
$$947$$
$$412$$
$$119$$

etc.

Of these we accept 754, 264, 412, and 119; we reject 957. If we come to a number already accepted, we pass on to the next one. The procedure is continued until 20 nonrepeated acceptable numbers are found. If this exhausts a particular set of columns on a page, the procedure for selecting new columns, or new rows, is repeated.

A useful and sometimes necessary device for obtaining a random sample in the statistical sense (one in which probabilities of selection are in fact equal and independent for all elements in the universe of interest) is a *frame*. A sampling frame is a convenient listing or arrangement of the elements in the universe so that one can assure that probabilities of selection are as they should be for the sampling type devised. Normally this may be a simple listing and serial numbering of all elements (such as employees in the firm), or it may be a set of cards, also serially numbered, such as invoices or customers, and so on. For unnumbered elements it may be difficult to assure equal probabilities of selection, since some elements may be hidden from view or by their very nature unlikely to be selected. For example, crates at the bottom of piles may have a larger number of broken components because of inaccessibility. An inventory of such items based on the selection of a sample confined to the easy-to-grab upper crates will not properly represent these higher-breakage crates.

Many samples are taken in a haphazard manner and then analyzed as if they were "random." Some of these have led to rather tragic results when the true universe values were subsequently revealed (such as when an election follows a sampling polling), and some have no doubt led to unfortunate conclusions without ever being correctly detected and blamed. The sampling principles on which this chapter is based require that randomization is in fact operating. There is a case for imaginary randomization in a process, but the degree of randomization present usually requires an appropriate inquiry to establish the fact.

2.10. Summary

This chapter has introduced the concepts of a universe and its population. The universe is a set of physical elements, such as stores, people, or entries on a ledger. We are presumably interested in at least one property or observable characteristic of these elements. Taken collectively, these observed characteristics are called a population. We may be interested in the mean or variance of this population or in some other parameter characterizing it.

A sample is some fraction of the universe. Our observations on this sample are to be used to infer something about the population parameters. The simplest way of doing this is to use the sample mean to estimate the population mean, the sample variance to estimate the population variance. If our sample is random, we know that on the average the sample mean and variance equal the population mean and variance exactly.

The variance of the sample mean was given. It contains a term called the "finite population correction." Two procedures for selecting random samples were described: replacement and nonreplacement. It was shown that in general nonreplacement sampling is the more "efficient."

S^2 **versus** σ^2. S^2 and σ^2, from classical statistics, have somewhat different meanings, although they appear to be related. S^2 is defined as the sum of squares of $(Y_i - \bar{Y})$ divided by $(N-1)$. It is a characteristic of a finite set of N observations, in our case a population. S^2 exists because we have a complete census, actual or conceptual, of a population. No sampling is involved and none may even be contemplated.

On the other hand, σ^2 comes into existence because of the existence of some probability mechanism, either imposed or imagined. In our case, the probability mechanism was imposed by the adoption of a specified sampling procedure. Therefore, σ^2 is an expected value, $E[(y - \bar{Y})^2]$, when random sampling of single elements is carried out, and becomes numerically equivalent to

$$(2.31) \qquad \sigma^2 = \frac{\sum (y - \bar{Y})^2}{N} .$$

2.11. Review Illustrations

1. A coin is to be examined for the number of "heads" it contains.

 (a) What is a plausible universe of elements in this case?
 (b) What is the corresponding population of head counts?
Similarly, for a die:
 (c) What is a plausible universe of elements?
 (d) What is a corresponding population of *dots*?

Solution

 (a) Each "face" of the coin is an "element," thus U_1: 1 2.
 (b) The number of heads on each face is the characteristic of interest. Thus, Y_i: 0, 1; $Y = 1$.

(c) Each of the six faces: U_i: 1 2 3 4 5 6.

(d) The number of dots on each face is the characteristic of interest. Thus, Y_i = 1, 2, 3, 4, 5, 6; $Y = 21$.

2. Suppose a universe consists of the four elements A, B, C, D with y values 2, 6, 8, and 9, respectively.

(a) What is the variance of the means of samples of size three if sampling is without replacement (RNR)?

(b) If with replacement (RR)?

(c) What is the relative efficiency of *nonreplacement* to *replacement* in this case?

Solution

(a) $\text{Var}(\bar{y}) = \left(1 - \dfrac{n}{N}\right)\dfrac{S^2}{n} = \left(1 - \dfrac{3}{4}\right)\dfrac{9.583}{3} = 0.7986.$

(b) $\text{Var}(\bar{y}) = \left(1 - \dfrac{1}{N}\right)\dfrac{S^2}{n'} = \left(1 - \dfrac{1}{4}\right)\dfrac{9.583}{3} = 2.39575.$

(c) $\text{RE (RNR/RR)} = \dfrac{N-1}{N-n} = \dfrac{4-1}{4-3} = 3 \text{ or } 300\%.$

3. Suppose a city comprises a large number of blocks of equal size, of which half contain 80 apartment dwellers (renters) each and half contain 10 single-family-unit homes (owners) each. A sample is drawn by putting 100 points on a map at random and interviewing a household selected at random on each block containing a random point. (If a block contains k random points, then k households are selected at random for interviewing.) Suppose the fraction of households in the city that are homeowners is being estimated.

(a) What is the expected value of the sample proportion, p, in this case? (Give the actual value.)

(b) What is the variance of p in (a)?

(c) What is the MSE (mean square error) of p in (a)?

Solution

(a) Assuming all blocks are equal in area, a random point has a probability of $\frac{1}{2}$ of falling on an apartment block and $\frac{1}{2}$ of falling on a single-housing-unit block. In the former case the interviewer ends up with a home renter and in the latter with a homeowner. Hence the sample will contain an expected 50 percent homeowners. Hence $E[p] = .5$.

(b) Since the population sampled contains a probability of homeowner = .5, the var $(p) = E[(p-1/2)^2] = pq/n = (.5)(.5)/100 = .0025.$

(c) Bias $= E[p] - P$. But P, the actual proportion of homeowners, is $10/(10 + 80) = .11$. Hence, bias in this case is $.50 - .11 = .39$.

2.12. References

Cochran, W. G. 1953, 1963   *Sampling Techniques.* (2nd ed.) New York: John Wiley & Sons, Inc. (Chapter 2. Excellent mathematical presentation.)

Deming, W. E. 1950   *Some Theory of Sampling.* New York: John Wiley & Sons, Inc. (Chapters 3 and 4. Good presentation of basic random sampling.)

Hansen, M. H., Hurwitz, W. N., Madow, W. G. 1953   *Sample Survey Methods and Theory.* Vol. 1. New York: John Wiley & Sons, Inc. (Chapter 4. An excellent and thorough technical presentation of basic random sampling.)

McCarthy, P. J. 1957   *Introduction to Statistical Reasoning.* McGraw-Hill Book Company. (Chapter 6. An elementary but sophisticated presentation of random sampling.)

Wilks, S. S. 1949   *Elementary Statistical Analysis.* Princeton, N.J.: Princeton University Press. (Especially Chapter 9. A good presentation of basic concepts of random sampling.

2.13. Mathematical Notes

Proof that $P_i = n/N$: Under RNR sampling P_i, the probability that element i appears in a sample of n, is n/N. For samples of $n = 1, 2, \ldots$, we have

$$n = 1: \quad P_i = \frac{1}{N}.$$

$$n = 2: \quad P_i = \frac{1}{N} + \frac{N-1}{N} \cdot \frac{1}{N-1} = \frac{2}{N}.$$

$$n = n: \quad P_i = \frac{1}{N} + \frac{N-1}{N} \cdot \frac{1}{N-1} + \frac{N-2}{N} \cdot \frac{1}{N-2} + \frac{N-3}{N} \cdot \frac{1}{N-3} + \frac{N-n}{N} \cdot \frac{1}{N-n}$$

$$= n \text{ terms of } \frac{1}{N}$$

$$= \frac{n}{N}.$$

Proof that $E[X_i^2] = [(N-1)/N]S^2 + \bar{X}^2$ (see Eq. 2.8):

$$E[X_i^2] = \sum_{i}^{N} X_i^2 P_i = \sum X_i^2 \frac{1}{N}.$$

From the definition of S^2, Eq. 2.4, we obtain

$$E[X_i^2] = \frac{N-1}{N} S^2 + \bar{X}^2.$$

Proof that $E[\bar{x}_{\mathrm{RNR}}] = \bar{X}$ (see Eqs. 2.12 and 2.13):

$$E[\bar{x}] = E\left[\frac{1}{n} \sum_{}^{n} X_i\right] = \left[\frac{1}{n} \sum_{}^{N} X_i I_i\right]$$

where $I_i = 1$ with P_i or 0 with $1 - P_i$.

$$E[\bar{x}] = \frac{1}{n} \sum_{}^{N} X_i P_i = \frac{1}{n} \sum X_i \frac{n}{N} = \frac{X}{n} \cdot \frac{n}{N} = \bar{X}.$$

Proof that $E[\bar{x}_{\mathrm{RR}}]$ (see Eq. 2.15):

$$E[\bar{x}] = E\left[\frac{1}{n} \sum_{}^{} X_i\right] = \frac{1}{n} \sum_{}^{n} E[x_i] = \frac{1}{n} \sum_{}^{n} \bar{X} = \bar{X}.$$

Proof that $E[\bar{x}_{\mathrm{RNR}}^2] = [(N-n)/N]S^2/n + \bar{X}^2$:

$$E[\bar{x}^2] = \frac{1}{n^2}\left[\sum_{}^{N} P_i X_i^2 + \sum_{j \neq i} P_{ij} X_i X_j\right]$$

$$= \frac{1}{n^2}\left[\frac{n}{N}\sum_{}^{N} X_i^2 + \frac{n(n-1)}{N(N-1)}\sum_{j \neq i}^{N} X_i X_j\right]$$

since

$$P_i = \frac{n}{N} \quad \text{and} \quad P_{ij} = \frac{n(n-1)}{N(N-1)},$$

$$\left(\sum X_i\right)^2 = \sum_{}^{N} X_i^2 + \sum_{j \neq i} X_i X_j,$$

$$S^2 = \left[-\sum_{j \neq i} X_i X_j + N(N-1)\bar{X}^2\right].$$

and

$$\sum_{}^{N}(X_i - \bar{X})^2 = -\sum_{j \neq i}(X_i - \bar{X})(X_j - \bar{X}),$$

whence

$$E[\bar{x}_{\mathrm{RNR}}^2] = \left(\frac{N-n}{N}\right)\frac{S^2}{n} + \bar{X}^2.$$

Proof that $\text{Var}(\bar{x}_{\text{RNR}}) = [(N-n)/N]S^2/n$ (see Eq. 2.19):

$$\text{Var}(\bar{x}) = E[(\bar{x}-\bar{X})^2]$$

$$= E[\bar{x}^2 - 2\bar{x}\bar{X} + \bar{X}^2]$$

$$= E[\bar{x}^2] - \bar{X}^2$$

$$= \left(\frac{N-n}{N}\right)\frac{S^2}{n}.$$

Proof that $E[s^2] = S^2$ (see Eq. 2.14 for RNR case):

$$E[s^2] = \frac{1}{n-1}\left[\sum_{i}^{n} X_i^2 - n\bar{x}^2\right]$$

$$= \frac{1}{n-1}\left[\sum_{i}^{n} E[X_i^2] - nE[\bar{x}^2]\right]$$

$$= \frac{1}{n-1}\left[n\left(\frac{N-1}{N}\right)S^2 + n\bar{X}^2 - n\left(\frac{N-n}{N}\right)\frac{S^2}{n} - n\bar{X}^2\right]$$

$$= S^2.$$

Proof that $[(N-n)/Nn]\,s^2$ unbiasedly estimates $\text{Var}(\bar{x}_{\text{RNR}})$:

$$E\left[\left(\frac{N-n}{Nn}\right)s^2\right] = \left(\frac{N-n}{Nn}\right)E[s^2]$$

$$= \left(\frac{N-n}{Nn}\right)S^2,$$

which is $\text{Var}(\bar{x}_{\text{RNR}})$; hence the estimator is unbiased.

Proof that $E[\bar{x}_{\text{RR}}^2] = [(N-1)/N]S^2/n' + \bar{X}^2$:

$$E[\bar{x}_{\text{RR}}^2] = \frac{1}{(n')^2}E\left[\sum X_i^2 + \sum_{j\neq i} X_i X_j\right]$$

$$= \frac{1}{(n')^2}\left\{\sum^{n'} E[X_i^2] + \sum_{j\neq i}^{n'} E[X_i X_j]\right\}$$

$$= \frac{1}{(n')^2}\left[n'\left(\frac{N-1}{N}\right)S^2 + n'\bar{X}^2 + n'(n'-1)\bar{X}^2\right]$$

$$= \left(\frac{N-1}{N}\right)\frac{S^2}{n'} + \bar{X}^2.$$

Proof that $\text{var}(\bar{x}_{RR}) = [(N-1)/N]S^2/n'$:

$$\text{var}(\bar{x}_{RR}) = E[(\bar{x}_{RR} - \bar{X})^2]$$
$$= E[\bar{x}_{RR}^2 - 2\bar{x}_{RR}\bar{X} + \bar{X}^2]$$
$$= E[\bar{x}_{RR}^2] - \bar{X}^2$$
$$= \left(\frac{N-1}{N}\right)\frac{S^2}{n'} \ .$$

Proof that $E[s_0^2] = [N-1)/N]S^2$: By definition

$$s_0^2 = \frac{1}{n'-1} \sum^{n'} (X_i - \bar{x}_{RR})^2$$

$$E[s_0^2] = \frac{1}{n'-1} E\left[\sum (X_i - \bar{x}_{RR})^2\right]$$

$$= \frac{1}{n'-1}\left\{\sum E[X_i^2] - n' E[\bar{x}_{RR}^2]\right\}$$

$$= \frac{1}{n'-1}\left[n'\left(\frac{N-1}{N}\right)S^2 + n'\bar{X}^2 - n'\left(\frac{N-1}{N}\right)\frac{S^2}{n'} - n'\bar{X}^2\right]$$

$$= \left(\frac{N-1}{N}\right)S^2$$

Proof that s_0^2/n' unbiasedly estimates $\text{var}(\bar{x}_{RR})$:

$$E\left[\frac{s_0^2}{n'}\right] = \frac{1}{n'}E[s_0^2]$$

$$= \left(\frac{N-1}{N}\right)\frac{S^2}{n'} \ ,$$

which is $\text{Var}(\bar{x}_{RR})$; hence the estimator is unbiased.

2.14. Exercises

1. The number of words in a book is to be determined by selecting a sample of pages and counting the number of words on those pages.

    (a)   What is the universe?
    (b)   The population?
    (c)   Define the population mean and variance in this case.

2. Suppose the probability of finding specified persons at home on a random

call is .7 whether it is the first or any subsequent call. We have a sample of 500 people to be interviewed.

(a)   How many people can be expected to be found at home on one, two, three, and four random calls?

(b)   How many calls should we make in order to expect interviews from 95% of the designated sample?

3.   For the following population list all possible RNR samples of $n = 2$ and, assuming selection is RNR, show that $E[\hat{p}] = P$ and $E[s^2] = S^2$.

$$Y_i: 0\ 0\ 0\ 1\ 1$$

4.   A survey is to be made of certain characteristics of public accounting firms in a given city. A list of certified public accountants from which to select a sample is available. For the purpose of selecting accounting firms for the sample, the N accountants on the list are numbered, n random numbers from a table of random numbers are selected, and the accountants associated with those random numbers are chosen. The accounting firms to which those accountants belong are then used as the desired sample. Is this a satisfactory way of taking the sample?

5.   How can one obtain a random sample of (a) names from a telephone directory? (b) grains of wheat in a bag?

6.   Suppose we have a universe of six elements, and an RNR sample of size two gives the values four and six for the observations on y.

(a)   Estimate $\bar{Y}$.
(b)   Estimate the standard error of the estimate in (a).
(c)   Estimate the relative efficiency of RNR to RR sampling in this universe.

CHAPTER 3

Why Variance?

3.1   Introduction

The preceding chapter introduced some elementary concepts of *universe*, *population*, *samples*, and *selection methods*. Definitions of *random*, *mathematical expectation, bias, variance*, and other concepts were given in the hope of developing a more concise language.

The uses of such estimates as $\bar{y}$ (estimated mean of y, where $\bar{y}$ is sample mean) and $\hat{Y}$ (the estimated total of Y, where $\hat{Y} = N\bar{y}$) are quite evident; they are estimates of population parameters that are usually of interest. The use of var($\bar{y}$) was briefly mentioned as a measure of precision of $\bar{y}$. But var ($\bar{y}$), and other quantities derived from it, will receive much more attention later in this book. Why variance? Mainly because variance is useful to compare sample designs, to determine the confidence to be placed in $\bar{y}$, to determine required sample size, and to test statistical hypotheses. These uses will be discussed later in this chapter.

51

For the moment we shall continue with a description of more basic concepts. In order to take advantage of the reader's presumed knowledge of basic statistics, we shall briefly review the principles and concepts of sampling a population in which the distributions of the population values of y (or x) are known (that is, the normal and binomial). The extension to the case where these distributions are not known (very common in applied sampling) will be made using a very important principle in statistics, the *central limit theorem*. In this discussion we shall adopt the more traditional symbol x rather than y in order to help the reader's recognition of these partially forgotten concepts. In Section 3.6 we shall return to the finite world again, and an attempt will be made to link the usual mathematical models to problems of the finite populations.

3.2. The Normal Distribution

A theoretical distribution that has proved useful in dealing with many natural distributions generally and particularly with sampling distributions is called the *normal distribution*. The graph of a general normal distribution is given in Fig. 3.1. Although a normal distribution is defined by the equation of its curve, this equation is not used explicitly in subsequent chapters, so it is not written down. The curve itself can be thought of as defining the distribution.

If the Greek letters μ and σ are used to represent the mean and standard deviation of a normal distribution, then it can be shown by advanced mathematical methods that σ, the limiting value of s, has the following geometrical interpretation with respect to that normal curve:

1. The area under the normal curve between $\mu - \sigma$ and $\mu + \sigma$ is 68% of the total area, to the nearest 1%.

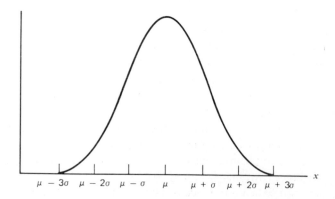

Fig. 3.1. Typical normal distribution. *Source.* Hoel and Jessen (1971). Copyright 1971 John Wiley & Sons, Inc. Reprinted by permission.

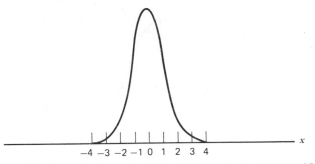

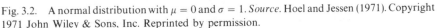

Fig. 3.2.   A normal distribution with $\mu = 0$ and $\sigma = 1$. *Source.* Hoel and Jessen (1971). Copyright 1971 John Wiley & Sons, Inc. Reprinted by permission.

2.   The area under the normal curve between $\mu - 2\sigma$ and $\mu + 2\sigma$ is 95% of the total area, to the nearest 1%.
3.   The area under the normal curve between $\mu - 3\sigma$ and $\mu + 3\sigma$ is 99.7% of the total area, to the nearest 0.1%.

The axis in Fig. 3.1 has been marked off in units of σ, starting with the mean μ. It is clear from this sketch that there is almost no area under the curve beyond 3σ units from μ; however, the equation of the curve would show that the curve actually extends from $-\infty$ to $+\infty$.

An interesting property of the normal curve is that its location and shape are completely determined by its value of μ and σ. The value of μ, of course, centers the curve, whereas the value of σ determines the extent of the spread. Since all normal curves representing theoretical frequency distributions have a total area of 1, then as σ increases, the curve must decrease in height and spread out. This is illustrated in Figs. 3.2 and 3.3, which give sketches of two normal curves with the same mean, namely 0, and standard deviations of 1 and 3, respectively. The fact that the shape of a normal curve is completely determined by its standard deviation enables one to reduce all normal curves to a standard one by a simple change of variable. To any point on the x axis of a normal curve there corresponds a point on the x axis of the standard normal curve, and its value can be determined by stating how many standard deviations it is away from the mean point of the curve. Thus the point $x = 6$ on Fig. 3.3 corresponds to the point $z = 2$ on the standard normal curve given by Fig. 3.4; therefore the value $x = 6$ can be obtained from Fig. 3.3 by stating that it is 2 standard deviations to the right of its mean 0.

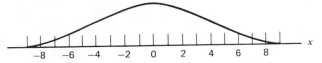

Fig. 3.3.   A normal distribution with $\mu = 0$ and $\sigma = 3$. *Source.* Hoel and Jessen (1971). Copyright 1971 John Wiley & Sons, Inc. Reprinted by permission.

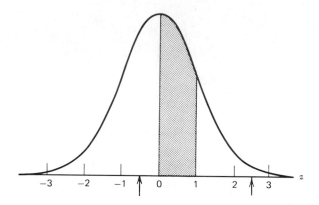

Fig. 3.4. Standard normal distribution. *Source.* Hoel and Jessen (1971). Copyright 1971 John Wiley & Sons, Inc. Reprinted by permission.

In general, if a point x on the axis of a normal curve with mean μ and standard deviation σ corresponds to a point z on the standard normal curve, then the point x is z standard deviations to the right of μ. The relationship between these corresponding points is therefore given by the formula

$$x = \mu + z\sigma.$$

Or, if z is expressed in terms of x,

(3.1)
$$z = \frac{x - \mu}{\sigma}.$$

This formula enables one to find the point z on the standard normal curve that corresponds to any point x on a nonstandard normal curve. Thus, the point $x = 4$ on Fig. 3.3 corresponds to the point $z = (4-0)/3 = 1\frac{1}{3}$ on Fig. 3.4. By this device of expressing all x values on a normal curve in terms of corresponding values on the standard normal curve, all normal curves can be reduced to a single standard one. Key values are given in appendix Table B.

As an illustration of the use of Table B, suppose one wishes to find the area between $x = 220$ and $x = 280$ for x possessing a normal distribution with $\mu = 230$ and $\sigma = 20$. The desired area is shown in Fig. 3.5. First it is necessary to calculate the corresponding z values by means of Eq. 3.1. These are

$$z_1 = \frac{220 - 230}{20} = -0.50$$

and

$$z_2 = \frac{280 - 230}{20} = 2.50$$

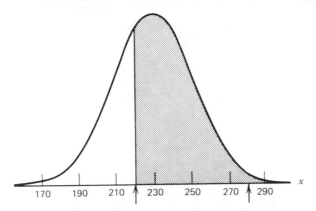

Fig. 3.5.   A particular normal distribution. *Source*. Hoel and Jessen (1971). Copyright 1971 John Wiley & Sons, Inc. Reprinted by permission.

These two z values are indicated by means of vertical arrows in Fig. 3.4. Now, the desired area is given by the area from -0.50 to 2.50 under the standard normal curve. From Table B, the area from $z = 0$ to $z = 2.50$ is 0.4938. By symmetry, the area from $z = -0.50$ to $z = 0$ is the same as that from $z = 0$ to $z = 0.50$. The latter area is found in Table B to be 0.1915; consequently, the desired area is the sum of these two areas, or 0.6853. Although the two normal curves in Figs. 3.4 and 3.5 would look quite different if the same scale were used on both axes, they purposely have been drawn to look alike so that the equivalence of corresponding areas will be apparent.

3.3.   The Distribution of $\bar{x}$ for a Normal x

Section 3.2 was concerned with the characteristics of a random variable that follows the normal distribution. In this section we shall study the characteristics of the sample mean, $\bar{x}$, when sampling from a normal distribution. Since many real-life random variables possess normal distributions, at least approximately, results based on sampling a normal distribution should prove to be very useful.

Any measuring device is of limited accuracy; therefore, it is possible to construct only approximate distributions for continuous variables. In order to simplify the experiment, a rather rough approximation to the distribution of a standard normal variable will be used. This approximation was obtained by using Table B to calculate the percentage of area under the normal curve shown in Fig. 3.4 corresponding to unit intervals with the middle interval centered at the origin. The resulting percentages written as decimal fractions, are shown in Fig. 3.6, which gives the histogram for this approximate normal distribution.

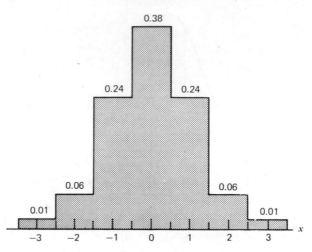

Fig. 3.6.   An approximation to the standard normal distribution. *Source*. Hoel and Jessen (1971).
Copyright 1971 John Wiley & Sons, Inc. Reprinted by permission.

Incidentally, the percentages obtained from Fig. 3.6, when the two end
interval percentages are combined on each side, give the percentages often used
by instructors who "grade on the curve" to determine the percentage of letter
grades A, B, C, D, and F to assign.

Samples of size 4 were taken from this discrete distribution by means of the
random numbers found in Table A. The values of $\bar{x}$ were next tabulated to
yield the frequency table shown in Table 3.3, in which the third row gives the
percentages in decimal-fraction form of the corresponding absolute
frequencies.

Table 3.3.   *Frequencies of values of $\bar{x}$*

$\bar{x}$:	$-\frac{5}{4}$	$-\frac{4}{4}$	$-\frac{3}{4}$	$-\frac{2}{4}$	$-\frac{1}{4}$	0	$\frac{1}{4}$	$\frac{2}{4}$	$\frac{3}{4}$	$\frac{4}{4}$	$\frac{5}{4}$
f:	1	1	3	8	5	14	9	6	2	0	1
$f/50$:	.02	.02	.06	.16	.10	.28	.18	.12	.04	.00	.02

Finally, this frequency table was graphed as a histogram, as shown in
Fig. 3.7.

When we compare Figs. 3.6 and 3.7, it appears that the sample means based
on four measurements each vary about one half as much as do individual
sample values. Furthermore, it appears that the $\bar{x}$ distribution possesses a
mean that is close to 0, which is the mean of the x distribution. Finally, except
for one rather pronounced irregularity for the interval centred at $-\frac{1}{4}$, it
appears that the distribution of $\bar{x}$, except for the difference in spread, possess a
distribution of the same approximate normal type as the x distribution.

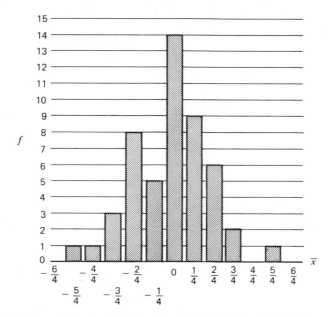

Fig. 3.7.   Distribution of $\bar{x}$ for samples of size 4 from the distribution of Fig. 3.6. *Source.* Hoel and Jessen (1971). Copyright 1971 John Wiley & Sons, Inc. Reprinted by permission.

If this sampling experiment had been carried out, say, 500 times rather than just 50, irregularities such as those in Fig. 3.7 would disappear and the properties of the $\bar{x}$ distribution just discussed would become increasingly apparent. Thus, it would be found that the histogram could be fitted very well with a normal curve, that the mean of the $\bar{x}$ distribution would be very close to 0, and that the variance of the $\bar{x}$ distribution would have a value very close to one-fourth the value of the variance of the x distribution. Although the samples here were taken from an approximate normal distribution as given by Fig. 3.6 rather than from an exact normal distribution, similar results would be obtained if we made the approximation increasingly good by choosing a very small class interval.

Fortunately, it is not necessary to carry out such repeated sampling experiments to arrive at the theoretical frequency distribution for $\bar{x}$. By using the rules of probability and advanced mathematical methods, we can derive the equation of the curve representing the distribution of $\bar{x}$ when the sampling is from the exact normal distribution rather than an approximation. It turns out that $\bar{x}$ will possess a normal distribution if x does, with the same mean but with a standard deviation that is $1/\sqrt{n}$ times the standard deviation of x. These mathematical results are expressed in the form of a theorem:

(3.2) *Theorem.* If x possesses a normal distribution with mean μ and standard deviation σ, then the sample mean $\bar{x}$, based on a random sample of size n, will also possess a normal distribution with mean μ and standard deviation $\sigma/\sqrt{n}$.

The distribution of $\bar{x}$ given by this theorem is called the sampling distribution of $\bar{x}$ because of its connection with repeated sampling experiments, even though it is derived by purely mathematical methods.

This theorem is a purely mathematical theorem about ideal distributions corresponding to smooth curves; however, the conclusions can be expected to hold well for actual populations, provided the population is large and provided the population histogram can be fitted well by a normal curve. This means that if one starts with a population whose histogram can be fitted well with a normal curve and if one takes a large number of samples each of size n and calculates the value of $\bar{x}$ for each sample, then the histogram for the $\bar{x}$ values will be fitted well by the normal curve specified in the theorem.

The results of the sampling experiment just completed appear to be in agreement with this theorem. This histogram of Fig. 3.7 looks like the type that one gets from samples from a normal population, and, since $\sigma = 1$ and $n = 4$ here, its mean, -0.035, and standard deviation, $\sqrt{23} = 0.48$, are in good agreement with the theoretical values of 0 and $\sigma/\sqrt{4} \doteq \frac{1}{2}$ given by the theorem. From this theorem one can draw the conclusion that the means of samples of size 4 from a normal population possess only one-half the variability about the mean of the population that the individual measurements do, as measured by the standard deviation.

3.4. The Distribution of $\bar{x}$ for a Non-normal x

Suppose now that the variable x does not possess a normal distribution. What then can be said about the distribution of $\bar{x}$? Since the unbiasedness of $\bar{x}$, Eq. 2.12, and the variance of $\bar{x}$, Eq. 2.19, do not depend upon the nature of the distribution of x, it follows that the $\bar{x}$ distribution, whatever the distribution of x, will possess the mean μ and the standard deviation $\sigma/\sqrt{n}$. The only unanswered question therefore is that of determining the nature of the distribution.

A number of statisticians have conducted sampling experiments with different kinds of nonnormal distributions for x to see what effect the nonnormality would have on the distribution for $\bar{x}$. The surprising result has always been that if n is larger than about 25, the distribution of $\bar{x}$ will appear to be normal in spite of the population distribution chosen for x. Several years ago an instructor of an advanced statistics course challenged his students to construct a distribution as nonnormal as they could and wagered them that if he took samples of size 25 from their population, the resulting $\bar{x}$ distribution would check out as a normal distribution. After agreement on the rules of the contest, the experiment was conducted. That evening the instructor ate in style at the students' expense. If the distribution for x does not differ too widely from a normal distribution, the distribution of $\bar{x}$ will often appear to be normal for n as small as 5. This remarkable property of $\bar{x}$ is of much practical importance, because a large share of practical problems involve samples sufficiently large to permit one to assume that $\bar{x}$ is normally distributed and thus to use familiar normal-curve methods to solve problems related to means without being concerned about the nature of the population distribution.

A well-known mathematical theorem, known as a *central limit theorem*, essentially states that under very mild assumptions the distribution of $\bar{x}$ will approach a normal distribution as the sample size, n, increases. This theorem, together with the results of sampling experiments of the type already discussed, can be expressed in the following manner:

(3.3) *Theorem.*   If x possesses a distribution with mean μ and standard deviation σ, then the sample mean $\bar{x}$, based on a random sample of size n, will possess an approximate normal distribution with mean μ and standard deviation $\sigma/\sqrt{n}$, the approximation becoming increasingly good as n increases.

The preceding theorem permits one to calculate the probability that $\bar{x}$ will lie in any specified interval, if the population mean and variance are known. Suppose it is known that $\mu = 68$ inches and $\sigma = 3$ inches for the distribution of stature of adult males. If a random sample of size $n = 100$ is taken, what is the probability that the resulting value of $\bar{x}$ will differ by more than $\frac{1}{2}$ inch from the population mean? Since $\bar{x}$ may be treated as a normal variable with mean $\mu_{\bar{x}}$ $= 68$ and standard deviation $\sigma_{\bar{x}} = 3/\sqrt{100} = 0.3$, it is necessary to calculate the area under a normal curve with this mean and standard deviation that lies to the right of 68.5 and to the left of 67.5. By symmetry this is equal to twice the probability that $\bar{x}$ will lie to the right of 68.5. Thus, it suffices to calculate

$$z = \frac{68.5 - 68}{0.3} = 1.67.$$

From Table B it will be found that this probability is .05; therefore the probability that $\bar{x}$ will differ from 68 by more than $\frac{1}{2}$ inch is equal to .10.

The preceding theorems will be the foundation for solving various types of statistical inference problems in the following chapters. They are extremely important and useful theorems.

3.5. Point and Interval Estimates of μ and $\bar{Y}$

Two types of estimates of parameters are in common use in statistics: the point estimate and the interval estimate. A *point estimate* is the familiar kind of estimate; that is, it is a number obtained from computations on the sample values that serves as an approximation to the parameter being estimated. For example, the sample proportion, $p = x/n$, of voters favoring a certain candidate is a point estimate of the population proportion P. Similarly, the sample mean $\bar{x}$ is a point estimate of the population mean μ. An *interval estimate* for a parameter is an interval, determined by two numbers obtained from computations on the sample values, that is expected to contain the value of the parameter in its interior. The interval estimate is usually constructed in such a manner that the probability of the interval's containing the parameter can be specified. The advantage of the interval estimate is that it shows how accurately the parameter is being estimated. If the length of the interval is very small, high precision has been achieved. Such interval estimates are called *confidence intervals*. Both point and interval estimates are determined for normal distribution parameters in this section and for the binomial in the next.

In Section 3.3 it was stated that the means of samples drawn from a distribution that is normal will also be normally distributed but with a standard deviation given by $\sigma_{\bar{x}} = \sigma/\sqrt{n}$. Hence the properties of the normal distribution can be applied to the sample means. For example, for a particular sample mean, $\bar{x}$, we can state (before the sample is actually drawn) that

$$P\{\mu - 1\sigma_{\bar{x}} < \bar{x} < \mu + 1\sigma_{\bar{x}}\} = .68$$

—that is, the probability that this particular sample mean lies in the region between one standard error less and one standard error more than the true mean is .68. Likewise,

$$P\{\mu - 2\sigma_{\bar{x}} < \bar{x} < \mu + 2\sigma_{\bar{x}}\} = .95$$

or

$$P\{\mu - 1.5\sigma_{\bar{x}} < \bar{x} < \mu + 1.5\sigma_{\bar{x}}\} = .87,$$

or, for any length of symmetrical region straddling the true mean, the probability is

(3.4) $$P\{\mu - z_{\theta}\sigma_{\bar{x}} < \bar{x} < \mu + z_{\theta}\sigma_{\bar{x}}\} = \theta.$$

Here θ is the probability that corresponds with z_{θ}, the number of standard errors on each side of the true mean that defines the region in which $\bar{x}$ may lie. If z_{θ} is 1, then θ is .68; if $z_{\theta} = 2$, θ is .95 approximately—the usual well-known relationships.

To be more useful, the probability statement can be turned around with a little algebra, yielding the form

(3.5) $$P\{\bar{x} - z_\theta \sigma_{\bar{x}} < \mu < \bar{x} + z_\theta \sigma_{\bar{x}}\} = \theta,$$

which now reads, "The probability that the interval between $\bar{x} - z_\theta \sigma_{\bar{x}}$ and $\bar{x} + z_\theta \sigma_{\bar{x}}$ includes the true mean, μ, is θ." If the statement is made after the sample has been drawn, which is the usual circumstance in which it is used, statisticians prefer to call it a "confidence" rather than a "probability statement," and θ is called a *confidence coefficient*. The interval defined in Eq. 3.5 is an *interval estimate* of μ.

Since the means of samples drawn from a normal population have a normal distribution with mean μ and standard deviation $\sigma/\sqrt{n}$, then z, the standard normal deviate, can be calculated by

(3.6) $$z = \frac{\bar{x} - \mu}{\sigma_{\bar{x}}}.$$

By letting $|\bar{x} - \mu| = e$, where e is the distance $\bar{x}$ may be from the true mean, then if we set a limit to e, say e_θ, such that we have a confidence of θ that it is not exceeded, we have the relationship

(3.7) $$e_\theta = z_\theta \sigma_{\bar{x}} = \frac{z_\theta \sigma}{\sqrt{n}}$$

between tolerable error, e_θ, and degree of confidence, θ. Note that the calculations for a confidence interval estimate are essentially the calculation of e_θ and the addition and subtraction of that value from the point estimate. Hence the limits on the population mean with confidence θ are given by

(3.8) $$\bar{x} \pm z_\theta \sigma_{\bar{x}}.$$

Example 3.1. A sample of size $n = 4$ is drawn from a normal population with a standard deviation, $\sigma = 10$, yielding a mean, $\bar{x} = 36$. Calculate the 95% confidence limits for the population mean, μ.

In this case $\bar{x} \pm z_\theta \sigma_{\bar{x}}$ is given by

$$36 \pm (1.96) \left(\frac{10}{\sqrt{4}} \right), \qquad 36 \pm 9.8$$

or

$$P\{26.2 < \mu < 45.8\} = .95.$$

When σ is Unknown. In many, if not most, cases of sampling we will not know the value of σ but may have to rely on an estimate of it from our sample. In this case we can use small-sample theory, where we replace z with t and σ with s and proceed as before. Now we write Eq. 3.6 in terms of t; thus

(3.9) $$t = \frac{\bar{x} - \mu}{s_{\bar{x}}}.$$

When we substitute $s_{\bar{x}}$ for $\sigma_{\bar{x}}$ in Eqs. 3.7 and 3.8, the confidence limits, with estimated standard errors replacing known standard errors, become

(3.10) $$e_\theta = t_\theta s_{\bar{x}} = \frac{t_\theta s}{\sqrt{n}}$$

and

(3.11) $$\bar{x} \pm t_\theta s_{\bar{x}},$$

respectively. Tabulated values of t, like z, are in Table D in appendix. Suppose in our example, $\bar{x} = 36$ and $n = 4$ as before, and we do not know σ, but $s = 12$ and we wish to calculate the 95 % confidence limits for μ—that is, $\theta = .95$. Now $t_\theta = t$ for $1 - (\alpha = .05)$ with three degrees of freedom, which according to Table D is 3.18; hence our required confidence limits are

$$36 \pm \left[\frac{(3.18)\,(12)}{\sqrt{4}} \doteq 19 \right]$$

or 17 and 55. When σ is known, the limits work out to be

$$36 \pm \left[\frac{(1.96)\,(10)}{\sqrt{4}} = 9.8 \right]$$

or 26.2 and 45.8. On the average the calculated confidence limits will be wider when σ must be estimated, and particularly so when s is based on very small samples. When n is 25 or larger, the "t" method is nearly equivalent to the "z" method. The advantage of the t method is that it is always correct if σ is known, use t for ∞ degrees of freedom.

The relative frequency base for the "confidence" that the calculated limits enclose the parameter being estimated is illustrated by an experiment in which 100 samples of size 4 and another 100 of size 16 were drawn from a normal population with $\mu = 40$ and $\sigma = 40$. For each sample $\bar{x}$ and s were calculated and from these the 50 % confidence limits. Figure 3.8 presents a plot of $\bar{x}$s for both sizes of samples together with the 50 % error bounds; that is,

$$\mu \pm (z_{.50}) \frac{\sigma}{\sqrt{n}}$$

for each sample size. Note that, as expected, about one-half of the point estimates fall within the limits in both cases. Also note how the limits can be narrowed when sample size is increased from four to 16, in which case we make

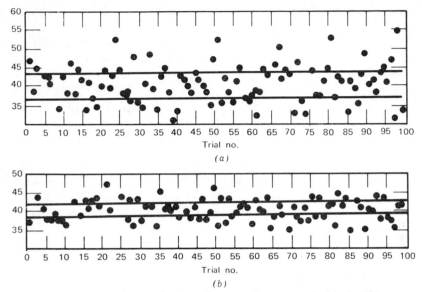

Fig. 3.8. Plotted means of each of 100 random samples compared with the 50% accuracy interval. (a) Samples of size 4. (b) Samples of size 16.

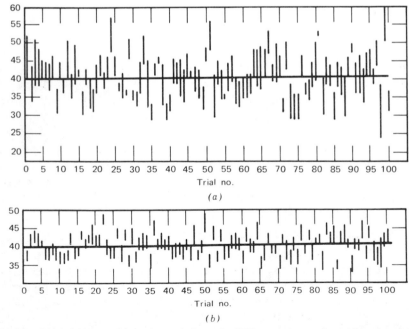

Fig. 3.9. Plotted 50% confidence intervals for each of 100 random samples. (a) Samples of size 4. (b) Samples of size 16.

the limits half as wide while retaining the same confidence level, which in this case is 50%.

Figure 3.9 presents the results from the 200 trials in the form of intervals rather than point estimates and also under the assumption that σ is unknown. The *confidence intervals*, the lines connecting the confidence limits, are plotted for each sample. Note how the length of the interval varies from sample to sample. When a line "cuts the solid line," the statement that the limits "enclose" or "contain" μ is "correct." Note that our 50% "confidence" in each statement is correct, since about one-half the intervals enclose the mean.

When N is Finite. In the case of applied sampling, it has been found that small departures from normality do not disturb the validity of either the z or t methods of interval estimation seriously even if sample sizes are small. However, if samples are large enough so that the central limit theorem can be depended upon, then both methods are approximately correct. The approximation is even found useful if sampling is from a finite population where the FPC corrected estimated variance, $\widehat{\text{var}}\,(\bar{y})$, may be substituted for σ^2/n. Hence the limits are given by

$$(3.12) \qquad\qquad\qquad \bar{x} \pm t_\theta \left(\frac{N-n}{N}\right)^{1/2} \frac{s}{\sqrt{n}}$$

instead of Eq. 3.11.

3.6. Point and Interval Estimates of π and P

Although such matters as sampling procedures and point estimation are unaffected by type of population dealt with, some simplifications can be made of previous formulas if the population is of a binomial type. Also, when confidence limits are under consideration, it is possible, under certain conditions, to use some of the simpler results from the normal populations.

For exact results, using RNR sampling, the appropriate theory is that for the hypergeometric distribution. If sampling is RR, the appropriate theory is that for the binomial distribution. The former may be computationally burdensome, and the latter is rarely used. Hence some approximate methods are generally found to be very satisfactory.

When $n_1 = n\pi$ is large and n/N is small. If the binomial type population is large and the sample, whether RNR or RR, is small relative to it, then var (p) will behave like a binomial proportion of successes in n trials, where the probability of a success on each trial is π, and the theory and procedures for the binomial are appropriate. If the number of successes, $x = n_1$, in a sample of n is large enough, then the simpler procedures used in dealing with the normal population can be used. In this case, the simpler forms for computing $\bar{X} = P$, S^2, $\sigma_{\bar{x}}^2 = \sigma_p^2$ and their corresponding sample values are given in Section 2.7.

(We can regard the limiting case of $P = N_1/N$ to be π when N is infinitely large.) Then for precision, and confidence intervals, we have by replacing σ with $\pi(1 - \pi)$ the following adaptations:

(3.13)
$$e_\theta = z_\theta \frac{\pi(1 - \pi)}{n},$$

(3.14)
$$P\{p - e_\theta < \pi < p + e_\theta\} = \theta,$$

and, since π, or P, must be estimated from the sample by p,

(3.15)
$$e_\theta = z_\theta \sqrt{\frac{pq}{n}}.$$

The FPC can be added to the standard error term, but care should be given to its effect on the confidence level, θ, being claimed.

A useful rule for deciding whether the normal approximation will be suitable is given in Table 3.2 [from Cochran (1963)].

Table 3.4. *Smallest values of np for use of the normal approximation*

P	$nP =$ number observed in the smaller class	$n =$ sample size
.5	15	30
.4	20	50
.3	24	80
.2	40	200
.1	60	600
.05	70	1400
$\sim 0^a$	80	∞

[a] When P is extremely small, nP follows the Poisson distribution.

To illustrate use of Table 3.2, suppose we have a sample of 50 persons of whom 10 are in favor of a certain proposal. What are the 95 % confidence limits for P? In this case np for the smaller class is 10, whereas 20 are required, hence we do not use normal approximation but should use tables (see references).

Suppose 20 instead of 10 are in favor. Using the normal approximation, we have Eq. 3.11,

$$\bar{x} \pm \frac{t_\theta s}{\sqrt{n}},$$

where $\bar{x} = p = 0.40$, $s = \sqrt{npq/(n-1)} = \sqrt{(50)\left(\frac{2}{5}\right)\left(\frac{3}{5}\right)/(50-1)} = 0.495$, $\sqrt{n}$

$= \sqrt{50} = 7.07$, $\theta = .95$, and $t_{.95}$ for 49 df $\doteq 2.0$. Hence, we have

$$0.40 \pm \frac{(2.0)\,(0.495)}{7.07} = 0.40 \pm 0.14,$$

or the true proportion, P, lies somewhat between 0.26 and 0.54.

3.7. Tolerable Error and Required Sample Size

Solving Eq. 3.7 for n, we obtain

(3.16) $$n = \left(\frac{z_\theta \sigma}{e_\theta}\right)^2,$$

and by choosing some *tolerable error*, e_θ, with level of confidence θ, and knowing σ, we can determine the size of sample that will satisfy our stated needs. If σ is unknown but a sample estimate of it is available, then we can use Eq. 3.10 and solve for n, whence

(3.17) $$n = \left(\frac{t_\theta s}{e_\theta}\right)^2.$$

The appropriate degrees of freedom for t_θ will depend on n, the required sample size, rather than on the size of sample on which s is calculated. As a consequence, t_θ depends on n. A solution can usually be obtained easily by iterative methods if n appears to be less than 25, where t_θ is most sensitive to the value of n.

Example 3.2.   Suppose we have drawn a sample from a normal population and obtained an s^2 of 16. We would be willing to tolerate an error of ± 2 units, but want to have 95 % confidence that the error is no greater than 2. How large a sample would be required?

In this case, $s^2 = 16$, $e_\theta^2 = 4$, and the probability level for t is .95, but we do not know the proper degrees of freedom to use, since we do not know n. We can proceed iteratively, however, by assuming the sample size is large, in which case $t(.95; \infty)$ (read as "t for 95 % confidence with ∞ degrees of freedom") is approximately 2.0 (Table D). Then for our first approximation, using Eq. 3.17.,

$$n = \frac{(16)\,(2.0)^2}{(2)^2} = 16.$$

Since n is in the neighbourhood of 16, we can try $t(.95; 15) = 2.13$ (Table D), and for our second approximation,

$$n = \frac{(16)\,(2.13)^2}{(2)^2} = 18.1,$$

or more precisely, using $t(.95; 17) = 2.11$, our third approximation is

$$n = 17.8.$$

Hence, the required sample size is 18.

Finite Population Correction for t. In order to take into account a finite as well as an infinite population, we can rewrite Eq. 3.9, the expression for t, by replacing $s/\sqrt{n}$ with the estimated standard error of $\bar{x}$ when a population is finite. Then

(3.18)
$$t = \frac{\bar{x} - \mu}{(s/\sqrt{n})\,[(N-n)/N]^{1/2}}.$$

The use of Eq. 3.18 will generally be practicable when samples are large even for large departures from normality and when samples are small for departures that are only moderate. [See Cochran (1963) for further guidance on these matters.]

Solving Eq. 3.18 for n, we obtain

(3.19)
$$n = \left(\frac{N-n}{N}\right)\left(\frac{st_\theta}{e_\theta}\right)^2$$

as an expression relating the size of sample required to provide a tolerable error, e_θ, with confidence, θ, when the population variance, S^2, is estimated by s^2 from a sample. Since n also appears on the right-hand side, Eq. 3.19 does not provide an explicit solution. However, Eq. 3.19 becomes

(3.20)
$$n = \frac{(st_\theta/e_\theta)^2}{1 + (st_\theta/e_\theta)^2/N},$$

or, if the required sample size ignoring the FPC is denoted by n_0,

(3.20a)
$$n = \frac{n_0}{1 + n_0/N}.$$

It may be convenient for many practical situations to ignore the FPC where $n/N \le 0.05$. Also, when s is based on samples where $n \ge 25$, we may use z_θ for t_θ.

3.8. Relative Standard Error, RSE

The *coefficient of variation*, or relative standard deviation, of a population may be defined as

(3.21)
$$V = \frac{S}{\bar{Y}}$$

and for the sample

(3.22)
$$v = \frac{s}{\bar{y}}.$$

Similarly, the *relative standard error* of some estimator, say $\bar{y}$, is defined as

(3.23)
$$\text{RSE}(\bar{y}) = \frac{SE(\bar{y})}{\bar{Y}}.$$

In many instances we are interested in attaining a precision within a certain percent of the parameter rather than within absolute limits. For example, we may desire a tolerance interval of $\pm 5\%$ of the mean, regardless of the size of the mean, rather than one of predetermined length. In the case where we wish relative rather than absolute sizes of intervals, Eq. 3.10 becomes

(3.24)
$$e_\theta / \bar{Y} = \frac{(s/\bar{y})t_\theta}{\sqrt{n}}.$$

Since $\bar{Y}$ usually has to be estimated from the same sample or source as s, using $\bar{y}$ to estimate $\bar{Y}$ and letting $s/\bar{y}$ be denoted by v, the coefficient of variation of the sample, we have an approximation

(3.25)
$$\frac{e_\theta}{\bar{y}} \doteq \frac{vt_\theta}{\sqrt{n}},$$

and solving for n we obtain

(3.26)
$$n = \left(\frac{vt_\theta}{e_\theta / \bar{y}}\right)^2.$$

As we shall see in Section 3.10, these formulas in practice frequently have an advantage over those involving s^2, since coefficients of variation frequently can be guessed more accurately than absolute variances.

Example 3.3. A preliminary sample of a large population gives a coefficient of variation of 1.00 (or 100%). How large a sample would be required to estimate the total of this population, Y, with an accuracy of 5% with a confidence of 95%?

Our estimator will be $\hat{Y} = N\bar{y}$, and the desired confidence limits will be $N\bar{y} \pm Ne_\theta$. To estimate sample size, we choose Eq. 3.26 (since it is stated that the population is "large," no FPC is needed):

$$n = \left(\frac{vt}{e/\bar{y}}\right)^2.$$

In our problem, we are told that

$$Ne_\theta = (0.05)Y = (0.05)N\bar{Y};$$

hence

$$e_\theta = \frac{(0.05)N\bar{Y}}{N} = (0.05)\bar{Y},$$

and, using $\bar{y}$ to replace $\bar{Y}$, 1.00 for v, and 2.1 (from Table D) for t, we have

$$n = \left\{ \frac{(1.00)\,(2.1)}{(0.05)\,(\bar{y})/\bar{y}} \right\}^2 = \left(\frac{2.1}{0.05} \right)^2 = 1764.$$

3.9.   Schemes for Conjecturing S^2, V^2, and $\bar{Y}$

It is often quite desirable when a sample is being planned to have some idea of the size of the S^2, even if only a rough estimate. Several aids will be given here that have proved quite helpful when the picture otherwise looks bleak.

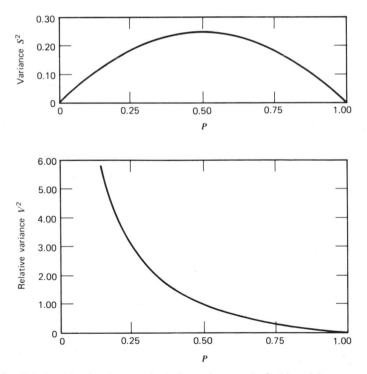

Fig. 3.10.   Relationship of variance and relative variance to P of a binomial type population.

In the case of a binomial type population, both S^2 and V^2 are functions of the mean, P, since $S^2 = PQ$ and $V^2 = Q/P$ when the FPC can be ignored. Figure 3.10 presents these functions graphically. If P can be estimated, then S^2 can be determined. If there is no clue of what P might be, it may be best to assume it to be .5, in which case variance is at a maximum. If relative error is to be controlled, the problem is to estimate V^2, for which the binomial type population gets infinitely large for small values of P. If accuracy is required for estimates of V^2, in the absence of knowledge of P, a preliminary sample may be required for that purpose.

Coefficients of variation seem to be much more amenable to classification than variances or means, and reasoning by analogy may be far less disappointing. Geography and time seem to have less effect on V^2, so experience with a given type of population in one location at a certain time may be utilized for another geographic area at another time.

3.10. Summary

The role of variance, or its square root, standard deviation, is an important one in evaluating samples for accuracy, for determining the goodness of an estimate in hitting near its target, and in determining sample size. When the underlying populations are truly normal, the methods presented in this chapter are exact. When, as is usually the case, this is not so, the methods are approximate, but in general they will serve satisfactorily for many practical purposes. This assurance is based on the facts that (1) if samples are large enough, the distribution of their means tends to be normal, thanks to the central limit theorem, even when the parent population is quite nonnormal, and (2) "t" is not very sensitive to nonnormalities. However, some care should be used in deciding when conditions do warrant the use of these approximations. In dealing with binomial type data (zeros and ones), it is suggested that the approximate methods be used because of their simplicity, unity, and generally satisfactory performance when required conditions are met—which they usually can be without difficulty. Otherwise the exact methods for the binomial and hypergeometric distributions suggested, although they are not presented here.

3.11. Review Illustrations

1. A company has 100 sales districts. In a marketing survey of customers in the overall market (all 100 districts) it has been found that a sampling rate of $\frac{1}{2}$ of 1 % provides the precision desired. The managers of each of the sales districts decided that they would like a similar survey for each of their areas and with the same precision. The $\frac{1}{2}$ of 1 % sampling rate made the proposal seem quite modest. Suppose the average sales district contained 2000 customers and had

an S^2-variance 80% of that for the whole market; how large a sample (in percent of customers) would be required for each district?

Solution

$$(1-0.005)\frac{S^2}{(100)(2000)(0.005)} = \left(1-\frac{n}{2000}\right)\frac{.8S^2}{n}.$$

$2n = 1600 - 0.8n$; hence $n = 573$, and $f = n/2000 = 28.6\%$.

2. In a study of women over 30 years of age residing in Iowa, a random sample of 1203 households contained a total of 1073 women over 30 years of age, distributed as follows:

Number of women over 30 years of age in a household	Number of households
0	232
1	876
2	88
3	7
4	0

(*a*) Do the 1073 women over 30 comprise a random sample of all such women in the state of Iowa? Why or why not?

(*b*) Make a 95% confidence statement on the fraction of all households in the state containing women over 30 years of age.

Solution

(*a*) No, the sample consists of random groups of women rather than random individual women.

(*b*) According to the sample the proportion of households in Iowa containing women over 30 years of age is $p = (876 + 88 + 7)/1203 = 971/1203 = 0.807$. Here, ignoring FPC,

$$e_\theta = z_\theta \sqrt{\frac{pq}{n}},$$

$$= (1.96)\sqrt{\frac{(0.807)(1-0.807)}{1203}} = 0.0223,$$

and a 95% confidence statement is given by

$$P\{p - e_\theta < P < p + e_\theta\} \qquad\qquad = .95,$$
$$P\{0.807 - 0.022 < P < 0.807$$
$$+ 0.022\} \qquad\qquad\qquad = .95,$$

or

$$P\{0.785 < P < 0.829\} \qquad\qquad = .95.$$

3. Four automobiles of a certain make when driven over a test road yielded the following results on gas mileage: 15, 18, 15, 20. Assuming these four cars are a random sample of all cars of this make this year:

(a) What is the estimated gas mileage of cars of this make?

(b) Does this test disprove the manufacturer's claim that this car averages 20 miles per gallon?

(c) How many cars should be tested to obtain an estimate within .2 miles per gallons with 99% confidence?

Solution

(a) $\bar{y} = 17.0$.

(b) $e = (2.45)(3.182)/\sqrt{4} = 3.9$ [here we take $t_\theta(.95; 3 \text{ df}) = 3.182$]. Our 95% confidence limits are 17.0 ± 3.9, or $P\{13.1 < \mu < 20.9\} = .95$. Since claimed mileage of > 20 is inside our interval, we regard the claim as correct with at least 95% confidence.

(c) $n = s^2 t^2 / e^2 = (6)(3.18)^2 / (0.2)^2 = 1.517,$ $\quad$ or $\quad$ with $\quad$ $t(0.05; \quad \infty)$
$= 6(2.0)^2 / (0.2)^2 = 600.$

4. A survey of quail (bobwhite) on a random sample of 40-acre tracts of land is being considered. The plan is to send trained wildlife biologists to the selected 40-acre units to determine if one or more coveys of quail appear to be occupying the areas. Knowing the fraction of all the 40s that are occupied can be helpful in determining the total number of quail in the region. Suppose about 20% of the 40s (sampling units) are expected to be occupied and the total number of 40s in the universe is very large.

(a) How large a sample of 40s should be surveyed in order to obtain an estimate of the population proportion occupied within 3% of the true proportion with 95% confidence?

(b) How large a sample if the confidence interval is 3 percentage points and with 95% confidence?

(c) Is the "t" method a satisfactory approximation for constructing confidence limits in (a)? In (b)?

Solution

(a) $\hat{P} = 20$; $e = (0.03)(\hat{P}) = 0.006$; $s^2 = (0.2)(0.8) = 0.16$; $t_\theta(.95, n$ large$)$ $= 1.96$. Then $n = (0.16)(1.96)^2/(0.006)^2 = 17,100$ units.

(b) In this case $e = 0.3$, hence $n = (0.16)(1.96)^2/(0.03)^2 = 683$ units.

(c) Yes, since Table 3.2 indicates for $P = 0.2$ a required n of at least 200. This is met in both (a) and (b).

5. Suppose the coefficient of variation of income of farms is 100% (that is, $V = 1.00$). How large a random sample of farms is required for an allowable error of $\pm 5\%$ with 95% confidence for a:

(a) County with 2000 farms?

(b) State with 200,000 farms?

(c) Nation with 5,000,000 farms?

Solution

(a) $n_0 = \sigma^2 z^2/e^2$, where $\sigma^2 = V^2/\mu^2 = (1)^2(\mu)^2 = 1$; $d^2 = (0.05)^2\mu^2$ and $z = 1.96$; hence $n_0 = \mu^2(1.96)^2/(0.05)^2\mu^2 = 3.84/(25/10,000) = 1535$ and $n = n_0/(1 + n_0/N) = 1535/(1 + 1535/2000) = 868$.

(b) Here n_0 is the same as in (a) but N is now 200,000; hence $n = 1535/(1 + 1535/200,000) = 1525$.

(c) Now $N = 5,000,000$; hence $n = 1535/(1 + 1535/5,000,000) = 1535$.

6. A sample survey was made of the adults in a town to determine how they stood in regard to a certain civic issue. A random sample of 200 of the town's 10,000 adults gave the following data:

Group	No. in sample
For	90
Against	80
No decision	30
Total	200

(a) What are the 67% confidence limits for the "pro" estimate?

(b) Is the "pro" group significantly (statistically) smaller than 50%? (Use 5% level.)

(c) How large should our sample be in order to provide 95% confidence that an error no larger than ± 2 percentage points takes place?

Solution

Here $N = 10,000$, $n = 200$, $p = 90/200 = 0.45$.

(a) $e = z\sigma/\sqrt{n} \doteq z\sqrt{pq/n}$, where $z = 1$, hence $e = (1)\sqrt{(0.45)(0.55)/200}$ = 0.035 (FPC can be ignored since $n/N = 200/10,000 \doteq 0.02$). The 67% confidence limits are given by $p + e = 0.45 \pm 0.035$ or $(0.415, 0.485)$.

(b) Here we wish to test the hypothesis that $P \geq 0.5$ against the alternative that $P < 0.5$, hence a one-tail test. Ignoring FPC, $z = (p - P)/\sigma_p = (0.45 - 0.50)/\sqrt{pq/n} = (-0.05)/\sqrt{(0.45)(0.55)/200} = -1.421$. Since $z(\alpha = 0.05)$ = 1.645 for a single tail, accept the hypothesis that $P \geq 0.5$. The evidence is insufficient to conclude that P is less than the majority.

(c) $n_0 = \sigma^2 z^2/e^2 = (0.45)(0.55)(1.96)^2/(0.02)^2 = 2377$, but $n_0 > 0.05N$; then $n = n_0/(1 + n_0/N) = 2377/(1 + 2377/10,000) \doteq 1921$.

3.12. References

Cochran, W. G. 1953, 1963	*Sampling Techniques.* (2nd ed.) New York: (Chapters 2, 3, and 4. An extensive discussion of confidence intervals.)
Hoel, P. G., Jessen, R. J. 1971	*Basic Statistics for Business and Economics.* New York: John Wiley & Sons, Inc.
McCarthy, P. J. 1957	*Introduction to Statistical Reasoning.* McGraw-Hill Book Company. New York: (A clear presentation of the nature of confidence intervals.)
Pearson, E. S., Hartley, H. O. (eds.) 1954	*Biometrika Tables for Statisticians.* Vol. 1. London: The Syndics of the Cambridge University Press. (Table 27, p. 174. Table gives values for n = 2 to 500 and by tens from 500 to 1000.)

3.13. Mathematical Notes

Proof that $\sigma_{\bar{x}}^2 = \sigma^2/n$:

$$\sigma_{\bar{x}}^2 = E[(\bar{x} - \mu)^2]$$

$$= E[\bar{x}^2] - \mu^2$$

$$= \frac{1}{n^2}[\sum^n x^2 + \sum_{j \neq i} x_i x_j] - \mu^2$$

$$= \frac{1}{n^2}[\sum^n E[x^2] + \sum_{j \neq 1} E[x_i x_j]] - \mu^2,$$

since

$$\sigma^2 = E[(x - \mu^2)] = E[x^2] - \mu^2$$

and
$$E[x_i x_j] = n(n-1)\mu^2$$

$$\sigma_{\bar{x}}^2 = \frac{1}{n^2}[n\sigma^2 + n\mu^2 + n(n-1)\mu^2] - \mu^2$$

$$= \frac{\sigma^2}{n}.$$

Derivation of Eq. 3.5: Given $P\{\mu + e < \bar{x} < \mu - e\} = 0$. Subtracting $\mu + \bar{x}$ from each segment,

$$P\{-\bar{x} + e < -\mu < -\bar{x} - e\} = 0.$$

Multiplying by -1,

$$P\{\bar{x} - e < \mu < \bar{x} + e\} = 0.$$

3.14. Exercises

1. Given that x is normally distributed with mean 20 and standard deviation 4, calculate the probability that the sample mean, $\bar{x}$, based on a sample of size 64, will (a) exceed 21, (b) exceed 20.5, (c) lie between 19 and 21, (d) exceed 25, (e) exceed 18.

2. For the purpose of saving time, Jones decides to find the total weight of 16 packages in a case rather than weigh each package separately. He finds the mean and standard deviation of case weights to be 24.2 and 1.2 pounds, respectively, for a large sample of such cases.

(a) Assuming that the cases are composed of random samples of individual packages, what is the standard deviation of weight per package?

(b) Suppose that it was found that the result in (a) did not agree with earlier results when packages were weighed individually. What explanation would you give?

3. To determine how many families in a community of 100,000 families qualify for the U.S. Department of Agriculture's food stamps, a random sample of 360 families was taken. It was found the 98 of those 360 families qualified. Calculate the 90% confidence limits for the total number of families in that community who will qualify for the stamps.

4. Let us assume that the average county in Iowa has 2000 farms and that there are 200,000 in the state. Suppose we wish to determine by a survey whether the fraction of farms owned by the operators has changed since the last farm census when, let us say, it was $\frac{1}{2}$. How large a random sample would be required to detect a change of $\pm 5\%$ (of 0.50) with a chance of error of 1 in 20 (a) for the state? (b) for an average county?

5. In the U.S. Presidential election of 1948, the Gallup Poll made the following

vote predictions by candidate. For comparison the actual outcome in terms of the popular vote is listed in column 3.

Candidate (1)	Gallup Poll (2)	Actual results (3)
Dewey	49.5	45.1
Truman	44.5	49.5
Thurmond	2.0	2.4
Wallace	4.0	2.4
Other	—	0.6
Total	100.0	100.0

Assume that the poll used a random sample of 2500 U.S. voters.

(*a*) Compute the 99% confidence intervals for Dewey. What do you conclude in view of the results?

(*b*) A sample of what size is required to obtain the Dewey vote within $\frac{1}{10}$ of 1 percentage point with 99% confidence?

CHAPTER 4

Choice of Sampling Unit

4.1. Introduction

Up to this point the set of elements comprising the universe has been assumed given and the problem of sampling has been how to select a fraction of those elements that would best represent the total. The set of elements in the universe and the set of elements available for sampling have been regarded as identical. However, now we shall consider situations in which: (i) the set of elements in the universe is not necessarily finite nor mutually exclusive and (ii) sampling need not be confined to any one concept of the universe.

    The set of elements used for performing the sampling operation shall be called a *sampling frame*, and in many practical situations a given universe may conceivably contain a number of alternative sampling frames. The choice of frame is often an important aspect of sample design, and it is one in which such criteria as statistical efficiency, costs, bias and practicality must be studied and evaluated.

4.2. The Universe and the Frame

Suppose our universe is Bigtown and we are interested in estimating the total number of people in it (its "population") by means of a sample survey. One *frame* for Bigtown might be the set of all its families, from which a sample could be selected to obtain the number of persons in each. Another frame might be the set of all its blocks (as given by a street plan map). A sample of blocks could be selected and the number of persons living in each ascertained by survey. These are *two* possible *frames* for a *single universe*. The total number of inhabitants in Bigtown is the same whether we total them up by the block on which they live or by the family to which they belong. For sampling, however, these two frames may be vastly different in their properties.

In agriculture, a universe of interest may be the area under cotton in Gin County. Here are some possible frames:

1. Farms.
2. Fields.
3. A grid of one-mile-square sections covering the county.
4. Lines across the county.
5. Points.

The first three frames have finite numbers of elements or *sampling units* (SUs), whereas the last two have an infinity of SUs in each. The line SU is observed by measuring the proportion of its length which lies across cotton fields; the point SU is observed by noting whether it lies in a cotton field or not. (To estimate total land in cotton we simply determine the proportion of random points in the sample falling in cotton and multiply it by the total land area of the county. In the case of the line SU the average proportion in cotton is multiplied by total land area.)

Some frames may not be satisfactory for sampling because the units are not *mutually exclusive*, a fact that can cause a bias in estimates made from samples containing these units. For example, in the Gin County universe, suppose our frame is the set of cotton buyers (those who buy cotton from farmers). The survey is to be carried out by obtaining the names of cotton farmers from a random sample of buyers. The farmers thus named will then be sought for information on amount of land they had planted to cotton. Since some farmers may have sold to more than one buyer, we do not have a frame of mutually exclusive units. Biases would be incurred if estimates were made from samples drawn from this frame unless appropriate estimators were used.

Although many frames suitable for sampling are lists of some sort, such as a list of subscribers to *Looklife* magazine, a list of dealers, a list of doctors practicing medicine in a given city, a list of telephone families (phone directory), yet in many cases lists are not available at all, or if they are, they are unsatisfactory because of omissions, errors, etc. In these cases, alternatives must be found—and here is where the sampler's ingenuity can be very useful.

4.3. Comparison of Efficiencies of SUs (RE and NRE)

Suppose for a given universe we have two frames, α and β, each comprised of mutually exclusive units, N_α and N_β, the aggregate sums of which have the same total, Y. The parameters are:

		Population			
	Total frame				
Frame	units	Total	Mean	Variance	Coefficient of variation
α	N_α	$Y_\alpha = Y$	$\bar{Y}_\alpha$	S_α^2	V_α
β	N_β	$Y_\beta = Y$	$\bar{Y}_\beta$	S_β^2	V_β

Suppose we are estimating Y, the total of y in the universe, by drawing RNR samples of sizes n_α and n_β from the N_α and N_β units in the frames, respectively. Then we have the respective estimators of Y,

$$\hat{Y}_\alpha = N_\alpha \bar{y}_\alpha \quad \text{and} \quad \hat{Y}_\beta = N_\beta \bar{y}_\beta,$$

each providing unbiased estimates of Y. How do these two cases compare in effectiveness?

Relative Precision.   One measure of effectiveness is precision. Using the concept of relative precision presented in Section 2.8, we find that the precision of sampling from frame β relative to that from sampling frame α is given by

$$(4.1) \qquad \text{RP}(\beta/\alpha) = \frac{1/\text{var}(\hat{Y}_\beta)}{1/\text{var}(\hat{Y}_\alpha)} = \frac{N_\alpha^2(1 - n_\alpha/N_\alpha)S_\alpha^2/n_\alpha}{N_\beta^2(1 - n_\beta/N_\beta)S_\beta^2/n_\beta}$$

$$(4.1a) \qquad\qquad = \frac{1 - n_\alpha/N_\alpha}{1 - n_\beta/N_\beta} \cdot \frac{V_\alpha^2}{V_\beta^2} \cdot \frac{n_\beta}{n_\alpha},$$

since var $(\hat{Y}) = N^2$ var $(\bar{y})$, and var $(\bar{y})$ for RNR sampling $= (1 - n/N)S^2/n$. Eq. 4.1a follows because the coefficient of variation is defined as $V = S/\bar{Y}$.

Relative Efficiency.   In order to have a standard of comparison, we need to fix certain conditions so that a comparison is meaningful. Since the sizes of the units in each frame may be quite different (one might be an individual person, whereas the other may be all the persons in a city block), let us fix the condition that the *sampling fraction*—the fraction of all units contained in the sample—is to be the same for both frames. With this condition, the efficiency of frame β relative to frame α is given by

$$(4.2) \qquad \text{RE } (\beta/\alpha) = \frac{N_\alpha S_\alpha^2}{N_\beta S_\beta^2} = \frac{\bar{Y}_\alpha V_\alpha^2}{\bar{Y}_\beta V_\beta^2},$$

where Eqs. 4.1 and 4.1a are solved under the condition of $n_\alpha/N_\alpha = n_\beta/N_\beta$.

Net Relative Efficiency. The costs of surveying different types or sizes of sampling units may be quite different. For example, the cost of interviewing a farmer to obtain information on the amount of land he has planted to barley may be quite different than that required to measure the amount of barley land within a square-mile area. Suppose, in the above example, that the cost of interviewing 265 farms is less than the cost of measuring 72 sections of land. Then clearly we would prefer the farm sample to the section sample, because for the same precision it simply costs less.

Suppose we have a total budget for field expenditures of C and the cost of surveying (interviewing, measuring, counting, traveling, etc.) is c_α and c_β per SU (sampling or survey unit) of frames α and β, respectively. The total budget is to be spent either on n_α SUs of α or on n_β SUs of β; hence

$$(4.3) \qquad n_\alpha = \frac{C}{c_\alpha} \quad \text{and} \quad n_\beta = \frac{C}{c_\beta}.$$

Starting with relative precision, Eq. 4.1a, and using the conditions of Eq. 4.3, the assumption that the rates of sampling n_α/N_α and n_β/N_β are small and the fact that $N_\alpha \bar{Y}_\alpha = N_\beta \bar{Y}_\beta = Y$, we obtain the *net relative efficiency*, which is approximately

$$(4.4) \qquad \text{NRE } (\beta/\alpha) \doteq \frac{N_\alpha^2 S_\alpha^2 c_\alpha}{N_\beta^2 S_\beta^2 c_\beta} \doteq \frac{V_\alpha^2 c_\alpha}{V_\beta^2 c_\beta}$$

or, alternatively,

$$(4.4a) \qquad \text{NRE } (\beta/\alpha) \doteq \frac{\text{RE } (\beta/\alpha)}{\text{RC } (\beta/\alpha)},$$

where RE and RC refer to relative efficiency and relative costs per SU, respectively.

4.4   The Cluster Sampling Unit; Estimation of $\bar{Y}$ and $\bar{\bar{Y}}$

The *cluster sampling unit* is simply a group of elements that in the sampling process is treated like a single unit. In the simple case the elements comprising a cluster are either in the sample as a group or not in the sample at all. For example, if a cluster is all the households in a city block, then the households in that block are either in the sample or not in the sample depending on whether the block happens to be selected. In no case will some of the households in a block be in a sample while others of the block are not—which would happen if a random sample of individual households in the city were selected.

To pursue the study of this case, let us assume we have a universe of M elements that are put into N clusters of exactly M_0 elements each. Since we have two sampling frames, we can draw a sample of m elements either of two ways: (i) a simple random sample of m elements drawn individually or (ii) a simple random sample of n clusters such that $nM_0 = m$. We wish to estimate $\bar{Y}$, the population mean per element. For example, consider the "year" as a frame of 364 individual days in a frame of 52 weeks of 7 days each. The day is an element, the week a cluster of seven days; and let the characteristic of interest be the amount of rainfall. Suppose we wish to compare a sample of 14 days taken as individual days on one hand or one taken as two weeks of seven days each on the other.

Notation.   The rainfall example will illustrate the notation to be used. See table on next page.

In both elemental and cluster sampling, the sample means per element, $\bar{y}_e$ and $\bar{\bar{y}}_c$, respectively, can be shown to be unbiased estimators of the population mean per element, $\bar{Y}$. They do not necessarily have the same sampling variability, however. These work out to be

$$(4.5) \qquad \text{var}(\bar{y}_e) = \left(1 - \frac{m}{M}\right)\frac{S^2}{m} = \left(1 - \frac{m}{M}\right)\frac{S^2}{nM_0},$$

since $m = nM_0$, and

$$(4.6) \qquad \text{var}(\bar{\bar{y}}_c) = \text{var}\left(\frac{\bar{y}_c}{M_0}\right) = \left(1 - \frac{n}{N}\right)\frac{S_c^2}{nM_0^2},$$

since $\bar{\bar{y}}_c = \bar{y}_c/M_0$ and $\bar{y}_c$ can be regarded as the mean of a sample of n taken from a population of N Y_is.

The relative efficiency of the cluster to elemental SUs can be obtained by comparing the relative precision of the two procedures when sampling fractions are held constant. Hence

$$\text{RE}(C/E) = \frac{\text{var}(\bar{y}_e)}{\text{var}(\bar{\bar{y}}_c)}, \qquad \text{with } \frac{n}{N} = \frac{m}{M},$$

$$(4.7) \qquad\qquad = \frac{M_0 S^2}{S_c^2}.$$

If these quantities are known, the relative efficiency is calculable. If not, some estimates can be made. Some help in determining policy can be obtained by

Item	Elemental frame (days)	Cluster frame (weeks)	Universe (the year)
Size of frame	$M_0 = 1$ (day)	$M_0 = 7$ (days)	—
No. of frame units	$M = 364$ (days) $M = NM_0$ $m = 14$ (days)	$N = 52$ (weeks) $n = 2$ weeks)	—
Observations	Y_{ij} (rainfall on jth day in ith week)	Y_i (rainfall for the ith week) $Y_i = \sum_j Y_{ij}$	Y (rainfall for the year) $Y = \sum_i Y_i$
Means	$\bar{\bar{Y}}$ (per day) $\bar{\bar{Y}} = Y/M$ $\bar{\bar{y}}_e = \dfrac{\sum_{i=1}^{n'} \sum_{j=1}^{m_i} Y_{ij}}{m}$ where[a] $m_i = 1, \ldots, M,$ $n' = 1, \ldots, n$	$\bar{Y}_c$ (per week) $\bar{Y}_c = Y/N$ $\bar{y}_c \doteq \sum^n Y_i/n$ $\bar{\bar{Y}}_c$ per day) $\bar{\bar{Y}}_c = \bar{Y}_c/M_0$ $\bar{\bar{y}}_c = \bar{y}_c/M_0$	— — — — — —
Variances	S^2 (per day) $S^2 = \dfrac{\sum \sum (Y_{ij} - \bar{\bar{Y}})^2}{NM_0 - 1}$ $s_e^2 = \dfrac{\sum^{n'} \sum^{m_i} (Y_{ij} - \bar{\bar{y}}_e)^2}{m - 1}$	S_c^2 (per week) $S_c^2 = \dfrac{\sum (Y_i - \bar{Y})^2}{N - 1}$ $s_c^2 = \dfrac{\sum (Y_i - \bar{y}_c)^2}{n - 1}$	—

[a] If m elements are selected RNR from M, the number of clusters selected, n', will range from 1 to n and the number of elements selected from the ith selected cluster will range from 1 to M_0. The indicated double summation is simply the sum over the m elements taken in the two stages.

studying the relationship between S^2 and S_c^2 for real populations and, if circumstances are favorable, to see if some principles can be developed.

Example 4.1. Beall (1939) made a detailed census of the distribution of Colorado potato beetle (*Leptinotarsa decemlineata*) in a heavily infested potato field. The field consisted of 48 rows 96 feet long. Counts were made of beetles found on each two feet of row; on 48 segments, for each of the 48 rows, a total of 10,913 beetles were found, an average of 4,737 beetles per segment. The counts, segment by segment, are shown in Fig. 4.1. Dividing the field into "square" clusters of various sizes and computing the appropriate variances and mean squares between clusters, we obtain the following data:

	1	2	3	4	5	6	7	8	9	10	11	12	13	14	15	16	17	18	19	20	21	22	23	24
1	2	4	2	4	2	3	4	6	6	7	11	1	3	2	14	3	5	1	6	4	5	4	3	11
2	0	2	0	2	7	4	2	3	9	10	5	13	7	10	11	4	5	3	6	5	1	0	14	5
3	16	9	2	8	5	10	8	7	9	3	1	9	17	10	7	7	16	10	4	12	4	13	11	12
4	5	7	7	14	20	5	9	6	8	4	7	13	1	7	18	7	5	13	7	1	4	6	8	10
5	12	5	7	0	4	10	13	5	13	3	13	3	14	5	16	7	16	9	15	11	7	1	6	6
6	11	6	17	1	9	9	5	17	6	24	9	23	16	9	23	12	5	10	9	18	12	15	18	6
7	11	10	13	21	10	11	2	20	7	16	12	9	5	13	12	10	7	11	18	3	3	22	9	11
8	2	3	5	7	4	1	0	14	11	4	5	4	0	1	7	14	10	8	4	9	2	18	7	18
9	6	8	0	0	2	1	7	1	4	9	10	6	0	3	18	11	8	10	9	0	2	23	8	14
10	8	14	10	7	3	3	17	8	5	9	1	10	14	4	8	11	9	10	8	11	13	16	13	4
11	7	12	13	16	11	8	9	1	11	5	12	19	12	11	13	5	9	12	20	20	5	8	14	22
12	13	2	10	10	7	8	15	28	8	3	9	13	4	10	14	7	18	7	13	14	12	6	10	8
13	4	7	3	2	4	1	3	7	9	7	3	3	0	3	3	9	3	8	5	6	7	4	5	3
14	2	1	7	8	2	6	7	1	5	3	13	2	6	6	3	13	10	5	4	4	5	4	4	2
15	5	7	5	2	5	6	8	2	10	4	6	2	8	3	7	2	8	4	2	8	8	7	4	2
16	6	1	3	5	6	3	5	7	8	6	5	4	4	2	1	2	5	4	11	9	6	3	9	5
17	5	4	15	6	8	5	10	6	9	6	5	3	12	5	8	8	4	0	5	8	3	1	4	6
18	6	5	1	6	6	7	5	3	2	11	2	6	5	3	8	5	6	4	4	3	2	4	4	11
19	4	3	7	10	5	5	2	2	1	5	9	10	3	6	7	11	9	2	2	7	7	4	3	1
20	6	3	7	6	3	5	6	1	4	6	1	0	2	3	2	1	5	0	4	3	6	6	7	7
21	5	4	1	5	3	6	1	2	0	5	6	3	2	2	6	1	3	2	5	1	7	5	6	2
22	3	7	1	4	0	2	4	0	3	4	5	2	2	4	4	1	2	0	4	4	4	12	8	5
23	6	5	5	6	2	0	1	5	3	3	2	2	0	1	7	2	4	2	4	7	1	2	4	2
24	5	4	6	3	10	0	9	3	3	10	1	4	5	4	2	3	1	1	3	4	1	5	0	1
25	5	6	10	6	13	3	14	7	6	8	7	10	1	6	3	0	4	0	9	15	1	4	2	13
26	3	3	10	3	10	1	1	4	2	6	1	12	2	1	2	5	7	5	8	3	4	6	8	1
27	3	8	8	6	4	7	8	7	9	8	13	8	6	1	9	13	0	3	9	2	5	4	2	3
28	9	7	10	3	5	6	6	2	6	10	4	4	1	2	0	1	6	9	3	2	4	1	1	9
29	9	11	9	3	10	2	6	6	5	22	5	6	7	4	2	4	5	13	4	7	3	2	3	4
30	4	3	9	9	9	4	6	7	2	1	8	3	5	1	6	6	15	1	7	0	4	7	3	3
31	7	4	4	8	14	5	4	9	4	11	0	4	6	4	6	8	5	7	6	11	9	2	7	4
32	6	8	10	4	9	2	0	13	9	8	1	4	1	2	2	6	9	4	11	12	8	7	12	8
33	2	8	4	9	3	4	9	5	3	4	8	3	0	1	9	6	7	8	5	12	7	4	6	7
34	4	10	6	8	7	1	8	2	4	5	6	6	11	3	5	12	11	7	8	5	2	4	3	2
35	2	4	5	9	11	10	3	4	3	5	7	1	8	8	6	2	6	7	13	5	8	5	5	2
36	4	13	4	2	8	0	3	2	2	8	6	5	1	6	4	15	5	5	3	4	15	8	13	16
37	2	6	4	11	0	3	0	0	3	2	2	7	6	6	2	4	2	4	4	7	5	3	5	5
38	9	7	10	0	2	4	5	3	2	7	4	1	1	0	1	5	8	3	4	3	5	1	2	2
39	1	6	4	0	4	2	0	5	1	6	2	2	1	0	4	8	5	1	4	2	9	5	5	4
40	4	5	0	3	4	3	4	8	1	7	7	5	5	0	12	4	0	2	4	7	2	1	2	5
41	9	6	3	6	13	5	2	4	2	2	3	1	2	2	5	4	6	5	2	3	5	0	5	4
42	6	9	3	0	2	4	7	16	11	7	5	1	7	1	9	5	4	7	8	2	0	3	3	5
43	5	5	1	10	5	4	1	4	0	1	4	7	6	1	7	5	3	4	2	2	5	4	4	3
44	5	4	5	2	4	3	4	1	2	4	1	4	2	3	2	0	1	4	3	2	1	2	5	2
45	1	0	1	8	2	5	1	2	2	1	7	10	1	5	2	1	2	4	2	1	2	0	3	1
46	0	1	0	4	0	0	0	0	0	4	1	2	4	6	1	4	2	6	4	3	1	2	6	0
47	2	1	0	0	0	15	0	8	4	2	6	2	11	1	0	1	4	4	1	4	5	2	6	3
48	2	2	4	0	0	2	11	2	5	4	2	6	2	2	1	2	5	1	2	4	6	5	2	3

	25	26	27	28	29	30	31	32	33	34	35	36	37	38	39	40	41	42	43	44	45	46	47	48
1	4	6	2	12	1	1	1	2	4	4	0	4	6	5	1	3	1	0	4	3	2	6	0	3
2	0	2	0	0	2	3	2	4	4	1	1	1	0	2	8	1	0	0	1	6	1	2	0	0
3	0	2	4	1	4	1	2	2	1	2	1	1	3	7	3	2	1	1	3	4	0	0	1	0
4	3	2	5	0	3	3	2	1	6	6	4	1	3	11	7	0	3	5	4	7	0	2	0	3
5	5	7	3	4	7	2	3	5	3	1	4	6	9	0	5	3	1	1	2	4	2	0	1	0
6	14	11	6	6	3	2	3	7	13	1	4	9	4	5	0	3	5	4	0	1	6	1	7	2
7	10	7	4	7	2	6	11	1	5	6	0	3	5	7	4	8	4	1	3	2	6	5	1	0
8	6	4	2	4	7	6	4	3	1	0	0	0	4	2	0	1	2	7	1	3	0	1	2	1
9	5	9	5	4	5	14	1	7	11	4	3	5	2	1	0	4	0	3	9	0	0	0	0	6
10	6	7	4	10	7	2	5	9	7	9	2	8	0	4	3	5	3	3	2	4	3	5	3	2
11	4	9	5	9	5	10	8	4	3	4	11	8	1	9	4	5	8	2	4	9	8	0	3	2
12	3	9	10	10	6	1	3	1	2	0	2	6	10	5	11	3	4	5	2	0	5	2	4	1
13	25	9	8	1	17	9	11	23	9	2	12	13	4	9	5	12	5	5	3	15	8	7	5	4
14	13	9	3	17	12	15	11	18	4	3	3	4	7	6	8	8	2	9	5	2	5	7	3	1
15	6	8	7	4	10	13	4	8	6	7	14	5	10	16	0	5	3	7	2	5	7	2	1	4
16	11	6	6	9	3	6	0	8	11	2	8	10	7	6	8	3	5	6	9	3	7	7	2	6
17	7	2	5	5	11	3	10	9	2	10	9	7	7	7	4	4	9	2	1	12	3	5	0	1
18	1	19	9	4	10	3	6	5	4	8	0	8	6	7	3	9	7	9	2	6	2	4	6	0
19	1	2	5	11	5	4	4	3	0	1	1	1	3	5	4	4	4	8	5	4	4	7	4	3
20	7	3	6	4	2	14	1	3	6	3	7	7	8	0	7	2	12	6	7	1	13	2	1	2
21	4	5	3	7	2	6	7	4	12	2	6	8	3	4	7	4	6	6	5	1	3	4	5	4
22	5	9	10	8	3	7	2	13	4	5	6	18	6	11	14	6	6	8	8	3	1	8	5	0
23	3	6	7	2	4	4	6	8	2	6	4	1	5	7	2	2	11	5	3	6	5	10	3	5
24	6	1	10	3	5	4	6	6	15	6	9	8	9	3	3	8	7	6	1	6	3	7	4	1
25	6	6	2	0	3	5	1	4	3	2	4	1	0	2	3	0	8	0	1	5	2	0	0	2
26	0	1	2	4	2	1	2	4	5	1	3	4	3	2	1	1	0	2	2	2	4	1	0	1
27	4	6	5	7	1	6	5	5	2	4	2	5	3	5	3	2	0	3	2	3	3	0	13	3
28	6	0	2	1	3	0	4	2	5	4	3	4	3	4	4	2	6	2	0	2	1	1	1	1
29	8	4	5	3	0	6	3	1	3	6	5	1	6	4	3	2	3	3	3	5	1	4	2	3
30	1	2	2	5	5	3	4	7	4	4	0	5	2	3	2	5	2	3	5	2	2	3	3	4
31	8	6	4	6	9	3	6	3	0	7	2	4	3	4	0	6	8	7	6	2	2	3	0	4
32	6	4	5	0	7	3	4	4	2	9	4	4	2	1	2	3	3	6	2	1	4	8	0	3
33	5	5	4	2	4	2	3	2	1	2	5	5	5	3	6	7	1	3	1	3	5	6	4	9
34	1	7	6	1	7	1	4	5	6	5	3	4	3	2	0	4	4	2	2	4	2	9	0	4
35	1	8	5	2	6	10	6	10	4	6	9	1	4	1	4	3	3	2	2	2	1	4	5	0
36	4	5	1	4	3	4	1	4	3	4	8	12	1	4	1	4	1	0	5	1	2	7	0	1
37	4	4	0	1	3	3	3	4	0	3	1	3	4	8	0	0	8	7	1	1	10	7	6	6
38	6	1	1	1	2	1	1	3	2	3	0	2	13	6	6	2	5	5	7	9	4	8	7	2
39	3	1	1	0	3	6	3	3	1	4	1	0	2	3	8	3	7	8	2	12	2	3	9	3
40	1	1	2	3	3	3	0	0	0	2	0	1	4	2	6	10	0	2	3	3	4	4	5	7
41	3	1	3	4	1	2	1	1	1	0	2	2	2	2	4	2	2	3	6	5	2	5	4	4
42	3	3	1	1	0	2	0	3	3	3	0	0	5	1	4	1	2	6	7	2	5	0	7	1
43	0	1	1	0	0	0	2	0	4	2	1	2	0	2	2	3	1	4	4	7	3	4	5	0
44	2	1	0	1	1	1	1	1	0	2	0	0	1	5	1	2	3	6	2	4	2	2	1	3
45	3	1	0	1	2	1	3	5	0	1	2	1	7	5	2	0	2	4	1	6	5	1	1	0
46	1	1	1	0	4	1	1	0	0	1	0	0	7	4	2	3	1	2	2	5	4	3	4	0
47	2	0	0	1	0	0	0	0	3	2	2	2	3	2	3	1	3	2	3	3	2	4	1	1
48	2	1	1	0	0	0	1	1	3	1	2	2	0	3	5	2	2	4	2	4	1	0	2	0

← Direction of rows of potatoes →

*Beall, G. Methods of estimating the population of insects in a field. Biometrika. 30: 422-439. 1939. Table VI. Figure 4.1 Distribution of Beetles in a Field.

Fig. 4.1. Distribution of beetles in a field.

Cluster structure	Cluster Size (M_0)	Cluster variance S_c^2
1×1	1	15.00
2×2	4	115.60
4×4	16	1,263.36
6×6	36	5,810.76
8×8	64	14,491.52

What are the relative efficiencies of each of these clusters as compares with individual segments? Using Eq. 4.7 on the beetle data, we obtain:

Cluster structure	:	1×1	2×2	4×4	6×6	8×8
Cluster size	M_0:	1	4	16	36	64
Cluster variance, S_c^2:		15.00	115.60	1263.4	5810.8	14,492
RE (C/E), percent:		100	52.7	19.0	9.3	6.6

In this case, the efficiency of clusters is lower than that of the individual segments. One must count twice as many segments in clusters of four, for example, in order to get the same amount of information in a given number of individually sampled segments.

It is not unusual for clusters in natural populations to be less efficient than elemental units in sampling. Some principles to summarize this phenomenon will be presented later—mainly in Section 4.8.

4.5.　Alternative Models for Clusters; ANOVA

A statistical tool useful for isolating and studying the sources of variation is that of analysis of variance (ANOVA). The ANOVA table is a compact arrangement for the various computations required; it also provides a useful stage for presenting alternative representations or models of sources of variation in the population and to estimate their contributions. In order to make ANOVA most useful in the analysis of survey data, some alterations may be made in the forms in which it is frequently found in other applications. For example, the finite population itself will be summarized in an ANOVA table, as well as that of the sample. Moreover, we shall be concerned with estimation rather than testing hypotheses. And three models instead of just one model will be presented in order to display the main statistical dialects used in survey literature. The three models are: (i) mean square model, (ii) components-of-variance model, and (iii) intraclass correlation model. All three will be represented in an ANOVA table format.

Mean Square Model. Table 4.1 presents the basic ANOVA table for a population of N clusters of M_0 elements each. The mean squares indicated in the table are the usual for ANOVA, except here they refer to a finite population. Hence,

$$MS\ (T) = \frac{\sum_{i=1}^{N} \sum_{j=1}^{M_0} (Y_{ij} - \bar{\bar{Y}})^2}{NM_0 - 1},$$

$$MS\ (B) = \frac{\sum_{i=1}^{N} \sum_{j=1}^{M_0} (\bar{Y}_i - \bar{\bar{Y}})^2}{N - 1},$$

$$MS\ (W) = \frac{\sum_{i=1}^{N} \sum_{j=1}^{M_0} (Y_{ij} - \bar{Y}_i)^2}{N(M_o - 1)},$$

where $\bar{Y}_i$ is mean of y per element in cluster i.

It can be shown that $MS\ (T) = S^2$ and $MS(B) = S_c^2/M_0$, and therefore the relative efficiency of the cluster can be expressed as the ratio of $MS(B)$ to $MS\ (T)$—that is,

(4.8) $$RE\ (C/E) = \frac{MS\ (T)}{MS\ (B)}.$$

If clusters were formed by a random process, we would expect the three mean squares to be equal (if the population is infinitely large) and therefore the cluster and elements to be of equal efficiency in estimating the population mean. But the natural populations that have been examined appear to have various desgrees of nonrandomness in the behavior of the groups or clusters. This behavior, however, may be altered by the sampler if the formation of clusters is under his control, a topic that will be dealt with later in this chapter.

Table 4.1. *Analysis of variance of a universe; mean square model* (*on a per-element basis*)

Source of variation	Degrees of freedom	Mean square[a]	Defined mean square
Total	$NM_0 - 1$	$MS\ (T)$	$\dfrac{N(M_0 - 1)}{NM_0 - 1} MS\ (W) + \dfrac{N - 1}{NM_0 - 1} MS\ B)$
Clusters	$N - 1$	$MS\ (B)$	$MS\ (B)$
Elements/clusters	$N(M_0 - 1)$	$MS\ (W)$	$MS\ (W)$

[a] The abbreviations have the obvious meanings: total, between, and within mean square, respectively.

Note that with this model we can express the variance of a sample mean per element by substituting MS (B) for S_c^2/M_0 in Eq. 4.6, obtaining

$$(4.9) \qquad \text{var}(\bar{y}_c) = \left(1 - \frac{n}{N}\right) \frac{\text{MS}(B)}{nM_0}.$$

Estimating MS (B), S_C^2, **and** S^2 **from a Cluster Sample.** In Table 4.2 an ANOVA table is presented for an RNR sample of n clusters of M_0 elements each taken from a population of N clusters of M_0 elements each. The column labeled "expected mean squares" indicates what the sample mean squares are estimating. With this information we can easily see that direct estimates of population mean squares for between and within clusters are given by

$$(4.10) \qquad \text{MS}(\hat{B}) = \text{MS}(b), \qquad \text{MS}(\hat{W}) = \text{MS}(w).$$

With a little algebra it can be shown that the total mean square of the population can be expressed in terms of between and within mean squares, yielding the relationship

$$(4.11) \qquad S^2 = \text{MS}(T) = \frac{N(M_0 - 1)}{NM_0 - 1} \text{MS}(W) + \frac{N - 1}{NM_0 - 1} \text{MS}(B),$$

which is estimated unbiasedly by

$$(4.12) \qquad \hat{S}^2 = \frac{N(M_0 - 1)}{NM_0 - 1} \text{MS}(w) + \frac{N - 1}{NM_0 - 1} \text{MS}(b).$$

Or, if n is large relative to M_0, a good approximation is given by the sample mean square,

$$(4.13) \qquad \hat{S}^2 \doteq \text{MS}(t).$$

Table 4.2. ANOVA of a sample; mean square model (*on a per-element basis*)

Source of variation	Degrees of freedom	Mean square	Expected mean square
Total	$nM_0 - 1$	MS (t)	$\dfrac{n(M_0 - 1)}{nM_0 - 1}$ MS $(W) + \dfrac{n - 1}{nM_0 - 1}$ MS (B)
Clusters	$n - 1$	MS (b)	MS (B)
Elements/clusters	$n(M_0 - 1)$	MS (w)	MS (W)

Components-of-Variance Model. The relationship of mean squares and components of variance for this case is given in Table 4.3.

We are now defining the between mean square as a sum with two parts: one, the within-cluster variance, S_W^2; the other, a term that under most situations

Table 4.3. ANOVA of a universe; components-of-variance model (on a per-element basis)

Source of variation	Degrees of freedom	Mean square	Defined mean square
Total	$NM_0 - 1$	MS (T)	$S_W^2 + M_0 \dfrac{N-1}{M_0 N - 1} S_B^2$
Clusters	$N - 1$	MS (B)	$S_W^2 + M_0 S_B^2$
Elements/clusters	$N(M_0 - 1)$	MS (W)	S_W^2

can be called a *component of variance*, S_B^2 (which is due *more* to clusters' varying among themselves than to the variation occurring within clusters). If the MS (B) is smaller than MS (W), which can occur for some populations, then S_B^2 will have a negative value and can no longer be regarded as a component of variance. However, in general, we may regard S_B^2 as a quantity defined as

$$(4.14) \qquad S_B^2 = \frac{\text{MS } (B) - \text{MS } (W)}{M_0},$$

which can be either positive or negative. If S_B^2 is positive, it can be regarded as a component of variance; if negative, simply as a number.

With this model we can express the variance of a sample mean (per element) by

$$(4.15) \qquad \text{var } (\bar{y}_c) = \left(\frac{N - n}{N} \right) \frac{S_W^2 + M_0 S_B^2}{n M_0}$$

by merely replacing S_c^2 / M_0 in Eq. 4.6 by its equivalent, $S_W^2 + M_0 S_B^2$.

Estimating S_W^2, S_B^2, and S^2 from a Cluster Sample. Again putting the relevant computations of an RNR sample into an ANOVA table, we have Table 4.4.

Table 4.4. ANOVA of a sample components-of-variance model (on a per-element basis)

Source of variation	Degrees of freedom	Mean square	Defined mean square	Expected mean square
Total	$nM_0 - 1$	MS (t)		$S_W^2 + \dfrac{M_0(n-1)}{M_0 n - 1} S_B^2$
Clusters	$n - 1$	MS (b)	$s_W^2 + M_0 s_B^2$	$S_W^2 + M_0 S_B^2$
Elements/clusters	$n(M_0 - 1)$	MS (w)	s_W^2	S_W^2

Examining Table 4.4, we can see that s_W^2 is an unbiased estimate of S_W^2 and hence is consistent with MS (w) as an unbiased estimate of MS (W) (Table 4.2). Since $s_W^2 + M_0 s_B^2$ estimates $S_W^2 + M_0 S_B^2$ unbiasedly [because MS(b) is an unbiased estimate of MS (B), see Table 4.2], then it follows that s_B^2 is an unbiased estimate of S_B^2. To calculate s_B^2, we have

$$(4.16) \qquad\qquad s_B^2 = \frac{MS(b) - MS(w)}{M_0}.$$

Likewise, an unbiased estimate of S^2 can be obtained by substituting s_W^2 and s_B^2 as estimates of S_W^2 and S_B^2, respectively, in Table 4.3, obtaining

$$(4.17) \qquad\qquad \hat{S}^2 = s_W^2 + \frac{M_0(N-1)}{M_0 n - 1} s_B^2.$$

Generally it will be found that MS (t) is an underestimate of S^2, but if n is large relative to M_0, the bias is negligible and MS (t) will be a good estimate of S^2.

It may appear that this model has no advantage over the simpler mean square model for dealing with cluster sampling; in fact, it is just more cumbersome. This is true for this case where clusters are surveyed intact. When subsampling is imposed (see Chapter 9), this model will be more useful.

Correlation Model. A method alternative to that of mean squares for expressing the variance of estimates made from cluster sampling, and their efficiencies relative to that of simple random sampling, employs the concept of *intraclass correlation*, which is the correlation between pairs of observations taken from the same class (in our case, cluster).

Suppose we have, as in Table 4.1, a universe of N clusters of M_0 elements each. We can represent the various sources of variation in an ANOVA-type table such as Table 4.5.

In this table, the between and within mean squares are represented as total mean squares multiplied by factors involving δ, called the intraclass

Table 4.5. ANOVA of a universe correlation model (on a per-element basis)

Source of variation	Degrees of freedom	Mean square	Defined[a] mean square
Total	$NM_0 - 1$	MS (T)	S^2
Clusters	$N - 1$	MS (B)	$S^2[1 + \delta(M_0 - 1)]$
Elements/clusters	$N(M_0 - 1)$	MS (W)	$S^2(1 - \delta)$

[a] When difference between $N - 1$ and N can be ignored.

correlation coefficient. It is defined as

$$(4.18) \qquad \delta = \frac{\sum_i \sum_{j<k} (Y_{ij} - \bar{\bar{Y}})(Y_{ik} - \bar{\bar{Y}})}{NM_0(M_0 - 1)S^2/2} \left(\frac{NM_0}{NM_0 - 1}\right)$$

where Y_{ij} = the jth element in the ith cluster. Hence the numerator is the covariance of all possible pairs of elements within groups. A simple yet exact computation form for δ in terms of mean squares is given by

$$(4.19) \qquad \delta = \frac{[(N-1)/N] \; MS \; (B) - MS \; (W)}{(M_0) \; MS \; (T)}.$$

The values of δ can range from $-(1/M_0 - 1)$ to $+1$.

If conditions are such that $N/(N-1)$ can be regarded as 1, then we can express the variance of the mean per element of a cluster sample as

$$(4.20) \qquad \text{var} \; (\bar{\bar{y}}) = \left(\frac{N-n}{N}\right) \frac{S^2}{nM_0} [1 + (M_0 - 1)\delta].$$

If δ is zero—that is, clusters are random groupings—then the variance of $\bar{\bar{y}}$ becomes simply that for simple random sample of elements.

Hence, whenever the intraclass correlation is negative, the cluster is more efficient than the individual element as a sampling unit. Although δ can range from $-(1/M_0 - 1)$ to $+1$, ordinarily it will be found to be small and positive and, in a few rare cases, negative. When $\delta = 0$, it can be said that the clusters are essentially random groupings of the elements in the universe. When δ is positive, the elements within clusters can be said to be "more alike" than elements in different clusters, and likewise are more "unalike" if δ is negative. Hence, intracluster correlation is a measure of the degree to which clusters depart from a "random formation."

Both S_W^2 and S_B^2 can be expressed in terms of δ. Thus, ignoring the case of small N,

$$(4.21) \qquad S_W^2 = S^2(1 - \delta)$$

and

$$(4.21a) \qquad S_B^2 \doteq S^2 \delta.$$

Moreover, in terms of components of variance,

$$(4.22) \qquad \delta \doteq \frac{S_B^2}{S_W^2 + S_B^2} \doteq \frac{S_B^2}{S^2},$$

hence, δ can be interpreted as the portion of total variance ascribable to cluster.

Estimating δ and S^2 from a Cluster Sample. The logic for the estimators of δ can usually be seen most easily in an ANOVA setting. Table 4.6 presents the sample results and what the various mean squares are estimating where the sample is n clusters of M_0 elements each.

Table 4.6.　ANOVA of a sample; correlation model (*on a per-element basis*)

Source of variation	Degrees of freedom	Mean square	Defined[a] mean square	Expected[b] mean square
Total	$nM_0 - 1$	MS (t)	s_T^2	S^2
Clusters	$n - 1$	MS (b)	$s_T^2[1 + (M_0 - 1)\hat\delta]$	$S^2[1 - (M_0 - 1)\delta]$
Elements/clusters	$n(M_0 - 1)$	MS (w)	$s_T^2(1 - \hat\delta)$	$S^2(1 - \delta)$

[a] Provided differences between n and $n - 1$ can be ignored.
[b] Provided differences between N and $N - 1$ can be ignored.

It may be noted in Table 4.6 that both s_T^2 and $\hat\delta$ are derivable from the sample between and within mean squares. Solving the two equations for the two unknowns, we obtain

$$(4.23) \qquad \hat\delta \doteq \frac{\text{MS }(b) - \text{MS }(w)}{\text{MS }(b) + (M - 1)\text{MS }(w)}.$$

This approximate estimate is unbiased if $N/(N - 1)$ can be regarded as 1. By solving for S^2, we obtain as an estimate of S^2,

$$(4.24) \qquad \hat{S}^2 \doteq \frac{\text{MS }(b) + (M_0 - 1)\,\text{MS }(w)}{M_0}.$$

We can estimate intraclass correlation coefficient in terms of components of variance by substituting unbiased estimates in Eq. 4.22:

$$(4.25) \qquad \hat\delta \doteq \frac{s_B^2}{s_W^2 + s_B^2}.$$

Example 4.2. In the beetle data of Example 4.1, an analysis of variance of clusters of four segments (2×2) gives the results below. Compute the intraclass correlation coefficient for this case by using the exact method for the mean square model and the approximation for the component-of-variance model.

Source of variation	Degrees of freedom	Sum of squares	Mean square	Mean square
Total	2303	34,549	15.00	MS (T)
Clusters	575	16,615	28.90	MS (B)
Segments/clusters	1728	17,934	10.38	MS (W)

From Eq. 4.19, we have the exact intraclass correlation using mean squares:

$$\delta = \frac{[(N-1)/N]\ \text{MS }(B) - \text{MS }(W)}{M_0\ \text{MS }(T)} = \frac{[(576-1)/576]\ (28.90) - 10.38}{4(15.00)} = 0.309$$

And from Eq. 4.22, an approximate value of δ is given by

$$\delta \doteq \frac{S_B^2}{S_W^2 + S_B^2} = \frac{4.63}{10.38 + 4.63} = 0.308.$$

Some Principles of Cluster Formation.   As an illustration, suppose we have a universe consisting of eight balls, of which four are black and four are white. The variate of interest is the color of the ball, $Y_i = 0$ if black and 1 if white. We wish to estimate $\bar{Y}$, which is the proportion of white balls, by sampling. We shall consider three sampling frames:

1. Eight individual balls.
2. Four clusters of two balls each, wherein "alike" colors occur together.
3. Four clusters of two balls each, wherein "unlike" colors occur together.

Pictorially the three frames are:

Frame unit	"Alike" pairs		"Unlike" pairs		"Random" pairs	
1	O	O	O	●	O	O
2	O	O	O	●	O	●
3	●	●	O	●	●	O
4	●	●	O	●	●	●

An ANOVA table prepared for each frame is shown in Table 4.7.

It is easy to see the effects on sampling variance brought about by the different schemes of pairing. When "unalikes" are put together in clusters, the MS (B) goes to zero, in which case there is no sampling variance. But when "alikes" are put together in clusters, the MS (B) and hence sampling variance becomes a maximum. The MS (B) for random sampling lies in between the two extremes. The principle to derive from this example is that sampling variance in cluster sampling can be reduced by putting population heterogeneity *within* the cluster.

Summary of the Models.   The relationships between the three variance models are summarized in Table 4.8, in which both the exact and approximate functions are presented. Since the three are algebraically equivalent, none has any technical advantage over the others. Each may have an advantage in specific circumstances, depending on the problem being dealt with and the kinds of statistics available. Because all three models are used, usually by

Table 4.7.   ANOVA of the eight-ball universe (on a per-ball basis)

Frame	Source of variation	Degrees of freedom	Sums of squares	Mean square
	Gross	8	4	
	Mean	<u>1</u>	<u>2</u>	
	Total	7	2	$2/7 = 0.286$
"Alikes" paired	Pairs	3	2	$2/3 = 0.667$
	Within pairs	4	0	0
"Unalikes" paired	Pairs	3	0	0
	Within pairs	4	2	$2/4 = 0.500$
"Random" pairs	Pairs	3	$6/7^a$	$2/7^a = 0.286$
	Within pairs	4	$8/7^a$	$2/7^a = 0.286$

[a] Average over the three possible pair configurations. Of the 28 possible random pairs, six are ○ ○, six are ● ●, and 16 are ○ ●. The imputed MS (B) is given by $[N/(N-n)]E[(\bar{y}_t - \bar{Y})^2]/M_0$ and imputed SS $(B) = $ (df) $[MS(B)]$. The imputed SS (W) is obtained by subtraction.

different "schools" of survey statisticians, it is sometimes helpful to have a convenient means of translation. Perhaps Table 4.8 can serve as a sort of Rosetta stone for this purpose.

4.6   Systematic Sampling

Suppose we have a universe of 12 elements and we wish to draw from it samples of size three. The selection is made by the following procedure. Divide universe size, M, by proposed sample size, m; the result, M/m, is called the *sampling interval*. In our case this interval is $\frac{12}{3} = 4$. Next choose a random number from 1 to the size of the sampling interval. This, the *start number*, designates the first element in the sample. The start number plus the sampling interval designates the second element in the sample. If our start number is 2, then the second sample number is $2+4 = 6$; likewise, by repeatedly adding the sampling interval we obtain the set of numbers designating the sample. Hence in our case, the sample consists of elements 2, 6, and 10. The structure of this type of selection, called *systematic sampling*, may be represented as follows for each possible sample:

```
Universe:  1  2  3  4  5  6  7  8  9  10  11  12
SU No. 1:  1           5           9
       2:     2              6           10
       3:        3              7            11
       4:           4              8             12
```

Table 4.8. *Basic interrelationships of the three variance models (for equal-sized clusters)*

Item	Mean squares	Components of variance	Intraclass correlation
MS (T): Exact	$\left[\dfrac{N(M_0-1)}{NM_0-1}\right]$ MS (W) + $\left(\dfrac{N-1}{NM_0-1}\right)$ MS (B)	$S_W^2 + \left(\dfrac{N-1}{NM_0-1}\right)M_0 S_B^2$	S^2
Approx.[a]	MS (W) + MS $(B)/M_0$	$S_W^2 + S_B^2$	—
MS (B): Exact	$\left(\dfrac{NM_0-1}{N-1}\right)S^2 - \left(\dfrac{N(M_0-1)}{N-1}\right)$ MS (W)	$S_W^2 + M S_B^2$	$S^2\left[\dfrac{NM_0-1}{M_0(N-1)} + \dfrac{N(M_0-1)}{N-1}\,\delta\right]$
Approx.[a]	$M_0[S^2 - \text{MS }(W)]$	—	$S^2[1+(M_0-1)\delta]$
MS (W): Exact	$\left[\dfrac{NM_0-1}{N(M_0-1)}\right]S^2 - \left[\dfrac{N-1}{M(M_0-1)}\right]$ MS (B)	S_W^2	$S^2\left(\dfrac{NM_0-1}{NM_0} - \delta\right)$
Approx.[a]	$S^2 - \text{MS }(B)/M_0$	—	$S^2(1-\delta)$

[a]If $(N-1)$ is approximately N and (M_0-1) is approximately M_0.

Notice that this kind of selection generated four possible samples:

$$1,5,9; \quad 2,6,10; \quad 3,7,11; \quad 4,8,12;$$

and that each has an equal chance of being drawn, because the start numbers 1, 2, 3, and 4 have equal probabilities of selection. Actually each of the four samples can be regarded as a cluster, since each element is included in one and only one group and all elements in a group are drawn into the sample by a single random draw. Hence the effect of systematic selection is to cause the creation of a sampling frame of $M/m = k$ clusters of m sampling units each. In simple systematic sampling, such as that described above, only one of these clusters is selected for the sample. The variance of the sample mean per element is given by Eq.4.6:

$$(4.26) \qquad \text{var}\,(\bar{y}_{\text{sys}}) = \left(\frac{N-n}{N}\right)\frac{S_c^2}{nM_0^2},$$

where now N = the total number of clusters = the sampling interval = k,

$\qquad n = 1$ unless more than one *start number* is selected,

$\qquad M_0$ = the number of elements in the cluster, $M_0 = M/N = m$,

$\qquad S_c^2$ = the variance of the population of k cluster totals.

In the case where only one start number is used, no estimate of sample variance can be made, because an s_c^2 cannot be computed for samples of one sampling unit—that is, one cluster. Thus if an unbiased estimate of sampling variance is desired, two or more start numbers may be drawn. An approximate estimate can be made if systematic samples are regarded as stratified random samples (see Chapter 7).

The efficiency of systematic samples compared with random is identical to that given by Eq. 4.7:

$$\text{RE (sys/random)} = \frac{M_0 S^2}{S_c^2},$$

or, in terms of analysis of variance,

$$\text{RE (sys/random)} = \frac{\text{MS }(T)}{\text{MS }(B)},$$

where S^2 = variance of the individual elements,

$\qquad S_c^2$ = variance of cluster totals,

$\qquad$ MS (T) = total mean square (of elements),

$\qquad$ MS (B) = between cluster mean square (per element).

If the method is to be efficient, the variation between clusters of a systematic sample must be less than the variation within clusters. Because of the wide

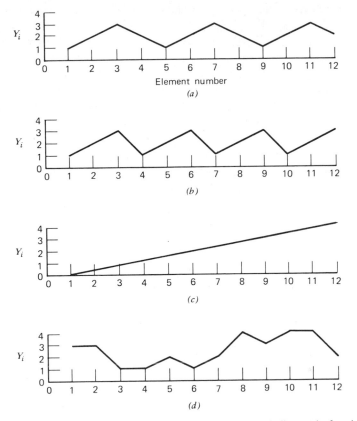

Fig. 4.2. Illustrative trends for a universe of 12 elements. (*a*) Periodic trend of period 4. (*b*) Periodic trend of period 3. (*c*) Linear trend. (*d*) Random trend.

spacing of the elements in a cluster, occasioned by the usual choices of sampling interval, this is frequently found to be the case in a wide variety of applications. Systematic sampling is not only usually more efficient than simple random sampling, but it is usually simpler to prepare.

There are, however, two main faults: (i) difficulty in estimating the sampling variance unbiasedly (unless replication is adopted—in which case there is usually some loss in efficiency over the unreplicated) and (ii) the possibility that conditions are such that a very great loss in efficiency may be suffered without adequate warning. Losses in efficiency may occur when conditions exist such as those shown in Fig. 4.2. A periodic or cyclic trend where the period is the same as the sampling interval is the worst possible situation for systematic sampling. In case (a), a sampling interval of four will put all the "alike" elements into the same samples; hence, all variability will be concentrated between samples and none within. The MS (B) will be very great; in fact, it will be $\frac{11}{3} \times$ MS (T), hence the RE (sys/random) in this case is $\frac{3}{11}$ or 27%—a severe loss. A slight modification of case (a), where the period is changed from four to three, makes the efficiency of systematic sampling a maximum for a sampling interval of four. As shown in case (b), the clusters have the same means and therefore no sampling variance. Hence, the systematic sample is quite sensitive to the character of the periodicity relative to the sampling interval.

In the case where the population values occur as a linear trend [case (c) of Fig. 4.2], systematic sampling will be more efficient than random. If k is the sampling interval, then the relative efficiency of systematic to random sampling for the case of a *linear trend* is given by

$$(4.27) \qquad \text{RE(sys/random)} = \frac{M+1}{k+1},$$

which in our case is

$$\text{RE (sys/random)} = \frac{12+1}{4+1} = \frac{13}{5} \text{ or } 260\%.$$

Random trends such as case (d) are sampled equally well by systematic or random selection. There is no essential difference in the two methods when the series is random. More on systematic sampling will be given in Chapter 12.

4.7. Some Cost Functions

As indicated in Section 4.3, if costs of surveying—that is, observing or measuring—different types or sizes of SUs differ, and if costs are important in the conduct of the survey, then they should be considered when deciding on the best SU. When the SU is a cluster, survey costs per element generally depend on the number of elements in the cluster, particularly if there is a fixed cost in getting access to the cluster itself. If the cluster is a page in a ledger, a crate of gidgets, a group of households in a city block, or a group of plants in a field, there is usually some cost in time and travel in getting the investigator to the page, crate, block, or field before he can start examining the individual elements there. A simple and yet useful function is the linear function for cluster SUs:

$$(4.28) \qquad C = nc_B + nM_0 c_W,$$

where C is the total cost, c_B is the cost of attaining access to a cluster, which is constant regardless of the size of cluster, and c_W is the cost associated with an element within a cluster, also independent of cluster size. For a cluster size $M_0 = 1$, the cost per SU would be $c_1 = c_B + c_W$; for $M_0 = 5$, $c_5 = c_B + 5c_W$.

Surveys in which travel is an important component of costs and where such costs differ significantly among different choices of type and size of SU (cluster) present special problems in constructing cost estimates. Many agricultural and farm surveys are of this sort. Some common travel situations, although somewhat abstracted, are presented in Section 4.13.

Interview surveys, rural or urban, have costs that can be easily categorized and put into simple but useful mathematical function form, where size of cluster is considered a continuous variable. An example is presented below. In the next section, variance, as a function of cluster size will be considered in order to facilitate a more general approach to obtaining an optimal cluster size.

A Cost Function for an Interview Survey.   Consider a survey of a sample of m elements (farms, households, etc.) consisting of n clusters of M_0 elements each. The information is to be obtained by a direct visit to each element (house, farm, store, etc.) by automobile. The various components of cost are denoted as follows:

A = area of the universe,

n = number of SUs (clusters) in the sample,

M_0 = number of elements in each cluster,

C = total budget for the survey except for overhead,

t_E = time (in hours) required for observing an element (interview),

t_c = time (in hours) required to "canvas" the cluster and perform other work at a cluster site that is essentially independent of cluster size,

t_T = time (in hours) required to travel one mile ($= 1/\text{mph}$),

C_T = cost per mile of travel (in dollars per mile),

C_H = average cost per hour of time for the investigator (including his meals and lodging) and the processor and his machines,

t_p = time required to edit and process a cluster as such that is essentially independent of the size of cluster,

t_q = time required to edit and process each element (questionnaire).

The costs by type and place of operation are:

		Place of operation	
Types of operations	Between SUs	At SUs between elements	At elements
Canvassing and interviewing		$C_H t_c n$	$C_H t_E n M_0$
Traveling (time)	$C_H t_T \sqrt{An}$		
Transport	$C_T \sqrt{An}$		
Processing		$C_H t_p n$	$C_H t_q n M_0$
Total	$(C_H t_T + C_T)\sqrt{An}$	$(t_c + t_p)C_H n$	$(t_E + t_q)C_H n M_0$

By letting $C_1 = (t_c + t_p)C_H$ plus other costs per cluster essentially independent of M_0,

$C_2 = (t_E + t_q)C_H$ plus other costs per element,

$C_3 = (C_H t_T + C_T)\sqrt{A}$ plus other costs per cluster due to distance,

we have for total cost function

(4.28a) $$C = C_1 n + C_2 n M_0 + C_3 \sqrt{n}.$$

Although estimates going into this function may be fairly rough, it may still be found useful for determining not only the costs of a survey but the optimum size of cluster (see Section 4.9).

Paths and Distances in Survey Travel.   In many surveys, the cost of travel is an important element of cost and may be an additional factor to consider in the overall design. We shall consider here briefly some of the problems of determining these costs.

The basic factors to be considered are: (i) the position of field points relative to the home base of the investigator, (ii) whether he must return home at the end of the day or week or not till the end of the survey, (iii) the size and shape of the area traveled, (iv) the nature of the road network, and (v) his choice of *path* (that is, does he attempt to minimize travel distance or time, or does he use some other principle). Some common path modes and the formulas for the expected distances involved are given in Figure 4.3. Some explanations for each of the eight cases follow.

Case (a): Systematic Arrangement; Home in Corner.   In this case the formula is exact if we have integral values for n_c and n_r, where $n_c n_r = n$. Here $\gamma = 1$ if n_c is odd and 0 if n_c is even.

Case (b): Random Arrangement; Home in Corner.   This is an approximate but quite useful formula. The values of D obtained will vary from sample to sample even if n is fixed, hence $E[D]$ is an *expected value*, or the average. Here $A = ab$.

Case (c): Straight Line.   Given a line of length L with n randomly located points along it. The expected travel in this case is the distance to the farthest point from one end of the line. The expected distance to the nearest point is $L/(n+1)$.

Case (d): Home in Center; One Point.   With home base in the center of a rectangle the expected distance to a single randomly located point and back is $(a+b)/2$.

Case (e): Home in Center; n Points, One Trip.   A total of n random points scattered over the rectangle.

Case (f): Home in Center; One of N Trips.   In this case it is assumed that there is a total of m points over the whole rectangle and only m_0 can be dealt with on

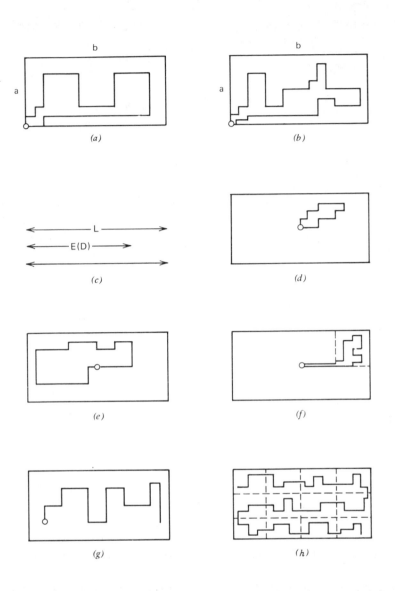

Fig. 4.3. Expected travel distances for some common cases. (a) Systematic arrangement; home in corner. $E[D] = \sqrt{abn} + b + \gamma(a - \sqrt{ab/n}$. (b) Random arrangement; home in corner. $E[D] = \sqrt{A}(1 + \sqrt{n})$. (c) Random points on line; home at end. $E[D] = nL/(n+1)$. (d) Random point; home in center. $E[D] = (a+b)/2 = \sqrt{A}$. (e) Random arrangement, home in center. $E[D] = \sqrt{An}$. (f) Random trip's worth, home in center. $E[D] = \sqrt{A} + \sqrt{A_0}(\sqrt{m_0} - 1)$. (g) Random arrangement; nonreverse path. $E[D] = (n-1)[a/(n+1) + b/3]$. (h) "Poststratified"; nonreverse path. $E[D] = [(\bar{m}+3)/3\sqrt{\bar{m}}\sqrt{Am}$

99

a single trip (such as a day's work, for example). Hence N trips will be required to complete the whole rectangle ($N = M/m_0$). The area covered on a single trip is denoted by A_0 ($A_0 = A/N$). The formula given is for a single trip. For N trips multiply by N.

Case (g): Nonreversing Path; Simple Random.   With n randomly located points scattered in a rectangle, where $b > a$, it is assumed an investigator begins at the "leftmost" point and proceeds to the right, visiting each point in order of leftmost position. Hence, once started, the investigator never moves to the left, only up, down, and to the right. Clearly this is not always the shortest route through the points, but it is mathematically simple and perhaps quite realistic for many practical situations.

Case (h): Nonreversing Path; "Post-Stratified."   In this case the path is carried out along "strips," the total path having a serpentine configuration. The optimum number of strips is determined by targeting $m/N_cN_r = \bar{m}$ at about 3. Hence, $N_c = (mb/a)^{1/2}$ and $N_R = (ma/b)^{1/2}$, where m is the total number of points in the whole rectangle, N_r is the number of row strips, and N_c is the number of column strips, whichever direction the strips are to be taken. At "optimum stratification," where $\bar{m} = 3$ and $N = m/3$,

$$E[D] = \frac{2}{\sqrt{3}}\sqrt{Am} = 1.155\sqrt{Am}$$

In the paths described above it is assumed that the "dogleg" is followed in going from point to point. If beeline distance is desired, it can be obtained by use of the factor θ for the appropriate shape of the grid unit.

$$\text{Beeline distance} = \frac{\text{Dogleg distance}}{\theta}.$$

Some illustrative values are:

Ratio of b to a of rectangle	1 : 1	2 : 1	4 : 1	∞ : 1
Factor θ	1.274	1.254	1.205	1.000

An additional factor affecting both beeline and dogleg distances is lack of rectangularity of the road net (such as in areas not in the Public Land Survey). If for the square it is set at one, it may be 1.2 or greater for some road nets, especially in hilly areas. If high precision is required, the values must be determined experimentally.

4.8. Some Variance Functions

If the relationship of the variance among clusters to the size of the cluster can be put in the form of a simple mathematical function, a number of advantages may result. For example, such a variance function may provide us estimates of between-cluster variances over a wide range of cluster sizes rather than just those chosen for experimentation, thus permitting a simpler and more precise determination of optimal cluster size. Three functions have appeared in the literature and are briefly described below. The three functions linking cluster and element variances by size of cluster differ from each other primarily with regard to the parameter on which each is structured. The parameters used—(i) the between-cluster variance, (ii) the within-cluster variance, and (iii) the intraclass correlation coefficient—will constitute a good way to introduce them.

Between-Cluster Variance. This method, proposed by Smith (1938),* assumes

$$(4.29) \qquad \frac{S_c^2}{M_0} = \text{MS } (B) = A_S M_0^s,$$

where ordinarily $s \geq 0$, and when clusters are random groups of elements, $s = 0$. Since $A_S = s^2$ when $M_0 = 1$, an estimate of s can be obtained if MS (B) can be observed for some $M_0 > 1$. Note that if the case $M_0 = 1$ is accepted, then MS $(B) = S^2 M_0^s$, with only one parameter to be determined.

Within-Cluster Variance. This method, proposed by Jessen (1942), assumes

$$(4.30) \qquad S_W^2 = \text{MS } (W) = A_j M_0^j,$$

where ordinarily $j \geq 0$, and when clusters are random groups of elements, $j = 0$. An estimate of j can be determined with an observation on only one level of M_0 if it is assumed that the universe is itself a cluster of size M and MS (B) of S^2. It may be noted that if $M_0 = 1$, then $A_J = S_W^2$, the "implicit" within-cluster variance for clusters of size 1.

Intraclass Correlation Coefficient. This method, presented by Hansen et al. (1953), assumes that the intraclass correlation

$$(4.31) \qquad \delta = A_H M_0^h,$$

where A_H is a constant to be determined and h can be either positive or negative but ordinarily will be negative and less than unity.† When clusters are random,

---

* About the same time, Mahalanobis (1939) presented a variance function similar to Smith's, in which the independent variable is between-cluster mean square (or variance). Mahalanobis' function contained a PQ in the place of S^2 in the numerator, where P is the fraction of area (or elements) in the universe containing, say, a given crop. It has the same general characteristics as Smith's function.

† This method can also be regarded as essentially one in which S_B^2, the component of variance due to cluster, is a function of cluster size. Thus

$$(4.31a) \qquad S_B^2 \doteq S^2 A_H M_0^h,$$

since when N is large, $\delta \doteq S_B^2/S^2$. Thus each of three methods directs its attention to one of the three variances involved: the between, the within, or the weighted average of the two—that is, the between mean square.

both A_H and $h = 0$. An estimate of A_H and h requires observations on at least two levels of M_0.

The appropriate function for a given situation will depend on what kind of data are available [MS (W), MS (B), etc.], the precision required, and the peculiarities of the phenomena dealt with. The S and H methods are somewhat simpler to carry out, although according to Hendricks (1944), the J method may be a little more accurate than the S method. (He did not compare the H method.) For most practical purposes the S and H methods are recommended, primarily because of their simplicity, although in every case an examination should be made to see how well the function fits the data.

Illustrations and Comparisons.   Sets of real data will be examined to study the nature of the three functions. In the biological and agricultural area we shall use the beetle data (mentioned in Section 4.4), and some data collected in 1937 to look into the problem of sampling unit choice in estimating areas in crops. The study covered 19 midwestern counties, but only Brookings County, S.D., will be used here. The county consisted of 720 one-mile-square sections, into which the county is mapped by the U.S. Public Land Survey. Each field was identified on aerial photographs, and the type of crop growing in it was determined by on-site inspection and its area measured, hence providing a set of data useful for testing sampling techniques. Basic ANOVA computations on acreages in barley and corn are shown, together with the beetle data, in Table 4.9A for several sizes of clusters. Note for the three cases shown that δ is positive, and it declines as M_0 increases; moreover that S_W^2 increases and S_B^2 decreases (usually) as M_0 increases, and the sum of the two is approximately constant and equal to S^2. Table 4.9B shows similar calculations of characteristics of some human populations. They are direct and derived results of studies carried out by the U.S. Bureau of the Census to study the sampling efficiency of different cluster sizes. The data were from the 1940 Census, confined to cities of 100,000 population or more.

Table 4.10A presents δs for some bio-agricultural characteristics for various cluster (plot) sizes. Note the rather high δs for yields of oranges among neighboring trees. Table 4.10B shows values of δ for some household characteristics in U.S. metropolitan areas. In both cases the δs decrease with increasing M_0.

Table 4.11 presents the calculated coefficients for each of the three variance functions for the items presented in Table 4.10A and B. For these data the S function seems to fit best and the H function the poorest. However, the data are sketchy, and in the case of the household items the original computations were not available, so a valid comparison of goodness-of-fit is not attainable.

Principle of Proximity; Gradients.   When clusters are formed by putting elements (farms, households, fields, etc.) that are in close proximity to each

Table 4.9A. *Values of mean squares, components of variance, and intraclass correlation coefficients for selected cluster sizes, for selected bio-agricultural items*

Item (1)	Cluster Size M_0 (2)	N (3)	MS (T) S^2 (4)	MS (W) S_W^2 (5)	MS (B) $S_W^2 + M_0 S_B^2$ (6)	S_B^2 (7)	$S_W^2 + S_B^2$ (8)	δ (9)
Beetle counts[a] $Y = 10{,}913$	1	2304	15.00	—	15.00	—	15.00	—
	4	576	15.00	10.38	28.90	4.63	15.01	0.309
	16	144	15.00	10.77	78.96	4.26	15.03	0.284
	36	64	15.00	10.88	161.41	4.18	15.06	0.279
	64	32	15.00	11.74	226.43	3.35	15.09	0.224
Barley acreage[b] $Y = 49{,}326$	1	720	1647	—	1647	—	1647	—
	4	180	1647	1492	2568	269.0	1761	0.153
	36	20	1647	1506	6902	269.8	1776	0.152
Corn acreage[b] $Y = 120{,}582$	1	720	3400	—	3400	—	3400	—
	4	180	3400	2077	7391	1329	3406	0.390
	36	20	3400	2339	42,500	1116	3455	0.323

[a] Basic element a segment of row 2 feet long. All clusters are square.
[b] Basic element a plot one mile square. All clusters are square.

Table 4.9B. *Values of mean squares, components of variance, and intraclass correlation coefficients for selected cluster sizes, for selected econo-demographic items**

Item (1)	Cluster size M_0 (2)	N (3)	MS $(T)^a$ S^2 (4)	MS $(W)^b$ S_W^2 (5)	MS $(B)^c$ $S_W^2 + M_0 S_B^2$ (6)	S_B^2 ^{d} (7)	$S_W^2 + S_B^2$ (8)	δ^e (9)
Unemployed males/HH	1	11,106	0.208	—	0.208	—	0.208	—
$Y = 2150$	3	3646	0.208	0.196	0.233	0.0123	0.208	0.060
$Y/M = 0.1936$	9	1157	0.208	0.193	0.324	0.0146	0.208	0.070
	27	323	0.208	0.199	0.451	0.0093	0.208	0.045
	62	180	0.208	0.201	0.639	0.0071	0.208	0.034
Population/HH	1	11,106	3.49	—	3.49	—	3.49	—
$Y = 37{,}515$	3	3646	3.49	2.69	5.10	0.803	3.49	0.230
$Y/M = 3{,}378$	9	1157	3.49	2.84	8.68	0.649	3.49	0.186
	27	323	3.49	2.99	16.38	0.496	3.49	0.142
	62	180	3.49	3.26	17.54	0.230	3.49	0.066
Percent home-owned HHs	1	11,106	0.123	—	0.123	—	0.123	—
$Y = 2185$	3	3646	0.123	0.102	0.165	0.0210	0.123	0.170
$Y/M = 0.1967$	9	1157	0.123	0.102	0.291	0.0210	0.123	0.171
	27	323	0.123	0.103	0.654	0.0204	0.123	0.166
	62	180	0.123	0.111	0.843	0.0118	0.123	0.096

*Based on data in Hansen et al. (1953). Vol. I, pp. 597 and 264.

[a] From Hansen, Hurwitz and Madow (1953), p. 597.

[b] $S_W^2 = S^2(1 - \delta)$.

[c] MS $(B) = S^2[1 + \delta(M_0 - 1)]$.

[d] $S_B^2 = [\text{MS}(B) - S_W^2]/M_0$.

[e] From Hansen, Hurwitz, and Madow (1963), p. 264.

Table 4.10A. *Values of intraclass correlation coefficients, δ, for selected bio-agricultural items in clusters of selected sizes*

	$\bar{Y}$				Cluster Size, M^0			
Item	(Y/M)	S^2	2	4	9	16	36	64
1. Corn acreage[a]	168.0	3400	—	0.390	—	—	0.323	—
2. Barley acreage[a]	68.5	1647	—	0.153	—	—	0.152	—
3. Flax acreage[a]	4.47	—	—	0.041	—	—	0.038	—
4. Potato acreage[a]	2.44	—	—	0.284	—	—	0.150	—
5. Orange yields[b]	—	38,500	0.460	0.442	0.402	—	0.381	—
6. Beetle counts[c]	4.737	15.00	—	0.309	0.302	0.284	0.279	0.224

[a] Brookings County, S.D. $M = 720$ sections (square-mile plots).
[b] Orange grove in Florida. $M = ?$ (not available) trees.
[c] Beall's Potato Field. $M = 2304$ 2-foot row segments.

other into the same cluster, an almost universal result is that the between-cluster variance will be greater than the variance among elements in the whole universe. This phenomenon is sometimes called the *principle of proximity*; proximity somehow either "makes" elements more alike or "attracts" elements that are somewhat "alike." Hence, owing to the principle of proximity, it follows that random samples of clusters of neighboring elements are usually less efficient than random samples of the same number of individuals.

To lend a little support for this contention, we shall consider a measure for

Table 4.10B *Values of intraclass correlation coefficients, δ, for selected econo-demographic items in clusters of selected sizes* (basic unit: neighboring households)*

	$\bar{Y}$			M_0		
Item[a]	(Y/M)	S^2	3	9	27	62
1. Males, 25–34	0.2802	0.265	0.045	0.026	0.018	0.008
2. Males, unemployed	0.1937	0.208	0.060	0.070	0.045	0.034
3. Males in labor force	1.0617	0.703	0.12	0.10	0.07	0.03
4. Housing units, $10–$14 rent	0.0525	0.048	0.235	0.169	0.107	0.062
5. Persons (per HH)	3.378	3.491	0.230	0.186	0.142	0.066
6. Home-owned HHs	0.1967	0.123	0.170	0.171	0.166	0.096
7. Average rental value	—	—	0.45	0.36	0.25	0.12

*From Hansen et al. (1953), Vol.I, pp. 308, 597, and 264. Data from 1940 Census for selected cities over 100,000 population.
[a] Based on 11,106 HHs ($= M$) comprising 37,515 persons.

Table 4.11. Illustrative coefficients of the three functions

Item*	S function		J function		H function	
	A_S	$\hat{s}$	A_J	$\hat{j}$	A_H	$\hat{h}$
Beetle counts	10.04	0.7587	9.749	0.0385	0.3645	−0.0982
Barley acreage[a]	1576	0.4044	1434	0.0197	0.1456	−0.0030
Corn acreage[a]	3137	0.7131	1755	0.0967	0.4393	−0.0858
. . .						
Unemployed males	2.168	0.336	0.190	0.0212	0.087	−0.206
Population (per HH)	3.448	0.423	2.682	0.0396	0.397	−0.385
Percent home-owned HHs	0.087	0.569	0.098	0.0157	0.226	−0.162

*See Tables 4.9 and 4.10 for data summaries.
[a] In the S and J functions, three data points were used; that is, $M_0 = 1$ and $M_0 = 720$ were used here. Otherwise the fittings excluded these points.

proximity and examine its relationship with observed variances. For such a measure consider a rectangular area $a \times b$ within which two points are randomly located. Then it can be shown [Jessen (1942)] that the squared distance (beeline) between them, D^2, has the expectation

$$(4.32) \qquad E[D^2] = \left(\frac{a^2 + b^2}{6}\right).$$

Separating the area and shape effects, we have

$$(4.33) \qquad E[D^2] = \left(\frac{ab}{3}\right)\left(\frac{a/b + b/a}{2}\right)$$

$$= \left(\frac{\text{area}}{3}\right)(\text{shape}),$$

where the shape factor, $(a/b + b/a)/2$, has the value 1 for a square and increasing values for rectangles with the greater departure from a square. It may be noted that for plots of the same area the slender plot would have a higher $E[D^2]$ than a square. If proximity and low variability are associated, then the between-plot variance should have high variability (see Section 4.5). Relationships like those of Eqs. 4.29, 4.30, or 4.31 could be adapted by replacing cluster size M_0 by $E[D^2]$. As an example, Eq. 4.29 becomes

$$(4.34) \qquad \log \text{MS}(B) = \log\left(\frac{ab}{3}\right)\left(\frac{a/b + b/a}{2}\right)$$

or

(4.34a) $$\text{MS } (B) = AB^b C^c,$$

where B and C are the size and shape factors, respectively, and A is a constant.

In some cases variability depends on the orientation of the plot shape. For example, households situated along a particular affluent street may be more alike than those on another street that cuts across poor and rich neighborhoods. Such trends of characteristics of elements, sometimes called *gradients*, may show monotonic, undulating, or other nonrandom patterns.

If these characteristics, as in the case of systematic sampling, can be anticipated, then appropriate choice of shape orientation is that where the long slender plot runs along the gradient. An application of the above variance function to the beetle data is given in Section 4.11, where it appears that beetle density in that field has a gradient running parallel to rows.

4.9. Optimal Size of Sampling Unit

The best size of cluster would be that which produces for a given cost a minimum variance of sample mean, or that which for a fixed variance of the sample mean has a minimum cost. If the variances and costs were known for all relevant values of M_0, then the net relative efficiency equations of Section 4.3 could be used to calculate all such values, and that which yielded the highest NRE would be the optimal choice. Moreover, if variances and costs can be put into functions of M_0, as indicated in Sections 4.7 and 4.8, we could perhaps obtain a mathematical solution for best value of M_0, provided we could determine the values of the parameters involved. However, a simple but approximate method may be quite useful here.

It may be noted in Table 4.9A and B that S_W^2 and S_B^2 do not vary much with small changes in M_0, and δ only moderately. This suggests that over a short but perhaps relevant range of M_0 they can be regarded as constant. Let us consider the variance of a sample mean with the components-of-variance model:

(4.35) $$\text{var } (\bar{y}) = \left(1 - \frac{n}{N}\right) \frac{S_W^2 + M_0 S_B^2}{n M_0}.$$

Also let us assume a simple cost function, where total costs are

(4.36) $$C = n c_B + n M_0 c_W.$$

Here c_B is a fixed cost associated with a cluster regardless of its size and c_W is a fixed cost associated with each element regardless of cluster size. Hence to observe a cluster of size 1 (i.e., $M_0 = 1$) costs $c_B + c_W$ units. With a little algebra and calculus, it can be shown that under the condition that S_W^2 and S_B^2 in Eq.

4.35 are constants, and with a fixed budget C and constant costs as shown in Eq. 4.36, then var $(\bar{y})$ is a minimum, hence M_0 is an optimum, when

$$(4.37) \qquad M_{opt} = \sqrt{\frac{c_B}{c_W} \cdot \frac{S_W^2}{S_B^2}}$$

or its approximate equivalent where $\delta \doteq S_B^2/(S_B^2 + S_W^2)$:

$$(4.37a) \qquad M_{opt} = \sqrt{\frac{c_B}{c_W} \cdot \frac{1 - \delta}{\delta}} .$$

For an example, suppose a characteristic under study has $\delta = .20$ and $c_B/c_W = 4$. Then $M_{opt} = \sqrt{(4)\,(.80)/(.20)} = 4$. If, at size 4, we decide c_B, c_W, and δ should be modified from our original conjecture, then a new value of M_{opt} can be computed from the revised information. Such iterations usually are quite few because of the stability of the coefficients involved.

4.10.   Summing Up on Choice of SU

The foregoing discussion was confined to cases where $\bar{Y}$ (and in the case where clusters are of equal size, also $\bar{\bar{Y}}$) is being estimated, but the conclusions reached on efficiency will be essentially the same for both situations. Recapitulated, these are:

Type: The choice of type of SU (or frame) will depend heavily on the nature of the investigation, on what sorts of frames are available, and a number of other practical considerations. Efficiency and costs are important criteria to use when alternative choices of SU type are available. Efficiency can be increased by putting unlike elements together in clusters. Costs are usually decreased by putting geographically contiguous elements together in clusters.

Size: Where size can be varied at will, then a problem of optimum size arises. Ordinarily, when costs are ignored, the smaller the size, the more efficient the unit. Consideration of costs and other practical matters usually indicate that the optimum size becomes somewhat larger.

Practical matters: These are many and differ markedly from one type of universe to another. Later we shall consider problems of errors on the part of investigators as an important practical matter in SU choice. Accuracy of available frames and the question of whether more accurate ones can be economically constructed will be mentioned in the next chapter. There is a great opportunity for ingenuity in the choice of the best SU for a sample survey.

4.11. Review Illustrations

1. The general nature of the data collected in Brookings County was discussed in Section 4.8. The location of each field was assigned to one of the 720 plots (square-mile Public Land Survey "sections") and also to one of 2190 farms in the county. The relevant data summaries for acreage in barley are:

	Plots (sections)		Farms	
Total frame units	$N_\alpha =$	720	$N_\beta =$	2190
Barley acreage, total	$Y_\alpha =$	49,326	$Y_\beta =$	49,326
Barley variance	$S_\alpha^2 =$	1647	$S_\beta^2 =$	651

(a) Determine which is the more efficient frame unit, the plot or the farm.

(b) The two frame units involve different costs to obtain observations. Suppose the cost of locating and interviewing a farmer is $10 and the cost of locating a plot, identifying and measuring crop acreages, and taking aerial photographs is $25. Which is the more efficient frame considering costs?

(c) Another possible SU in this case is the point. A point randomly located within the county will either fall on barley or not. An estimate of total acreage in barley is obtained by multiplying the total area of county (512,640 acres) by the fraction of points falling on barley. What is the precision of a point relative to a one-square-mile plot in surveying barley acreage?

Solution

(a) The relative efficiency of the farm to the section is given by

$$RE\ (F/S) = \frac{N_s S_s^2}{N_F S_F^2} = \frac{(720)\ (1647)}{(2190)\ (651)} = 0.83.$$

(b) Considering costs (where $C_s = 25$ and $C_F = 10$):

$$NRE\ (F/S) = \frac{N_s^2 S_s^2 \cdot C_s}{N_F^2 S_F^2 \cdot C_F} = \frac{(720)^2(1647)\ (25)}{(2190)^2\ (651)\ (10)} = 0.68$$

hence the farm as an SU is even worse than the section when costs are considered.

(c) Our two estimators of Y are

$$\hat{Y}_s = N\bar{y}_s \quad \text{and} \quad \hat{Y}_p = A_p,$$

and the relative precision of point to section is (ignoring FPC)

$$RP\ (P/S) = \frac{\text{Var } (\hat{Y}_s)}{\text{Var } (\hat{Y}_p)} = \frac{N_s^2 S_s^2 / n_s}{A^2 S_p^2 / n_p} = \frac{N_s^2 S_s^2 n_p}{A^2 PQ n_s}$$

$$= \frac{(720)^2 (1647) n_p}{(512{,}640)^2 (49{,}326/512{,}640) n_s} = \frac{0.0338 n_p}{n_s},$$

or about 30 points are equivalent to one section in sampling efficiency here. Note that $S_p^2 \doteq PQ$.

2.   A survey organization is given the job by an advertising agency to determine the average number of households in the market area that turns its TV sets onto a particular station at a particular hour during a week. The survey organization obtained the following information on a small pilot survey of 10 households. A problem confronting the survey organization is whether to use a panel type sample, where the same household is questioned whether he had the station of interest turned on each day of the week (such as by asking him to keep a diary), or to question different households each day regarding their watching performance during one day only. The questions to be asked, assuming the sample of 10 to be reliable, are (a) what is the degree of persistent watching or "station loyalty" evidenced by the sample using δ as a measure? and (b) how many individual day respondents would be equivalent to a single panel respondent for estimating average daily viewingship?

| Household | Viewing (1 denotes viewing, 0 otherwise) | | | | | |
no.	Mon.	Tues.	Wed.	Thurs.	Fri.	Total
1	0	1	0	0	1	2
2	1	1	1	1	1	5
3	1	0	1	0	1	3
4	0	0	0	0	0	0
5	0	1	0	1	0	2
6	1	1	0	0	1	3
7	0	1	1	1	0	3
8	0	0	0	0	0	0
9	0	0	0	0	0	0
10	1	1	1	1	0	4
Total	4	6	4	4	4	22

Solution

(a) The computations can be summarized in the ANOVA table:

Source	df	SS	MS	Mean square as components of variance
Total	49	12.32	0.2514	
HHs	9	5.52	0.6133	$s_W^2 + 5s_B^2$
Day/HHs	40	6.80	0.1700	s_W^2

Using Eq. 4.25, and since $s_B^2 = (0.6133 - 0.1700)/5 = 0.08866$,

$$\hat{\delta} \doteq \frac{s_B^2}{s_W^2 + s_B^2} = \frac{0.0887}{0.1700 + 0.0887} = 0.339.$$

Hence there appears to be a significant amount of "station loyalty."

(b) The efficiency of the cluster unit relative to the individual HH for a single day is given by Eq. 4.8, where

$$RE\ (C/E) = \frac{MS\ (T)}{MS\ (B)}.$$

Since n appears to be "reasonably" large here, we can use MS (t) to estimate MS (T), and MS (b) is always a good estimator of MS (B). Hence,

$$\widehat{RE}\ (C/E) = \frac{MS\ (t)}{MS\ (b)} = \frac{0.2514}{0.6133} = 0.408$$

or, inversely, 2.44 (i.e., 1/0.408) households in a panel are required to give the same information on daily station watching as one taken at random each day. (However, since each "panel household" reports for five days, it is still "worth" about two nonpanel households for its five-day record!)

3. Suppose an investigator can complete five elements per day including the travel time from his home base out to the cluster and return each day. Which has the highest NRE, a five-element cluster or a 10-element cluster?

The travel argument is shown in Fig. 4.4. In case A, the expected travel for one round trip is a miles per day or $2a$ miles for the 10 elements. In case B it is also a miles for each day and $2a$ for the two days. Hence, the relative cost *per element* for the two cases is the same, since 10 elements were completed in both cases for the same mileage. The MS (B) for the size 10 cluster is expected to be larger than the MS (B) for the size 5 cluster, for the following reasons. As cluster size increases, the mean square *within*-cluster is expected to increase.

From Eq. 4.29 we see that if Smith's variance holds, MS $(B) = S^2 M_0^s$, where $s \geq 0$, and is most commonly greater than zero. Hence MS (B) must increase with any increase in cluster size M_0. This being the case, and since R3 $(M_0/1)$ = MS (T)/MS (W), then for this variance function

$$\text{RE } (M_0/1 = \frac{1}{M_0^s} ;$$

hence relative efficiency continues to drop off continuously as cluster size increases. Since NTR = RE/RC—that is, relative efficiency divided by relative cost—we conclude in our case that since RS/RC is constant, then NRE falls continuously with increasing M_0, therefore the two five-element clusters are more efficient than the one 10-element cluster, considering costs.

4. The following mean squares (on a per-element basis) were computed from the beetle data of Fig. 4.1 for eight different sizes of square plots and 10 different shapes for nonsquare plots, all consisting of 48 elements. (Structure is given in number of rows times number of columns.)

Square plots			Nonsquare plots		
Size (M_0)	Structure	MS (B)	Size (M^0)	Structure	MS (B)
4	2×2	28.896	48	1×48	112.53
9	3×3	51.286	48	2×24	210.44
16	4×4	78.958	48	3×16	182.19
36	6×6	161.413	48	4×12	219.56
(48)	—	—	48	6×8	210.49
64	8×8	226.429	48	8×6	173.39
144	12×12	565.623	48	12×4	189.48
256	16×16	600.690	48	16×3	132.84
576	24×24	1219.000	48	24×2	100.58
			48	48×1	50.79

(a) Using the square-plot results, plot MS (B) against expected squared distance due to $ab/3$, the size factor, on double-log paper. Is the relationship reasonably linear?

(b) By inspecting the results on the nonsquare plots, does there appear to be a gradient present? If so, in what direction?

(c) Calculate the values of the shape factor, $(a/b + b/a)/2$, and for those plots having identical values, obtain the simple average of their MS (B)s. Calculate the $E[D^2]$, using both factors for these combined plots, and plot the results— that is the MS (B)s—against $E[D^2]$ on the graph used in (a). Is this relationship reasonably linear? Comment.

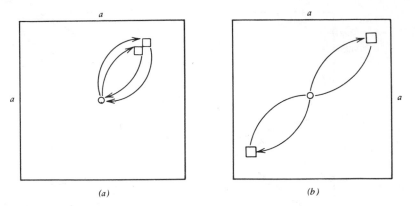

Fig. 4.4. Comparison of two travel plans. (*a*) Two trips of work in one combined unit. (*b*) One trip of work in each of two units.

(*d*) Using a least-squares fit on the logarithms of the results above, the following relationship was determined:

$$\text{MS } (B) = V = 12.0(ab)^{0.76}\left(\frac{a}{b} + \frac{b}{a}\right)^{-0.26}.$$

In quantitative terms discuss how the factors size and shape of plot affect beetle density variability through the proximity principle. How was the presence of a gradient dealt with here?

Solution

(*a*) Calculations of $E[D^2] = ab/3$ are

M_0:	4	9	1 6	36	64	144.	256	576
$E[D^2]$:	1.33	3.00	5.33	12.0	21.3	48.0	85.3	192.0
MS (B):	28.9	51.3	79.0	161	226	565	601	1219

And the plot is shown in Fig. 4.5. Yes, the relationship is reasonably linear.

(*b*) Yes, there is a gradient running along columns [since plots stretched along the columns have a lower MS (B), they must contain more heterogeneity].

(*c*) Combining appropriate pairs—that is, 1×48 with 48×1, and so on—and averaging MS (B)s, we obtain

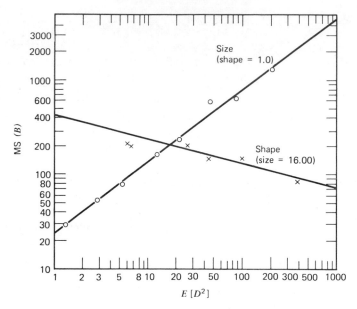

Fig. 4.5.   Plot of MS (B) on $E[D^2]$, beetle data.

Structure:	1/48	2/24	3/16	4/12	6/8	(7/7)
$(a/b + b/a)/2$:	24.01	6.04	2.76	1.67	1.04	(1.00)
$ab/3$:	16	16	16	16	16	(16)
$E[D^2]$:	384.0	96.7	44.2	26.7	16.7	(16.0)
MS (B):	81.7	156	158	205	192	(194)

The plot is shown above in (a). Yes, the plot can be regarded as linear, but the fit is rather poor. However, the average MS (B) does appear to decline as the plot is stretched out.

(d) Between-plot mean square increases with size and decreases when the plot shape departs from a square. The function gives at least a general representation of heterogeneity in the field. When proximity of elements is lessened, for a fixed size, a gain in efficiency is obtained through lower between-plot mean square. The proximity effect is obscured when plot size is increased, owing to the mass effect of M_0 (this will be dealt with in Chapter 9 by subsampling within plots). By averaging the two plot configurations with identical size values, we are in effect assuming that plot orientation along rows or along columns is to be randomly determined for each sample plot.

5.   A survey is to be taken to determine family expenditures for food in Metroville. A personal interview is planned and since an area-list type frame is

to be used, it is expected that costs associated with the cluster will be $8 and those associated with each household about $4. A budget of $15,000 is provided but only $10,000 will be available for operations. It is expected that δ will be about 0.20 and the coefficient of variation of food expenditures per household will be around 0.80. Determine (a) the optimal size cluster, (b) the number of interviews that will be made, and (c) the relative standard error of the sample mean.

Solution

(a) For M_{opt} we use Eq. 4.37a, since N appears large.

$$M_{opt} \doteq \sqrt{\frac{C_B}{C_W} \cdot \frac{1-\delta}{\delta}} = \sqrt{\frac{8}{4} \left(\frac{1-0.20}{0.20}\right)} = 2.8 \doteq 3.$$

Hence our best cluster size is 3.

(b) Total costs, $C = nC_B + nM_0 C_W = \$10,000$. To determine n we have

$$n = \frac{C}{C_B + M_0 C_W} = \frac{10,000}{8 + (3)(4)} = \frac{10,000}{20} = 500.$$

Total interviews will be $m = nM_0 = (500)(3) = 1500$.

(c) The variance of the sample mean when δ is known is given by Eq. 4.20:

$$\text{var } (\bar{y}) = \left(\frac{N-n}{N}\right) \frac{S^2}{nM_0} [1 + (M_0 - 1)\delta],$$

but the relative variance will be (ignoring FPC)

$$\text{rel var } (\bar{y}) = \left(\frac{N-n}{N}\right) \frac{V^2}{nM_0} [1 + (M_0 - 1)\delta]$$

$$= \frac{(0.8)^2}{(500)(3)} [1 + (3-1)(0.20)]$$

$$= \frac{0.64}{1500} (1.40) = 0.000,597$$

whence RSE $(\bar{y}) = \sqrt{0,000,597} = 0.024$ or 2.4%.

6. A survey is to be undertaken of the 48 capitals of the continental United States. The plan is to have an investigator drive his automobile to each of the capitals in one continuous trip.

(a) About how many miles of travel are required?

(b) What path (visiting order of the 48 cities) should he choose to minimize travel?

Solution

(a) The area of the continental United States is 3,022,387 square miles and is roughly a rectangle of about 1500 by 2000 miles. The case (a) travel path seems appropriate here. The traveler begins in a corner and returns there, after visiting a systematic arrangement (6×8) of 48 points. Here

$$E[D] = \sqrt{abn} + b + \gamma(a - \sqrt{ab/n})$$

where $a = 1500$, $b = 2000$, $n = 48$, $\gamma = 0$ (since 6 and 8 are even. Then $E[D]$ $= \sqrt{3,022,387(48)} + 2000 = 14,000$ miles. Note that this is dogleg and not beeline mileage.

(b) This is a version of what is sometimes called the "traveling salesman" problem. It is more difficult than (a) unless one has access to high-speed computers and appropriate programs. However, we may approximate a path by following the principles of case (h). Here we wish to put transcepts across the United States that will divide the 48 points into about 16 cells of three each—that is, forming three east-west strips across and five north-south strips. This results in the 15 cells with the distribution as shown.

4	4	1	1	5
3	4	4	4	4
1	3	6	3	0

Applying the formula:

$$E[D] = \left(\frac{\bar{m}+3}{3\sqrt{\bar{m}}}\right)\sqrt{Am}$$

$$= \left(\frac{48/15+3}{3}\right)\sqrt{48/15]}\,\sqrt{3.02 \times 10^6(48)}$$

$$= \left(\frac{6.2}{(5.3666)}\right) 12,040$$

$$= 13,910 \text{ miles,}$$

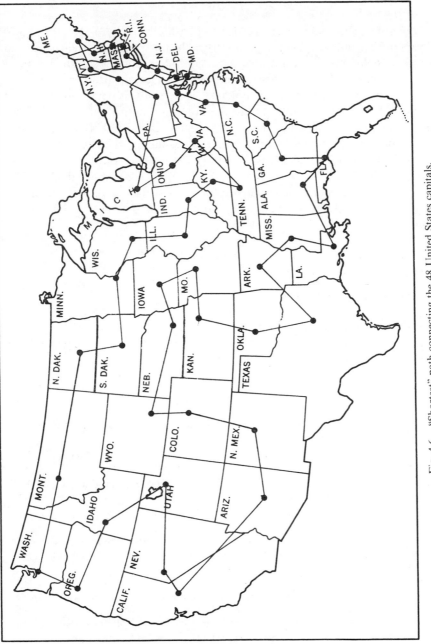

Fig. 4.6. "Shortest" path connecting the 48 United States capitals.

117

to which we must add about $(2.5/3)a+(4.5/5)b$ or $(2.5/3)$ $(1500)+(4.5/5)$ $(2000) = 3050$ miles to bring him home from the northeast cell. Hence the total travel is 16,960 miles, using this general path-determining scheme. With at most six points in a cell to choose a minimum path for, the traveler can usually pick the best path without much difficulty. However, this path is much longer $(16,960/14,000 = 1.21$ or $21\%)$ than the minimum path implied in (a). That path can be approximated by the results of Danzig et al. (1954) dealing with a similar problem involving 48 cities. The approximate path inferred from their results is shown in Fig.4.6. Measuring the length of this route on the map, we obtain 11,300 miles beeline, or $(1,254)$ $(11,300) = 14,200$ miles dogleg. This does not involve a "home" in the corner, for which we say $(\frac{1}{2}\cdot\frac{1}{3})a+(\frac{1}{2}\cdot\frac{1}{5})b$ or $\frac{1500}{6}$ $+\frac{2000}{10} = 450$ might be added, giving a total of 14,650 miles. In summary we have

Case (a) (unknown route)	14,000
Case (h) (zigzag route)	16,960
Specified "minimum" route	14,650

The zigzag route, case (h), provides a simple method for selecting a route or path, but it is unnecessarily costly. To obtain the best route, some dependence on high-speed computers is justifiable.

4.12. References

Beall, G.
1939
Methods of Estimating the Population of Insects in a Field. *Biometrika*, **30**: 422–439.

Cochran, W. G.
1953, 1963
Sampling Techniques. (2nd ed.) New York: John Wiley & Sons, Inc. (Chapter 9. Good on sampling unit efficiency.)

Danzig, G.
Fulkerson, R.
Johnson, S.
1954
Solution of a Large-Scale Traveling Salesman Problem. *Journal of the Operations Research Society of America*, **2**: 393–410 (November).

Hansen, M. H.
Hurwitz, W. N.
Madow, W. G.
1953
Sample Survey Methods and Theory. Vol. I. New York: John Wiley & Sons, Inc. (Chapter 6. Good on sampling unit efficiency.)

Hendricks, W. A.
1944
The Relative Efficiencies of Groups of Farms as Sampling Units. *Journal of the American Statistical Association*, **39**: 367—376. (A comparison of the J and S variance functions for agricultural populations.)

Homeyer, P. G.
Black, C. A.
1946
Sampling Replicated Field Experiments on Oats for Yield Determination. *Soil Science Proceedings*, **11**: 341–344.

Jessen, R. J.
1942
Statistical Investigation of a Sample Survey for Obtaining Farm Facts. *Iowa Agricultural Experiment Station Research Bulletin* 304.

Mahalanobis, P. C.   A Sample Survey of the Acreage Under Jute in Bengal. *Sankhȳa* (India),
1940   **4**: 511–530. (An excellent early paper on cost and variance functions.)

McCreary, G. E.   *Cost Functions for Sample Surveys.* Ph.D. Thesis, Iowa State College,
1950   Ames, Iowa. (A comprehensive study of expected distances between
                     points in an area.)

Smith, H. F.   An Empirical Law Describing Heterogeneity in Yields of Agricultural
1938   Crops. *Journal of Agricultural Science*, **28**: 1–23.

Sukhatme, P. V.   The Problem of Plot Size in Large-Scale Yield Surveys. *Journal of the*
1947   *American Statistical Association*, **42**: 297–310.

Sukhatme, P. V.   *Sampling Theory of Surveys with Applications*, Ames, Iowa: Iowa State
1954   College Press. (Chapter 6. Good on sampling unit efficiency, especially
                     crops.)

Waksberg, J.   *Reliability of Estimates with Alternative Cluster Sizes in the Health*
Hanson, R. H.   *Interview Survey.* DPEW Publication No. (HSM) 73-1326. Washington,
Jacobs, C. A.   D.C.: U.S. Government Printing Office. [A good source of δs and design
1973   effects of various health items is given for cluster sizes 3, 6, and 9
                     (households).]

4.13. Mathematical Notes

ANOVA Notation. A universe consists of M elements clustered into N groups of M_0 elements each, where

Y_{ij} = the value of some characteristic y on the jth element in the ith group,

$$Y_i = \sum_j^{M_0} Y_{ij} = \text{total } y \text{ for group } i,$$

$$Y = \sum_i^N Y_i = \text{total } y \text{ for the } M \text{ elements,}$$

$\bar{Y}_i = Y_i/M_0$ = mean of y per element in group i,

$\bar{Y} = Y/N$ = mean of y per *group* in the universe,

$\bar{\bar{Y}} = Y/M$ = mean of y per *element* in the universe,

$$\delta = \frac{\underset{i,j<k}{E}\left[(Y_{ij}-\bar{\bar{Y}})(Y_{ik}-\bar{\bar{Y}})\right]}{E[(Y_{ij}-\bar{\bar{Y}})^2]} = \text{intraclass correlation coefficient.}$$

The Mean Square Model. Consider the identity
$$Y_{ij} = Y_{ij},$$

$$Y_{ij} - \bar{\bar{Y}} = (\bar{Y}_i - \bar{\bar{Y}}) + (Y_{ij} - \bar{Y}_i).$$

Squaring each side and summing over all j and i,

$$\sum_i \sum_j (Y_{ij} - \bar{\bar{Y}})^2 = \sum_i \sum_j \left[(\bar{Y}_i - \bar{\bar{Y}})^2 + 2(\bar{Y}_i - \bar{\bar{Y}})(Y_{ij} - \bar{Y}_i) + (Y_{ij} - \bar{Y}_i)^2\right]$$

(4.13.1)
$$= \sum_i \sum_j (\bar{Y}_i - \bar{\bar{Y}})^2 + \sum_i \sum_j (Y_{ij} - \bar{Y}_i)^2.$$

Hence

$$SS\ (T) = SS\ (B) + SS\ (W)$$

with degrees of freedom

$$NM_0 - 1 = N - 1 + N(M_0 - 1).$$

The Components-of-Variance Model. Let us regard the finite universe as having Y_{ij}s generated by a linear sum

$$Y_{ij} = \bar{\bar{Y}} + G_i + E_{ij},$$

where G_i is a component common to all elements in group i and E_{ij} is simply a residual. Then

$$Y_{ij} - \bar{\bar{Y}} = G_i + E_{ij}.$$

Summing over j,

$$Y_i - M_0 \bar{\bar{Y}} = M_0 G_i + E_i,$$

$$Y_i - \bar{Y} = M_0 G_i + E_i.$$

Squaring and summing over i,

(4.13.2) $$\sum_i (Y_i - \bar{Y})^2 = \sum_i M_0^2 G_i^2 + 2 \sum_i M_0 G_i + \sum_i E_i^2.$$

But

$$\sum G_i = 0, \quad \sum_i (Y_i - \bar{y})^2 = M_0(N - 1)\ \text{MS}\ (B),$$

$$\sum G_i^2 = (N - 1)S_B^2, \quad \text{say and} \quad \sum E_i^2 = (N - 1)M_0 S_W^2;$$

then

$$M_0(N - 1)\ \text{MS}\ (B) = M_0^2(N - 1)S_B^2 + (N - 1)M_0 S_W^2,$$

(4.13.2a) $$\text{MS}\ (B) = M_0 S_B^2 + S_W^2,$$

and

$$\text{MS}\ (W) = S_W^2.$$

The Correlation Model

$$\sum_j (Y_{ij} - \bar{\bar{Y}}) = Y_i - M_0 \bar{\bar{Y}} = Y_i - \bar{Y}.$$

Hence for the ith group,

$$Y_i - \bar{Y} = \sum_j (Y_{ij} - \bar{\bar{Y}}) = Y_{i1} - \bar{Y}) + (Y_{i2} - \bar{Y} + \cdots + (Y_{iM_0} - \bar{Y}).$$

Squaring and summing both sides,

$$(4.13.3) \qquad \sum_i (Y_i - \bar{Y})^2 = \sum_i \sum_j (Y_i - \bar{\bar{Y}})^2 + 2 \sum_i \sum_{j<k} (Y_{i,j} - \bar{\bar{Y}})(Y_{ik} - \bar{\bar{Y}}).$$

$$M_0 \sum_i \sum_j (\bar{Y}_i - \bar{\bar{Y}})^2 = \sum_i \sum_j (Y_{ij} - \bar{\bar{Y}})^2 + 2 \sum_i \sum_{j<k} (Y_{ij} - \bar{\bar{Y}})(Y_{ik} - \bar{\bar{Y}}).$$

Hence

$$(4.13.3a) \qquad (M_0)\, \text{SS}\,(B) = \text{SS}\,(T) + \text{SS}\,(T) \left(\frac{NM_0(M_0 - 1)}{NM_0 - 1}\right) \delta$$

since

$$\delta = \frac{2 \sum_i \sum_{j<k} (Y_{ij} - \bar{\bar{Y}})(Y_{ik} - \bar{\bar{Y}})}{NM_0(M_0 - 1)S^2}$$

From Eq. 4.13.3a we can derive δ in terms of MS (W) and MS (B). Thus

$$(M_0)\, \text{SS}\,(B) = \text{SS}\,(T) + \text{SS}\,(T) \left[\frac{NM_0(M_0 - 1)}{NM_0 - 1}\right] \delta$$

$$\delta = \frac{(M_0 - 1)\, \text{SS}\,(B) - \text{SS}\,(W)}{S^2[NM_0(M_0 - 1)]}$$

$$= \frac{(1/M_0)\{[(N-1)/N]\, \text{MS}\,(B) - \text{MS}\,W)\}}{S^2}$$

And MS (B) and MS (W) can be put in terms of S^2 and δ. Thus from Eq. 4.13.3a,

$$\text{BMS} = \frac{(NM_0 - 1)S^2}{M_0(N-1)} + \frac{(NM_0 - 1)S^2}{M_0(N-1)} \left[\frac{NM_0(M_0 - 1)}{NM_0 - 1}\right] \delta$$

$$= S^2 \left[\frac{NM_0 - 1}{M_0(N-1)} + \frac{N(M_0 - 1)}{N-1} \delta\right],$$

and similarly from Eq. 4.13.3a, noting that SS $(W) = $ SS $(T) - $ SS (B),

$$\text{MS}\,(W) = S^2 \left(\frac{NM_0 - 1}{NM_0} - \delta\right).$$

Alternative exact expression for δ are given by

$$\delta = \frac{[(N-1)/N]\, S_B^2 - (1/NM_0)\, S_W^2}{[M_0(N-1)/(NM_0-1)]\, S_B^2 + S_W^2}.$$

Unbiased estimates of δ are given by

$$\delta = \frac{MS(b) - MS(w)}{[M_0(N-1)/(NM_0-1)]\,MS(b) - [NM_0(M_0-1)/(NM_0-1)]\,MS(w)}$$

and

$$\hat{\delta} = \frac{[(N-1)/N]\,s_B^2 - (1/NM_0)\,s_W^2}{[M_0(N-1)/(NM_0-1)]\,s_B^2 + s_W^2}.$$

Optimal Size of SU.   Let C, C_B, and C_W be total unit costs associated with clusters (other than elements within a cluster) and element, respectively; then

$$C = nC_B + nM_0C_W$$

and

$$\text{var}\,(\bar{y}) = \frac{MS\,(B)}{nM_0} = \frac{S_W^2 + M_0 S_B^2}{nM_0}$$

or

$$\text{var}\,(\bar{y}) = \frac{S_W^2(C_B + M_0C_W)}{M_0C} + \frac{M_0 S_B^2(C_B + M_0C_W)}{M_0C},$$

$$\frac{dV}{dM_0} = \frac{-C_B S_W^2 C}{(M_0C)^2} + \frac{C_W S_B^2}{C} = 0,$$

whence

$$M_{\text{opt}} = \sqrt{(C_B/C_W)\,(S_W^2/S_B^2)} \doteq \sqrt{(C_B/C_W)\,[(1-\delta)/\delta]}.$$

4.14.   Exercises

1.   We have data of the following sort for every household in Statisburg:

Household no.	Person no.	Age	Income 1966	Possess TV?
1	1	40	5000 ⎫	Yes
	2	35	0 ⎬	
2	1	25	1000	No
3	1	25	4000 ⎫	
	2	25	500 ⎬	No
	3	5	0 ⎪	
	4	3	0 ⎭	
4	1	70	2000	Yes
.	.	.	.	.
.	.	.	.	.
.	.	.	.	.
N	1	40	4000	Yes

(a) Identify two universes.

(b) Identify two populations.

(c) Identify the population of household incomes for the five households shown.

2. Out of a consignment of 1000 boxes of eggs, each containing 40 eggs, 20 boxes were drawn RNR, and among these nine were found to contain no cracked eggs, eight were found to contain 1 cracked egg each, and three were found to contain two cracked eggs each.

(a) Estimate the total number of cracked eggs in the consignment and its variance.

(b) Compare the variance in (a) with that computed from the "binomial variance" and discuss the discrepancy between these variances in this and similar situations.

3. Here is a synthetic micropopulation of $N = 3$ groups of $M_0 = 4$ observations in each

Group, i:	1	2	3
Observation, ij:	4,7,2,7	10,6,2,10	8,17,18,5

(a) Summarize the relevant computations in an ANOVA table.

(b) Determine MS (T), MS (B), and MS (W).

(c) Determine S^2, S_W^2, and S_B^2.

(d) Calculate the exact intraclass correlation coefficient, δ, and compare it with the approximate (Eq. 4.22) value.

4. Nadir Enterprises has 8800 employees. A sample of 100 is desired for a survey. Suppose the employee list is alphabetized from A to Z.

(a) Explain carefully how you would select a systematic sample of 100 employees.

(b) Are any benefits of stratification present in your sample? Explain.

(c) Are any dangers of periodic arrangements present? Discuss.

(d) Would there likely be any advantage in using systematic rather than simple random selection in this case? Discuss.

5. A sample of families in Bigtown is to be selected for a survey of newspaper readership. Two schemes are being considered: (i) selecting individual families at random and (ii) selecting, at random, clusters of five neighboring families.

(*a*) Ignoring travel costs, which scheme is likely to be more efficient? Why?

(*b*) What effect would the introduction of heavy travel costs have on your answer? Explain.

(*c*) Suppose the optimal size of cluster under certain conditions is two. Now let us suppose the expected "at-home" probability is half what was initially expected. What effect will this have on our optimum? Explain.

CHAPTER 5

Estimation

5.1. Introduction

In earlier chapters we have been concerned primarily with the problems of estimating the simple functions of a y-population, such as its total, Y, its mean per SU, $\bar{Y}$ (or P in case the Y_i are 0s and 1s), or in the case of cluster samples, the mean per element, $\bar{\bar{Y}}$. Since these parameters are closely related—that is, $Y = N\bar{Y}$ and $\bar{\bar{Y}} = \bar{Y}/M_0$ (where M_0 is a constant number of elements in each cluster)—it is not surprising that their estimators and the variances of those estimators are closely related. Hence $\hat{Y} = N\bar{y}$ and $\bar{\bar{y}} = \bar{y}/M_0$ and var $(\hat{Y}) = N^2$ var $(\bar{y})$ and var $(\bar{\bar{y}}) = (1/M_0^2)$ var $(\bar{y})$. Moreover, these estimators are unbiased,

125

as are the estimators of their variances. Our attention has been confined to a single population, y.

Now we shall consider two populations, x and y, usually related in some manner, and we wish to estimate not only the usual characteristics of each separately but also some joint relationships, such as the ratio of y to x, $\bar{Y}/\bar{X}$, the sum of y and x, $\bar{Y}+\bar{X}$, the regression of y on x, say B, or their correlation, say R. The problems of estimating these and other two-parameter functions will be considered here.

Having looked into the properties of these two-parameter estimators we shall reconsider the problem of estimating $\bar{Y}$ again, with the object now of improving our estimates by somehow using our knowledge of the x-population, and perhaps even by utilizing extrasample information about the y-population.

As we get involved in estimation, we shall be confronted with new problems. We shall find that unbiasedness is not an attainable property in estimating some parameters, and that even if it is, some biased estimators may be preferred. Expressions for variances will become more complex; in some cases only approximations will be available, and we will need some guidance as to how good the approximations are. On the other hand, we will discover how important the choice of appropriate estimator is in many practical situations and how it provides a new power to the survey designer.

Throughout this chapter we shall be concerned with a finite universe consisting of N elements on each of which there is a y and x characteristic. Thus

$$U_i: 1, 2, 3, \ldots, N;$$
$$Y_i: Y_1, Y_2, Y_3, \ldots, Y_N;$$
$$X_i: X_1, X_2, X_3, \ldots, X_N.$$

A sample of n elements (SUs) are selected with RNR unless otherwise specified.

5.2.   Estimating $\bar{Y}+\bar{X}$ and $\bar{Y}-\bar{X}$, the Sum and Difference of Two Population Means

When we are dealing with two populations rather than one, it may be of interest to estimate from the sample the sum of the two populations, $\bar{Y}+\bar{X}$, or the difference, $\bar{Y}-\bar{X}$. The logical estimators in this case turn out to be quite useful. Thus to estimate the sum, $\bar{Y}+\bar{X}$, we have

(5.1) $$w_s = \bar{y}+\bar{x},$$

which is unbiased and has the variance

$$\text{var}(w_s) = \text{var}(\bar{y})+\text{var}(\bar{x})+2R[\text{var}(\bar{y})\,\text{var}(\bar{x})]^{1/2}$$

(5.2) $$= \left(\frac{N-n}{Nn}\right)(S_y^2+S_x^2+2RS_yS_x),$$

where the y- and x-populations belong to the same universe of N elements from which an RNR sample of n has been drawn.

Similarly, to estimate the difference $\bar{Y}-\bar{X}$, we have

(5.3)
$$w_d = \bar{y} - \bar{x},$$

which is also unbiased and has the variance

$$\text{var}(w_d) = \text{var}(\bar{y}) + \text{var}(\bar{x}) - 2R[\text{var}(\bar{y})\,\text{var}(\bar{x})]^{1/2}$$

(5.4)
$$= \left(\frac{N-n}{Nn}\right)(S_y^2 + S_x^2 - 2RS_yS_x),$$

where, as before, both Y_i and X_i are both observed as the same n elements. Note that Eqs. 5.2 and 5.4 are identical except for the sign on the covariance term. In the case where R is 0, or if two samples are drawn independently of each other, each observing only Y_i or only X_i, the variance of the sum or difference is equal to the sum of variances. A positive correlation increases Var (w_s) but decreases Var (w_d).

To estimate Eqs. 5.2 and 5.4 from a sample we can simply replace S_y^2, S_x^2, and R by their sample estimates, s_y^2, s_x^2, and r.

If separate samples are taken to observe x and y, say n_x for the X_i and n_y for the Y_i, the estimator for $\bar{Y}-\bar{X}$ will remain the same (see Eq. 5.3), but the variance will be

(5.4a)
$$\text{var}(\bar{y}-\bar{x}) = \left(\frac{N-n_y}{N}\right)\frac{S_y^2}{n_y} + \left(\frac{N-n_x}{N}\right)\frac{S_x^2}{n_x},$$

or, if FPCs can be ignored,

(5.4b)
$$\text{var}(\bar{y}-\bar{x}) = \frac{S_y^2}{n_y} + \frac{S_x^2}{n_x}.$$

Equations 5.4a and 5.4b also are appropriate when one is estimating the sum $\bar{Y}+\bar{X}$.

5.3. Estimating $\bar{Y}/\bar{X}$, the Ratio of Two Population Means

Suppose we wish to estimate $\bar{Y}/\bar{X}$ from the sample. An obvious estimator is $\bar{y}/\bar{x}$, and for most practical cases it is satisfactory. But it is biased, since its expectation is

(5.5)
$$E[\bar{y}/\bar{x}] \doteq (\bar{Y}/\bar{X})[1 + \left(\frac{N-n}{Nn}\right)(V_x^2 - RV_xV_y)],$$

where V_x = coefficient of variation of $x = S_x/\bar{X}$,
V_y = coefficient of variation of $y = S_y/\bar{Y}$,
R = correlation coefficient of y and x, which is defined as

(5.6)
$$R = \frac{\sum(Y_i - \bar{Y})(X_i - \bar{X})}{[\sum(Y_i - \bar{Y})^2 \cdot \sum(X_i - \bar{X})^2]^{1/2}} = \frac{S_{xy}}{S_xS_y}.$$

The *bias* of $\bar{y}/\bar{x}$ in estimating $\bar{Y}/\bar{X}$ is the difference of the expected value of the estimator and the quantity being estimated. Using Eq. 5.5,

(5.7)
$$\text{bias }(\bar{y}/\bar{x}) = E[\bar{y}/\bar{x}] - \bar{Y}/\bar{X}$$

$$= (\bar{Y}/\bar{X})\left(\frac{N-n}{Nn}\right)(V_x^2 - RV_xV_y).$$

Note that the bias:

1.   Goes down as sample size increases.
2.   Goes down as the sampling rate increases.
3.   Goes down as the correlation goes to $+1$.
4.   Goes down as the variation of x (the denominator) decreases.
5.   Is zero if the regression of y on x is a straight line going through the origin (in which case $V_x^2 = RV_xV_y$).

Example 5.1.   A survey was taken in two successive years of the same 100 randomly selected Iowa farms to determine net cash income each year. Letting x be the net cash income for the f rst year and y for the second years, it was desired to estimate $\bar{Y}/\bar{X}$, the relative year-to-year change in net cash income for all farms in the state. Suppose $\bar{Y}/\bar{X} = 1.20$, $V_y = V_x = 1.80$, and $R = .78$; what is the approximate bias of $\bar{y}/\bar{x}$, the sample ratio, in estimating the change?

Since the bias is given by Eq. 5.7, we have

$$\text{bias }(\bar{y}/\bar{x}) = (\bar{Y}/\bar{X})\left(\frac{N-n}{Nn}\right)(V_x^2 - RV_xV_y)$$

$$= (1.20)\left(\frac{1}{100}\right)[(1.80)^2 - (.78)(1.80)(1.80)]$$

(the FPC is ignored, since N for all farms in the state is more than 2000; hence the sampling rate is less than 5%).

$$\text{bias }(\bar{y}/\bar{x}) = \frac{0.856}{100} = 0.00856,$$

or, as a percent of the parameter,

$$(100)\frac{\text{bias}}{\bar{Y}/\bar{X}} = \frac{(100)(0.00856)}{1.20} = 0.713\%.$$

Hence for samples of this size from this universe the sample ratio would tend to overestimate the true change by about 0.7 of 1%.

The variance of $\bar{y}/\bar{x}$ is given by an approximate formula (since no exact formula is known):

$$(5.8) \qquad \text{var } (\bar{y}/\bar{x}) \doteq (\bar{Y}/\bar{X})^2 \left(\frac{N-n}{Nn}\right) \left(\frac{S_y^2}{\bar{Y}^2} + \frac{S_x^2}{\bar{X}^2} - 2R\frac{S_y S_x}{\bar{Y}\bar{X}}\right),$$

or, in terms of coefficients of variation,

$$(5.8a) \qquad \text{var } (\bar{y}/\bar{x}) \doteq (\bar{Y}/\bar{X})^2 \left(\frac{N-n}{Nn}\right) \left[(V_y^2 + V_x^2 - 2RV_y V_x)\right].$$

The variance of a ratio depends not only on the variance of y but also on that of x and the correlation of the two. It may be noted that the variance of $\bar{y}/\bar{x}$ decreases as expected as sample size increases, but it also:

1. Goes down as the correlation goes to $+1$.
2. Goes down when the variability of either y or x goes down.
3. Goes down when the sampling rate goes up.

The formulation is also general enough so that when $\bar{x}$ is a constant, equaling $\bar{X}$, then Eq. 5.8 becomes (since $S_x = 0$)

$$(5.8b) \qquad \text{var } (\bar{y}/\bar{X}) = \left(\frac{1}{\bar{X}^2}\right) \left(\frac{N-n}{N}\right) \frac{S_y^2}{n} = \left(\frac{1}{\bar{X}^2}\right) \text{var } (\bar{y}),$$

as we might expect. A special case of this situation would be that of cluster sampling, where the cluster sizes constitute the x-population. If the mean per element is being estimated and the clusters are of equal size (i.e., $M_i = \bar{X}$), Eq. 5.8b is appropriate; if the clusters vary in size, Eq. 5.8 is appropriate for the variance, substituting M_i for X_i, and so on.

The Ratio as a Difference. An alternative approximate variance expression is sometimes more convenient to use. Let the error, the difference between the estimate and the parameter being estimated, be written as the identity.

$$\frac{\bar{y}}{\bar{x}} - \frac{\bar{Y}}{\bar{X}} = \frac{\bar{y} - (\bar{Y}/\bar{X})\bar{x}}{\bar{x}} = \frac{(1/n)\sum^n[Y_i - (\bar{Y}/\bar{X})X_i]}{\bar{x}}$$

$$(5.9) \qquad\qquad = \frac{1}{\bar{x}}\frac{1}{n}\sum^n D_i,$$

where

$$D_i = Y_i - \left(\frac{\bar{Y}}{\bar{X}}\right)X_i;$$

and where $\bar{x}$ is stable enough to approximate $\bar{X}$, the variance of Eq. 5.9 is given approximately by

$$\operatorname{var}\left(\frac{\bar{y}}{\bar{x}}\right) = \operatorname{var}\left(\frac{\bar{d}}{\bar{X}}\right) \doteq \frac{1}{\bar{X}^2}\operatorname{var}(\bar{d})$$

(5.10)
$$\doteq \left(\frac{1}{\bar{X}}\right)^2\left(\frac{N-n}{N}\right)\frac{S_d^2}{n},$$

where, since $E[D_1] = 0$,

(5.11)
$$S_d^2 = \frac{\sum D_i^2}{N-1} \quad \text{and} \quad i = 1, \ldots, N.$$

The two forms are equivalent approximations; a preference may depend on convenience or what one may wish to emphasize.

If we divide the variance of an estimate of some quantity by the square of that quantity, we obtain a useful parameter called *relative variance*, which is a squared coefficient of variation. Thus the relative variance of $\bar{y}/\bar{x}$ is given by removing the factor $(\bar{Y}/\bar{X})^2$ from Eq. 5.8a:

$$\operatorname{rel\ var}(\bar{y}/\bar{x}) = \frac{\operatorname{var}(\bar{y}/\bar{x})}{(\bar{Y}/\bar{X})^2}$$

(5.12)
$$= \left(\frac{N-n}{Nn}\right)(V_x^2 + V_y^2 - 2RV_xV_y).$$

Estimates of var $(\bar{y}/\bar{x})$ can be obtained from RNR samples by replacing the parameters in Eqs. 5.8 and 5.10 by their sample estimates. Thus we have

(5.13)
$$\widehat{\operatorname{var}}(\bar{y}/\bar{x}) = \left(\frac{\bar{y}}{\bar{x}}\right)^2\left(\frac{N-n}{Nn}\right)(v_x^2 + v_y^2 - 2rv_xv_y)$$

or

(5.14)
$$\widehat{\operatorname{var}}(\bar{y}/\bar{x}) = \left(\frac{1}{\bar{x}}\right)^2\left(\frac{N-n}{Nn}\right)s_d^2,$$

where we have substituted s_x^2 for S_x^2, s_y^2 for S_y^2, $\bar{x}$ for $\bar{X}$, $\bar{y}$ for $\bar{Y}$, v_x for V_x, v_y for V_y, r for R, $\bar{y}/\bar{x}$ for $\bar{Y}/\bar{X}$, s_d^2 for S_d^2, and where

$$v_x = \frac{s_x}{\bar{x}}, \qquad v_y = \frac{s_y}{\bar{y}},$$

$$r = \frac{\sum(\bar{X}_i - \bar{x})(\bar{Y}_i - \bar{y})}{[\sum(\bar{X}_i - \bar{x})^2 \cdot \sum(\bar{Y}_i - \bar{y})^2]^{1/2}}, \qquad i = 1, \ldots, n,$$

$$r = \frac{s_{xy}}{s_xs_y},$$

$$S_{xy} = \frac{1}{n-1} \sum (\bar{X}_i - \bar{x})(\bar{Y}_i - \bar{y}),$$

$$v_{xy} = \frac{S_{xy}}{\bar{x}\bar{y}},$$

$$s_d^2 = \frac{1}{n-1} \sum [Y_i - (\bar{y}/\bar{x})X_i]^2.$$

For ease of computation s_{xy} can be obtained by

$$S_{xy} = \frac{1}{n-1} \left[\sum X_i Y_i - \frac{(\sum X_i)(\sum Y_i)}{n} \right].$$

Example 5.2.   A survey on a sample of 132 area SUs from the frame total of 9460 in Des Moines, Iowa (see Section 5.13) obtained data on the number of TV sets and number of households for each SU such as the following:

SU no.	Total households (X_i)	No. possessing sets (Y_i)
1	3	1
2	8	0
3	8	1
4	7	0
5	7	4
.	.	.
.	.	.
.	.	.
132	4	2
Total	691	232

The proportion of households having TV sets works out to 0.3358. What is the standard error of this estimate of the population proportion? (The following statistics are available: $v_y = 1.0038$; $v_y^2 = 1.0076$; $v_x = 0.3812$; $v_x^2 = 0.1453$; $r = 0.4964$.)

The estimated variance of $\bar{y}/\bar{x}$ is given by Eq. 5.13:

$$\widehat{\text{var}}\,(\bar{y}/\bar{x}) = (\bar{y}/\bar{x})^2 \left(\frac{N-n}{Nn}\right)(v_y^2 + v_x^2 - 2rv_y v_x)$$

$$= (0.3358)^2 \left[\frac{9460-132}{(9460)(132)}\right]\left[1.0076 + 0.1453\right.$$

$$\left. - 2(0.4964)(1.0038)(0.3812)\right]$$

$$= (0.01127) \left[\frac{1}{132} (1.1529 - 0.3700) \right]$$

$$= (0.1127) (0.005930) = 0.000668.$$

The estimated standard error is simply the square root of the variance or

$$\widehat{SE} \ (\bar{y}/\bar{x}) = \sqrt{\widehat{var} \ (\bar{y}/\bar{x})} = \sqrt{0.000668} = 0.0259,$$

and the approximate 67 % confidence limits are 0.3358 ± 0.0259. Hence, unless a one-in-three chance has come off, the true proportion of sets is between 31.0 % and 36.2 %.

5.4. Estimating $\bar{Y}\bar{X}$, the Product of Two Population Means

The product estimator, although similar to the ratio estimator, is much less frequently used in statistical surveys. To estimate $\bar{Y}X$ from an RNR sample of n from a universe of N we have as an estimator

(5.15) $$w_p = \bar{y}\bar{x},$$

which, although biased, is for many purposes quite useful. The bias is given by

$$\text{bias} \ (\bar{x}\bar{y}) = \left(\frac{N-n}{Nn} \right) S_{xy}$$

(5.16) $$= \left(\frac{N-n}{Nn} \right) RS_x S_y.$$

An unbiased variant of Eq. 5.15 is given in Section 5.11. Since bias decreases with sample size, it is generally ignorable in many practical situations.

The variance of w_p is given approximately by

(5.17) $$\text{var} \ (w_p) \doteq \bar{X}^2 \ \text{var} \ (\bar{y}) + \bar{Y}^2 \ \text{var} \ (\bar{x}) \ + 2\bar{Y}\bar{X}R[\text{var} \ (\bar{y}) \ \text{var} \ (\bar{x})]^{1/2}$$

$$\doteq \left(\frac{N-n}{Nn} \right) (\bar{X}^2 S_y^2 + \bar{Y}^2 S_x^2 + 2\bar{Y}\bar{X}RS_y S_x),$$

and by replacing the parameters with sample estimates we have as the variance estimated from a sample

(5.18) $$\widehat{var} \ (w_p) \doteq \left(\frac{N-n}{Nn} \right) (\bar{x}^2 s_y^2 + \bar{y}^2 s_x^2 + 2\bar{y}\bar{x}s_{xy}).$$

5.5. Estimating $\bar{Y}$ with the Ratio Estimator

An estimate of $\bar{Y}$ can be obtained by multiplying the average amount of y per unit of x observed in the sample by the average amount of x for the entire universe, which is assumed to be known. Thus we define the ratio estimator,

$$(5.19) \qquad \bar{y}_{rat} = \bar{X}(\bar{y}/\bar{x}).$$

The logic of this estimator may be seen in Fig. 5.1, where a line is drawn from the origin through the point representing the sample mean of $\bar{y}$ and $\bar{x}$. The true mean of x—that is, $\bar{X}$—is known and in this case is shown to be larger than the mean obtained in the sample. For this reason we expect the true mean of y to be larger than the sample mean $\bar{y}$. Hence we choose $\bar{y}_{rat}$ as an improved estimate of $\bar{Y}$, obtaining it by the proportional relationship:

$$(5.19a) \qquad \frac{\bar{y}_{rat}}{\bar{y}} = \frac{\bar{X}}{\bar{x}},$$

from which Eq. 5.19 can be derived.

The variance of the ratio estimator is given by

$$(5.20) \qquad var\,(\bar{y}_{rat}) = \bar{X}^2\,var\,(\bar{y}/\bar{x}),$$

whence, from Eq. 5.8,

$$(5.20a) \qquad var\,(\bar{y}_{rat}) \doteq \bar{Y}^2 \left(\frac{N-n}{Nn}\right) \left(\frac{S_y^2}{\bar{Y}^2} + \frac{S_x^2}{\bar{X}^2} - 2R\,\frac{S_y S_x}{\bar{Y}\bar{X}}\right),$$

and from Eq. 5.8a, in terms of relative variances,

$$(5.20b) \qquad var\,(\bar{y}_{rat}) \doteq \bar{Y}^2 \left(\frac{N-n}{Nn}\right)(V_y^2 + V_x^2 - 2RV_y V_x).$$

To obtain Y_{rat}, an estimate of Y with the help of x, we simply multiply Eq. 5.19 by N, and to obtain var $(\hat{Y}_{rat})$, its variance, we multiply Eq. 5.20a by N^2.

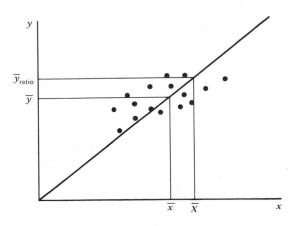

Fig. 5.1. Diagrammatic representation of the ratio estimator.

Since both of the estimators in this section contain the ratio $(\bar{y}/\bar{x})$, differing only by the constant by which it is multiplied, both are thereby *biased* by the moment given in Eq. 5.7 except for the appropriate multiplier.

With two estimators of the same parameter, $\bar{Y}$, we have the opportunity for a choice. In many cases we would prefer the estimator with the greater accuracy, hence the smaller variance or, if biased, the smaller MSE. Ignoring bias, which is usually small if n is large, we find the relative efficiency of the ratio to simple estimator by

(5.21) $$\text{RE (ratio/simple)} \doteq \frac{\text{var }(\bar{y})}{\text{var }(\bar{y}_{\text{rat}})} = \frac{[(N-n)/Nn]S_y^2}{\text{Eq. 5.20b}}$$

$$\doteq \frac{[(N-n)/Nn]\ \bar{Y}^2 V_y^2}{[(N-n)/Nn]\ (\bar{Y}^2)\ (V_x^2 + V_y^2 - 2RV_x V_y)}$$

(5.21a) $$\doteq \frac{V_y^2}{V_x^2 + V_y^2 - 2RV_x V_y}.$$

Hence $\bar{y}_{\text{rat}}$ is a better (more precise) estimator than $\bar{y}$ whenever

(5.22) $$R > \frac{1}{2}\frac{V_x}{V_y}$$

or where $V_x = V_y$ (which is often the case) and

$$R > \tfrac{1}{2}.$$

5.6. Estimating $\bar{Y}$ with the Regression Estimator

Another estimator is based on the possible regression relationship of y on x. Thus to estimate $\bar{Y}$ we have

(5.23) $$\bar{y}_{\text{regr}} = \bar{y} + b(\bar{X} - \bar{x}),$$

where

(5.24) $$b = \frac{\sum^n (X_i - \bar{x})(Y_i - \bar{y})}{\sum^n (X_i - \bar{x})^2} = \frac{s_{xy}}{s_x^2},$$

the ordinary (least-squares) regression coefficient computed from the sample. The nature of this estimator may be seen in Fig. 5.2. As in Fig. 5.1, we represent the relationship of y on x by a straight line through the sample means $\bar{y}$ and $\bar{x}$, but in this case the line does not have to go through the origin. The slope of the line, b, is given by

$$b = \frac{\bar{y}_{\text{regr}} - \bar{y}}{\bar{X} - \bar{x}},$$

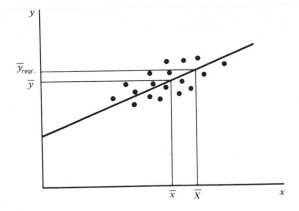

Fig. 5.2.   Diagrammatic representation of the regression estimator.

hence

$$\bar{y}_{regr} = \bar{y} + b(\bar{X} - \bar{x}).$$

The regression estimator, like the ratio estimator, is also generally biased except under certain special conditions. The bias in $\bar{y}_{regr}$ is given by Cochran (1942):

$$\text{bias } (\bar{y}_{regr}) = E[\bar{y}_{regr}] - \bar{Y}$$

(5.25)
$$\doteq \frac{-(N-n)}{(n-1)NS_x^2} \frac{\sum^N E_i(X_i - \bar{X})^2}{N-1}$$

where

(5.26)
$$E_i = Y_i - [\bar{Y} + B(X_i - \bar{X})]$$

and B is the true regression coefficient for a finite population; thus

(5.27)
$$B = \frac{\sum (Y_i - \bar{Y})(X_i - \bar{X})}{\sum (X_i - \bar{X})^2} = \frac{S_{xy}}{S_x^2}.$$

Note that the bias:

1. Goes down as sample size increases.
2. Goes down as the sampling rate increases.
3. Goes down as the correlation goes to ± 1 (since in this case E_i goes to 0).
4. Goes down as variation of x *increases*.
5. Is zero if the regression of y on x is linear [since $\sum E_i(X_i - \bar{X})^2$ will then be zero].

Since the bias goes down as n increases, the bias in the estimator can usually be disregarded when samples are reasonably large. To make sure, however, Eq. 5.25 can be used as a check using sample estimates of S_x^2 and the "error covariance."

For any *specific sample* the sampling variance is given by the *conditional variance* (that is, from the set of xs observed), which is approximately

$$(5.28) \qquad \text{var} \, (\bar{y}_{\text{regr}}) \doteq \left(\frac{N-n}{N} \right) S_y^2 (1-R^2) \left[\frac{1}{n} + \frac{(\bar{X} - \bar{x})^2}{\sum^n (X_i - \bar{x})^2} \right] = Var(cond)$$

and which can be estimated from the sample by

$$(5.29) \qquad \widehat{\text{var}} \, (\bar{y}_{\text{regr}}) \doteq \left(\frac{N-n}{N} \right) s_y^2 \, (1-r^2) \left[\frac{1}{n} + \frac{(\bar{X} - \bar{x})^2}{\sum^n (X_i - \bar{x})^2} \right].$$

[handwritten: Var about reg.]

The *average* variance of the regression estimator—that is, the variance over repeated samples of size n—first given by Cochran (1942), is

$$(5.30) \qquad \overline{\text{var}} \, (\bar{y}_{\text{regr}}) \doteq \left(\frac{N-n}{Nn} \right) S_y^2 (1-R^2) \left(1 + \frac{1}{n}\right),$$

or where $1/n$ is sufficiently small

$$(5.30a) \qquad \overline{\text{var}} \, (\bar{y}_{\text{regr}}) \doteq \left(\frac{N-n}{Nn} \right) S_y^2 (1-R^2), = Var(uncond.)$$

where R is the correlation coefficient between y and x, as defined by Eq. 5.6. To obtain $\hat{Y}_{\text{regr}}$, the regression estimate of Y, and its variance $\overline{\text{var}} \, (\hat{Y}_{\text{regr}})$, Eqs. 5.23 and 5.30a are multiplied by N and N^2, respectively. To estimate Eq. 5.30a from a sample

$$(5.31) \qquad \widehat{\overline{\text{var}}} \, (\bar{y}_{\text{regr}}) \doteq \left(\frac{N-n}{Nn} \right) s_y^2 (1-r^2).$$

The variances given in Eqs. 5.30 and 5.30a are those that would be obtained on the average by the use of the regression estimator.

5.7. Estimating $\bar{Y}$ with the Difference Estimator

A variant of the regression estimator is one where the estimate b of the parameter B is not obtained from the sample, such as by least squares, but is some arbitrary constant, say b_0. In this case the estimator of $\bar{Y}$ is given by a formula identical in form to Eq. 5.23:

$$(5.32) \qquad \bar{y}_{\text{diff}} = \bar{y} + b_0 (\bar{X} - \bar{x}).$$

Note that Eq. 5.32 can be written as

$$(5.32a) \qquad \bar{y}_{\text{diff}} = (\bar{y} - b_0 \bar{x}) + b_0 \bar{X}.$$

Since b_0 is an arbitrary constant, a simple and in many cases a reasonable value to choose is $b_0 = 1$. In this case it simplifies to

$$(5.32b) \qquad \bar{y}_{\text{diff}} = (\bar{y} - \bar{x}) + \bar{X},$$

which is simply a difference between two random variables. Hansen et al. (1953), who pioneered the idea and use of the $\bar{y}_{\text{diff}}$ estimator, called it a "difference" estimator because of this property. When b_0 is not unity, the term in parentheses of Eq. 5.32a can be regarded as a *weighted difference*. The estimator is always unbiased, since

$$E[\bar{y}_{\text{diff}}] = E[\bar{y} - b_0(\bar{x} - \bar{X})]$$
$$(5.33) \qquad\qquad = \bar{Y} - b_0\bar{x} - b_0\bar{X} = \bar{Y}.$$

A result that may be surprising inasmuch as b_0 is an arbitrary choice.

The variance of $\bar{y}_{\text{diff}}$ is given by

$$(5.34) \qquad \text{var}(\bar{y}_{\text{diff}}) = \left(\frac{N-n}{Nn}\right)(S_y^2 + b_0^2 S_x^2 - 2b_0 R S_x S_y) \qquad ※$$

which is somewhat similar in form to var $(\bar{y}_{\text{rat}})$, Eq. 5.20b. The reliability of this estimator depends, like the ratio and regular regression estimators, on the correlation between y and x. It depends also on the choice of b_0. It can be found mathematically that the best choice is

$$(5.35) \qquad b_0 = \frac{S_{xy}}{S_x^2} = B$$

—that is, when one chooses the least-squares value of B for the population of x and y.

To estimate Eq. 5.34 from a sample, we simply replace the parameters with appropriate estimates, thus

$$(5.36) \qquad \widehat{\text{var}}(\bar{y}_{\text{diff}}) = \left(\frac{N-n}{Nn}\right)(s_y^2 + b_0^2 s_x^2 - 2b_0 r s_x s_y).$$

The efficiency of $\bar{y}_{\text{diff}}$ relative to the simple estimator is given by

$$(5.37) \qquad \text{RE}(\bar{y}_{\text{diff}}/\bar{y}) = \frac{S_y^2}{S_y^2 + b_0^2 S_x^2 - 2b_0 R S_x S_y}.$$

Hence, whenever $R > b_0 S_x/2S_y$, the $\bar{y}_{\text{diff}}$ is more accurate. If $b_0 = B$, the true regression coefficient, then whenever $R > \frac{1}{2}$ the $\bar{y}_{\text{diff}}$ estimator is more accurate.

As direct Differences. An alternative way of looking at this estimator is to express $Y_i - b_0 X_i$ as a new variate, say D_i, a difference between Y_i and X_i with b_0 as an arbitrary constant. If

$$(5.38) \qquad D_i = Y_i - b_0 X_i,$$

then

(5.38a) $$\bar{D} = \bar{Y} - b_0 \bar{X},$$

and for a sample of n,

(5.38b) $$\bar{d} = \bar{y} - b_0 \bar{x}.$$

The estimator, Eq. 5.32a, can be rewritten as

$$\bar{y}_{\text{diff}} = (\bar{y} - b_0 \bar{x}) + (\bar{Y} - \bar{D})$$

(5.39) $$= \bar{d} + (\bar{Y} - \bar{D}),$$

the variance of which is

(5.40) $$\text{var}(\bar{y}_{\text{diff}}) = \left(\frac{N-n}{Nn}\right) \frac{S_d^2}{n},$$

where

(5.41) $$S_d^2 = \frac{\sum (D_i - \bar{D})^2}{N-1}, \qquad i = 1, \ldots, N.$$

If only sample data are available, an unbiased estimate of Eq. 5.39 is given by

(5.42) $$\widehat{\text{var}}(\bar{y}_{\text{diff}}) = \left(\frac{N-n}{Nn}\right) s_d^2,$$

where

(5.43) $$s_d^2 = \frac{\sum (D_i - \bar{d})^2}{n-1}, \qquad i = 1, \ldots, n.$$

A comparison of this estimator, the regular or least-squares regression estimator, and the ratio estimator will be considered in Section 5.9.

5.8. Estimating $\bar{Y}$ with the Product Estimator

To estimate $\bar{Y}$ with the help of x we have the product estimator,

(5.44) $$\bar{y}_{\text{prod}} = \bar{y}\left(\frac{\bar{x}}{\bar{X}}\right)$$

(5.44a) $$= \left(\frac{1}{\bar{X}}\right) w_p,$$

in which information on x, namely the population mean, $\bar{X}$, is included in an attempt to improve the estimate of $\bar{Y}$. In contrast to the ratio estimator, in this case the positions of $\bar{x}$ and $\bar{X}$ are reversed, $\bar{X}$ now being in the denominator. Equation 5.44 is biased because of the bias in $\bar{y}\bar{x}$, but it is generally ignorable for large n.

The variance approximated by

$$\text{var}\,(\bar{y}_{\text{prod}}) \doteq \frac{1}{\bar{X}^2}\,\text{var}\,(w_p)$$

(5.45)
$$\doteq \left(\frac{N-n}{Nn}\right)\left[S_y^2 + \left(\frac{\bar{Y}}{\bar{X}}\right)^2 S_x^2 + 2\left(\frac{\bar{Y}}{\bar{X}}\right)RS_xS_y\right],$$

which can be established from a sample by using analogous sample estimators for the parameters.

Note that, to be considered seriously as a higher accuracy alternative to estimate $\bar{Y}$, the product estimator depends on a negative correlation between y and x.

5.9. Estimating Y and $\bar{Y}$: Comparison of Estimators

The common structure and the special features of the simple, ratio, product, difference, and regression estimators for estimating the total, Y, with the help of x are presented below. To facilitate the comparison, the estimators are modified slightly in form from their presentations above, but they are algebraically identical.

Simple: $N\bar{y} = N[(1)\,(\bar{y}) + 0(\bar{X} - 0)]$

Ratio: $N\bar{y}_{\text{rat}} = N[(0)\,(\bar{y}) + \dfrac{\bar{y}}{\bar{x}}\,(\bar{X} - 0)]$

Regression: $N\bar{y}_{\text{regr}} = N[(1)\,(\bar{y}) + b(\bar{X} - \bar{x})]$

Difference: $N\bar{y}_{\text{diff}} = N[(1)\,(\bar{y}) + b_0(\bar{X} - \bar{x})]$

Product: $N\bar{y}_{\text{prod}} = N[(0)\,(\bar{y}) + \bar{y}\bar{x}\left(\dfrac{1}{\bar{X}} - 0\right)]$

Each estimator utilizes information about $\bar{y}, \bar{X}$, and $\bar{x}$, but they differ in the weights assigned to each. In some cases these weights are zeros. Note that the simple estimator gives zero weights to $\bar{X}$ and $\bar{x}$, relying completely on $\bar{y}$, whereas the regression and difference estimators give weight to all three.

The actual accuracy of the estimators, however, depends on the nature of the y- and x-populations and their relationship with each other. These matters generally must be determined empirically. However, in order to get some general ideas of how each may behave, we can make some plausible assumptions and see what happens. First, let us consider the problem of estimating the total Y from a sample. How do they compare? We note that besides $\bar{y}$:

1. The simpler estimator requires we know only N.
2. The ratio estimator requires we know only $\bar{X}$ (and not N).
3. The regression estimator requires we know both N and X (or $\bar{X}$).

To compare the variances we have the following summary, which omits the factor $(N-n)/Nn$:

Simple: $\quad\quad\quad\quad \mathrm{var}\ (\bar{y}) = S_y^2$

Ratio: $\quad\quad\quad\quad \mathrm{var}\ (\bar{y}_{\mathrm{rat}}) = S_y^2 + \left(\dfrac{\bar{Y}}{\bar{X}}\right)^2 S_x^2 - 2\left(\dfrac{\bar{Y}}{\bar{X}}\right) RS_y S_x$

Regression: $\quad\quad \mathrm{var}\ (\bar{y}_{\mathrm{regr}}) = S_y^2 + B^2 S_x^2 - 2BRS_y S_x$

Difference: $\quad\quad \mathrm{var}\ (\bar{y}_{\mathrm{diff}}) = S_y^2 + b_0^2 S_x^2 - 2b_0 RS_y S_x$

Product: $\quad\quad\quad \mathrm{var}\ (\bar{y}_{\mathrm{prod}}) = S_y^2 + \left(\dfrac{\bar{Y}}{\bar{X}}\right)^2 S_x^2 + 2\left(\dfrac{\bar{Y}}{\bar{X}}\right) RS_y S_x$

Since efficiency—aside from biases, which we shall ignore—depends on the magnitude of these variances, we can compare efficiency by comparing these variances algebraically to determine conditions under which one is to be preferred over the other.

The relative efficiency of ratio to simple is

$$(5.46) \quad\quad \mathrm{RE\ (ratio/simple)} = \frac{\mathrm{var}\ (\bar{y})}{\mathrm{var}\ (\bar{y}_{\mathrm{rat}})} = \frac{\mathrm{var}\ (\bar{y})}{\mathrm{var}\ (\bar{X}\bar{y}/\bar{x})},$$

and, substituting from Eq. 5.20b, we obtain

$$(5.47) \quad\quad \mathrm{RE\ (ratio/simple)} = \frac{1}{1 + (V_x/V_y)\ (V_x/V_y - 2R)},$$

whence we can say the ratio estimator is more efficient than the simple whenever

$$(5.48) \quad\quad\quad\quad\quad\quad R > \left(\frac{1}{2}\right)\frac{V_x}{V_y}.$$

In cases where $V_x \doteq V_y$, such as those when x and y are the same population but at two different periods of time, then ratio is better than simple when R is greater than .5. The relative efficiency in this case is

$$(5.49) \quad\quad\quad\quad\quad \mathrm{RE\ (ratio/simple)} = \frac{1}{2(1-R)}.$$

In the case of the regression estimator we have

$$\mathrm{RE\ (regr/simple)} = \frac{\mathrm{var}\ (\bar{y})}{\mathrm{var}\ (\bar{y}_{\mathrm{regr}})}$$

$$(5.50) \quad\quad\quad\quad\quad\quad\quad \doteq \frac{1}{1-R^2};$$

hence the regression estimator is always more efficient than the simple if R is greater than zero, either negatively or positively.

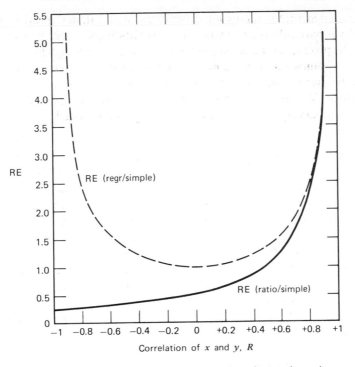

Correlation of x and y, R

Fig. 5.3.   Graph of the approximate effeciencies of the ratio and regression estimators, relative to the simple, when $V_x = V_y$.

A graph has been prepared in Fig. 5.3 to show how the three estimators compare in efficiency for all values of R for the case where $V_x = V_y$.

Table for Fig. 5.3

R	RE (ratio/simple)[a]	RE (regr/simple)[b]
-1.00	0.25	∞
$-.90$	0.263	5.27
$-.75$	0.2857	2.28
$-.50$	0.33	1.33
$-.25$	0.40	1.067
0	0.50	1.000
$+.25$	0.67	1.067
$+.50$	1.00	1.33
$+.75$	2.00	2.28
$+.90$	5.00	5.27
$+1.00$	∞	∞

[a] RE (ratio/simple) $= 1/[2(1-R)]$.
[b] RE (regr/simple) $= 1/(1-R^2)$.

Note that in general the regression estimator is better than the ratio, and the superiority is particularly great when the correlation coefficient nears -1. (In this case the product estimator should be considered.) Although the regression estimator is somewhat more cumbersome to compute, serious consideration ought to be given it because of its safety (no chance of accuracy loss) and possible high efficiency. Also notice the rather handsome gains that can be made if a helping variate can be found that is highly correlated with our object of direct interest, y.

Example 5.3. After completion of the experiment on judgment versus random sampling of a universe of 126 stones described in Section 1.6, each selector was asked to observe and guess the weight of each stone in the universe. The results for one of the selectors are recorded as X_i, and the true weight of each stone is recorded as Y_i. Suppose we regard these "eye estimates" of weights as our auxiliary population, which we would like to use to improve our estimate of $\bar{Y}$ from RNR samples of n stones. What gain in efficiency can be expected in random sampling if these eye estimates are used?

The necessary computations from these data are: $\bar{Y} = 98.4$, $\bar{X} = 115.3$, $S_y^2 = 15,605$, $S_x^2 = 21,869$, $S_y = 124.9$, $S_x = 147.9$, $V_y = 1.27$, $V_x = 1.28$, $R = .895$.

The efficiency of the ratio estimator relative to the simple in this case is given by Eq. 5.47:

$$\text{RE (ratio/simple)} = \frac{1}{1 + (V_x/V_y)\,(V_x/V_y - 2R)}$$

$$= \frac{1}{1 + (1.28/1.27)\,[1.28/1.27 - 2(.895)]}$$

$$= 4.29,$$

a rather large gain, indeed.

The regression estimator may be somewhat better. From Eq. 5.50

$$\text{RE (regr/simple)} = \frac{1}{1 - R^2}$$

$$= \frac{1}{1 - (.895)^2} = \frac{1}{1 - .80} = 5.00.$$

Hence the use of eye-estimated weights results in a rather large gain in sampling efficiency through the use of ratio and regression estimators. If sample sizes are so small that bias is troublesome, it is suggested that the unbiased ratio estimator Eq. 5.63 be used.

5.10. Estimating $\bar{\bar{Y}}$: Clusters of Unequal M_i

If our object is to estimate the mean of y per element, rather than per cluster, the appropriate estimator, presented in Section 4.4, is

$$(5.51) \qquad \bar{y}_c = \frac{\sum^n Y_i}{nM_0} = \frac{\sum^n M_i \bar{Y}_i}{nM_0} = \frac{\bar{y}}{M_0},$$

and its variance, given by Eq. 4.6, is

$$(5.52) \qquad \text{var}(\bar{y}_c) = \left(\frac{N-n}{N}\right) \frac{S_c^2}{nM_0^2}.$$

If $M_i \neq M_0$, we have three potentially useful estimators of $\bar{\bar{Y}}$:

$$(5.53) \qquad \bar{\bar{y}}_{(1)} = \frac{\sum \bar{Y}_i}{n},$$

$$(5.54) \qquad \bar{\bar{y}}_{(2)} = \frac{\sum M_i \bar{Y}_i}{n\bar{M}} = \frac{\bar{y}}{\bar{M}},$$

$$(5.55) \qquad \bar{\bar{y}}_{(3)} = \frac{\sum M_i \bar{Y}_i}{\sum M_i} = \frac{\bar{y}}{\bar{m}}.$$

The first and third are generally biased, whereas the second is unbiased. However, the second requires that $\bar{M}$ be known. Although the bias in the first is constant over all sample sizes, that in the third decreases with increases in sample size.

To compare the three estimates we should use MSE rather than variance to measure accuracy, since two of the estimators are biased. The MSEs are:

$$(5.56) \qquad \text{MSE}(\bar{\bar{y}}_1) = \left(\frac{N-n}{Nn}\right) \frac{\sum(\bar{Y}_i - \bar{\bar{Y}}_{(1)})^2}{N-1} + (\bar{\bar{Y}}_{(1)} - \bar{\bar{Y}})^2$$

$$(5.57) \qquad \text{MSE}(\bar{\bar{y}}_2) = \left(\frac{N-n}{Nn}\right) \frac{\sum(Y_i - \bar{Y})^2}{\bar{M}^2(N-1)},$$

$$(5.58) \qquad \text{MSE}(\bar{\bar{y}}_3) \doteq \left(\frac{N-n}{Nn}\right) \frac{\sum M_i^2(\bar{Y}_i - \bar{\bar{Y}})^2}{\bar{M}^2(N-1)},$$

where $\bar{\bar{Y}}_{(1)} = \sum^n \bar{Y}_i/N$, the expected value of $\bar{\bar{y}}_{(1)}$.

As usual, none of these estimators is best—that is, has the smallest MSE—under all circumstances. Inspection of the estimators shows that the MSE depends on the relationship between $\bar{Y}_i$ and M_i, types 1, 2, and 3 being best when the regression of $\bar{Y}_i$ on M_i is zero, negative, and positive, respectively.

5.11. Other Estimators

A number of other estimators have been in use or proposed for use because of their particular properties. Of the five to be described here, one is an alternative estimator (but biased) of $\bar{Y}/\bar{X}$; the remaining four are estimators of $\bar{Y}$ with the help of x, including unbiased versions of the ratio, regression, and product estimators. Since these schemes are usually somewhat more complex in structure and sometimes less accurate than their biased counterparts, they are less frequently used. However, they are presented here briefly for the light they shed on the estimation problem.

Means-of-ratios Estimator: $\bar{r}$. Suppose we define the Y_i/X_i for each SU as R_i and

$$(5.59) \qquad \bar{R} = \frac{1}{N} \sum^{N} R_i.$$

If we define

$$(5.60) \qquad \bar{r} = \frac{1}{n} \sum_i^n (Y_i/X_i) = \frac{1}{n} \sum_i^n R_i,$$

then $\bar{r}$ is an unbiased estimator of $\bar{R}$, but it is a biased estimator of $\bar{Y}/\bar{X}$. The bias of $\bar{r}$ as an estimator of $\bar{Y}/\bar{X}$ is defined as

$$(5.61) \qquad \text{bias } (\bar{r}) = E[\bar{r}] - \bar{Y}/\bar{X}$$

and is given by

$$(5.62) \qquad \text{bias } (\bar{r}) \doteq -(\bar{Y}/\bar{X})(V_x^2 - RV_xV_y)$$

—that is, the bias of the ratio estimator when the sample size is 1. Therefore, this estimator has the fault that its bias does not decrease with an increase in sample size, nor does it go to zero when we take a complete census. Hence, bias is the most important characteristic to watch in this estimator.

Unbiased Ratio Estimator, $\bar{y}_{\text{u-rat}}$. Although a general unbiased estimator for $\bar{Y}/\bar{X}$ is not available, one is for estimating $\bar{Y}$ using the ratio. The estimator,

$$(5.63) \qquad \bar{y}_{\text{u-rat}} = \bar{X}\bar{r} + \frac{n}{n-1} \cdot \frac{N-1}{N} (\bar{y} - \bar{r}\bar{x}),$$

is due to Hartley and Ross (1954). Some properties of this estimator are given in Goodman and Hartley (1958) together with comparisons with the ratio estimator. Its precision is usually less than that of the simple ratio estimator, and it is generally useful only when bias is a particular problem; see Section 7.9.

Unbiased Regression Estimator: $\bar{y}_{\text{mick}}$. A class of unbiased ratio and regression estimators (of $\bar{Y}$ and Y) has been proposed and described by Mickey (1959). Since these are fairly sophisticated and as yet not in general use, they will not be considered here.

Unbiased Product Estimator: $\bar{y}_{\text{u-prod}}$ or w_{up}. An unbiased estimator of $\bar{Y}\bar{X}$ is given by

(5.64)
$$w_{up} = \bar{x}\bar{y} - \left(\frac{N-n}{Nn}\right) s_{xy}$$

(5.65)
$$= \left(\frac{N-1}{N}\right)\left(\frac{n}{n-1}\right) \overline{yx} - \left[\frac{N-n}{N(n-1)}\right] \overline{yx},$$

where $\overline{yx}$ is the mean of yx product in the sample—that is, $\sum Y_i X_i/n$—and s_{xy} is sample covariance,

(5.66)
$$s_{xy} = \frac{\sum (Y_i - \bar{y})(X_i - \bar{x})}{n-1}.$$

Equation 5.64 was obtained by simply supplying an estimate of the bias in w_p (Eq. 5.16) Since this estimate is unbiased, the result is an unbiased estimator. Equation 5.65 may be a more convenient computational form.

An unbiased product estimator of $\bar{Y}$ can be made by simply substituting w_{up} in Eq. 5.44a, whence we obtain

(5.67)
$$\bar{y}_{\text{u-prod}} = \left(\frac{1}{\bar{X}}\right) w_{up}$$

(5.67a)
$$= \left(\frac{1}{\bar{X}}\right)\left[\bar{x}\bar{y} - \left(\frac{N-n}{Nn}\right) s_{xy}\right].$$

The occasion where this estimator is to be preferred over the biased one seems fairly limited. However, here it is if the need arises.

Use of the Known Coefficient of Variation, V_y. If one wishes to estimate $\bar{Y}$ and happens to know the relative variability of y—that is, V_y—he can consider the estimator,

(5.68)
$$\bar{y}_{\text{coef}} = \left(\frac{n}{n+V_y^2}\right)\bar{y},$$

which, although biased, has a lower mean square error (MSE) than $\bar{y}$. The estimator, due to Searls (1964), is particularly attractive when one is dealing with small samples.

5.12. Summary

There are many opportunities for making rather substantial gains in efficiency through the choice of appropriate auxiliary populations, extracting information from them by the choice of appropriate estimators. Only the more common ones have been presented. A few suggestions on choice of auxiliary populations will be given below.

There are many opportunities for using judgment in providing an auxiliary population. The costs are usually low, and it is frequently surprising how high a correlation can be achieved. In dealing with human populations in towns and cities it is quite common to make quick eye estimates while driving a car around the block. Another technique is to make these block population estimates from a quick inspection of aerial photographs of a city. In forest surveys it is sometimes possible to make quick eye estimates of tree stands by direct observation or by looking at aerial photographs. In agricultural surveys similar techniques have been used.

Old census data are another good source of auxiliary populations. The new survey becomes simply a method for estimating a change to obtain the present status. Marked reductions in the size of sample required can be effected by this method. It is not necessary to be able to obtain the X_i exactly as it was originally for each SU. For example, if we have the x data for block totals only, when actually we have subdivided blocks into SUs, we need only divide the X for the total block by the number of SUs in the block to obtain a usable X_i. Other unbiased schemes of this sort can be constructed.

Sometimes it is not convenient to determine the total number of elements in a universe. For example, suppose we wish to determine the total area Y of all the leaves on a tree. If the tree is stripped of leaves and the leaves weighed, we obtain the value of X. If a small sample of leaves is selected for both leaf area measurements and weight, we can obtain values of $\bar{y}$ and $\bar{x}$. An estimator of Y, the total leaf area, can then be made by the ratio estimator. This would not be possible with either the simple or regression estimators.

It is hoped that further advances in theory and application will be forthcoming, making possible even greater opportunities for improvements in sample design through more efficient estimation schemes.

5.13. Review Illustrations

1. A survey of 691 households in Des Moines, Iowa, was carried out to study TV viewing (October 1951). The sample consisted of 132 area-list sampling units. There were 56,296 households in Des Moines in April of 1950, according to the U.S. Bureau of the Census, and the total number of SUs in the area-list frame was 9460. Following are the total number of households, X_i, and of those, the number possessing TV sets, Y_i, for each of the 132 SUs.

SU no.	X_i	Y_i	SU no.	X_i	Y_i	SU no.	X_i	Y_i	SU no.	X_i	Y_i	SU no.	X_i	Y_i
1	3	1	28	6	4	55	6	2	82	44	0	109	1	3
2	8	0	29	4	4	56	5	1	83	2	0	110	3	0
3	8	1	30	5	2	57	7	2	84	3	0	111	5	0
4	7	0	31	5	2	58	6	1	85	7	1	112	0	0
5	7	4	32	5	1	59	5	1	86	11	3	113	1	0
6	7	6	33	5	3	60	7	1	87	4	0	114	4	0
7	3	0	34	6	3	61	5	4	88	4	1	115	3	0
8	7	2	35	6	1	62	3	1	89	4	0	116	6	1
9	7	1	36	6	3	63	7	3	90	6	1	117	5	1
10	7	4	37	4	1	64	4	1	91	4	0	118	5	3
11	7	4	38	4	2	65	4	3	92	4	2	119	5	1
12	8	0	39	4	0	66	6	3	93	6	2	120	4	1
13	7	3	40	9	1	67	4	2	94	5	1	121	4	1
14	5	2	41	6	1	68	3	2	95	5	3	122	5	1
15	5	2	42	6	2	69	7	3	96	5	3	123	6	2
16	5	2	43	4	1	70	8	4	97	9	2	124	5	1
17	8	1	44	5	0	71	4	0	98	5	1	125	5	0
18	4	1	45	3	2	72	2	0	99	4	1	126	7	3
19	7	3	46	7	2	73	5	2	100	6	3	127	8	4
20	14	8	47	7	2	74	0	0	101	6	2	128	5	2
21	6	2	48	3	1	75	4	2	102	5	1	129	5	2
22	5	3	49	7	2	76	3	1	103	5	4	130	6	4
23	8	4	50	6	4	77	3	0	104	4	1	131	3	2
24	6	6	51	9	1	78	9	0	105	5	2	132	4	2
25	4	2	52	7	3	79	4	2	106	4	3			
26	7	4	52	5	2	80	5	1	107	5	2			
27	3	2	54	3	1	81	3	1	108	5	2			

Total   691   232

The following statistics were computed from the data (where y and x are the number of TV sets and number of households on each SU, respectively):

$$\bar{y} = 1.7576, \qquad \bar{y}^2 = 3.0892,$$
$$\bar{x} = 5.2348, \qquad \bar{x}^2 = 27.4031,$$
$$s_y = 1.7643, \qquad s_y^2 = 3.1127,$$
$$s_x = 1.9956, \qquad s_x^2 = 3.9826,$$
$$v_y = 1.0038, \qquad v_y^2 = 1.0076,$$
$$v_x = 0.3812, \qquad v_x^2 = 0.1453,$$
$$r = 0.4964, \qquad r^2 = 0.2464,$$
$$\bar{y}/\bar{x} = 0.3358, \quad (\bar{y}/\bar{x})^2 = 0.1127,$$
$$b = 0.4389.$$

Assuming the 132 SUs are a random sample from a total of 9460:

(a) What are the variance and standard error of $\bar{y}/\bar{x}$, the proportion of HHs with TV?

(b) What is the estimated bias in $\bar{y}/\bar{x}$ when we are estimating Y/X?

(c) What is the estimated variance in $\bar{y}/\bar{x}$ if the 691 households were selected individually at random (RNR)? [*Hint:* In this case $\bar{y}/\bar{x} = p$.]

(d) Discuss the comparison of (a) and (c). [*Hint:* Compute the implicit intraclass correlation coefficient, $\hat{\delta}$.]

Solution

(a) $\widehat{\text{var}}\ (\bar{y}/\bar{x}) = (\bar{y}/\bar{x})^2 \left(\dfrac{v_x^2 + v_y^2 - 2v_x v_y r}{n} \right), \qquad$ ignoring FPC

$$= \left(\frac{1.7576}{5.2348} \right)^2 \frac{1}{132} \ (0.1453 + 1.0076$$

$$- (2)\ (0.3812)\ (1.0038)\ (0.4964))$$

$$= \frac{0.1127}{132}\ (0.7729) = 0.000,660,$$

$\widehat{\text{SE}}\ (\bar{y}/\bar{x}) = (0.000,660)^{1/2} = 0.0257.$

Hence $(\bar{y}/\bar{x}) \pm \text{SE}\ (\bar{y}/\bar{x}) = 0.3358 \pm 0.0257.$

(b) Estimated bias in $\bar{y}/\bar{x}$: From Eq. 5.7:

$$\left(\frac{N-n}{Nn} \right) \frac{\bar{Y}}{\bar{X}}\ (v_x^2 - r v_x v_y)$$

$$= \frac{9460 - 132}{(9460)\ (132)}\ (0.3358)\ [0.1453$$

$$- (0.4964)\ (0.3812)\ (1.0038)]$$

$$= \frac{0.0153}{132} = 0.000,115 \quad \text{or} \quad 0.0343\% \text{ of } \bar{Y}/\bar{X}.$$

(c) $\widehat{\text{var}}$ of p (if HHs were selected at random):

$$\widehat{\text{var}}\ (p) \doteq \frac{pq}{n} = \frac{(0.3358)\ (0.6642)}{691} = 0.000,323.$$

(d) Discussion of (a) and (c):

$$\text{RE (cluster of 5/1)} = \frac{0.000,323}{0.000,660} = 0.490 \qquad \text{or about half as good.}$$

Since $RE \, (C/E) = \dfrac{1}{[1+(M_0-1)\delta]},$

$$\delta = \frac{1/RE - 1}{M_0 - 1} = \frac{1/0.490 - 1}{5.2348 - 1} = \frac{1.04}{4.2348} = +0.245.$$

2.   Here is a microuniverse with y- and x-populations as follows:

U_i:	1	2	3	4	5
Y_i:	6	3	4	8	4
X_i:	4	4	6	8	3

(a) Compute: Y, $\bar{Y}$, S_y^2; X, $\bar{X}$, S_x^2; $\bar{Y}/X$, S_{xy}, B, R.

(b) Enumerate all possible RNR samples for $n = 1, 2, 3,$ and 4 and determine the $\bar{y}/\bar{x}$ and $(\bar{y}/\bar{x} - \bar{Y}/\bar{X})^2$ for each. Compute $E[\bar{y}/\bar{x}]$ and $E[(\bar{y}/\bar{x} - \bar{Y}/\bar{X})^2]$.

(c) Check the validity for each sample size of the approximate formula for bias in $(\bar{y}/\bar{x})$ for estimating $\bar{Y}/\bar{X}$. Does it behave as you expect?

(d) Check the validity for each sample size of the approximate formula for var $(\bar{y}/\bar{x})$. Does it behave as you expect?

Solution:

(a) $Y = \sum Y_i = 25,$ $\bar{Y} = \frac{25}{5} = 5,$ $S_y^2 = \dfrac{\sum Y_i^2 - Y^2/N}{N-1} = 4.0$

$X = \sum X_i = 25,$ $\bar{X} = \frac{25}{5} = 5,$ $S_x^2 = \dfrac{\sum X_i^2 - X^2/N}{N-1} = 4.0$

$\bar{Y}/\bar{X} = \frac{5}{5} = 1,$ $S_{xy} = \dfrac{\sum Y_i X_i - YX/N}{N-1} = \dfrac{136 - 125}{4} = 2.75,$

$B = \dfrac{S_{xy}}{S_x^2} = \dfrac{2.75}{4} = .6875,$ $R = \dfrac{S_{xy}}{S_x S_y} = \dfrac{2.75}{\sqrt{4}\sqrt{4}} = .6875.$

(b) The $(\bar{y}/\bar{x})$s and $(\bar{y}/\bar{x} - \bar{Y}/\bar{X})$s for each n are

$n = 1$: $\bar{y}/\bar{x}$: 1.50, 0.75, 0.67, 1.00, 1.33
       $(\bar{y}/x - \bar{Y}/X)^2$: 0.2500, 0.0625, 0.1111, 0, 0.1111
       $E[\bar{y}/x] = 1.0500$
       $E(\bar{y}/\bar{x} - \bar{Y}X)^2 = 0.10694$

$n = 2$: $\bar{y}/\bar{x}$: 1.125, 1.000, 1.167, 1.429, 0.700, 0.917, 1.000, 0.857,
         0.889, 1.091
     $(\bar{y}/\bar{x} - \bar{Y}/\bar{X})^2$: 0.1563, 0, 0.02780, 0.18404, 0.09000, 0.00689, 0,
         0.02024, 0.01232, 0.00828
     $E[\bar{y}/\bar{x}] = 1.0175$
     $E(\bar{y}/\bar{x} - \bar{Y}\bar{X})^2 = 0.036{,}55$

$n = 3$: $\bar{y}/\bar{x}$: 0.929, 1.063, 1.182, 1.000, 1.077, 1.200, 0.833, 0.847,
         1.000, 0.942
     $(\bar{y}/\bar{x} - \bar{Y}/\bar{X})^2$: 0.00504, 0.00397, 0.03310, 0; 0.00593, 0.04000,
         0.02780, 0.002340, 0, 0.00580.
     $E[\bar{y}/\bar{x}] = 1.0073$
     $E(\bar{y}/\bar{x} - \bar{Y}/\bar{X})^2 = 0.014{,}504$

$n = 4$: $\bar{y}/\bar{x}$: 0.905, 1.408, 1.105, 1.000, 0.955
     $(\bar{y}/\bar{x} - \bar{Y}/\bar{X})^2$: 0.009,025, 0.002,320, 0.011,300, 0,
         0.002,025
     $E[\bar{y}/\bar{x}] = 1.0026$
     $E(\bar{y}/\bar{x} - \bar{Y}/\bar{X})^2 = 0.004{,}934$

(c)   bias $(\bar{y}/\bar{x}) = (\bar{Y}/\bar{X}) [(N-n)/Nn] (V_x^2 - RV_xV_y) = (5.0/5.0) [(5$
$-n)/5n] [4/25 - (0.6875) (\tfrac{2}{5}\cdot\tfrac{2}{5})]$
           $= (5-n)/5n) (0.05)$.
For $n = 1$, 2, 3, and 4 we obtain

| | Sample size, n | | | |
Bias $(\bar{y}/\bar{x})$	1	2	3	4
Theoretical bias	0.040,0	0.015,0	0.006,67	0.002,50
Actual bias	0.050,0	0.017,5	0.007,30	0.002,60
Percent error in theoretical bias	−20.0%	−13.5%	−8.6%	−3.8%

As expected, bias gets smaller as n increases, and the error in theoretical bias gets smaller as n increases, however irregularly.

(d)   var $(\bar{y}/\bar{x}) = \left(\dfrac{N-n}{Nn}\right)\left(\dfrac{\bar{Y}}{\bar{X}}\right)^2\left(\dfrac{S_y^2}{\bar{Y}^2} + \dfrac{S_x^2}{\bar{X}^2} - \dfrac{2RS_yS_x}{\bar{Y}\bar{X}}\right)$

$$= \left(\frac{5-n}{5n}\right)\left(\frac{5.0}{5.0}\right)^2\left\{\frac{4.0}{(5.0)^2} + \frac{4.0}{(5.0)^2} - 2\left[\frac{2.75}{(5.0)(5.0)}\right]\right\}$$

$$= \frac{5-n}{5n} (0.1).$$

For $n = 1, 2, 3$, and 4 we obtain

	Sample Size, n			
	1	2	3	4
Theoretical var $(\bar{y}/\bar{x})$	0.080,00	0.033,00	0.013,30	.005,00
Actual MSE $(\bar{y}/\bar{x})$	0.106,94	0.020,14	0.014,50	.004,93
Percent error in var $(\bar{y}/\bar{x})$	-25%	$+64\%$	-8%	$+1\%$

As expected, the error in var $(\bar{y}/\bar{x})$ decreases as n increases toward N. It is also a poor estimator of MSE when n is small, as in this case.

3. Following is a method sometimes used to estimate the total fish of a given species in a lake. A certain number of "tagged" fish are dumped into the lake, and after some time has elapsed (so that the tagged fish can resume more or less normal life) a sampling of fish is obtained by nets or by electric shock and the number of tagged and total fish netted or shocked to the surface is observed. The fraction of fish tagged is used to estimate the total fish in the lake. Suppose each haul of the net is regarded as a sampling unit and that the net is placed at random points in the lake. If the total fish tagged is X, the total number of fish and the number tagged on the ith haul are denoted by Y_i and X_i, respectively; consider the estimator $\hat{Y} = X\bar{y}/\bar{x}$ for estimating the total fish in the lake.

The following data were obtained from a creel census of Ventura Marsh, Iowa, in October 1953 on Northern Pike. Number of fish tagged $X = 1146$. (Data obtained from R. J. Muncy and R. L. Ridenhour.)

Haul no.	Total catch (Y_i)	No. tagged (X_i)	Haul no.	Total catch (Y_i)	No. tagged (X_i)
1	493	11	9	722	41
2	1584	57	10	1056	48
3	1275	80	11	640	25
4	1488	60	12	490	15
5	1070	32	13	610	15
6	1443	54	14	1810	67
7	437	15	15	1089	39
8	657	27	16	331	10

$$\sum y_i = 15{,}192, \qquad \sum x_i = 596,$$
$$\bar{y} = 949.50, \qquad \bar{x} = 37.25,$$
$$s_y^2 = 216{,}180, \qquad s_x^2 = 482.20,$$
$$v_y^2 = 0.2398, \qquad v_x^2 = 0.3475,$$
$$v_y = 0.4897, \qquad v_x = 0.5895,$$
$$r = 0.8965, \qquad \bar{y}/\bar{x} = 25.4811.$$

(a) Assuming the 16 hauls are a random sample of all possible hauls that could be made in the lake, estimate the total number of Northern Pike in the lake and its standard error.

(b) What assumptions are required to make the estimator $X\bar{y}/\bar{x}$ an unbiased (except technical bias) estimator in problems of this sort?

(c) Suppose we express the estimator of Y as

$$\hat{Y}_p = \frac{X}{p} \quad \text{rather than} \quad \hat{Y}_r = X\frac{\bar{y}}{\bar{x}},$$

where p is the observed fraction of the fish netted that are tagged. Regarding p as a binomial p, what is the estimated standard deviation of $\hat{Y}_p$? [Use $\widehat{SE}(\hat{Y}_p) = X[\widehat{SE}(p)]/p^2$.]

(d) What evidence do we have on the degree to which the tagged and untagged fish have been thoroughly intermixed?

Solution

(a) $\hat{Y}_r = X\dfrac{\bar{y}}{\bar{x}},$ where $X = 1146$ and $\bar{y}/\bar{x} = 25.4811,$

           $= (1146)\,(25.4811)$

           $= 29{,}212$ fish in the lake.

   $\widehat{SE}(\hat{Y}_r) = X[SE(\bar{y}/\bar{x})]$

           $= X\left(\dfrac{N-n}{N}\right)^{1/2}\left(\dfrac{\bar{y}}{\bar{x}}\right)\left(\dfrac{v_y^2 + v_x^2 - 2rv_xv_y}{n}\right)^{1/2}$

           $= (1146)\,(1)\,\dfrac{25.4811}{\sqrt{16}}\,[0.2398 + 0.3475$

                                   $- 2(0.4897)\,(0.5895)\,(0.8965)]^{1/2}$

           $= (1146)\,(6.3725)\,(0.2642)$

           $= (1146)\,(1.6836) = 1929.$

Hence our interval is $\hat{Y} = 29{,}212 \pm 1929.$

(b) (1) Samples must be "random," or if the tagged and untagged fish should "mix" at random, the randomness of sampling would be unnecessary. (2) Tagged and untagged fish equally "catchable." (3) X must be known; that is, if there is mortality in the tagged fish, it must be known.

(c) $SE(\hat{Y}_p) = X[SE(1/p)],$ $p = \dfrac{1}{\bar{y}/\bar{x}} = \dfrac{1}{25.48} = 0.0392,$

           $= \dfrac{X}{p^2}\left(\dfrac{n}{n-1}\cdot\dfrac{pq}{n}\right)^{1/2}$ $n = 15{,}192,$

           $= \dfrac{1146}{0.001537}\left[\dfrac{(0.0392)\,(0.9608)}{15{,}192}\right]^{1/2}$

$$= \frac{(1146)}{0.001537} (0.000,002,479)^{1/2}$$

$$= (1146) \left(\frac{0.001575}{0.001537} = 1.0247 \right)$$

$$= 1174.$$

(d)   If the tagged and untagged fish were thoroughly mixed, the variance of the ratio $\bar{y}/\bar{x}$ would be the same as the variance of $1/p$ (apart from errors in the approximations). But

$$\widehat{\text{var}}\ (\bar{y}/\bar{x}) = (1.6836)^2 = 2.8345,$$

and

$$\widehat{\text{var}}\ (1/p) = (1.0247)^2 = 1.0500;$$

hence there is a large "cluster" effect.

4.   In the Venice area of Los Angeles City a sample of 14 blocks was selected from those that existed in the 1950 census and matched with the 1960 census block maps. The area consisted of three tracts, 248 blocks, and 7368 households in the 1950 census and of seven tracts and 10,072 households in the 1960 census. For each sample block, the number of households found in each census is given below.

SU no.	Tract and block numbers		No. of HHs		Diff. 1960–1950
	1950	1960	1950	1960	
1	190:7	2731:7	0	0	0
2	190:16	2731:31,32	66	76	10
3	190:28	2737:8	18	23	5
4	190:41	2737:23	24	27	3
5	191:8	2736:6,7	85	34	− 51
6	191:23	2732:19	10	20	10
7	191:38	2732:4	44	52	8
8	191:53	2736:21,22,23	78	76	− 2
9	191:75	2733:20	34	78	44
10	191:91	2733:5	43	93	50
11	192:4	2734:2	45	62	17
12	192:19	2734:17	63	64	1
13	192:57	2735:3	73	92	19
14	192:67	2735:18	41	48	7
Total			624	745	121
Mean			44.57	53.21	8.64

(*a*)   Plot the 1960 observations against those of 1950.

(*b*)   Does it appear that the ratio or regression estimators may be more accurate in estimating the 1960 mean than the simple estimator? Why or why not? Confirm by calculations.

Solution

(*a*)   The plot of the data:

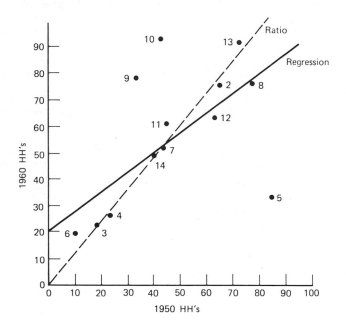

(*b*)   Yes, it appears that although scatter is quite large, both ratio and regression estimators may be effective. Computations are:

$$s_y^2 = 835.874, \qquad \bar{y} = 53.214, \qquad v_y^2 = 0.2952;$$

$$s_x^2 = 676.725, \qquad \bar{x} = 44.571, \qquad v_x^2 = 0.3406;$$

$$s_{xy} = 487.330, \qquad r = 0.6480,$$

$$\text{RE (ratio/simple)} = \frac{v_y^2}{v_y^2 + v_x^2 - 2v_{xy}} = 1.312$$

and

$$\text{RE (regr/simple)} = \frac{1}{1 - r^2} = 1.724.$$

Therefore although both are effective in obtaining a more precise estimate of $\bar{Y}$ (i.e., the average number of HHs per block in 1960), the regression estimator is much more than the ratio. In fact the regression estimator is nearly twice as effective as a simple mean of $\bar{y}$.

5.14. References

Cochran, W. G.
1953, 1963
Sampling Techniques. New York: John Wiley & Sons, Inc. (Pages 29–33 and Chapters 6 and 7 for a good discussion of theory.)

Cochran, W. G.
1942
Sampling Theory When Sampling Units Are of Unequal Sizes. *Journal of the American Statistical Association*, **37**:199–212. (Basic paper on the regression estimator.)

Goodman, L.
Hartley, H. O.
1958
The Precision of Unbiased Ratio-type Estimators. *Journal of the American Statistical Association*, **53**: 491–508.

Hansen, M. H.
Hurwitz, W. N.
Madow, W. G.
1953
Sample Survey Methods and Theory. Vol. I. New York: John Wiley & Sons, Inc. (Good general discussion with excellent examples.)

Hartley, H. O.
Ross, A.
1954
Unbiased Ratio Estimators. *Nature*, **174**: 270.

Hasel, A. A.
1942
Estimation of Volume in Timber Stands by Strip Sampling. *Annals of Mathematical Statistics*, **13**: 179–206. (Presentation of regression theory with example based on timber data.)

Jessen, R. J.
Kempthorne, O.
Daly, J.
Deming, W. E.
1947
On a Population Sample for Greece. *Journal of the American Statistical Association*, **42**: 357–384. (An example of the use of the mean-of-ratios estimator.)

Jessen, R. J.
1942
Statistical Investigation of a Sample Survey for Obtaining Farm Facts. *Iowa Agricultural Experiment Station Bulletin* 304. Ames, Iowa: Iowa State University.

Kish, L.
1965
Survey Sampling. New York: John Wiley & Sons, Inc.

Mickey, M. R.
1959
Some Finite Population Unbiased Ratio and Regression Estimators. *Journal of the American Statistical Association*, **54**: 594–612.

Raj, D.
1958
Sampling Theory. New York: McGraw-Hill Book Company. (Chapter 5.)

Ross, A.
1960
Ph.D. Dissertation, Iowa State University, Ames, Iowa.

Searls, D. T.
1964
The Utilization of a Known Coefficient of Variation in the Estimation Procedure. *Journal of the American Statistical Association*, **59** (308): 1225–1226.

Watson, D. J.
1937
The Estimation of Leaf Areas. *Journal of Agricultural Science*, **27**: 474. (Interesting example of early use of regression estimator.)

Yates, F.   The Estimation of the Efficiency of Sampling with Special Reference to
Zacopanay, I.   Sampling in Cereal Experiments. *Journal of Agricultural Science* **25**:
    1935   545 577. (Early use of the ratio estimator.)

5.15.   Mathematical Notes

$E[Y_iX_i]$; *Definition of* S_{xy} *and R.*   Define the covariance of y and x in a finite universe as

$$S_{xy} = \frac{\sum (Y_i - \bar{Y})(X_i - \bar{X})}{N-1}, \qquad i = 1, \ldots, N,$$

whence

$$R = \frac{S_{xy}}{S_x S_y}.$$

To obtain $E[Y_iX_i]$, when samples of $n = 1$ are selected at random from a universe of size N,

$$Y_iX_i = [(Y_i - \bar{Y}) + \bar{Y}][(X_i - \bar{X}) + \bar{X}]$$

$$= (Y_i - \bar{Y})(X_i - \bar{X}) + \bar{Y}(X_i - \bar{X} + \bar{X}(Y_i - \bar{Y}) + \bar{Y}\bar{X}$$

$$E[Y_iX_i] = \left(\frac{N-1}{N}\right) S_{yx} + \bar{Y}\bar{X}.$$

Hence

$$\text{bias } (Y_iX_i) = \left(\frac{N-1}{N}\right) S_{yx}.$$

$E[\bar{y}\bar{x}]$

$$\bar{y}\bar{x} = [(\bar{y} - \bar{Y}) + \bar{Y}][(\bar{x} - \bar{X}) + \bar{X}]$$

$$= (\bar{y} - \bar{Y})(\bar{x} - \bar{X}) + \bar{X}(\bar{y} - \bar{Y}) + \bar{Y}(\bar{x} - \bar{X}) + \bar{X}\bar{Y}$$

$$E[\bar{y}\bar{x}] = E[(\bar{y} - \bar{Y})(\bar{x} - \bar{X})] + \bar{X}\bar{Y}$$

$$= \left(\frac{N-n}{Nn}\right) S_{yx}^2 + \bar{Y}\bar{X}.$$

Hence

$$\text{bias } (\bar{y}\bar{x} = \left(\frac{N-n}{Nn}\right) S_{yx}^2 \quad \text{or} \quad \left(\frac{N-n}{Nn}\right) RS_y S_x.$$

$E[\bar{y}/\bar{x}]$

$$\frac{\bar{y}}{\bar{x}} - \frac{\bar{Y}}{\bar{X}} = \frac{\bar{y} - \bar{Y}/\bar{X}\,\bar{x}}{\bar{x}}$$

$$\doteq \frac{\bar{y} - \bar{Y}/\bar{X}}{\bar{X}} \qquad \text{when } \bar{x} \text{ is replaced with } \bar{X}$$

$$\doteq \frac{\bar{y} - \bar{Y}/\bar{X}\bar{x}}{\bar{X}} \left(1 - \frac{\bar{x} - \bar{x}}{\bar{X}} + \cdots\right) \text{ by using the power series.}$$

$$E\left(\frac{\bar{y}}{\bar{x}} - \frac{\bar{Y}}{\bar{X}}\right) \doteq \left(\frac{1-f}{n\bar{X}^2}\right)\left(\frac{\bar{Y}}{\bar{X}} \cdot S_x^2 - RS_y S_x\right).$$

var $(\bar{y}/\bar{x})$

$$\frac{\bar{y}}{\bar{x}} - \frac{\bar{Y}}{\bar{X}} = \frac{\bar{y} - (\bar{Y}/\bar{X})\bar{x}}{\bar{X}},$$

$$\left(\frac{\bar{y}}{\bar{x}} - \frac{\bar{Y}}{\bar{X}}\right)^2 = \frac{1}{\bar{X}^2}\left[\bar{y} - \left(\bar{Y}/\bar{X}\right)\bar{x}\right]^2.$$

$$E\left[\left(\frac{\bar{y}}{\bar{x}} - \frac{\bar{Y}}{\bar{X}}\right)^2\right] = \frac{1}{\bar{X}^2}\left(\frac{1-f}{n}\right)S_d^2 \quad \text{where } D_i = Y_i - (\bar{Y}/\bar{X})X_i.$$

Now let

$$Y_i - (\bar{Y}/\bar{X})X_i = (Y_i - \bar{Y}) - (\bar{Y}/\bar{X})(X_i - \bar{X}).$$

Squaring and taking expectations on each side,

$$\left(\frac{N-1}{N}\right)S_d^2 = \left(\frac{N-1}{N}\right)\left[S_y^2 + \left(\frac{\bar{Y}}{\bar{X}}\right)^2 S_x^2 - 2\left(\frac{\bar{Y}}{\bar{X}}\right)RS_x S_y\right],$$

$$\text{var } (\bar{y}/\bar{x}) = \left(\frac{1-f}{n}\right)\left(\frac{\bar{Y}}{\bar{X}}\right)^2\left(\frac{S_y^2}{\bar{Y}^2} + \frac{S_x^2}{\bar{X}^2} - 2R\frac{S_x}{\bar{X}}\frac{S_y}{\bar{Y}}\right).$$

MSE $(\bar{\bar{y}}_{(1)})$, MSE $(y_{(2)})$, MSE $(\check{y}_{(3)})$

$$\bar{\bar{y}}_{(1)} = \frac{\sum \bar{Y}_i}{n}, \qquad \bar{\bar{y}}_{(2)} = \frac{\bar{y}}{\bar{M}}, \qquad \bar{\bar{y}}_{(3)} = \frac{\bar{y}}{\bar{m}}.$$

$$\text{MSE } (\bar{\bar{y}}_{(1)}) = E[(\bar{\bar{y}}_{(1)} - \bar{\bar{Y}}_{(1)} + \bar{\bar{Y}}_{(1)} - \bar{Y})^2]$$

$$= E[(\bar{\bar{y}}_{(1)} - \bar{\bar{Y}}_{(1)})^2 + (\bar{\bar{Y}}_{(1)} - \bar{Y})^2]$$

$$= \left(\frac{N-n}{Nn}\right)\frac{\sum (\bar{Y}_i - \bar{\bar{Y}}_{(1)})^2}{N-1} + (\bar{\bar{Y}}_{(1)} - \bar{Y})^2.$$

$$\text{MSE } (\bar{\bar{y}}_{(2)}) = \text{MSE } (\bar{y}/\bar{M}) = \text{var } (\bar{y}/\bar{M})$$

$$= \left(\frac{N-n}{Nn}\right)\frac{\sum (Y_i - \bar{Y})^2}{\bar{M}^2(N-1)}$$

$$\text{MSE } (\bar{\bar{y}}_{(3)}) = \text{MSE } (\bar{y}/\bar{m})$$

$$= \frac{1}{\bar{X}^2}\left(\frac{1-f}{n}\right)S_d^2$$

$$= \left(\frac{N-n}{Nn}\right) \frac{\sum M_i^2 \, (\bar{Y}_i - \bar{Y})^2}{\bar{M}^2(N-1)},$$

since $\bar{X} = \bar{M}$, $D_i = Y_i - (\bar{Y}/\bar{M})M_i$, $D_i^2 = M_i^2(\bar{Y}_i - \bar{Y})^2$.

5.16. Exercises

1. A survey was undertaken to determine the population mean of two components, A and B, of some characteristic. Owing to the nature of the components, both could not be measured on the same element. A sample of 1000 elements in a large universe was selected with the following results:

Component A: $n = 500$, $\bar{x}_A = 630$, $s_A = 120$.
Component B: $n = 500$, $\bar{x}_B = 475$, $s_B = 110$.

Calculate the 95% confidence limits on $\mu_A + \mu_B$.

2. A survey was undertaken to determine the number of ranch-style houses in Urbantown. The SU was a block or portion of a block containing a variable number of houses ranging from about 0 to 10. Suppose a random sample of SUs is selected RNR from a universe of 1000 SUs, and we have the entire results from the sample of three SUs as follows:

SU no.	House no.	Ranch style?
1	1	Yes
	2	Yes
	3	No
2	1	No
3	1	No
	2	No
	3	No
	4	Yes

(a) Estimate the total number of ranch-style houses in Urbantown and its standard error.

(b) Suppose we know that the total of houses of all styles is 2500. Does this information provide any help? Discuss.

3. Here is a small universe with y and x populations as follows:

$$U_i: \quad 1 \quad 2 \quad 3 \quad 4 \quad 5$$
$$Y_i: \quad 6 \quad 3 \quad 4 \quad 8 \quad 4$$
$$X_i: \quad 4 \quad 4 \quad 6 \quad 8 \quad 3$$

RNR

(a)   List all possible samples of $n = 2$ and calculate the regression estimates of $\bar{Y}$ for each. Obtain their average (expected) value. Is $\bar{y}_{regr}$ an unbiased estimator here? Evaluate bias $(\bar{y}_{regr})$ by the formula. How does this result compare with $E[\bar{y}_{regr} - \bar{Y}]$? Comment.

(b)   Calculate $E[(\bar{y}_{regr} - \bar{Y})^2]$ from the data in (a). Evaluate var $(\bar{y}_{regr})$ by the formula. How well does the formula perform here? Comment.

4.   (a)   Show that the precision of the regression estimator is always greater than that of the ratio estimator unless

$$\rho = V_x/V_y, \quad = R$$

and then they are the same. (Here V_x and V_y are the coefficients of variation of x and y, respectively, and ρ is the correlation coefficient between x and y.)

(b)   What sort of relationship between x and y must exist in order that $\rho = V_x/V_y$?

5.   Normally we regard the regression and ratio estimators as feasible alternatives.

(a)   Under what conditions, if any, will one be feasible and the other not? Consider estimating both $\bar{Y}$ and Y.

(b)   Suppose the true regression of y on x is curvilinear through the origin. Does this fact tell us that one or the other or both are inappropriate or inefficient? Comment.

CHAPTER 6

Building Sampling Frames;
Area Sampling

6.1. Introduction

The concept of a *frame* for sampling was discussed briefly in previous chapters. In many practical situations the frame is a matter of choice to the survey planner, and sometimes a critical one, since frames not only should provide us with a clear specification of what collection of objects or phenomena we are undertaking to study, but they should also help us carry out our study efficiently and with determinable reliability. Some very worthwhile investigations are not undertaken at all because of the lack of an apparent frame; others, because of faulty frames, have ended in disaster or in a cloud of doubt.

In Chapter 4 we studied the properties of samples, where the SU consisted of a cluster of elements. It was assumed that these clusters or SUs were in a list or in some arrangement such that random samples of them could be obtained—in other words, these clusters existed in a frame of some sort.

In this chapter we shall consider some problems of conceptualizing feasible frames for a variety of real-life situations and some methods of constructing them to be operational. We shall see how to evaluate a proposed scheme and how to generate frames that utilize information available to the planner, thereby increasing the efficiency of the resulting survey.

In some field situations it is common practice to keep separate the operations of frame construction (e.g., listing) and sample selection (e.g., selecting HHs to be interviewed) in order to avoid possible biases in selection and other difficulties ascribable to human frailties. We shall see how some of these problems can be overcome, thereby reducing costs.

We shall continue with the case where we are estimating Y or, where the SU is a cluster of elements, estimating $\bar{Y}$. We shall deal with the construction of clusters of variable as well as fixed size.

6.2.   Types of Frames

One of the simplest of frames is a listing of each element in the universe of interest, numbering the entries serially from 1 to N. This frame has some of the more important features that are desired: (i) all the elements in the universe of interest are included, (ii) the total number of elements is known, and (iii) a random sample of any size may easily be obtained from it (e.g., by the use of random numbers). It may or may not be efficient, depending on the variance characteristics of various alternative groupings of the elements and the costs of making the subsequent observations on them.

Another common frame is a set of cards, vouchers, capsules, or the like, in which each card, voucher, or capsule identifies each element (such as employee, sales transaction, patient, etc.) in the universe of interest. In this case the *frame units* (FUs) are physical elements that are more or less alike and may be physically mixed to simulate a randomization of position or order. The situation is somewhat like that of selecting at random "a ball from an urn" or "a card from a shuffled deck." The FUs—the balls, cards, or whatever—need not be serially numbered to permit randomization (although experience has shown that random mixes are in fact very difficult to achieve), and the total can be obtained by a direct count.

A great number of universes of investigational interest have no "natural" frames. For example, there is no existing frame of adults living in Los Angeles County, of production items coming off a production line, of branches in a tree (where we may be interested in the amount of fruit they contain), of the area under some crop in Iowa, or of tourists in Hawaii. For these situations various frames can be conceptualized and constructed, sometimes quite simply, though frequently some degree of ingenuity and experience is required to develop a frame that is operational as well as valid and reasonably efficient.

These specially constructed, or synthesized, frames can be points, lines, or areas, such as are sometimes used in surveying areas under crops, or specially compiled lists of addresses, such as are sometimes used to obtain a frame for households living on a city block. In the latter case the frame is valid provided every household is associated with an address and every address on the block is listed.

Frame Universe versus Target Universe. Normally one wishes to have the universe of interest or *target universe* completely represented in the *frame universe.* For example, suppose we want to study the characteristics of all hunters in Montana and whether they are licensed or not. A list of persons issued hunting licenses would constitute a convenient frame universe, but it would not be our desired target universe. A telephone directory may be a convenient frame universe, but if we wished to study all households in the area, it would not be our target universe. Sometimes the two are close enough so that the difference in coverage can be safely ignored. When they are not, one can try to find another frame that is complete, or find or construct one that covers just the nonlisted households and combine it with the directory (see Section 6.9, multiple frames), or perhaps settle for the directory and tell your audience that you have nothing in the survey to tell you what is going on in the nondirectory households. The truncated universe may be of sufficient interest to justify the investigation.

Count Frames; Systematic Selection. Generally a complete listing is the preferred frame—particularly if the FUs are serially numbered because samples can then be selected by the use of random numbers. Neither the list nor the numbering is necessary, however, for assuring adequate control in the sample selection. If the units are ordered in some identifiable manner, then all we need to do is count off the units to obtain particular "numbered" units. A systematic sample of, say, 1 in 90 cards in a file drawer can be easily drawn this way with a given random start number. However, to avoid possible problems due to human frailties (inability to count correctly, making substitutions in selecting sample cards, etc.), it is wise to make the procedure verifiable—and then to verify the work to see if it is done correctly.

6.3. The FU, the SU, and the OU

In the foregoing section it was generally presumed that the frame unit (FU) and the sampling or survey unit (SU) were identical. If the frame unit was a point, the SU was a point, because it was the unit *individually* selected for the sample. Likewise, if the FU was a name on a list, the SU was an individual name drawn from that list, and then the person belonging to that name was observed (interviewed, or whatever). There is no need for this one-to-one correspondence between the unit of the frame, the unit selected for the sample, and the unit observed. As will be shown in Sections 6.4 and 6.5, the FU can consist of clusters of variable numbers of SUs, and in Chapter 4 we dealt with an SU comprising clusters of one or more elements, or what we shall call OUs, *observation units*. We shall generally regard the OU as having some role in the analysis—for example, determining the fraction of *households* having more than one adult present, or the average income *per household*, or the fraction of *persons* spending at least one day in a hospital last year.

We may regard the normal target universe as consisting of OUs, on all or some sample of which we wish to obtain one or more characteristics. The frame may consist of a set of elements that are or can be linked in some way with the

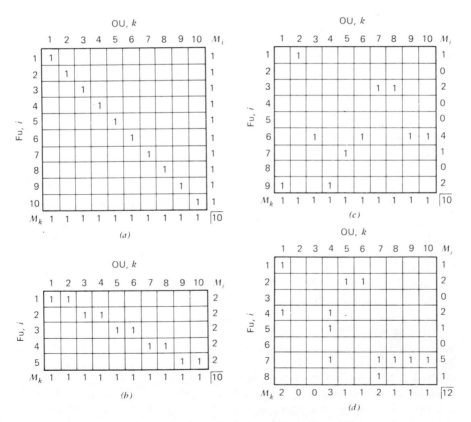

Fig. 6.1.   Examples of different FU-OU relationships.

OUs. For example, a city block can be linked with the households living on it. By selecting a city block, we thereby select a group of households. The linkages can be simple or complex. Regardless of the complexity it is usually possible, if the pattern is known or determinable, to make the frame useful. Figure 6.1 shows four common cases of linkages between FUs (of one SU each) and OUs in the target universe. Case (*a*) is that of one-to-one correspondence; case (*b*) is that where the FU consists of a cluster of two OUs each; case (*c*) is that of clusters of variable numbers of OUs; and case (*d*) is a complex case where an OU is linked with one or more FUs, or in some instances with no FUs at all. An example for case (*a*) is a complete list of OUs; for case (*b*), a list of duplex housing units; for case (*c*) a list of blocks of varying numbers of households; for case (*d*), a list of households used to obtain a sample of community leaders.

6.4.   Cluster FUs; Equalizing SU Size, M_i by M_h

Suppose we wish to draw a sample of households from Census Tract No. 7012 of Santa Monica. The tract consists of 87 blocks with a total of 2699 households in the 1960 Census. Since no list of households for this tract exists, we wish to construct a frame such that each SU contains an optimal number of households in close proximity—which, after considering all relevant variances and costs, we think is about 4 HHs. In a special publication the Census provides a map and listing of each block and its number of households. We wish to construct a feasible frame of SUs such that each SU contains four HHs and any number *n* of them can be easily selected at random.

If there has been no change since 1960, the desired frame should consist of $N = 675$ SUs ($= 2699 \div 4$). The block data are given in Table 6.1. A simple procedure for designating the 675 SUs is also indicated by Table 6.1. For each block h, $h = 1, 2, \ldots, 87$, we have M_h, the number of HHs. The number of desired SUs for block h is simply $M_h \div \bar{M}$ ($= M_0 = 4$). These are given in column 4. Hence, 10 of our desired 675 SUs are to be found in block 1, $4\frac{2}{4}$ in block 2, and so on. A simple way to designate the individual 10 SUs in block 1 is to specify either actually or conceptually a procedure for numbering the households serially from 1 to 40 and to regard the 10 SUs as comprising the 10 sets of numbers shown in column 6 of Table 6.1.

By adding another step we put the skeletal frame into a form convenient for selecting samples. By computing the cumulative totals of N_h, where $\sum_{h=1}^{h} N_h$, we have now, in column 5, a list of the 675 SU numbers, wherein we regard SUs 1–10 to be in block 1, SUs 11–14 to be in block 2, SU 15 to be $\frac{2}{4}$ in block 2 and $\frac{2}{4}$ in block 3, SUs 16–20 to be wholly in block 3, and so on. To select an SU at random, we draw a random number, RN, in the range 1–675. Suppose the RN is 12. Scanning column 5, we see that SU 12 is in block 2 and consists of HHs 5, 6, 7, and 8 of that block.

Table 6.1. A frame of 87FUs with 675 SUs of exactly 4 HHs each

			Tract 7012 of Santa Monica		
Block no.	No. of HHs, M_h		(Setting $\bar{M} = M_0 = 4$) No. of SUs, N_h		
(h)	Absolute	Cum.	Absolute	Cum.	HH nos. in each SU, i
(1)	(2)	(3)	(4)	(5)	(6)
1	40	40	10	10	# 1 = 1, 2, 3, 4, . . . #10 = 37, 38, 39, 40
2	18	58	$4\frac{2}{4}$	$14\frac{2}{4}$	#11 = 1, 2, 3, 4, . . . #14 = 13, 14, 15, 16 #15 = 17, 18(pt.)
3	24	82	6	$20\frac{2}{4}$	#15 = 1, 2(pt.), . . . #20 = 19, 20, 21, 22 #21 = 23, 24(pt.)
4	19	101	$4\frac{3}{4}$	$25\frac{1}{4}$	#21 = 1, 2(pt.)
5	11	112	$2\frac{3}{4}$	28	
6	15	127	$3\frac{3}{4}$	$31\frac{3}{4}$	
7	0	127	0	$31\frac{3}{4}$	
8	0	127	0	$31\frac{3}{4}$	
9	44	171	11	$42\frac{3}{4}$	#43 = 42, 43, 44(pt.)
10	16	187	4	$46\frac{3}{4}$	#43 = 1(pt.) #44 = 2, 3, 4, 5, . . . #47 = 14, 15, 16(pt.)
.	.	.	.	.	.
.	.	.	.	.	.
.	.	.	.	.	.
87	0	2699	0	$674\frac{3}{4}$	
Total	2699		$674\frac{3}{4}$		

If the sampling rate is relatively small, several shortcuts can be taken without any loss of validity. Note that the cumulative totals of N_h can also be obtained by the implied relationship

$$\sum_{h=1}^{h} N_h = \frac{\sum_{h=1}^{h} M_h}{\bar{M} = M_0},$$

since the case dealt with here requires $\bar{M}$ to be a constant $= M_0 = 4$. Hence, column 4 can be omitted, and columns need be completed only in those blocks within which an SU falls.

Table 6.2. *A frame of 87 FUs with 675 SUs of approximately 4 HHs each*

Block no. (h) (1)	No. of HHs, M_h		(Setting $M_i = 4$HHs) No. of SUs, N_h	
	Absolute (2)	Cumulative (3)	Cumulative (4)	Absolute (5)
1	40	40	10	10
2	18	58	15	5
3	24	82	21	6
4	19	101	26	5
5	11	112	28	2
6	15	127	32	4
7	0	127	32	0
8	0	127	32	0
9	44	171	43	11
10	16	187	47	4
.	.	.	.	.
.	.	.	.	.
.	.	.	.	.
87	0	2699	675	0
Total	2699			675

Other methods of designating the particular way an FU can be divided into SUs (and still retain certain desired properties) will be given in later chapters, particularly Chapter 9. The scheme presented here is purely illustrative.

Approximate equalization of M_i. The foregoing procedure may be a bit messy where an SU falls in more than one FU. If costs are involved in dealing with FUs (in this case, blocks) as well as with the SU itself (in this case, HHs), it may be desirable to eliminate fractional SUs by simply allowing M_i to vary a bit from a fixed $\bar{M} = M_0$. If this is preferred, we can keep the procedure essentially the same by examining Table 6.2. From the cumulative totals of M_h we calculate cum N_h directly by

$$\sum_{h=1}^{h} N_h = \frac{\sum_{h=1}^{h} M_h}{M_0}$$

and follow some rounding rule (in this case the nearest whole number) to make the N_h integers. Then we obtain the N_hs by taking the differences between successive cumulative totals. Hence

$$N_h = \sum_{h=1}^{h} N_h - \sum_{h=1}^{h-1} N_h$$

—that is, the reverse of cumulating totals. The N_hs so obtained specify the number of SUs to be made in each FU, where if desired the procedure in column 6 of Table 6.1 can be followed.

6.5. Cluster FUs; Equalizing SU Size by X_h

Section 6.4 presented procedures for equalizing the sizes of the SUs so that each contained exactly or approximately the same number of elements—that is, $M_i = M_0$ or $\doteq M_0$. Presumably M_0 would be chosen to be the optimal value, M_{opt}, as well as it can be determined. However, to equalize the number of elements in each SU requires that we have knowledge of where the elements are—that is, the M_h for each frame unit h. In cases where M_h are not known and it is still desirable to equalize M_i in the SUs, it may be possible to find some characteristic of each FU related to M_h on which M_i can be equalized. In the foregoing example, where we wish to equalize SUs to 4 HHs as they exist now, that related characteristic might be HHs in the last published census, or total inhabitants in the last census, or current eyeball-estimated HHs. These data, which are regarded as related to what we are really interested in, are commonly called *measures of size* (MOS) and may be used as substitutes.

Let us denote the measure of size, MOS, of the hth FU by X_h; then X_0, our desired X_i, replaces M_0. To determine a suitable X_0 to correspond with our desired M_0 we must be willing to make an estimate of M, say $\hat{M}$. Then the total frame units, N, will be given by

$$(6.1) \qquad N = \frac{\hat{M}}{M_0}$$

rounded to nearest integer. And X_0, the desired average cluster size, will be

$$(6.2) \qquad X_0 = \frac{X}{N}.$$

Suppose in the case of tract 7012 that we did not have available the number of households on each block, M_i, but did have the total number of inhabitants on each block, X_i. We wish to construct a frame containing clusters of approximately 4 HHs each. Here the desired $M_0 = 4$, $X = 7203$, and let $\hat{M} = M = 2699$ HHs. Then $N = 2699/4 = 674.75$ or 675, and $X_0 = X/N = 7203/675 = 10.6711$. Using the previous procedure based on M_i, Table 6.3 shows the adaptation using X_i. Again the cumulative N_h were rounded to the nearest whole number to obtain integer N_hs.

Note that we obtain a different distribution of N_h by block now, since the average persons per HH differs somewhat from block to block. However, we still obtain exactly 675 SUs in the frame, and, if current HHs are reasonably proportional to previous total persons, the SUs will be reasonably equal in

Table 6.3. A frame of 87FUs with 675 SUs of $\bar{M}$ estimated at 4 HHs and $M_i \doteq 4$

Block no. (h) (1)	No. of persons, X_h		No. of SUs, N_h	
	Absolute (2)	Cumulative (3)	Cumulative (4)	Absolute (5)
1	101	101	10	10
2	45	146	14	4
3	77	223	21	7
4	53	276	26	5
5	34	310	29	3
6	42	352	33	4
7	14	366	34	1
8	12	378	35	1
9	158	536	50	15
10	52	588	55	5
.	.	.	.	.
.	.	.	.	.
.	.	.	.	.
87	0	7203	675	0
Total	7203			675

Tract 7012 of Santa Monica
(Setting $X_i \doteq X_0 = 10.6711$ persons)

size. Note that the actual size of each SU is not determined until the block in which it is located is finally segmented into SUs, so it depends somewhat on the equalization policy followed within the block.

In the example above we assumed that it was desirable to equalize SU size on M_i because we had an optimal M_{opt} in mind. X_n, which was available, was used in place of the unavailable M_n, and therefore we equalized on X_0. This is appropriate if we are estimating Y, the total of y in the universe, or Y/X, the mean of y per unit of x (a case considered in Chapter 5) and if $\bar{X}$ is in fact the true optimum for this population. In the first case the appropriate estimator may be

(6.3) $$\hat{Y} = N\bar{y};$$

in the second, an estimator of Y/X may be

(6.4) $$\bar{y}/\bar{x} = \frac{\sum Y_i}{\sum X_i}, \qquad i = 1, 2, \ldots, n,$$

which, if X_i is completely equalized—that is, $X_i = X_0$—becomes simply $(1/X_0)\bar{y}$, the sample mean of y per SU multiplied by a constant. Where X_i is not completely equalized, Eq. 6.4 is a ratio of two random variables and therefore has a somewhat complicated variance. This is a more complex situation, as is the case where we wish to estimate, say, the mean of y per HH—that is, Y/M—by

(6.5)
$$\bar{y}/\bar{m} = \frac{\sum Y_i}{\sum X_i}, \qquad i = 1, 2, \ldots, n.$$

Equations 6.4 and 6.5 are called ratio estimators; their properties were dealt with in Chapter 5.

Since an MOS such as X_h is regarded as only approximately proportional to either M_h or Y_h, whichever we are trying to equalize, it is necessary to avoid excluding FUs constining $X_h = 0$ from the sample. Two procedures are commonly used: (i) attach all "zero FUs" to "nonzero FUs" so that when a nonzero FU comes into the sample, it, together with its attached zero FU, is appropriately divided into SUs, and (ii) select a separate sample of zero FUs (in other words, treat it as a separate stratum; see Chapter 7).

6.6. Multistage Frames; Several MOSs

The procedures presented in Sections 6.4 and 6.5 can be easily extended to more complex situations, where one may wish to construct his frame in stages and, depending on the kind and availability of information usable for MOS, perhaps use different MOSs at different stages or even over different FUs of the same stage.

Suppose we wish to construct a frame of SUs of approximately 4 HHs each for all of the city of Santa Monica. A convenient first-stage unit for many U.S. cities is the so-called Census Tract, or simply Tract, which typically includes a fairly compact cluster of about 50 blocks or about 3000 HHs, although wide departures from this average are not uncommon. The tract is mapped, and usually a lot of demographic and some economic data are collected about it and published by the U.S. Bureau of the Census. Every 10 years a new census is taken and the tract's characteristics are brought up to the new date. It may also be redefined from time to time.

In Santa Monica, in 1960, there were 12 tracts and a total of 35,070 HHs. Since we wish M_0 (current) = 4 HHs, and if we suspect that current HHs are, say, 45000, then N would calculate out to be $45,000/4 = 11,250$, whence $X_0 (1960) = 35,070/11,250 = 3.11$.

An allocation of SUs to each of the tracts can be carried out using the same procedure as that of Section 6.4. This is shown in Table 6.4. Note that tract 7012 gets an allocation of 675 SUs. Within each tract the procedures of Tables 6.1, 6.2, or 6.3 are applicable. The extension to more stages should be evident; for example, Santa Monica could be an FU in a listing of cities in California. Moreover, we can look upon the Table 6.2 as a two-stage scheme, where a listing of households within the block constitutes a second stage and can therefore be dealt with independently of whatever is done at the first stage—in

Table 6.4.   A first stage frame (tracts) for Santa Monica

| | City of Santa Monica [Setting X_i (1960) = 3.11 HHs] | | | |
| | No. of HHS, X_g | | No. of SUs. N_g | |
Tract no. (g) (1)	Absolute (2)	Cumulative (3)	Cumulative (4)	Absolute (5)
7012	2699	2699	675	675
7013	4081	6780	1695	1020
7014	2706	9486	2372	677
7015	3330	12,816	3,204	832
7016	2963	15,779	3945	741
7017	2423	18,202	4,551	606
7018	3817	22,019	5505	954
7019	2042	24,061	6,015	510
7020	2,123	26,184	6,546	531
7021	2701	28,885	7,221	675
7022	3583	32,468	8117	896
7023	2602	35,070	8768	651
Total	35,070			8768

this case, blocks. For example, we could use a list of addresses within the block as the second stage, and within an address a list of HHs as a third stage, and perhaps even a list of persons within HHs as a fourth stage—provided our desired SU was a person.

We may also select our sample SU in stages also. For example, to find out what tract it is in, we chose an RN from 1 to 8768. Any RN from 1 through 675 puts us into tract 7012, from 676 through 1695 into tract 7013, and so on. After n trials, some tracts may not contain any selected SUs and therefore may be omitted from further consideration (unless, of course, one finds at lower stages that the same SUs have been drawn more than once; and if nonreplacement selection is desired, new draws at the first stage are required). Then one can proceed through the next stage on those selected at the first stage.

6.7.   Area Frames—Grid and Plot Types

Area sampling is the term commonly used where the sampling units are areas, usually but not necessarily fairly small, and with either regular or irregular shapes. For example, the areas may be city blocks or political units such as counties, or they may be the square-mile sections in the U.S. Public Land Survey. Having selected a sample of these areas, we obtain the observations either on elements contained in the sample area or on the sample area itself.

Hence, two types of area SUs can be distinguished: (i) that in which the area contains a cluster of elements, which we shall call the *grid type*, and (ii) that in which the area itself is measured, the *plot type*. For example, suppose the area is a circle placed at random on a map and we wish to measure the area in cultivated crops. In the grid type we first determine the elements (farms) contained in the grid (by some rule, when farms extend beyond the grid), and then "measure" the amount of cultivated land in each farm associated with the grid. In the plot type, we measure the cultivated land contained within the grid directly.

The grid type is commonly used for selecting clusters of households, business establishments, farms, and so on, whereas the plot type is frequently used in agricultural and biological surveys and the like.

The area method of sampling—that is, the use of areas as frame units—is very useful when: (1) other suitable frames do not exist, (2) a check is desired on the accuracy of some other frame, and (3) a cluster of neighboring elements is desired. Area frames have been found useful in a wide range of applications in biological, physical, and social fields—wherever a universe can be defined in terms of areas.

Suppose we have a universe of 12 dots, which are located as shown in Fig. 6.2. The dots can be imagined to be households, farms, or whatever, and the lines can be regarded as streets or roads. In this case we can regard the four road-bound area segments as comprising an area frame for this universe of dots. Each dot is contained in one and only one area segment and, if an area segment is drawn at random, the particular dot or group of dots contained in it will be drawn at random; hence areas may make convenient frame units.

But "natural" area segments may not contain anywhere near the optimal number of elements, or whatever MOS is relevant. Hence we may alter the natural boundaries or adopt new boundaries that provide M_i nearer to an

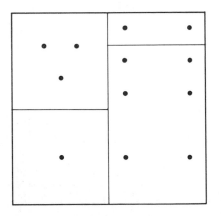

Fig. 62.   Universe of 12 dots with road net.

optimal M_0. This may be done a number of ways, depending on the existence of detailed maps (such as Sanborn maps in urban areas and aerial photographs and the like in rural areas). Using multistage framing as indicated in Section 6.6, we need these materials only in those frame units falling into the sample.

Criteria for equalization of area SUs depend on the nature of what is being surveyed. Sometimes equal areas may be most appropriate (as in crop surveys and certain ecological surveys), sometimes equal elements (e.g., surveys of farm characteristics, household surveys). Figure 6.3 illustrates the two kinds of segmentation—the plot type, equalized on area, and the grid type, equalized on elements—for the case of Fig. 6.2.

Area sampling units are quite useful in surveys involving people—generally where lists are not available and where their numbers as well as their characteristics are the objectives of the survey. The area may be the SU itself or

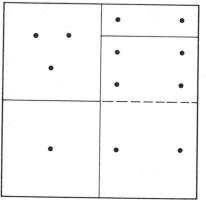

(a)

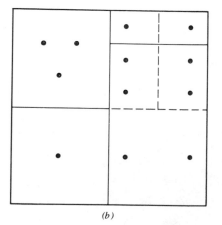

(b)

Fig. 6.3.   Equal-size SUs on areas or dots. (a) Equal areas. (b) Equal dots.

an FU containing several allocated SUs that require field work for final designation. This can be done by listings or, since accurate listings may be too expensive, by simple sketch-mapping, segmenting these into area SUs. Figure 6.4 is a sketch-map of a pair of city blocks to show the general location of buildings containing inhabitants and a rough estimate of numbers of HHs at each structure location. (If addresses were available, the sketch-mapping would probably not be worth the cost.) The resulting four area SUs are indicated. The exact determination of number of HHs will be made only on the SU falling into the sample.

Boundary Problems on Area SUs. Although area frames can be made to be very rigorous conceptually, in practice estimates based on area samples sometimes have been found to have serious biases. These biases are usually ascribable to faulty work by investigators in accounting for the elements in the sample area because of indefiniteness about where the boundaries actually are. Care should be taken to assure good boundaries in the first place and to provide maps and other aids (such as aerial photographs) to help the investigator establish the designated boundaries.

The problem is made more acute by choosing small areas (because of their usual high statistical efficiency), resulting in a high perimeter to unit of area. Hence, in many practical situations, it may be wise to accept a larger, less efficient area in order to reduce possible boundary bias.

The bias and its direction may be difficult to anticipate, depending on the type of survey being made and the training procedure. If a dwelling unit is

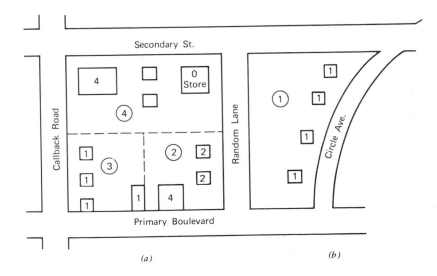

Fig. 6.4. Field sketches of blocks 2 and 3 for an area frame. (*a*) Block 2. (*b*) Block 3.

situated so that the investigator is in doubt whether it is in the SU or out, may tend to include it "in" if he was stongly warned to "get everything"; hence, overestimates will result. On the other hand, if no such warning is given, he may well ignore doubtful cases, so that underestimates result. Some of these problems will be dealt with in Chapter 13.

6.8.   Shortcut Designations; Noninteger Frames

In this chapter we have presented some working procedures for constructing frames that have an exact and predeterminable number of SUs, N, where an MOS is used to make M_i or X_i as near a chosen M_0 or X_0 as practicable and where the N_h are integers. To avoid possible difficulties with rounding errors, cumulative totals of the N_h were calculated to assure complete control and to illustrate the method.

Incomplete Designation.   Where only a few SUs are to be selected from a large number of FUs, a shortcut eliminating some of these steps can be used and still maintain complete control. This method requires the complete cumulative totals for M_h or X_h only. Table 6.5 displays the procedure for the data of tract

Table 6.5.   Incomplete designation of integer $N_h s$

			Tract 7012 of Santa Monica			
Block no. (h) (1)	No. of persons, X_h		Sample points		Cumulative SUs $\sum N_h$ (6)	Absolute SUs N_h (7)
	Absolute (2)	Cumulative (3)	RN_X (4)	RN_N (5)		
1	101	101				
2	45	146				
3	77	223				
4	53	276				
5	34	310			29	
6	42	352	364	35	33	4
7	14	366			34	1
8	12	378			35	1
9	158	536				
10	52	588				
.	.	.				
.	.	.				
.	.	.				
87	0	7203				
Total	7203					

7012, Santa Monica, given in Table 6.3 for the long method. Suppose we wish to select one SU from this frame of 675 with a probability of exactly 1/675. We draw a random number, RN, between 1 and X ($= 7203$), obtaining $RN_X = 364$. Looking at column 3 in Table 6.5, we may conclude that our selected SU must be in block 7. Converting our RN from the X_h scale to the N_h scale, we obtain $RN_N = 35$ ($= 364/10.67$ *rounded up to the next whole number*). Calculating cumulative N_h for blocks 5, 6, 7, and 8, we obtain the cumulative totals of N_h, rounded, to be 29, 33, 34, and 35, respectively, with the resulting N_hs of 4, 1, and 1 for blocks 6, 7, and 8, respectively. But it may be noted that according to column 6, our desired SU, 35, is actually in block 8 rather than 6. Since block 8 has been allotted one SU, the block and the SU are coextensive.

It can be shown that the procedure assures that the probability of selecting any of the 675 SUs, although only implicitly alloted by the procedure, is in fact $1/N$, or in this case $1/675$. If we adopt the rounding rule, *rounding up to the next whole number* in calculating cumulative X_h or M_h as well as in scaling RN_X to RN_N, we will always arrive at the correct FU directly, with only a minor loss in the equality of SU size.

It may be noted that by selecting blocks 7 and 8 (see Table 6.1), we would not obtain any HHs (in 1960), since there were none. Our measure of size X_h indicated some persons living there, but presumably they were not in households. This is an example of what may happen when we are guided by an MOS that is not perfectly accurate. In a survey of HHs those SUs would end up as "duds."

Noninteger Frames. The requirement that N be predetermined and fixed as part of the specifications for the frame may be conceptually useful but is certainly not necessary for good sampling. We can relax this requirement but hold to the one that the frame be such that we can select each element, OU, with equal probability.

Suppose we adopt the following procedure for selecting HHs in Tract 7012. (See Table 6.1.) To determine the block we want, we select an RN from 1 through X ($= 2699$ HHs) say 120. This puts us into block 6, which has a reported 15 HHs in it (in 1960). Let us now select at random four of these 15 HHs for our SU. In this scheme we can calculate the probability of selecting a block and selecting a household given the block; hence

$$(6.6) \qquad \text{prob} = \frac{X_h}{X} \cdot \frac{4}{X_h} = \frac{4}{X},$$

which is the same for all HHs. Hence our *sampling rate* is *constant*. We prefer to look at this case as part of the problem of multistage sampling, so we shall not consider it further here.

6.9. Special Techniques: "Co-Listing," "Random Fractions," Multiple Frames

In the design and construction of practicable frames, situations arise that require special treatment or special features to be devised. Some of these special requirments are one-time occurrences; others seem to appear frequently enough to justify the creation and to some extent standardization of accessories that can be attached to the basic design. Here are some.

"Co-listing."   In surveys where the sample consists of blocks on which households must be listed to provide a frame for selecting those households to be observed (e.g., interviewed), a common procedure is to send a person to the blocks to compile the lists and have him bring the list back to the office, where the sample selection takes place. The lister or another person may return to the block to make the required observations on the households. Dividing this work into two or perhaps three operations—(i) listing, (ii) sampling, and (iii) interviewing—permits the use of specially trained personnel for each task and assures that the sampling is *purely objective.* But it is expensive and may be impractical where travel is costly and where personnel are competent as both listers and interviewers.

The above procedure is sometimes referred to as "prelisting." An alternative method permits all operations to be done more or less simultaneously and yet assures control of the objectivity of the sample selection. In the case of listing households, or addresses, a fixed point may be specified as a reference point (e.g., northeast corner of block), and a specified order of listing from that point may be required (such as each address in clockwise direction around the block, although complex multiple dwelling units, scattered unaddressed dwellings, and the like may make a rigid order of listing difficult if not impossible). If sample selection is to be done simultaneously with listing, the simplest, if a fixed fraction is to be taken, is probably a systematic selection with a random start point.

Random Fraction.   An example of the usefulness of tying the listing and sampling together in a single operation is that where an individual is to be selected at random from a household at the time an interviewer calls on a household and lists its members. The following method should assure an objective selction of a random individual. Let each HH in the sample be assigned a *random fraction* RF, drawn say between 0.00 and 0.99. The individuals in the HH are to be listed in a rigid order, say in order of birthdates, and serially numbered $1, 2, 3, \ldots, k$. A random number RN is generated by the relationship $RN = k(RF)$, *rounded up* to the next whole number. The RN designates the selected individual. For example, if RF is 0.51 and k is 3, then $RN = 3(0.51) = 1.53 \longrightarrow 2$. Hence person number 2 is selected. (This method is exact for all even k and is approximately so for odd k, provided RF is taken to sufficient decimal places.) Note that the probability of an individual's being selected is $1/k$ that of a household's being selected.

Multiple Frames.   Sometimes the nature of the universe to be dealth with is such that it is more convenient to divide the OUs into two or more groups and to construct frames that are appropriate to each group. The manner in which sampling is carried out in thise case when the groups are mutually exclusive is called stratified sampling and will be dealt with in Chapter 7. Suppose one wishes a frame for business firms in California. It will be found that most of the larger firms appear on lists (trade organizations, financial statement compilations, tax reporting bodies, industry directories, and so on) and the smaller ones, or those of certain types, on none at all. Frames can be built on lists where they exist and are suitably accurate, and perhaps on area segments supplemented by lists for those units not on lists. When multiple frames are used, care must be taken to determine the extent OUs appear on more than one list and to adopt measures to deal with it. This is dealt with in Section 12.7.

Selecting FUs from a List with PPS (***Probability Proportional to Size***).   The problem of selecting from a list of FUs, or clusters, of variable sizes, N_h, such that the probability of selecting cluster h is N_h/N is one that occurred implicitly and explicitly in Sections 6.4–6.6 in selecting an SU in two stages or more. In the examples given, the selection was made by use of cumulative totals of the M_h. To make the selection without the use of cumulative totals, let us consider a rectangular frame of $K \times N_0$ cells, where K represents the number of clusters in

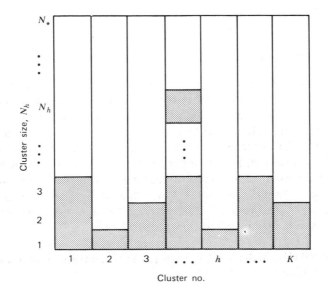

Cluster no.

the list and N_* is equal to or greater than the largest N_h. By selecting a pair of coordinates at random, where $1 \leq RN_K \leq K$, and $1 \leq RN_{N*} \leq N_*$, a cell is picked at random. If that cell occurs within the cluster, then the cluster is regarded as picked for the sample. If the random coordinates pick a cell outside the cluster's size N_h, regard it as a dud and repeat the selection of a new pair of random coordinates. The probability of picking cluster h can be seen to be $N_h/\sum N_h$ or N_h/N. This is sometimes called the *Lahiri scheme*. (See Section 8.10.)

6.10.   Summary

In this chapter we have considered the basic principles of constructing sampling frames that provide: (i) sampling units having certain desired properties and (ii) a means for selecting these units at random. We have examined the use of certain information correlated with what is being measured, called measure of size, to equalize the size of the SU. Two kinds of MOS are helpful: (i) that which is correlated with M_i the number of elements per cluster, and (ii) that which is correlated with Y_i. The first type is generally best for equalization when $\bar{\bar{Y}}$ is being estimated, and the second when $\bar{Y}$ is being estimated. Some illustrations are given on how equalization on MOS can be done in the case of an MOS correlated with M_i.

Our exploration of the principles and the nature of the problem should have made evident the vast opportunities for individual ingenuity. Although great progress has been made in the last 30 years, there is no indication that the supply of new and better ways of constructing efficient sampling frames has been exhausted.

6.11.   Review Illustrations

1.   The object is to obtain a sample of one-fourth of the HHs (households) in a city block, which may be regarded as picked at random. Six plans have been offered for doing this in a valid and efficient manner. Examine the following six schemes in regard to each of the three criteria: (a) possible bias in selection, (b) expected size of the variance of the sample mean, (c) expected cost. Make a two-way table with criteria (a), (b), and (c) in the heading and the six schemes in the stub. For criteria (b) and (c) determine relative ratings of the six schemes, such as "highest," "high," "medium," "low," "lowest," and enter them into the appropriate cells. Ties are likely. For criterion (a) state whether "biased" or "unbiased." Assume field work is done accurately.

(1) A list of all HHs in the block is made (20 are found) and a random sample of five HHs is taken.

(2) The block has four sides. One side is selected at random and all HHs having an address on that side are taken for the sample.

(3) A random corner (one of the four) is selected. The sample consists of the first and every fourth house (structure) encountered as one proceeds clockwise around the block. All HHs found in these houses are taken.

(4) A quick walk is taken around the block and a rough list of addresses is made in the order encountered. The list is divided into quarters, of which one is drawn at random. Field instructions are that all HHs actually having an address within the limits of the selected list segment are to be taken.

(5) The block is divided into four areas, of which one is selected at random. All HHs found in the selected area are to be taken.

(6) A random point is selected on the periphery of the block. The nearest house to that point and every fourth house around the block is taken.

Solution

Rough judgments are as follows:

Scheme	(a) Possible bias	(b) Variance		(c) Cost	
(1) Random individual HH	No	Low	4	Highest	1
(2) Random block side	No	Highest	1	Lowest	6
(3) Random corner/systematic	Yes	Least	5	Low-medium	4
(4) Random list segment	No	Medium	3	Medium	2
(5) Random area	No	High	2	Low	5
(6) Random point systematic	Yes	Least	5	Low-medium	3

2. A study was undertaken to determine, among other things, the average size of families from which college students come. The procedure was to select a random sample of students in colleges and ask them to fill out a questionnaire requesting information on the number of children born to their parents and to their grandparents. The simple mean of the sample was used to estimate the population means (average size of parents families say). Was the estimate biased or unbiased? Why?

Solution

Biased. Although the sample of students was random and the mean of students' characteristics would be unbiased, it does not follow that the sample of *families* having students in college is random. Actually, since families may have one or more students in college, then a family with two students will have twice the chance of being included as a family with but one student. Likewise a three-student family will have three times the chance. Since these are likely to be the larger families, family size in the sample is expected to be larger than the true size.

3.   A market researcher is given the job of determining the "market area" served by the XYZ Supermarket. He obtains a sample of customers by selecting for interview every tenth person leaving the check-out counters during all operating hours for a week. The addresses given in the interview are plotted on a map to indicate the "market area" of the store. Critically evaluate this scheme, bearing in mind the purpose of the survey.

Solution

The different frequencies of shopping by customers could cause a problem. Far-away customers may shop less frequently but buy more at a time—hence density of dots need not indicate density of sales volume.

4.   Determine a scheme for estimating the size of a crowd, such as: (*a*) number of persons in a crowded street or beach, (*b*) number of persons in a department store, (*c*) number of moving insects or animals in a given area. [*Hint:* See Yates (1949, pp. 43–44)]

Solution

For small areas (where all persons are visible in a line of sight across area): Use an observer moving from one end of area to other, counting all people he passes (regardless of direction they are moving) and deducting all people who overtake him. Repeat this in the reverse direction. The average of the two provides a correct result.
For larger areas: Use two or more observers moving in parallel across area, counting only in the space between each.
For very large areas: Use the above schemes for strips across the area, where strips are randomly chosen.

5.   Procedures are being considered for sampling the patients discharged from a large municipal hospital. An agreement has been reached that about 260 can be dealt with, although the total discharges during a year is about 100 times that number. Mr. A advocates the selection of a random sample of 1 % of the patients discharged during each week of the year. Mr. B wants to draw a random number between 1 and 100, start with the dischargee of that number (say 20), and take a systematic selection of every 100th patient through the files

for the year. Mr. C wants to pick a typical month and take 12 % of the patients at random from that month. Mr D wants to take all the patients discharged on four randomly selected days during the year. Mr E prefers to take the first 260 patients in the alphabetical file. Evaluate these schemes in regard to (*a*) bias in estimating a mean per dischargee for the year, and (*b*) variance of the estimated means relative to each other.

Solution

Scheme	Sampling Method	(a) Bias?	(b) Variance
Mr. A's	Strat-random	None	Low
Mr. B's	Systematic	None	Lowest, probably
Mr. C's	Purposive	Yes	Low, probably
Mr. D's	Day cluster	None (except Technically)	High, probably
Mr. E's	Top of alphabet	Yes	Low, like B, probably

If no estimate of sampling error is required, Mr. B's is likely to yield the best results here.

6.12. References

Bowley, A. L.
1913

Working Class Households in Reading. *Journal of the Royal Statistical Society*, **76**: 672–701. (An interesting pioneer work in selecting households in a relatively small town.)

Bureau of the Census
1947

How to Read Aerial Photographs for Census Work. Washington, D.C.: U.S. Government Printing Office. (44 pp.)

Deming, W. E.
1960

Sample Design in Business Research. New York: John Wiley & Sons, Inc. (A number of examples of preparing frames for sampling inventory materials, accounts, and other universes of interest to the business world.)

Hansen, M. H.
Hurwitz, W. N.
Madow, W. G.
1953

Sample Survey Methods and Theory. Vol. I. New York: John Wiley & Sons, Inc. (Extensive presentations of area frames for human populations.)

Jessen, R. J.
1945

The Master Sample of Agriculture: Design. *Journal of the American Statistical Association*, **40**: 45–56. (Description of an area frame for farms and agricultural characteristics.)

Jessen, R. J.
Strand, N. V.
1953

Statistical Methodology in Multi-Purpose Surveys of Crete, Greece. Appendix 1 (pp. 323–394) of Leland G. Allbaugh, *Crete: A Case Study of an Underdeveloped Area.* Princeton, N.J.: Princeton University Press. (Available as a monograph An early detailed description of methods used to sample a human population in an underdeveloped country.)

King, A. J.
1945

The Master Sample of Agriculture: Development and Use. *Journal of the American Statistical Association*, **40**: 38–45.

Lahiri, D. B.
1951

A method of Sample Selection Providing Unbiased Ratio Estimates. *Bulletin of the International Statistical Institute*, **33**(2): 133–140.

Mahalanobis, P. C.
1940

A Sample Survey of Acreage under Jute in Bengal. *Sankya*, **4**: 511–530. (Description of a frame for a crop.)

Monroe, J.
Finkner, A. L.
1959

Handbook of Area Sampling. Philadelphia: Chilton Company. (Paper cover, 55 pp.) (Some useful aids in preparing area frames and conducting field operations in agricultural surveys.)

Sukhatme, P.V. .
1954, 1970

Sampling Theory of Surveys with Applications. Ames, Iowa: Iowa State College Press. (2nd ed. with B. V. Sukhatme. London: Asia Publishing House.)

Yates, F.
1949

Sampling Methods for Censuses and Surveys. London: Charles Griffin Co. (Introducing schemes of point, line, and area sampling units.)

6.13.   Exercises

1.   A much-used listing for selecting people to be used in opinion surveys, political polls, marketing surveys, and the like is the telephone directory. As a frame for surveys of this sort critically evaluate the telephone directory.

2.   In a study of more than 400 wedding announcements published in a large metropolitan daily newspaper during the month of June over a ten-year period, it was erroneously concluded that such announcements avoided mention of a synagogue or any other association of a wedding with the Jewish faith. Why the error?

3.   Some bulk material is to be examined by sample. It is planned to use the tree frame method with 2^k splits and with two replications—that is, $n = 2$, where $k = 4$. Since $N = 16$, two random numbers, 3 and 8, were drawn in advance. Show how the sampler should proceed with a minimum of steps to obtain his sample without possible bias.

4.   The following sampling schemes are used in selecting households in a city block. Evaluate each scheme in regard to bias, statistical efficiency and costs relative to each other (e.g., low, lowest, high, etc.).

   (1)   A listing is made of all households in the block and a random sample of one-fourth of them is taken for interview.
   (2)   A count of households on the whole block is made. A random point on the perimeter of the block is selected and, proceeding clockwise, the next k households are taken, where $k = \frac{1}{4}$ of the total HHs in the block.
   (3)   The interviewers are told to select the most representative $\frac{1}{4}$ of the homes in the block.
   (4)   What appear to be the richest and poorest households are taken, and h

more are taken at random in order to obtain one-fourth of all households in the block.

(5) Starting at a given corner with a random number (selected from 1 to 4), the interviewer is requested to start listing households in a clockwise direction and interviewing the rth and every $(r+4)$th as he goes around the block, where r is the random number.

5. We wish to estimate the total number of "hospital admits" in California during the past year.

(a) Suggest at least two feasible frames for sampling.

(b) Choose one from (a) and indicate in some detail how the frame might be constructed.

CHAPTER 7

Stratification

7.1. Introduction

In the schemes of sampling discussed thus far, a selected sampling unit, whether a single element or cluster, could come from any part of the universe by pure chance. Such is a basic property of simple random sampling, whether RR or RNR. There is little intuitional appeal in the rather haphazard way that random sampling permits any part of the universe to be represented. In an attempt to control this behavior of random samples, a method called stratified-random sampling was developed. A considerable literature has been produced on its various ramifications, going back at least to Tschuprov (1923), Bowley (1926), and Neyman (1934). Only the more commonly used principles and procedures will be presented here. References are provided for further study of some aspects of stratification.

Stratification can be regarded as a procedure for selecting samples in which certain restrictions or constraints are imposed on the ordinary randomization procedure. When conditions are right and the constraints are imposed properly, a welcome gain in precision can be obtained—and usually with little or no additional cost. In another but related sense, stratification is a way to incorporate one's knowledge of a universe and its populations into a sampling procedure so as to increase its efficiency.

7.2. A Small Finite Case

Suppose we continue with the four-stone universe of Chapter 2 and the population of weights.

	Stratum 1		Stratum 2	
U_i	1	2	3	4
Y_i	2	5	7	14

Let us divide the universe into two groups such that elements 1 and 2 go into stratum 1 and elements 3 and 4 into stratum 2. Stratified random sampling differs from simple random by having the random selection of elements take place independently in each of the several strata; hence each stratum is treated as if it were itself a universe. Suppose we draw a stratified random sample of size two from the universe above. To do this, we must draw a random sample of one from each stratum. There will be four possible stratified random samples,

$$1,3; \quad 1,4; \quad 2,3; \quad 3,4$$

in contrast with six RNR samples. By looking at column 2 of Table 2.1 (Section 2.6), we can see that stratification has eliminated samples 1,2 (no. 1) and 3,4 (no. 6), which had the most divergent means of the RNR samples. The variance of the four stratified sample means would therefore be expected to be less than that of the means of the RNR samples. The average of the sample means turns out to be

$$E[\bar{y}_{\text{strat}}] = \frac{1}{4} \sum_{t=1}^{4} \bar{y}_t$$

$$= \frac{1}{4} (4.5 + 8.0 + 6.0 + 9.5) = 7.0$$

$$= \bar{Y},$$

hence the method, although restrictive, is in this case unbiased. The variance of the mean of our stratified sample,

$$\text{var } (\bar{y}_{\text{strat}}) = E[\bar{y}_{\text{strat}} - E[\bar{y}_{\text{strat}}]]^2$$

$$= \frac{1}{4} \sum_{t=1}^{4} (\bar{y}_t - \bar{Y})^2$$

$$= \frac{1}{4} [(-2.5)^2 + (1.0^2 + (-1.0)^2 + (2.5)^2]$$

$$= \frac{14.50}{4} = 3.625.$$

This can be compared with the variance of the mean from RNR samples, obtained by adding the squared deviations for samples 1,2 and 3,4 and changing the divisor to 6:

$$\text{var } (\bar{y}_{\text{RNR}}) = \frac{(14.50) + (-3.5)^2 + (3.5)^2}{6}$$

$$= \frac{39.00}{6} = 6.50,$$

which is considerably larger.

The gain in precision that stratification brought in this case may be attributed to the requirement that each stratum be represented by its proper number of sample elements, which in this case meant that for samples of size two, one element must be taken from each stratum. In this example, it was quite important to have each stratum properly represented, because the elements had much higher values in stratum 2 than in 1.

7.3. Stratified-Random Sampling

We shall now consider a rather general case of stratified random sampling, where a universe of N elements is divided into an arbitrary number of groups or strata. For this we shall find a set of symbols and definitions of help.

								Total
Stratum number,	U:	1	2	...	h	...	L	—
No. of elements,	U:	N_1	N_2	...	N_h	...	N_L	N
	S:	n_1	n_2	...	n_h	...	n_L	n
Totals of y,	U:	Y_1	Y_2	...	Y_h	...	Y_L	Y
Means of y,	U:	$\bar{Y}_1$	$\bar{Y}_2$	...	$\bar{Y}_h$	...	$\bar{Y}_L$	$\bar{Y}$
	S:	$\bar{y}_1$	$\bar{y}_2$	...	$\bar{y}_h$	...	$\bar{y}_L$	$\bar{y}$
Variances of y,	U:	S_1^2	S_2^2	...	S_h^2	...	S_L^2	S^2
	S:	s_1^2	s_2^2	...	s_h^2	...	s_L^2	s^2

In the above table "U" denotes universe and "S" denotes the sample values. In the general case of stratified random sampling a *random sample* of n_h elements is taken from the N_h in the hth *stratum*. In other words, each stratum is sampled as if it were a universe of interest. The sizes of the strata need not be equal, and the sampling rate, n_h/N_h, for the hth stratum need not be the same over all strata. Since the mean of the portion of the total sample taken in any stratum provides an unbiased estimate for the true mean of that stratum—that is, $\bar{y}_h$ is an unbiased estimate of $\bar{Y}_h$—for an estimate of Y_h we have the unbiased estimator

$$(7.1) \qquad\qquad \hat{Y}_h = N_h \bar{y}_h,$$

and for the y-population total we have the simple sum over all strata:

$$(7.2) \qquad \hat{Y}_{\text{strat}} = \sum \hat{Y}_h = \sum N_h \bar{y}_h, \qquad h = 1, 2, \dots, L.$$

To obtain an estimate of $\bar{Y}$ we simply divide Eq. 7.2a by N, obtaining

$$(7.3) \qquad\qquad \bar{y}_{\text{strat}} = \sum \frac{N_h}{N} \bar{y}_h = \sum W_h \bar{y}_h$$

—that is, $\bar{y}_{\text{strat}}$, which is an unbiased estimator of $\bar{Y}$, is a *weighted* average of the stratum means in the sample, where the weights are the relative sizes of the strata, N_h/N.

The variance of this estimator, since (i) it is a sum of independent variates (because randomization is independent in each stratum) and (ii) each term in the sum is a product of a constant and a variable, is given by

$$(7.4) \qquad\qquad \text{var}\,(\bar{y}_{\text{strat}}) = \sum \left[\left(\frac{N_h}{N} \right)^2 \text{var}\,(\bar{y}_h) \right].$$

Since the sample mean per stratum is

$$(7.5) \qquad\qquad \bar{y}_h = \frac{1}{n_h} \sum Y_{hi}, \qquad i = 1, \dots, n_h,$$

and its variance is

(7.6) $$\text{var } (\bar{y}_h) = \left(\frac{N_h - n_h}{N_h}\right)\frac{S_h^2}{n_h},$$

therefore

(7.4a) $$\text{var } (\bar{y}_{\text{strat}}) = \sum \left[\frac{N_h^2}{N^2}\left(\frac{N_h - n_h}{N_h}\right)\frac{S_h^2}{n_h}\right].$$

Example 7.1. Use Eq. 7.4a to compute the variance of the mean of stratified samples of size 2 (one from each stratum) for the population of stone weights given in Section 7.2.

In this case $N_1 = N_2 = 2; n_1 = n_2 = 1;$ in stratum 1, the y values being 2 and 5, then $S_1^2 = (2-5)^2/2 = 4.5$ and $S_2^2 = (7-14)^2/2 = 24.5;$

$$\text{var } (\bar{y}_{\text{strat}}) = \frac{N_1^2}{N^2}\left(\frac{N_1 - n_1}{N_1}\right)\frac{S_1^2}{n_1} + \frac{N_2^2}{N^2}\left(\frac{N_2 - n_2}{N_2}\right)\frac{S_2^2}{n_2}$$

$$= \frac{2^2}{4^2}\cdot\left(\frac{2-1}{2}\right)\cdot\frac{4.5}{1} + \frac{2^2}{4^2}\cdot\left(\frac{2-1}{2}\right)\cdot\frac{24.5}{1}$$

$$= \frac{2.25 + 12.25}{4} = 3.625.$$

(which is identical to the result given in Section 7.2).

7.4. Best Allocation of Sample

The previous section dealt with a very general case of stratified random sampling, where the amount of the total sample size, n, allocated to each of the several strata was arbitrary. We cannot compare this design with that of simple sampling unless some decision is made on how much of the total sample goes to each stratum. We consider a general solution first and then point out some useful special cases later.

Suppose the cost of observing an element from the hth stratum is c_h and we have a total budget of C units for the observations (exclusive of overhead). The sampling cost for the hth stratum is given by

(7.7) $$C_h = c_h n_h,$$

and total costs for all strata will be

(7.8) $$C = \sum C_h = \sum n_h c_h.$$

We wish to minimize the variance

(7.9) $$\text{var } (\bar{y}_{\text{strat}}) \doteq \sum_{h}^{L} \left(\frac{N_h^2}{N^2}\right)\left(\frac{S_h^2}{n_h}\right).$$

With the budget restriction of Eq. 7.8 this works out to be a minimum for all possible allocations when

$$(7.10) \qquad n_h = n \frac{N_h S_h / \sqrt{c_h}}{\sum (N_h S_h / \sqrt{c_h})}$$

—that is, when the number of elements allocated to the hth stratum is made proportional to the factor

$$(7.11) \qquad \frac{N_h S_h}{\sqrt{c_h}}.$$

Hence, the sample should be allocated most heavily to the larger, more variable, and cheaper strata.

The variance of the case of *net optimum stratification* turns out to be

$$(7.12) \qquad \text{var } (\bar{y}_{\text{net-opt}}) \doteq \frac{1}{n} \sum \left(\frac{N_h}{N}\right)^2 \left(\frac{S_h^2}{(N_h S_h / \sqrt{c_h}) / \sum (N_h S_h / \sqrt{c_h})}\right).$$

When the costs of observing elements are the same in all strata—that is, when $c_h = c$—then the *optimum allocation* of n is given by

$$(7.13) \qquad \text{opt } (n_h) = n \left(\frac{N_h S_h}{\sum N_h S_h}\right).$$

The variance of this case of *optimum stratification* is given by substituting from Eq. 7.13 into Eq. 7.4a, whence,

$$(7.14) \qquad \text{var } (\bar{y}_{\text{opt}}) = \frac{1}{N^2} \left[\frac{(\sum N_h S_h)^2}{n} - \sum N_h S_h^2\right].$$

In the case when both costs and variances are constant over strata, the optimum allocation of n, Eq. 7.13, reduces to

$$(7.15) \qquad n_h = n \left(\frac{N_h}{\sum N_h}\right) = n \left(\frac{N_h}{N}\right)$$

—that is, the allocation should be proportional to stratum sizes. This simple stratification is the most common case and will be dealt with in some detail in the next section.

In this case of simple allocation, whether costs and variance are equal or not, an unbiased estimate of $\bar{Y}$ is given by

$$(7.16) \qquad \bar{y}_{\text{simp}} = \sum \left(\frac{N_h}{N}\right) \bar{y}_h = \sum \left(\frac{n_h}{n}\right) \bar{y}_h = \sum_h \sum_i^{n_h} \left(\frac{Y_{hi}}{n}\right)$$

$$(7.16a) \qquad = \frac{\text{sum of the } ys \text{ in the sample}}{n}$$

—that is, the simple mean of the sample. Its variance is obtained from 7.4a by substituting 7.15, whence

$$(7.17) \qquad \text{var} \, (\bar{y}_{\text{simp}}) = \left(\frac{N-n}{N}\right) \left(\frac{\sum N_h S_h^2}{Nn}\right)$$

$$(7.17a) \qquad = \left(\frac{N-n}{N}\right) \frac{\bar{S}^2}{n}$$

where

$$(7.18) \qquad \bar{S}^2 = \sum \left(\frac{N_h}{N}\right) S_h^2.$$

Other forms of this formula will be given in Section 7.5.

While Eqs. 7.10 and 7.13 provide mathematical solutions to the problem of what is the best allocation of a sample under certain conditions of costs and variances, they have serious limitations in many situations met in practice, since (1) knowledge of c_h and S_h is required, (2) the estimators require weighting of sample results, (3) a good allocation for one population may be a poor one for another—a serious fault when several populations are dealt with simultaneously.

If reasonably good estimates of c_h and S_h can be made, and if good computing facilities exist, some of these difficulties are overcome. Good estimates of c_h are not very important because c_h enters only as the square root, hence rough estimates may be helpful, particularly where actual cost differences are large. But if cost differences are small and the cost estimates are rough, one probably should ignore differences in costs. The potential for improving efficiency in this case is small, and a slight loss of efficiency could occur. Good estimates of S_h can be more helpful in improving efficiency. The shortcut methods of Section 3.10 may be helpful. Another good scheme is to allocate n_h proportional to estimated Y_hs.

7.5. Simple Stratified-Random ("Representative") Sampling; ANOVA

The most common and generally the most useful type of stratification is *simple stratification*, where the allocation of the total sample to the various strata is made proportional to N_h. This method is sometimes called "representative" sampling. In the previous section it was shown that this scheme is optimum when costs and variances are the same in all strata.

A convenient way to compare the efficiency of simple stratified sampling with that of random sampling in estimating the mean is by the use of a technique known as the *analysis of variance*, often abbreviated to ANOVA. This powerful and useful technique was developed by R. A. Fisher (1950) during the 1920s and is briefly defined by him as a treatment of data "by the separation of the variance ascribable to one group of causes from the variance ascribable to other groups." One of the convenient features of ANOVA is its rather simple procedure for carrying out a variety of computations and organizing them into a compact and meaningful table for a number of uses. A number of models are available, and we shall use whichever one seems useful for the occasion. Table 7.1 presents a simple ANOVA table format for a universe of N elements grouped in L groups (strata) with unequal N_h in each.

Table 7.1.   Analysis of variance of a stratified universe (on a per-element basis)

Source of variance	Degrees of freedom	Mean square	Mean square is identical with
Total	$N-1 = \sum N_h - 1$	MS (T)	S^2
Strata	$L-1$	MS (B)	—
Elements/strata	$N-L = \sum (N_h - 1)$	MS (W)	$S_w^2 \doteq S^2$

Note that since the within-stratum mean square is

$$(7.19) \qquad MS\ (W) = \sum \left[\frac{N_h - 1}{\sum (N_h - 1)} \right] S_h^2,$$

then when the N_h are equal or when they are large so that $(N_h - 1)$ approximates N_h, Eq. 7.19 reduces to Eq. 7.18; hence

$$(7.20) \qquad MS\ (W) \doteq \bar{S}^2.$$

The relative efficiency of simple stratification to RNR sampling in the estimation of $\bar{Y}$ is

$$(7.21) \qquad RE\ (simple/random = \frac{var\ (\bar{y})}{var\ (\bar{y}_{simp})} = \frac{[(N-n)/N]\ (S^2/n)}{[(N-n)/N]\ (\bar{S}^2/n)}$$

$$(7.22) \qquad = \frac{S^2}{\bar{S}^2},$$

or, in terms of mean squares,

$$(7.22a) \qquad \doteq \frac{MS\ (T)}{MS\ (W)},$$

a comparison which can be made easily from an analysis-of-variance table. Unless the within-stratum mean square is more than the total mean square, a situation which rarely occurs in the real world, simple stratification will be more efficient than random sampling. In order for MS (W) to be small, MS (B) must be large. Hence, to achieve high efficiency in stratification, we must create strata that are quite different from each other in regard to their means, $\bar{Y}_h$. Strata should therefore be formed that contain "alike" elements, which is what we want to *avoid* in making up clusters. Units that are useful for clusters are useless for strata, and vice versa. (More on strata will be given in Section 7.7.)

To estimate the variance of the mean of a simple stratified sample we use the MS (w) from the sample, since it is an unbiased estimate of MS (W). Thus

$$(7.23) \qquad \text{var} \ (\bar{y}_{\text{simp}}) = \left(\frac{N-n}{N}\right) \frac{\text{MS} \ (w)}{n} \ .$$

To estimate the relative efficiency of a stratified to a random sample in estimating $\bar{Y}$, having drawn and observed a simple stratified sample, we may use the following procedure. First let us write out the ANOVA table for both a universe and a sample where strata are of equal size, as shown in Table 7.2.

Table 7.2.   *Analysis of variance of a stratified universe and sample; equal-sized strata (i.e., $N_h = N_0$ and $n_h = n_0$) (on a per-element basis)*

Source of variation	df	Universe Mean square	Mean square defined as	df	Sample Mean square	Mean square defined as
Total	$N-1$	MS (T)	$S_W^2 + N_0 \left(\dfrac{L-1}{N-1}\right) S_B^2$	$n-1$	MS (t)	$s_w^2 + \left(\dfrac{L-1}{n-1}\right) s_B^2$
Strata	$L-1$	MS (B)	$S_W^2 + N_0 S_B^2$	$L-1$	MS (b)	$s_w^2 + n_0 s_B^2$
Elements/strata	$L(N_0 - 1)$	MS (W)	S_w^2	$L(n_0 - 1)$	MS (w)	s_w^2

Note that we have L strata of N_0 elements each, and a sample of n_0 elements is drawn from each stratum, giving a total sample of $n = Ln_0$. We are now defining the mean squares in terms of *components of variance*, S_B^2 and s_B^2 for the universe and sample, respectively.

To estimate S^2 from the sample we use the fact that s_B^2 of the sample estimates s_B^2 of the universe unbiasedly. The corresponding sample and universe mean squares may not have this property, as is the case for MS (B) and MS (T). Since

$$(7.24) \qquad S^2 = \text{MS} \ (T) = S_W^2 + N_0 \left(\frac{L-1}{N-1}\right) S_B^2$$

$$\doteq \sum \left(\frac{N_h}{N}\right) S_h^2 + \sum \left(\frac{N_h}{N}\right)(\bar{Y}_h - \bar{Y})^2$$

then

(7.25)
$$\hat{S}^2 = s_w^2 + N_0 \left(\frac{L-1}{N-1}\right) s_B^2$$

(7.26)
$$= \text{MS}(w) + N_0 \left(\frac{L-1}{N-1}\right) \frac{\text{MS}(b) - \text{MS}(w)}{n_0}.$$

If strata sizes are not equal, an estimate of S^2 can be obtained by substituting N_* for N_0 and n_* for n_0 in Eqs. 7.25 and 7.26, where

(7.27)
$$n_* = \frac{1}{L-1} \left(\sum n_h - \frac{\sum n_h^2}{\sum n_h}\right)$$

and N_* is obtained from Eq. 9.27 by substituting N_h for n_h. In the case where n_h/N_h is constant,

(7.28)
$$\frac{N_*}{n_*} = \frac{N_0}{n_0}.$$

Example 7.2. In 1940 a survey of the acreage under corn in Iowa was carried out with a sample of areas, on each of which the area under corn was measured by the use of aerial photos and field inspection. The universe consisted of approximately 58,182 sections (square-mile areas), which were in 1617 strata (public land survey townships) of 36 sections each. A random sample of two sections was drawn from each of the 1617 townships. An ANOVA of the data is given in Table 7.3.

Table 7.3. *Analysis of variance of the sample (acres under corn, on a section basis)*

Source of variation	df	Mean square	Mean square denoted as
Total	3233	4232.4	MS (t)
Townships	1616	6511.9	MS (b)
Sections/townships	1617	1954.3	MS (w)

What gain in precision, if any, was obtained from township stratification over simple random sampling?

Here $58,182 = N$; $1617 = L$; $36 = N_0$; $2 = n_0$.

In this case of equal-sized strata we use Eq. 7.26:

$$\hat{S}^2 = \text{MS}(w) + \frac{N_0(L-1)\left[\text{MS}(b) - \text{MS}(w)\right]/n_0}{N-1}$$

$$= 1954.3 + \frac{36(1617-1) \; (6511.9 - 1954.3)/2}{58{,}181}$$

$$= 1954.3 + (0.999914) \; (2278.8)$$

$$= 4232.9.$$

Substituting $\hat{S}^2$ and MS (w) as estimators of S^2 and $\bar{S}^2$ in Eq. 7.22, we obtain

$$\widehat{RE} \; (\text{strat/random}) = \frac{\hat{S}^2}{MS \; (w)} = \frac{4232.9}{1954.3}$$

$$= 2.17,$$

or a gain of 117%.

In the simple stratification case the total mean square from the sample will be found to be a good estimate of S^2 when L, the number of strata, is large. Hence, in the above example the total mean square was 4232.4, or only a fraction of a percent different from the unbiased estimate of 4232.9.

7.6. Stratified: Purposive and/or Judgment Samples; Quota Sampling

So far in this chapter it has been assumed that the selection of elements within strata is random—either RNR or RR. The properties of such samples can be described by a fairly simple extension of the theory for simple random sampling. However, in practice, many samples have been drawn where selection of elements within strata has not been restricted to random. Although a great variety of such nonrandom procedures have been proposed and/or used, they can generally be classified as either the purposive or judgment types described in Chapter 1. Hence we have three types of stratified samples: random, purposive, and judgment. The common features are that (i) the universe is partitioned into groups, (ii) the sample is selected such that each group has at least one member drawn from each, and (iii) the objective is to obtain an estimate of the population mean of y or some related estimate (e.g., total ratio)—that is, a collective estimate.

To simplify a comparison, let us assume that the same strata are being considered for all three types and that selection of elements within each stratum is independent of selections in other strata. The stratified-random method would provide unbiased estimates of the population mean (Section 2.3). However, both the judgment and purposive methods are biased, and if the bias is in the same direction over all strata, then the weighted mean (of the stratum sample means) will also be biased. Conditions may be such that the sampler may be able to obtain a precision—say, the expectation

(7.29) $E[\sum (\bar{y}_h - \bar{Y}_h)^2]$

over a number of trials—that will be smaller for judgment or purposive samples than for random, but this is likely only when the total sample size is small. However, the properties of these types of samples are only partially known, and generalization is difficult. Because of the lack of theory and experience for these samples, one should be wary of them.

Quota Sampling.   A method of sampling widely used by many commercial and some governmental survey organizations (the American Institute of Public Opinion or Gallup Poll, the Social Survey in Great Britain, the Crossley Poll, to mention only a few) is commonly known as quota sampling. This method has some similarities with that of stratified random sampling, but it differs in several essential respects. Since proponents have in the past appealed to arguments supporting stratified random sampling as justification for the quota method, this seems to be a logical place to point out the essential differences.

In one of its most elementary forms the method may be described as follows. Suppose an advertising agency is interested in determining the characteristics of households that purchase a particular brand of food or drug product in some state. We will assume that this information has relevance to a projected advertising campaign. To determine the characteristics of interest it is proposed that a "representative" or "cross-section" sample of households in the state be selected and visited by interviewers to obtain records of the household purchases of the item under study. Since purchases of the item are thought to be related to various social and economic factors, it is further proposed that the households in the state be "stratified" on a number of social and economic factors, say "socioeconomic status," occupation of head of household, location of household (rural or urban), educational attainment of the homemaker or head of the household, color (white or black), and perhaps other characteristics. Information on the percentage of households in each classification established may be available from the last population census or similar source. The total sample of households is then to be selected so that the proportion of the sample households in each of the classifications matches the proportion for the state as a whole. A "geographic spread" of the sample households may be insured by first allocating the sample to regions within the state in proportion to the number of households in each region. At this stage in the proceedings each interviewer is instructed to obtain interviews from certain numbers of households in each category—this is probably why the word "quota" was adopted for this scheme.

A quota assignment for an interviewer to complete might be: "Obtain from your community the following families: two farm males, 30–40 years old; three urban males, 20–30 years old, socioeco class B; one urban female, 30–40 years old, socioeco class A; two urban females, 30–40 years old, socioeco class C."

The investigator is generally given considerable freedom in the manner in which the persons within a particular quota group are selected, although there are considerable differences between survey organizations in this respect. Few, if any, of the procedures provide equal (or even known) probabilities of selection to each person in the universe belonging to the quota class. Because of this lack of positive control on the selection, biases usually result. Most users of the method make an attempt to remove these biases through adjustments in the estimates. But this is not an easy task, and it is difficult to determine the precision of the final estimate in any probability sense. The method is therefore analogous in some respects to simple judgment selection (see Chapter 1) and, like that method, may be appropriate only for small samples, where the possible reduction in sampling error may offset the effects of the unknown biases.

Another source of possible bias is erroneous identification of quota classes. For example, a rather widely used factor for determining quotas is 'socioeconomic class." In some applications each household may be classified as a member of one of four classes. For example:*

CLASS A.   "Those who are sufficiently 'well-to-do' to have most of the things they want or to be able to afford any reasonable luxury sometimes enjoyed in your part of the country. These families live in large one-family houses of eight or more rooms, usually with garage for one and frequently two cars. Usually two or more bathrooms."

CLASS B.   "Those who are 'comfortably situated' with both the necessities of life and some of the luxuries, but who, if a luxury runs into considerable money, have to weight the having of one luxury against another. These families live in mainly moderate sized one-family houses, some of the best two-family and duplex houses, and moderately expensive apartment houses."

CLASS C.   "Those who are 'getting by' with the necessities of life, and who normally are able to save up for a few 'extras.' They live in small one-family houses fairly well kept, many two-family houses, and older, cheaper apartments."

CLASS D.   "Those who 'struggle along' with the bare necessities down to and including the 'reliefers.' These live in run-down one-family houses, poor two-family houses, and tenements."

Because this factor is (i) highly subjective, (ii) continuous rather than discrete in nature, it is difficult to obtain for a given set of households a consistent ranking by the several investigators or even by one from trial to trial. Furthermore, for the same reasons it is difficult to determine the correct sizes

*From instructions issued to interviewers by a polling organization using the "quota method." About 1947.

of the various quota classes in a given universe. Such information may be approximated by using the last U.S. population Census, but some quota classes are not available in the Census.

For example, in the case of socio economic classes, Blankenship (1943) gives the relative sizes of these classes for the United States as follows:

Socioeconomic class : A   B   C   D
Stated size (percent) : 10   30   40   20

In a national probability sample taken in the United States, the author asked the investigators to classify each household after the interview was completed into one of the four socioeconomic classes, following the rules set forth above. This survey, taken in January 1948, gave the following results:

Table 7.4.   Estimated from survey

Socioeconomic class	Total in U.S.	Percent and SE[a]	Stated by Blankenship (%)
A	1,529,000	4.1 ± 0.9	10
B	8,340,000	22.4 ± 2.7	30
C	18,974,000	51.1 ± 2.5	40
D	8,381,000	22.4 ± 2.2	20
Total	37,124,000	100.0	100

[a] Standard error of the percentage estimate expressed in percentage points.

The rather wide differences in this example between "suggested" sizes and "true sizes" can give rise to serious biases in the final estimates if the character under measurement differs greatly from "stratum to stratum." Although the quota method of sampling can be regarded as stratified random sampling in which relaxation has been permitted by not requiring (i) the strata sizes to be *known* and (ii) *positive* randomization in the selection of the elements within strata, the effects of these relaxations on the results are very difficult to determine. (See References.)

7.7.   Strata Design: Kind and Number of Strata: Single SU Strata

Information useful for the formation of strata is generally of two kinds: (1) that which is based on the *arrangement* of the elements in the universe, such as a listing structure either explicit or implicit, and (2) that based on some *knowledge* about individual elements themselves, such as on a variate X_i related in some manner with Y_i.

An example of the first type is the listing of the United States population by state, county, city, and minor civil division, such as township or precinct. This listing is useful for forming strata that are political or geographical units, and therefore it utilizes the principle of proximity to attain a lower within-strata variance. Another example of this type is a universe for which subdivisions are or can be shown on a map, such as a map showing major soils regions or medical care areas, or even just a plain map divided into arbitrary areas for strata.

For the second type we may use as an example a universe, say, of business establishments on each of which we have last year's volume of business and from which we wish to draw a sample to determine this year's volume. Here each element, a business establishment, has an X_i that may be sufficiently correlated with Y_i to justify its use for the formation of strata. Type of business, number of employees, and various kinds of other information may be available on each establishment, so that we may have several potential X_is.

Factors and Levels.   In some practical situations the sampler is confronted with several potential stratification "factors", each at a number of possible "levels." He may find he must choose among alternative factors and levels, because he may not be able to provide enough strata to accommodate them all. For example, in the case of the business establishments, the sampler may wish to use geographic location (at four levels), volume of last year's business (at five levels), number of employees (at five levels), and type of business (at five levels) simultaneously for strata, thus yielding a total of 500 ($=4 \times 5 \times 5 \times 5$) strata when the total sample is to be, say, only 200 establishments and, in order to have replication within strata, he would like no more than 100 strata and preferably, say, 10. He must drastically reduce either the number of factors or the number of levels, or both.

In this case some rough and simple rules for deciding on preference may be of help.

Rule 1.   In general, qualitative and nonmeasurable characteristics should be preferred over quantitative characteristics for use in stratification. Qualitative information is difficult to use anywhere except in stratification, whereas quantitative may be more fully utilized in the estimator (Chapter 5) or in selection probabilities (see Chapter 8).

Rule 2.   If the quantitative information, say X_i, is not related to Y_i in a simple manner (for example, linear), then it may be better to utilize it in stratification rather than in the estimator or selection probabilities.

Rule 3.   If more than one characteristic is being surveyed and each is roughly of equal importance, then it is better to forego use of quantitative information thought to be correlated with one or a few of the characteristics under measurement in deference to its use in ways other than for stratification.

If these rules are followed, the competition of various types of information for a place in the stratification scheme will be sharply reduced. In the example above, a good choice may be geographic at two levels and by type of business at five levels. (If there is still an "oversupply," a scheme called "multiple stratification," briefly described in Section 7.12, may be appropriate.) In fact, stratification may well be confined largely to area-geographic considerations and to such *classes* of qualitative information as "type," "ethnic," "status," "socioeconomic," "season," and "zone of residence" (e.g., rural versus urban).

Number of Strata.   Where "natural" classes are used, the number of strata may be governed by the number of classes, but in area and geographic stratification, or where quantitative information is used, the number of strata must be decided by the sampler. Since it is usually possible to form substrata within a given stratum that are more homogeneous than the original stratum, almost continuous reduction in $\bar{S}^2$ (the average within-stratum variance) can be obtained by simply increasing the number of strata. This is almost universally true of area and geographic strata and frequently true of strata based on some measure of size of element, X_i. One good rule is to try to make the strata equal in size, choosing the number of strata, L, to be some number less than or equal to the sample size, n. However, if unbiased estimates of sampling errors are desired, the number of strata may not exceed $n/2$, so that there will be two sample elements in each stratum with which to estimate the within stratum variance. Carrying stratification out to this extent should not be undertaken unless the gains appear to be worth the cost. Since it is usually possible to form strata rather cheaply, the formation of numerous small strata is usually worth while even if the gain is small. With such a scheme, however, the loss in degrees of freedom for estimating error may be great. In this case, some samplers prefer to accept some bias in estimating error to obtain a smaller error. This is achieved by grouping neighboring strata and regarding the group as a single stratum for purposes of estimating sampling variance.

Paper Strata.   If it is desirable to have equal-sized strata, even if it may mean a slight loss in efficiency, the following procedure may be helpful. Suppose we have a listing as presented in Table 7.5. The listing units (LUs) may be counties, cities, census tracts, or the like. The object is to form four equal-sized strata from this universe.

Since the universe contains 19,000 SUs, each stratum we construct must contain $19,000 \div 4$ or 4750 SUs. Suppose the order of the listing is such that contiguous LUs are most "alike" in the relevant characteristics. We can regard the cumulative total column as a numbering scheme or frame for the whole universe of 19,000 SUs, with stratum 1 consisting of SUs "numbered" 1 through 4750; stratum 2, those "numbered" 4751 through 9500, and so on. To identify stratum 3 we can regard it as consisting of SUs numbered 9501

Table 7.5.   Listing frame of counties of Parameter Valley

County (LU)	No. of SUs (N_h)	Cumulative SUs	Strata Limits on Cumulative total	Limits on Cumulative total	Stratum no.
A	10,000	10,000	4750	1– 4750	1
			9500	4751– 9500	2
				9501–10,000	
B	500	10,500		1– 500	
C	1500	12,000		1– 1500	3
D	3200	15,200	14,250	1– 2250	
				2251– 3200	
E	700	15,900		1– 700	
F	1200	17,100		1– 1200	4
G	1900	19,000		1– 1900	

through 10,000 of LU A, all of LUs B and C, and SUs numbered 1 through 2250 of LU D. To achieve maximum homogeneity within the "paper" strata, the numbering within LUs can be carried out so that the parts comprising each stratum are as alike as possible.

Paper strata are quite effective when the listing units are areas, political units, or some grouping based on a "continuous" variate (such as age, percent urban, percent white, gross sales, etc.). In these cases the dividing lines between strata will at best be set rather arbitrarily. When differences between LUs are discrete or sharp, this method is somewhat less efficient, and the anticipated gains in equality of stratum sizes must outweigh the apparent losses. Some of the advantages of equal-sized strata will be discussed in Section 14.5 under shortcut methods of estimating sampling variance.

Single SU Strata.   Some universes contain elements that are known to have very large Y_is relative to $\bar{Y}$, the parameter to be estimated. This knowledge is usually in the form of some X_is that are known from past history, by hearsay, or the like. In this case one may not wish to allow this element to appear in his sample by pure chance—the effect on the sample mean will be too pronounced, one way or the other. In this case the following rule, although rough, may be helpful.

Suppose we know the X_is, which we presume to be a reasonably trustworthy measure of size for the Y_is, and suppose a sample of n of the N elements is being considered, hence a sampling rate of n/N. Suppose also that we have plenty of flexibility in grouping the N elements into any number of strata, and we wish to have $L = n$ and therefore all $n_h = 1$—that is, one SU to be drawn from each stratum. Let us group elements into n groups of about equal aggregations of X_i. Hence $X_h \doteq n\bar{X}/N$. If any element has an $X_i \geq n\bar{X}/N$, it is a stratum unto itself. Its inclusion in the sample is a certainty. Hence these may be called "certainty SUs."

The theory behind this rule is as follows. If the coefficients of variation of Y_i within strata are constant over strata—that is,

$$(7.30) \qquad V_h = \frac{S_h}{\bar{Y}_h} = k$$

—then optimal allocation, ignoring FPC, is given from Eq. 7.13, substituting Eq. 7.30,

$$\text{opt } (n_h) = n \left(\frac{N_h S_h}{\sum N_h S_h} \right)$$

$$(7.31) \qquad = h \left(\frac{k N_h \bar{Y}_h}{k \sum H_h \bar{Y}_h} \right) = n \left(\frac{Y_h}{Y} \right) ;$$

hence the best allocation to strata is that proportional to the amount of Y_h in each, or in place of unknown Y_h we use the known X_h. Since X_h are to be made equal, $X_h = X/n$ and opt $(n_h) = 1$. If the assumption, $V_h = k$, holds and if R_{xy} is reasonably large, any $X_i \geq Xn/N$ should be made a single SU, or "self-representing" stratum.

7.8. Poststratification

Suppose a random sample is drawn and, after the observations are taken, we decide to regard the sample as a stratified-random one, where Eq. 7.3 is used as the estimator of $\bar{Y}$:

$$(7.32) \qquad \bar{y}_{\text{post}} = \sum \left(\frac{N_h}{N} \right) \bar{y}_h.$$

It is assumed, of course, that each stratum has at least one sample element in it, so that a $\bar{y}_h$ exists for each. The variance of $\bar{y}_{\text{post}}$ is not the same as for prestratification, because now the n_h are not fixed but are variables. Hence the sample mean for the hth stratum,

$$(7.33) \qquad \bar{y}_h = \left(\frac{1}{n_h} \right) \sum_i^{n_h} Y_{hi} = \frac{\sum Y_{hi}}{n_h} ,$$

is actually a ratio of two random variables. The appropriate variance of $\bar{y}_{\text{post}}$, Eq. 7.32, is given by the approximation (from Hansen et al., 1953)

$$(7.34) \qquad \text{var } (\bar{y}_{\text{post}}) \doteq \frac{N-n}{Nn} \left\{ \bar{S}^2 \left[1 + \frac{1}{\bar{n}} \left(\frac{L-1}{L} \right) \right] \right\},$$

where $\bar{n} = \dfrac{n}{L}$

and, if $\bar{n}$ is large,

(7.34a)
$$\operatorname{var}(\bar{y}_{\text{post}}) \doteq \left(\frac{N-n}{N}\right)\frac{\bar{S}^2}{n},$$

which means that poststratification tends in this case to become the same as that of simple prestratification. In fact the efficiency of post-relative to pre-stratification is given by

(7.35)
$$\operatorname{RE}(\text{post/pre}) \doteq \frac{\bar{n}}{\bar{n}+(L-1)/L}.$$

Example 7.3. Suppose a random sample of 100 was drawn from a universe. After the survey was completed it was discovered that information was available for the formation of two strata, and this information was in fact used in the analysis. (a) What loss was taken because the discovery was not made sooner? (b) What loss if information permitting 25 strata becomes available?

(a) $L = 2$, $n = 100$, therefore $\bar{n} = 50$.

$$\operatorname{RE}(\text{post/pre}) = \frac{50}{50 + \dfrac{2-1}{2}} = 0.99,$$

or a loss of only 1% approximately.
(b) $L = 25$, $n = 100$, therefore $\bar{n} = 4$.

$$\operatorname{RE}(\text{post/pre}) = \frac{4}{4 + \dfrac{24}{25}} = 0.806,$$

or a loss of about 19% approximately.

Notice in the example that when the number of strata is small, or what is the same, when $\bar{n}$ is large, there is little difference in efficiency between stratification in advance or after sampling. When strata are numerous relative to sample size, however, the loss can be important.

It is interesting to note that an important part of the gain in efficiency due to stratification is due to the estimator (poststratification). Further gains due to allocation (simple pre- rather than poststratification) are frequently modest relative to the gains from the estimator effect. Unless special conditions exist, such as large differential costs, large variance differences, or large number of strata, stratification can almost be ignored except for the use of the stratified estimator.

Poststratification may be used where the identification of the stratum cannot be done in advance of the survey. For example, the ages and education of persons in a general population survey may not be known when a sample is drawn, but the sizes of such strata may be found from U.S. Census reports.

7.9. Estimation of $\bar{Y}/\bar{X}$ and $\bar{Y}$ with Help of x in Stratification

Up to this point in stratification we have considered the simple unbiased estimates of the mean and total of a single population. In the preceding section a slightly more complex case of the same problem was looked at, where stratification is in a sense imposed on a random sample through the estimator. Now we shall go into the matter of estimating a ratio of the means (or totals) of two populations, such as $\bar{Y}/\bar{X}$ say, and also estimating $\bar{Y}$ with the help of x where the sample is stratified. The extension of the principles to the estimation of other parameters such as B and R (regression and correlation coefficients) will be briefly mentioned or suggested by the procedure developed.

Estimation of $\bar{Y}$ by Ratio.   To estimate y with the help of x, two estimators are in use. One employs a separate ratio for each stratum, the other a combined ratio over all strata. The first, the *separate* ratio estimator, is given by

$$\bar{y}_{\text{strat-rat(s)}} = \frac{\hat{Y}_{\text{strat-rat(s)}}}{N}$$

(7.36)
$$= \frac{\sum X_h(\bar{y}_n/\bar{x}_h)}{N}$$

It has a bias given approximately by

(7.37) $\text{bias}\,(\bar{y}_{\text{strat-rat(s)}}) \doteq \dfrac{1}{N}\sum\left(\dfrac{1-f_h}{n_h\bar{X}_h^2}\right)X_h[(\bar{Y}/\bar{X})_h S_{xh}^2 - R_h S_{yh} S_{xh}],$

which is a weighted mean of the biases associated with the ratios in each stratum. If the n_h are small and the biases are all in the same direction, this weighted mean bias may be troublesome, since it depends so heavily on the individual n_h. Here $f_h = n_h/N_h$, the fraction sampled in stratum h.

The variance of Eq. 7.36 is approximately

(7.38) $\text{var}\,(\bar{y}_{\text{strat-rat(s)}}) = \left(\dfrac{1}{N^2}\right)\sum N_h^2\left(\dfrac{1-f_h}{n_h}\right)S_{dh}^2,$

where S_d^2 is given by Eq. 5.11 and S_{dh}^2 is the variance of the D_is in stratum h; that is, $D_{hi} = Y_{hi} - (\bar{Y}_h/\bar{X}_h)X_{hi}$.

An alternative to Eq. 7.36 utilizes a single ratio for the whole universe. This *combined* ratio estimator is defined as

(7.39)
$$\bar{y}_{\text{strat-rat(c)}} = \frac{\hat{Y}_{\text{strat-rat(c)}}}{N}$$

$$= \frac{X}{N} \left(\frac{\sum N_h \bar{y}}{\sum N_h \bar{x}} \right) = \bar{X} \, \frac{\bar{y}_{\text{strat}}}{\bar{x}_{\text{strat}}}.$$

Its bias is given approximately by

(7.40) $$\text{bias}\,(\bar{y}_{\text{strat-rat(c)}}) \doteq \left(\frac{1-f}{n\bar{X}^2} \right) [(\bar{Y}/\bar{X})\bar{S}_{xh}^2 - \bar{R}_h \bar{S}_{yh} \bar{S}_{xh}],$$

where $\bar{S}_{xh}^2$, $\bar{S}_{yh}^2$, and $\bar{R}_h$ are the average within-stratum variances of x and y, and $\bar{R}_h$ is the average within-stratum correlation (see Table 7.6). Note that this bias depends heavily on n rather than on individual n_h and, since the term in variances is not likely to be much larger than the analogous term in Eq. 7.37, the bias is usually much smaller than that for the individual ratio estimator with the same n and n_h.

The variance of this estimator is approximately

(7.41) $$\text{var}\,(\bar{y}_{\text{strat-rat(c)}}) \doteq \left(\frac{1}{N^2} \right) \sum N_h^2 \left(\frac{1-f_h}{n_h} \right) S_{d'h}^2,$$

where $S_{d'h}^2$ is the variance of the d'-population in stratum h, in which $D_{hi} = Y_{hi} - (\bar{Y}/\bar{X})X_{hi}$. Note that the d'-population employs the overall ratio $\bar{Y}/\bar{X}$ for all Y_{hi}, whereas the d-population is calculated around separate ratios, $\bar{Y}_h/\bar{X}_h$, for each stratum. One would normally expect $S_{d'h}^2 > S_{dh}^2$ and therefore the separate ratio estimator to be more precise than that of the combined ratio. Except for the danger of bias and the added computational burden, the separate ratio would always be preferred.

Estimation of $\bar{Y}$ by Regression. The regression estimator can also be extended to stratified samples, and it will also have both separate and combined versions. For the separate version we have, following Eq. 5.23,

(7.42) $$\bar{y}_{\text{regr(s)}} = \frac{1}{N} \sum N_h [\bar{y}_h + b_h (\bar{X}_h - \bar{x}_h)],$$

where b_h is the estimated regression coefficient for stratum h. It has a bias given approximately by

(7.43) $$\text{bias}\,(\bar{y}_{\text{regr(s)}}) \doteq \sum \frac{N_h}{N} [\text{bias}\,(\bar{y}_{\text{regr}})_h],$$

where the structure of the term in brackets is given by Eq. 5.25 applied to stratum h. Since that term includes the factor $1/(n_h - 1)$, it is apparent that this estimator, like the separate ratio estimator, is sensitive to bias when the n_h are small.

The combined version is given by

$$(7.44) \qquad \bar{y}_{\text{regr}(c)} = \bar{y}_{\text{strat}} + \bar{b}(\bar{X} - \bar{x}_{\text{strat}}),$$

where $\bar{b}$ is the weighted within-stratum regression coefficient, given as

$$(7.45) \qquad \bar{b} = \frac{\sum W_h^2 s_{xyh}/n_h}{\sum W_h^2 s_{xh}^2/n_h},$$

and when stratification is simple, $\bar{b}$ simplifies to the pooled or self-weighted

$$(7.45a) \qquad \bar{b} = \frac{\bar{s}_{xy}}{\bar{s}_x^2}$$

obtained from the ordinary ANOCOVA computations. (See Table 7.6.)

The variances of the regression estimators, both separate and combined versions, ignoring FPC, are given by

$$(7.46) \qquad \bar{var}\,(\bar{y}_{\text{regr}(s)}) \doteq \frac{1}{N^2} \sum \left[\frac{N_h^2 S_{yh}^2}{n_h} (1 - R_h^2) \right]$$

and

$$(7.46a) \qquad \bar{var}\,(\bar{y}_{\text{regr}(c)}) \doteq \frac{1}{N^2} \sum \left[\frac{N_h^2 S_{yh}^2}{n_h} (1 - \bar{R}^2) \right],$$

where $\bar{R}$ is the pooled within-stratum correlation coefficient defined as

$$(7.47) \qquad \bar{R} = \frac{\sum (N_h - 1)\,(S_{xy})_h}{[\sum (N_h - 1)S_{yh}^2 \sum (N_h - 1)S_{xh}^2]^{1/2}}.$$

In general since ratios, $\bar{Y}_h/\bar{X}_h$, regressions, B_h, and correlation coefficients, R_h, in real populations are likely to vary from stratum to stratum, one would expect the separate to be more precise than the combined version of these two estimators. This is generally true, but it may be recalled that both the ratio and regression estimators are biased, and for small samples this bias may be a large portion of the MSE. Since this bias depends on n_h rather than n, and since the bias may be of the same sign over the strata, it is likely to be fairly large for small n_h even if n is quite large. Hence, if n_h are small, one should be cautious in choosing the version to be used.

Estimation of $\bar{Y}/\bar{X}$ ***by Ratio.***   An estimation of this ratio from a stratified random sample is given by

$$(7.48) \qquad w_{\text{strat}-\text{rat}} = \frac{\hat{Y}_{\text{strat}}}{\hat{X}_{\text{strat}}} = \frac{\sum N_h \bar{y}_h}{\sum N_h \bar{x}_h} = \frac{\bar{y}_{\text{strat}}}{\bar{x}_{\text{strat}}}.$$

This estimator is not unbiased, since it is a ratio of two random variables. However, the bias goes down with aggregate sample size, n; hence for reasonable sizes of n it is usually of little concern. (See discussion above on the combined ratio estimator.) The variance of Eq. 7.48 is given by

$$(7.49) \qquad \text{var}\,(w_{\text{strat-rat}}) = \frac{1}{\bar{X}^2} \sum \left(\frac{N_h}{N}\right)^2 \left(\frac{1-f_h}{n_h}\right) S_{d'h}^2,$$

where

$$S_{d'h}^2 = \frac{\sum_h^L \sum_i^{N_h} (D_{hi}')^2}{N-1}$$

and

$$D_{hi}' = Y_{hi} - (\bar{Y}/\bar{X})X_i.$$

In terms of ys and xs, Eq. 7.49 can be expressed as

$$(7.50) \;\; \text{var}\,(w_{\text{strat-rat}}) = \sum \frac{N_h^2}{N^2\bar{X}^2} \left(\frac{1-f_h}{n_h}\right) \left[S_{yh}^2 + (\bar{Y}/\bar{X})_h^2 S_{xh}^2 - 2(\bar{Y}/\bar{X})_h R_h S_{xh} S_{yh}\right]$$

to show its relationship to that given in Chapter 5, such as Eq. 5.8.

Simple Stratification; ANOVA.   In the simple stratification case, that is when $n_n \propto N_h$, some of the foregoing formulas can be simplified a bit and, for those who like an ANOVA format, the computational procedure can be somewhat standardized. In Eq. 7.18 we defined

$$\bar{S}^2 = \sum \left(\frac{N_h}{N}\right) S_h^2 \doteq \text{MS}\,(W),$$

where MS (W) is the within mean square in an ANOVA table for a universe. Similarly, we can substitute $\bar{S}_x^2$ for $\sum (N_h/N) S_{xh}^2$, $\bar{S}_{xy}^2$ for $\sum (N_h/N) S_{yh}^2$, and $\bar{S}_{xy}$ for $\sum (N_h/N) S_{xyh}$, the latter being a cross-product term involving both x and y that is analogous to the squared deviations.

With these weighted variances we can simplify Eq. 7.50, whence the variance of $w_{\text{strat-rat}}$ under simple stratification is approximately

$$(7.51) \qquad \text{var}\,(w_{\text{strat-rat}}) = \frac{1-f}{n\bar{X}^2} \left[\bar{S}_y^2 + (\bar{Y}/\bar{X})^2 \bar{S}_x^2 - 2\,(\bar{Y}/\bar{X})\,\bar{R}\bar{S}_x\bar{S}_y\right],$$

where $\bar{R} = \bar{S}_{xy}/\bar{S}_x\bar{S}_y$, the average within-stratum correlation coefficient.

The computations can be carried out following standard procedures for ANOVA and ANOCOVA (analysis of covariance) and the results conveniently presented in a format such as Table 7.6. The parameters obtained in this way can be used for various formulas in this chapter.

If data are available on a simple stratified sample, the ANOCOVA computations based thereon will result in a table identical to Table 7.6., except that lower-case symbols replace the capitals and $\bar{s}_y^2$, $\bar{s}_{xy}$, and $\bar{s}_x^2$ will be essentially unbiased estimators of $\bar{S}_y^2$, $\bar{S}_{xy}$ and $\bar{S}_x^2$, respectively.

Table 7.6. *Analysis of convariance* (*ANOCOVA*) *of a stratified universe of N elements* (*on a per element basis*)

Source of variation	df	General mean square (or cross-product) designation	Mean square (or cross-product) exactly (or approximately) defined as		
			y	xy	x
Gross	N	—			
Mean	1	—			
Total	$N-1$	MS (T), MCP (T)	S_y^2	S_{xy}	S_x^2
Strata	$L-1$	MS (B), MCP (B)	—	—	—
Elements/strata	$N-L$	MS (W), MCP (W)	$\bar{S}_y^2$	$\bar{S}_{xy}$	$\bar{S}_x^2$

Optimal Allocation for Ratios and Regressions. The general principles involved in determining the optimal way to allocate n_h to the L strata developed for the one-variate case will also apply for the ratio and regression cases, but both theory and application become more complex. In the case of the separate estimators one can, if the samples are large in each stratum, allocate the sample by the usual formula.

$$(7.52) \qquad n_h \propto \frac{N_h S_{dh}}{\sqrt{c_h}},$$

where S_{dh} is the standard deviation of the differences in stratum h, in which, as in Eq. 5.9 and Eq. 5.38, $D_{hi} = Y_{hi} - (\bar{Y}_h/\bar{X}_h)X_{hi}$ for the ratio case or $D_{hi} = Y_{hi} - B_0 X_{hi}$ for the regression case. Usually these quantities must be estimated from samples of some sort (unless arbitrary values such as used in the difference estimator are employed).

7.10. Stratification for Several Populations

When several populations are to be dealt with in a single survey, the appropriate policy for stratification becomes somewhat difficult. If the populations are related such that they all share approximately the same optimum allocation, then we have no new problems, since any one will suffice for computing the allocation for all. The population chosen might appear to be the more basic one for the group or perhaps the one for which greatest precision is desired. If no basic one is evident or all require the same precision, then the optima for several or all can be worked out and a simple average obtained for the group.

However, when the populations are unrelated and about equal precision is required for each, the allocation problem becomes tougher. Here are several suggestions:

1. Use a random sample and then choose poststratified estimators appropriate for each population dealt with. With this scheme a different X_i may be used with each population.

2. Use simple stratification based on area and geography and then use specific estimators (ratio, regression, etc.) for each population.

3. Split the sample into two parts with n_c units to be used for the "core" of the overall sample and n_s units to be used for supplements. For the n_c units in the core use either random or simple stratification based on such general information as area or geography. Special stratification is carried out for each of the several populations under study, and the supplemental sample is used to fill out the more serious deficiencies that show up when the n_h of the core sample are observed in the special strata. (This scheme was used in a national farm survey, where in addition to general farm information common to all farms certain information was desired from farms specializing in peanuts, potatoes, sugar beets, and so on. Such farms are concentrated in certain areas of the United States and therefore require considerably different allocations from each other and from the corn sample.)

4. Where two-way stratification is used (see Chapter 11), rows, columns, files, and so on can be used to designate the "strata" to be supplemented. A supplementation of sample selection in these rows, and so on, can be used to increase the accuracy of the desired subgroups—that is, subuniverses.

7.11. Stratification versus Regionalization; Collective versus Individual Estimates

In this chapter the term strata has been used to designate a set of partitions of a universe for the purpose of improving the precision of samples taken to estimate some overall population parameter such as the mean. In this context the function of the strata was to increase sampling efficiency. Other functions of partitioning may have little or nothing to do with sampling efficiency. For example, we wish to divide a universe into regions that make some sort of analytical sense and where each region is to have its own estimate with, let us suppose, equal precision. Or it may be that after a survey is taken, we wish to separate out different classes of elements on the basis of the value of some characteristic measured in the survey, such as age groups, or sex, or size of farm. It may be of interest to obtain separate estimates of the characteristics of these groups and to determine their precision so that we can compare them with some statistical confidence. Surveys with such objectives are sometimes called *comparative* surveys.

In these cases it may be confusing to call these groups strata. They were not designed for sampling purposes, although in some cases they could have been. Hence, to distinguish their different purpose, we should use a new label. The general term for a group used for analysis is *domain of study*. The common term for those groups where each is to stand on its own is *region* or *subuniverse*. In some cases a group will play the role of all three: of stratum, subuniverse, and domain.

Allocating Samples to Subuniverses.   The problem of sample allocation will depend on circumstances. Suppose our groups of interest are subuniverses, such as two regions, and we know N_h and S_h for each. If each is to have estimates of $\bar{Y}_h$ with equal accuracy, how should a sample of n be allocated?

In this case, since we wish var $(\bar{y}_1) =$ var $(\bar{y}_2)$,

$$(7.53) \qquad \left(\frac{N_1 - n_1}{N_1}\right)\frac{S_1^2}{n_1} = \left(\frac{N_2 - n_2}{N_2}\right)\frac{S_2^2}{n_2}.$$

Ignoring FPCs, then

$$(7.54) \qquad n_1 = n\left(\frac{S_1^2}{S_1^2 + S_2^2}\right):$$

hence allocation should be proportional to the population variances in the regions regardless of their sizes. If the variances are equal, then so should be sample sizes. Generalization of this result to L subuniverses with the precision requirements SE $(\bar{y}_h)$ for the hth subuniverse will be left to the reader.

Allocating Samples to Domains.   In some studies the object is not to estimate $\bar{Y}$ but to estimate the differences between the means of several groups. Suppose we wish to determine which group of students in State University spend the most money, men or women, and that there are three times as many men as women students. Assuming that S_h and c_h are the same for both groups, we would be inclined to allocate a stratified sample to these groups proportional to N_h—that is, to take three men students for every woman student. However, we wish to obtain a good estimate of $\bar{Y}_m - \bar{Y}_w$ rather than $\bar{Y}$. The variance of the difference between two means of independently drawn samples is (see Eq. 5.4)

$$(7.55) \qquad \text{var } (\bar{y}_1 - \bar{y}_2) = \left(\frac{N_1 - n_1}{N_1}\right)\left(\frac{S_1^2}{n_1}\right) + \left(\frac{N_2 - n_2}{N_2}\right)\left(\frac{S_2^2}{n_2}\right).$$

Assuming that the FPC can be ignored and that costs are the same in each group, var $(\bar{y}_1 - \bar{y}_2)$ will be minimized when

$$(7.56) \qquad n_1 = \frac{nS_1}{S_1 + S_2} \quad \text{and} \quad n_2 = \frac{nS_2}{S_1 + S_2}$$

—that is, when allocation is proportional to the population standard deviations. Where $S_1 = S_2$, the sample should be the same size in both strata, a result quite different from that of optimum stratification.

In the case where domain cannot be identified prior to sampling, the sample can be drawn with the hope that n_h will be sufficiently large in the postdetermined domains to permit $\bar{y}_h$ to be estimated with reliability adequate for the purposes at hand. It should be noted that the n_h are random variables, and therefore comparisons of any two postdomain means will have a variance somewhat larger than if the n_h were fixed. In Chapters 12 and 14 more will be given on this case.

7.12. Remarks

More elaborate and special types of stratification have been developed and may be used for problems suited to them. In Chapter 10 a method will be described where the sizes of strata will be estimated from a portion of the sampling resources. This method, a type of *double sampling*, makes it possible to use stratification where information for forming strata is not available.

Where it is desired to stratify a sample on more criteria than the size of sample will permit using the methods described in this chapter, methods of *multiple stratification* are available; see Chapter 11.

7.13. Summary

Stratification is a useful design device for sampling. It is a method of using prior knowledge of a universe and its populations without risk of loss. The gains are usually modest but in some cases may be quite large. The costs of carrying out stratification are usually quite low, and schemes can be adopted that keep the sample self-weighting. Therefore, the analysis can be kept simple.

Quota sampling is a type of stratified sampling in which selection of elements within the quota strata is sometimes quite nonrandom. The savings in costs due to this relaxation may not be as large as expected, because filling the final portions of the quotas properly may require costly searching— particularly if the quotas have much cross-classification in an attempt to achieve stratification efficiency.

The technique of stratification is only one of a number of related restrictions that can be imposed on the selection of sampling units. Later chapters discuss these related techniques, and we shall see they greatly enlarge the number of possibilities that should be considered by the sampler in arriving at a decision on best design for his problem. The whole group of techniques will be presented as alternative methods of utilizing available information about the universe for increasing the efficiency of samples. From this point of view these techniques are uses of prior information alternative to those such as (1) "equalizing" sampling unit sizes (Chapter 6) and (2) recovery through an estimator (Chapter 5) and alteration of the probabilities with which SUs are selected (Chapter 8).

7.14. Review Illustrations

1.   During World War II (March 1945) a survey was carried out to determine the number of new truck and bus tires in the inventories of U.S. dealers, in order to arrive at appropriate allocation policy (Deming and Simmons, *Journal of the American Statistical Association*, **41**: 16–33, 1946). From previous contacts a good deal was known about the universe of dealers. This enabled the samplers to classify the dealers into strata according to the inventories reported earlier. Also, they could estimate the standard deviations in the several strata with fair accuracy. Owing to wartime regulations, practically 100% returns could be obtained from mailed questionnaires. The prior knowledge available about the dealers included the following:

Stratum, h	No. of dealers, N_h	Estimated variance, S_h^2
1	19,850	34.8
2	3250	92.2
3	1007	174.2
4	606	320.4
	24,713	

For a sample of 4170, determine the best allocation among the four strata for estimating total tire stock in the United States.

Solution

Stratum:	1	2	3	4	$\sum$
N_h:	19,850	3250	1007	606	24,713
$\hat{S}_h$:	5.90	9.60	13.20	17.90	
$N_h S_h$:	117,115	31,200	13,292	10,847	172,454
n_h:	2832	754	321	262	4169

Although based on procedures for estimating $\bar{Y}$, the same results hold for estimating Y.

2.   Here is a synthetic microuniverse of four groups of four elements each,

with the 16 observations:

$$\frac{1}{9,8,6,7}, \quad \frac{2}{9,5,7,3}, \quad \frac{3}{10,1,7,4}, \quad \frac{4}{14,6,10,18}.$$

(a) Summarize the relevant results in an ANOVA table. Does it appear that stratification by group would be effective?

(b) Regarding the groups as strata, compute var ($\bar{y}$) for stratified random samples of four elements (one from each stratum).

(c) Compute the efficiency of (b) relative to simple random sampling (RNR).

Solution

(a) ANOVA:

Source	df	SS	MS
Gross	16	1216	
Mean	1	961	
Total	15	255	17
Group	3	105	35
Elements/group	12	150	12.5

Yes, stratification should be effective, since MS $(W) <$ MS (T).

(b) var $(\bar{y}_{\text{strat}}) = \left(\dfrac{M_0 - m_0}{M_0}\right)\left(\dfrac{\text{MS }(W)}{m}\right)$

$$= \frac{4-1}{4}\left(\frac{12.5}{4}\right) = \left(\frac{3}{4}\right)(3.125) = 2.34375.$$

(c) RE (strat/ran) $= \dfrac{\text{MS }(T)}{\text{MS }(W)} = \dfrac{17}{12.5} = 1.36.$

3. Some phenomena are associated with day of the week, day of the month, or both (for example, business activity). Suppose it is possible to select a sample of 52 days out of the 365 in a year and it is known that the characteristic being observed (say, sales) on Fridays is about five times that on non-Fridays, on the first of each month is about 10 times that of other days, and on Fridays that are also first days of the month about 15 times that of other days.

(a) Suggest a set of strata that you believe would be suitable for this problem. Do you think your strata would be better than, say, the 52 weeks? Why?

(b) Suggest an allocation policy for your strata and determine the allocations you recommend.

Solution

(a) It appears that at least these four strata should be formed:

Strata:	Friday Firsts	Non-Friday Firsts	Nonfirst Fridays	Other Days
N_h:	2	10	50	303

(The number of days that are both Friday and first-of-the-month will depend on the year.) This should be better than 52 weeks as strata, unless weeks differ greatly. (If this is the case, then the 303 days should be put into groups of contiguous or near-contiguous days to form additional strata for controlling this possible source of variation.)

(b) If we wish to assume that the coefficient of variation, V_h, is fairly constant over the strata, then we should allocate $n_h \propto Y_n$. In this case,

Stratum, h	N_h	Size factor	Index on Y_h	Unadjusted n_h	Adjusted n_h
Friday firsts	2	15	30	2.3	2
Non-Friday firsts	10	10	100	7.6	8
Nonfirst Fridays	50	5	250	19.0	19
Other days	303	1	303	23.1	23
Σ	365		683	52.0	52

Note that sampling is 100% in the first stratum, 80% in the second, about 40% in the third, and 8% in the fourth. If our assumption on V_h is reasonably correct, this should be a near optimal allocation (except for adjustments needed for the FPC).

4. A universe contains the following data:

Stratum, h	N_h	S_h	$\bar{Y}_h$
A	10,000	10	10
B	5,000	5	5

(a) Compute S of the combined population.

(b) What SE $(\bar{y})$ is expected for random samples of 150 drawn from the combined population?

(c) What SE $(\bar{y})$ is expected for a stratified sample of 150 drawn proportional to stratum size?

(d) What SE $(\bar{y})$ is expected for a stratified sample of 150 of which half are drawn from each stratum?

(e) What SE $(\bar{y}_B - \bar{y}_A)$ is expected of the difference between the observed means of the two strata if a sample of 150 is allocated as that of (c)?

(f) Of (d)?

Solution

(a) From Eq. 7.24

$$S^2 = \sum \left(\frac{N_h}{N}\right) S_h^2 + \sum \left(\frac{N_h}{N}\right)(\bar{Y}_h - \bar{Y})^2$$

and, since $\bar{Y} = \sum \left(\frac{N_h}{N}\right) \bar{Y}_h$, then

$$\frac{N_A}{N} = \frac{10,000}{15,000} = 0.67, \qquad \frac{N_B}{N} = \frac{5000}{15,000} = 0.33,$$

and
$$\bar{Y} = (0.67)\,(10) + (0.33)\,(5) = 6.70 + 1.65 = 8.35,$$
$$S^2 = [(0.67)\,(10)^2 + (0.33)\,(5)^2]$$
$$\qquad + [(0.67)\,(10 - 8.35)^2 + ((0.33)\,(5 - 8.35)^2]$$
$$\qquad = [75.25] + [5.52] = 80.77,$$

and
$$S = \sqrt{80.77} = 8.99.$$

(b) $\mathrm{SE}\,(\bar{y}_{\mathrm{ran}}) = \left(\frac{N-n}{N}\right)^{1/2} \left(\frac{S}{\sqrt{n}}\right) \doteq \frac{8.99}{\sqrt{150}} = \frac{8.99}{12.25} = 0.734.$

(c) Average within-stratum variance is given by $\sum (N_h/N)\, S_h^2 = 75.25$.

$$\mathrm{SE}\,(\bar{y}_{\mathrm{strat}}) = \left(\frac{N-n}{N}\right)^{1/2} \left(\frac{\bar{S}}{\sqrt{n}}\right) \doteq \frac{\sqrt{75.25}}{\sqrt{150}} = \frac{8.67}{12.25} = 0.708.$$

(d) $\mathrm{SE}\,(\bar{y}_{75/75}) = \left[\left(\frac{N_A}{N}\right)^2 \left(\frac{S_A^2}{n_A}\right) + \left(\frac{N_B}{N}\right)^2 \left(\frac{S_B^2}{n_B}\right)\right]^{1/2}$

$$\qquad = \left[(0.67)^2 \left(\frac{10^2}{75}\right) + (0.33)^2 \left(\frac{5^2}{75}\right)\right]^{1/2} = 0.797.$$

(e) $\text{SE } (\bar{y}_B - \bar{y}_A) = \sqrt{\dfrac{S_A^2}{n_A} + \dfrac{S_B^2}{n_B}} = \left(\dfrac{10^2}{100} + \dfrac{5^2}{50}\right)^{1/2} = (1 + 0.5)^{1/2} = 1.225.$

(f) $\text{SE } (\bar{y}_B - \bar{y}_A) = \sqrt{\dfrac{10^2}{75} + \dfrac{5^2}{75}} = (1.67)^{1/2} = 1.29.$

5.   The U.S. National Health Survey, begun in 1957, was designed originally with 42 subuniverses and each was allocated about the same size of sample (i.e., $n_h = $ constant). Using the following data on the sizes of the subuniverses (in 1950) determine the loss in accuracy suffered by the national estimates (because allocation was $n_h = $ constant rather than $n_h \propto N_h$) in order to obtain subuniverse estimates of equal accuracy. [Hint: Assume $S_h^2 = \bar{S}^2$.]

N_h (in millions of persons)

h	N_h	h	N_h	h	N_h	h	N_h
1	3.7	11	2.7	21	5.9	31	3.8
2	1.6	12	4.0	22	3.4	32	1.7
3	1.6	13	5.1	23	1.8	33	2.3
4	6.1	14	3.1	24	6.3	34	2.4
5	2.2	15	5.8	25	5.4	35	12.9
6	3.0	16	1.8	26	3.3	36	3.7
7	6.6	17	1.1	27	5.8	37	2.2
8	1.9	18	3.1	28	1.5	38	3.0
9	2.8	19	4.0	29	1.5	39	5.5
10	3.9	20	2.5	30	2.0	40	2.8
						41	4.4
						42	2.2

Total for U.S.: $\sum N_h = 150.4$, $\sum N_h^2 = 725.04$.

Solution

Ignoring PFCs and using Eq. 7.9 for the variance of each estimate:

$$\text{RE(sub/strat)} = \frac{\sum (N_h/N)^2 \, \bar{S}^2/n_h}{\sum (N_h/N)^2 \, \bar{S}/n_h} \, ;$$

but for strat, $n_h = nN_h/N$, and for sub, $n_h = \bar{n} = n/L$, then

$$\text{RE(sub/strat)} = \frac{\sum (N_h/N)^2/(nN_h/N)}{\sum (N_h/N)^2/(n/L)} = \frac{1/L}{(1/N)^2 \sum N_h^2}$$

$$= \frac{1/42}{725.04/(150.4)^2}$$

$$= 0.742 \text{ or } 26 \text{ percent loss.}$$

6.   Show that if the coefficients of variation of y are constant over strata, then optimal allocation will occur when n_h are made proportional to Y_h.

Solution

Optimal n_h, ignoring costs, is given by Eq. 7.13;

$$n_h = n\left(\frac{N_h S_h}{\sum N_h S_h}\right).$$

But $V_h = S_h/\bar{Y}_h = k$; hence $S_h = k\bar{Y}_h$, and

$$n_h = n\left(\frac{kN_h\bar{Y}_n}{\sum kN_h\bar{Y}_h}\right) = n\left(\frac{Y_h}{Y}\right)$$

or $n_h \propto Y_h$.

7.   An investigation on sources of Iowa farmers' information on agricultural subjects was conducted in 1951 on a stratified random sample of the 71,040 area SUs in the "open country" portion of the state. The stratification was accomplished by grouping geographically contiguous units to form 120 equal-sized strata (i.e., each stratum containing 592 SUs). The area sampling units were equalized on "dot counts" (see Section 6.7). Two sampling units were selected at random from each stratum. The data on numbers of households and radios on each sampling unit are presented below.

(a) Estimate the total number of households in the universe.

(b) Estimate the total number of radios in households in the universe.

(c) Estimate the average number of radios per household (1) using a "combined ratio" estimator, (2) using the individual stratum ratios (mean of strata ratios), (3) using the simple mean of the radio/HH ratios, $\bar{r}$.

(d) Estimate the variances of (a), (b), and (c1) above, and construct 95 percent confidence intervals for the quantities being estimated.

(e) Estimate the gain in efficiency due to stratification for estimating (1) total number of households, (2) total number of radios.

(f) Compute the sample coefficient of variation on the "size" of the sampling units measured in terms of households. Do you conclude that the attempt to equalize sampling unit size by dot count information was successful?

(g) A state agency reports the number of "farm households" as 241,000 for a comparable date. Compare this figure with results in (a) and (d) above and comment.

Column 1, stratum number, h Column 3, number of HHs per segment, M_{hi}
Column 2, segment number, h_i Column 4, number of radios per segment, Y_{hi}

h (1)	h_i (2)	M_{hi} (3)	Y_{hi} (4)	h (1)	h_i (2)	M_{hi} (3)	Y_{hi} (4)	h (1)	h_i (2)	M_{hi} (3)	Y_{hi} (4)
1	1	3	4	18	1	4	4	35	1	3	3
	2	4	6		2	3	3		2	3	4
2	1	2	2	19	1	3	4	36	1	3	5
	2	2	2		2	1	1		2	3	3
3	1	2	2	20	1	3	3	37	1	4	6
	2	3	2		2	3	4		2	25	49
4	1	3	4	21	1	2	2	38	1	3	4
	2	3	4		2	2	2		2	2	4
5	1	3	4	22	1	2	3	39	1	3	3
	2	3	5		2	2	2		2	2	3
6	1	3	3	23	1	3	2	40	1	9	17
	2	3	4		2	3	4		2	3	2
7	1	2	3	24	1	3	6	41	1	3	4
	2	3	3		2	3	4		2	3	4
8	1	3	3	25	1	3	5	42	1	3	4
	2	3	4		2	3	3		2	4	6
9	1	4	6	26	1	2	2	43	1	4	5
	2	7	7		2	4	6		2	3	3
10	1	3	3	27	1	2	3	44	1	3	3
	2	3	5		2	3	5		2	3	4
11	1	4	7	28	1	3	4	45	1	1	1
	2	3	5		2	3	4		2	3	4
12	1	3	5	29	1	3	6	46	1	3	3
	2	3	3		2	3	3		2	2	3
13	1	3	6	30	1	3	5	47	1	3	3
	2	3	7		2	3	3		2	3	5
14	1	2	1	31	1	4	7	48	1	2	5
	2	2	4		2	3	4		2	3	3
15	1	3	5	32	1	3	6	49	1	4	4
	2	3	6		2	3	5		2	2	3
16	1	5	7	33	1	4	3	50	1	2	3
	2	4	8		2	3	6		2	3	3
17	1	5	7	34	1	2	3				
	2	3	4		2	3	8				

adaptation of Eq. 5.60,

$$\bar{r} = \frac{1}{L}\sum(r_h) = \frac{1}{L}\sum(\bar{y}_h/\bar{m}_h)$$

$$= 164.81/120 = 1.3734.$$

(c3)   The simple mean of HH means is

$$\bar{r} = \sum_{h=1}^{120}\sum_{i=1}^{2} r_{hi} = \frac{326.87}{240} = 1.3620.$$

(d)   For (a):

$$\text{var}(\hat{M}) = \text{var}(N\bar{m}) = N^2\text{var}(\bar{m})$$

$$= N^2\left(\frac{N-n}{N}\right)\frac{\bar{S}_m^2}{n},$$

since the stratification is simple (i.e., $n_h \propto N_h$). $\bar{S}_m^2$ is obtained easily from an ANOVA of the M_i.

Source	df	SS	MS
Gross	240	2780.00	
Mean	1	2124.15	
Total	239	655.85	2.744
Strata	119	360.85	3.032
SUs/strata	120	295.00	2.458

$$\text{SE}(\bar{m}) = \left[\frac{(1-\frac{1}{296})\bar{S}_m^2}{240}\right]^{1/2} = \left(\frac{2.458}{240}\right)^{1/2} = (0.010242)^{1/2} = 0.1012,$$

$$\text{SE}(\hat{M}) = (71{,}040)(0.1012) = 7189 \text{ HHs},$$

$$\pm 2\text{SE}(\hat{M}) = \pm 14{,}378.$$

Hence, 95 percent CLs are

$$\hat{M} = (71{,}040)(714/240) \pm 14{,}378$$

$$= 211{,}344 \pm 14{,}378$$

$$= 196{,}966 \text{ to } 255{,}722.$$

For (b):

$$\text{var}(\hat{Y}) = \text{var}(N\bar{y}) = N^2\text{var}(\bar{y}) = N^2\left(\frac{N-n}{N}\right)\frac{\bar{S}_y^2}{n}.$$

$\bar{S}_y^2$ is obtained easily from an ANOVA of the Y_i.

Source	df	SS	MS
Gross	240	6939.00	
Mean	1	4158.34	
Total	239	2780.66	11.635
Strata	119	1483.16	12.464
SUs/strata	120	1297.50	10.813

$$\text{SE}(\bar{y}) = \left[\frac{(1-\frac{1}{296})\bar{S}_y^2}{240}\right]^{1/2} = \left(\frac{10.813}{240}\right)^{1/2} = (0.045054)^{1/2} = 0.2123,$$

$$\text{SE}(\hat{Y}) = (295,700)(0.2123) = 62,766 \text{ radios},$$

$$2\text{SE}(\hat{Y}) = \pm 125,530.$$

Hence

$$95\% \text{ CLs} = \hat{Y} \pm 125,530$$
$$= 295,700 \pm 125,530$$
$$= 170,170 \text{ to } 421,230.$$

For (c1):

$$\text{var}(w_{\text{strat-rat}}) = \left(\frac{1-f}{n\bar{X}^2}\right)[\bar{S}_y^2 + (\bar{Y}/\bar{X})^2\,\bar{S}_x^2 - 2(\bar{Y}/\bar{X})\bar{S}_{xy}].$$

$S_{ym} = \bar{S}_{xy}$ and is obtained easily from an ANOCOVA as the mean cross-products within strata:

Source	df	SCP	MCP
Gross	240	4225.00	
Mean	1	2972.03	
Total	239	1252.97	5.243
Strata	119	696.47	5.853
SUs w strata	120	556.50	4.638

Sukhatme, P. V.	*Sampling Theory of Surveys with Applications*, (2nd ed.) Ames, Iowa:
Sukhatme, B. V.	Iowa State College Press. (Chapter 3.)
1953, 1970	
Tschuprov, A. A.	On the Mathematical Expectation of the Moments of Frequency
1923	Distribution in the Case of Correlated Observations. *Metron*, **2**(3): 461–493; (4) 646–683. (Perhaps the first presentation of the conditions for optimal allocation.)
Williams, W. H.	The Variance of an Estimator with Post-Stratified Weighting. *Journal of*
1962	*the American Statistical Association*, **57**: 622–627 (March).

7.16. Mathematical Notes

Proof that optimal allocation is given by $n_h \propto N_h S_h$ when costs are ignored: Consider the simple two-strata case and ignore FPC. In general,

$$\text{var}(\bar{y}_{\text{strat}}) = \frac{\sum W_h^2 S_h^2}{n_h}.$$

Let $W_h^2 S_h^2 = k_h$; then

$$V = \frac{k_1}{n_1} + \frac{k_2}{n_2}, \qquad n = n_1 + n_2,$$

$$Z = \frac{k_1}{n_1} + \frac{k_2}{n_2} + \lambda(n - n_1 - n_2),$$

$$\frac{\delta z}{\delta n_1} = -\frac{k_1}{n_1^2} - \lambda = 0, \quad \frac{\delta z}{\delta n_2} = -\frac{k_2}{n_2^2} - \lambda = 0, \quad \frac{\delta z}{\delta \lambda} = n - n_1 - n_2 = 0,$$

$$\frac{k_1}{n_1^2} = \left[\frac{k_2}{n_2^2} = \frac{k_2}{(n - n_1)^2} \right], \qquad (n - n_1)\sqrt{k_1} = n_1\sqrt{k_2},$$

$$n\sqrt{k_1} = n_1\sqrt{k_1} + n_1\sqrt{k_2},$$

$$n_1 = \frac{n\sqrt{k_1}}{\sqrt{k_1} + \sqrt{k_2}} = n\left(\frac{W_h S_h}{\sum W_h S_h} \right);$$

therefore, $n_h \propto W_h S_h$ or $n_h \propto N_h S_h$.

var $(\bar{y}_{\text{strat-rat(s)}})$

$$\hat{Y} - Y = \sum (\hat{Y} - Y_h)$$
$$= \sum N_h[\bar{X}_h(\bar{y}_h/x_h) - \bar{Y}_h],$$
$$E[(\hat{Y} - Y)^2] \doteq \sum N_h^2 E[\bar{X}_h(\bar{y}_h/\bar{x}_h) - \bar{Y}_h]$$
$$= \sum N_h^2 \bar{X}_h^2 \, \text{var}(\bar{y}_h/\bar{x}_h),$$
$$\text{var}(\bar{y}_{\text{strat-rat(s)}}) = \frac{1}{N^2} \sum N_h^2 \left(\frac{1-f_h}{n_h}\right) S_{dh}^2.$$

Since

$$\text{var}(\bar{y}/\bar{x}) = \frac{1}{\bar{X}^2}\left(\frac{1-f}{n}\right) S_d^2,$$

then

$$\text{var}(\bar{y}_{\text{strat-rat(s)}}) = \frac{1}{N^2} \sum N_h \left(\frac{1-f_h}{n_h}\right) \left[S_{yh}^2 - (\bar{Y}/\bar{X})_h^2 S_{xh}^2 - 2(\bar{Y}/\bar{X})_h R_h S_{xh} S_{yh}\right].$$

var $(\bar{y}_{\text{strat-rat(s)}})$

$$\hat{Y} - Y = X\frac{\bar{y}_{\text{strat}}}{\bar{x}_{\text{strat}}} \doteq N\bar{X} - (\bar{y}_{\text{strat}} - (\bar{Y}/\bar{X})\bar{x}_{\text{strat}}).$$

Let $D'_{hi} = Y_{hi} - (\bar{Y}/\bar{X})X_{hi}$; then

$$\bar{d}'_h = \bar{y}_h - (\bar{Y}/\bar{X})\bar{x}_h$$

$$\doteq N[\sum\left(\frac{N_h}{N}\right)\bar{y}_h - (\bar{Y}/\bar{X})\sum\left(\frac{N_h}{N}\right)\bar{x}_h]$$

and

$$\bar{y} - \bar{Y} = \sum N_h \bar{d}'_h,$$

$$\text{var}(\bar{y}_{\text{strat-rat(c)}}) \doteq \frac{1}{N^2} \sum N_h^2 \left(\frac{1-f_h}{n_h}\right) S_{d'h}^2,$$

which in terms of xs and ys becomes

$$\text{var}(\bar{y}_{\text{strat-rat(c)}}) = \frac{1}{N^2} \sum N_h^2 \left(\frac{1-f_h}{n_h}\right)(S_{yh}^2 + (\bar{Y}/\bar{X})S_{xh}^2 - 2(\bar{Y}/\bar{X})R_h S_{xh} S_{yh}).$$

5.   An advertiser is interested in determining the number of persons watching the TV program he sponsors. He is planning to make a survey of TV viewers by means of the "telephone coincidental method," wherein a random sample of telephone subscribers is drawn from the directory and called during the showing of the program. Each respondent is asked whether or not he (or she) was watching the program at the time of the call.

(a) Suppose the advertiser would like to have 95% confidence that the fraction of families watching is within 5 percentage points of the true; how large a sample is required for a *single* presentation?

(b) If the program is presented weekly over a 13-week period, how large a random sample is required with the precision requirements of (a) if the *average* fraction for the period is being estimated?

(c) How large, if the sample is stratified by week—that is, $n/13$ calls made week?

CHAPTER 8

Elements of Probability Sampling

229

A numerical example may help fix this notion. Table 8.3 presents the calculations leading to the determination of the exact variance of estimates of such samples of size one drawn from the universe of Table 8.1, where var $(\hat{Y}_P)$ = $\sum P_i \{(\hat{Y})_i - Y\}^2 = 7.333$.

Table 8.3.   *Determination of var* $(\hat{Y})$ *where* $N = 4$, $n = 1$

Possible sample, U_i	Estimate $(\hat{Y})_i = Y_i/P_i$	Error $(\hat{Y})_i - Y$	(Error)² $[(\hat{Y})_i - Y]^2$	Probability of [ ]², P_i	P_i[ ]²
U_1	$(\hat{Y})_1 = 30$	$+6$	36	.1	3.600
U_2	$(\hat{Y})_2 = 20$	-4	16	.2	3.200
U_3	$(\hat{Y})_3 = 23\frac{1}{3}$	$-\frac{2}{3}$	$\frac{4}{9}$	.3	.133
U_4	$(\hat{Y})_4 = 25$	$+1$	1	.4	.400
					7.333

To estimate $\bar{Y}$ we can simply estimate $\hat{Y}$ and divide by N; hence

(8.4)
$$\bar{y}_P = \frac{\hat{Y}_P}{N} = \frac{1}{N}\left(\frac{Y_i}{P_i}\right)$$

and likewise its variance is simply $(1/N)^2$ that of var $(\hat{Y})$. Hence

$$\text{var }(\bar{y}_P) = \frac{1}{N^2}\text{var }(\hat{Y}_P)$$

(8.5)
$$= \frac{1}{N^2}\sum P_i\left(\frac{Y_i}{P_i} - Y\right)^2.$$

Comparing Probability with Random Sampling When $n = 1$. It will be convenient, particularly in making comparisons between equal and unequal probability sampling, to define a *weighted variance* of, say, the *y*-population, $\overset{*}{S}^2$, where

(8.6)
$$\overset{*}{S}^2 = \frac{\sum P_i (Y_i/P_i - Y)^2}{N(N-1)} = \frac{\sum (Y_i^2/P_i) - Y^2}{N(N-1)}.$$

Now we can write Eq. 8.3a as

(8.7)
$$\text{var }(\hat{Y}_P) = N^2\left(\frac{N-1}{N}\right)\overset{*}{S}^2 = N(N-1)\overset{*}{S}^2$$

and Eq. 8.5 as

(8.8)
$$\text{var }(\bar{y}_P) = \left(\frac{N-1}{N}\right)\overset{*}{S}^2.$$

To compare the efficiency of *probability sampling* with random sampling—

that is, samples using unequal versus equal probabilities of selection where samples are of size $n = 1$, for estimating either Y or $\bar{Y}$, we have

$$(8.9) \qquad \text{RE (prob/ran)} = \frac{S^2}{\overset{*}{S}^2}.$$

Example 8.1. Using the data in Table 8.1, determine the efficiency of sampling with the unequal probabilities given as compared to equal probabilities when samples are size $n = 1$. The data and the computations can be arranged as follows:

i	Y_i	P_i	Y_i^2	Y_i^2/P_i
1	3	.1	9	90
2	4	.2	16	80
3	7	.3	49	163.333
4	10	.4	100	250
$\sum$	24	1.0	174	583.333

From Eq. 2.4

$$S^2 = \frac{\sum Y_i^2 - Y^2/N}{N-1}$$

$$= \frac{174 - (24)^2/4}{4-1} = \frac{30}{3} = 10,$$

and from Eq. 8.6

$$\overset{*}{S}^2 = \frac{\sum Y_i^2/P_i - Y^2}{N(N-1)}$$

$$= \frac{583.333 - (24)^2}{4(4-1)} = \frac{7.333}{12} = 0.6111.$$

Then from Eq. 8.9

$$\text{RE (prob/ran)} = \frac{S^2}{\overset{*}{S}^2} = \frac{10.000}{0.611} = 16.36$$

or 1636%. In this case a tremendous gain in efficiency is obtained by using the unequal probabilities given. Unfortunately, this is not always the case. Later we shall examine the circumstances on which the efficiency of this method depend.

8.4. Optimal Probabilities; Measures of Size; Estimating Y and $\bar{Y}$ with PPX$_i$

In Example 8.1, S^2 and $\overset{*}{S}{}^2$ turned out to be 10 and 0.611, respectively. The efficiency of PR relative to RR sampling for any size of sample n' is given by

$$(8.21) \qquad \text{RE (PR/RR)} = \frac{\text{var (RR)}}{\text{var (PR)}}.$$

Using Eqs. 8.20 for PR and 2.20 for RR sampling for estimating $\bar{Y}$, we obtain

$$(8.21a) \qquad \text{RE (PR/RR)} = \frac{[(N-1)/Nn']\,S^2}{[(N-1)/Nn']\,\overset{*}{S}{}^2} = \frac{S^2}{\overset{*}{S}{}^2},$$

and for Example 8.1

$$= \frac{10}{0.6111} = 16.36,$$

the same as for the case where $n' = 1$.

The rather large advantage in efficiency of PR over RR sampling for this case suggests a potentially powerful device for sampling design. What are the conditions by which departure from equal probabilities of selections can bring about gains in efficiency? Since P_i are arbitrary, what is the optimal choice of P_i?

It will be found that the variance of $\hat{Y}_{PR}$ (in a computational form of Eq. 8.19a)

$$\text{var } (\hat{Y}_{PR}) = \frac{\sum (Y_i^2/P_i) - Y^2}{n'}$$

is a minimum—in fact, is equal to zero—when

$$(8.22) \qquad P_i = Y_i/Y$$

—that is, when the probabilities of selection are proportional to values being observed. This suggests that if we could determine a set of probabilities that are proportional to the amount of the y characteristic on each element, we would achieve a perfect sampling scheme. This is an interesting fact, but achieving this result in practice is difficult. How are we to determine such P_is unless we know the Y_is, which of course is what we are seeking by means of the survey. The principle is helpful, however, since we may be able to find some *measure of size*, X_i, that is correlated with Y_i, and use this information to determine the P_is. One source of such information may be previous census results. A listing form for utilizing the measure of size information may look like that of Table 8.4.

The selections are made from the cumulative total of X_i; in the example, element number 3 was drawn twice. Column 5 was presented merely to show that the probabilities of selection are $(P_i = X_i/X)$ in case they are to be used in computing the estimates. The specific P_is are not necessary for the selection of

Table 8.4.   Listing form for PR sampling

Element no. (1)	Previous census data (X_i) (2)	Cumulative total $(\sum X_i)$ (3)	Sample points (RN) (4)	Selection probabilities (P_i) (5)
1	20	20		.222
2	8	28	21	.089
3	32	60	29, 31	.355
4	16	76		.178
5	14	90		.156
				1.000

the elements when a measure of size is used. We shall call this "sampling with PPX$_i$,"—that is, with probabilities proportional to X_i.

Best MOS (Measure of Size).   The measure of size, X_i, useful for determining the P_i, in order to be efficient must be proportional to Y_i. A simple plot of the data may help us judge what might result from departures from proportionality. Using the data of Table 8.1 and regarding the P_i as based on some set of X_i/X, we have as a plot of Y_i on P_i (or X_i/X) the following:

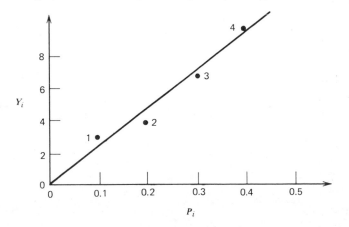

In this case, the strong relationship between Y_i and P_i is apparent, and so is the reasonableness that the line of relationship should go through the origin. If we compute the Y_i/P_i and the individual "errors," $(Y)_i - Y$, we obtain the results of Table 8.5. Column 7 gives the portion of the $\overset{*}{S}{}^2$ variance contributed by each element. The largest contributor is element 1, even though its probability of occurrence is smallest. Its large "error" denoted by $Y_i/P_i - Y$ (shown in column

Likewise, for estimators of the variances we have from Eq. 8.13,

$$(8.28) \qquad \widehat{\text{var}}\,(\hat{Y}_{\text{PR}}) = \frac{1}{n'(n'-1)} \sum_{}^{n'} [(\hat{Y})_i - \hat{Y}_{\text{PR}}]^2$$

$$(8.28a) \qquad = \frac{N^2}{n'}\, \overset{*}{s}{}^2_{y|x} = \frac{\bar{X}^2}{n'(n'-1)} \left[\sum (Y_i/X_i)^2 - \frac{1}{n'} (\sum Y_i/X_i)^2 \right],$$

where from Eq. 8.14, by replacing P_i with X_i/X,

$$(8.29) \qquad \overset{*}{s}{}^2_{y|x} = \frac{\bar{X}^2}{n'-1} \sum \left(\frac{Y_i}{X_i} - \frac{\hat{Y}}{X} \right)^2$$

$$(8.29a) \qquad = \frac{\bar{X}^2}{n'-1} \left[\sum \left(\frac{Y_i}{X_i} \right)^2 - \frac{1}{n'} \left(\sum \frac{Y_i}{X_i} \right)^2 \right]$$

and hence

$$(8.30) \qquad \widehat{\text{var}}\,(\bar{y}_{\text{PR}}) = \frac{1}{N^2}\, \widehat{\text{var}}\,(\hat{Y}_{\text{PR}}) = \frac{1}{n'}\, \overset{*}{s}{}^2_{y|x}.$$

The next three sections will consider estimators of other parameters and functions of them for two MOSs, M_i and X_i.

8.5. Estimating $\bar{\bar{Y}}$ (that is, Y/M) with PPM$_i$

Up to now we have considered the case of selecting sampling units with arbitrary probabilities and have shown that, if those probabilities are known, unbiased estimates of Y and $\bar{Y}$ can be made and unbiased estimates of the variances of such estimates can also be made. Now let us consider sampling units that are clusters of arbitrary sizes M_i, where we wish to estimate $\bar{\bar{Y}}$, the mean of y per element. In this section we shall select SUs with probabilities proportional to M_i (that is, PPM$_i$). In the next section we shall consider an arbitrary probability, $A_i = X_i/X$.

To estimate $\bar{\bar{Y}}$ we have from Eq. 8.11, substituting M_i/M for P_i and $\bar{Y}_i$ for Y_i/M_i and putting it on an element basis,

$$\bar{\bar{y}}_{PM1} = \frac{\hat{Y}_{\text{PR}}}{M} = \frac{1}{Mn'} \sum_{}^{n'} \left(\frac{Y_i}{P_i} \right) = \frac{1}{Mn'} \sum_{}^{n'} \left(\frac{Y_i}{M_i/M} \right)$$

$$(8.31) \qquad = \frac{1}{n'} \sum_{}^{n'} \bar{Y}_i.$$

This unbiased estimator is simply the mean of the element means of each cluster, a somewhat surprising but useful result. The variance of Eq. 8.31 can be derived from Eq. 8.12a by replacing P_i by M_i/M and dividing $\hat{Y}_{\text{PR}}$ by M_i, whence

$$(8.32) \qquad \text{var}\,(\bar{\bar{y}}_{PM1}) = \frac{1}{n'} \sum \left(\frac{M_i}{M} \right) (\bar{Y}_i - \bar{\bar{Y}})^2,$$

which is estimated unbiasedly from a sample by

$$(8.33) \qquad \widehat{\text{var}}\ (\bar{y}_{PM1}) = \frac{\sum^{n'}(\bar{Y}_i - \bar{y})^2}{n'(n'-1)}$$

by appropriately modifying Eq. 8.28.

8.6.   Estimating $\bar{\bar{Y}}$ and Y/X with PPX_i

If the selection probabilities are proportional to A_i, where $A_i = X_i/X$ (where X_i may be regarded as an approximate measure of M_i), and we wish to estimate $\bar{\bar{Y}}$ = Y/M, we have a choice of three reasonable estimators, such as

$$(8.34) \qquad \bar{y}_{PX1} = \frac{1}{n'}\sum \bar{Y}_i,$$

$$(8.35) \qquad \bar{y}_{PX2} = \frac{1}{n'M}\sum Y_i/A_i = \frac{1}{n'M}\sum\left(\frac{M_i}{A_i}\right)\bar{Y}_i,$$

$$(8.36) \qquad \bar{y}_{PX3} = \frac{\sum Y_i/A_i}{\sum M_i/A_i} = \frac{\sum (M_i/A_i)\,\bar{Y}_i}{\sum (M_i/A_i)}.$$

Note that these are analogous to the three types presented in Section 5.10, where clusters were being sampled with equal selection probabilities. As with PE selection—that is, probabilities equal—only type 2 is unbiased in the case of PPX_i—that is, probabilities proportional to X_i.

The MSEs (mean square errors) are given by

$$(8.37) \qquad \text{MSE}\ (\bar{y}_{PX1}) = \frac{1}{n'}\sum A_i(\bar{Y}_i - \bar{Y}_1)^2 + (\bar{\bar{Y}}_1 - \bar{\bar{Y}})^2,$$

$$(8.38) \qquad \text{MSE}\ (\bar{y}_{PX2}) = \frac{1}{n'M^2}\sum A_i\left(\frac{Y_i}{A_i} - Y\right)^2,$$

$$(8.39) \qquad \text{MSE}\ (\bar{y}_{PX3}) = \frac{1}{n'M^2}\sum\left(\frac{M_i^2}{A_i}\right)(\bar{Y}_i - \bar{\bar{Y}})^2,$$

where $\bar{\bar{Y}}_1 = A_i\bar{Y}_i$. Again the best—that is, the smallest MSE—estimator depends on circumstances. It can be shown that when $A_i = M_i/M$, Eqs. 8.37–8.39 become identical to Eq. 8.32.

To estimate the MSEs from a sample we have

$$(8.40) \qquad \widehat{\text{var}}\ (\bar{y}_{PX1}) \doteq \frac{\sum (\bar{Y}_i - \bar{y}_1)^2}{n'(n'-1)},$$

$$(8.41) \qquad \widehat{\text{var}}\ (\bar{y}_{PX2}) \doteq \frac{\sum (Y_1/A_i - M\bar{y}_2)^2}{M^2 n'(n'-1)},$$

$$(8.42) \qquad \widehat{\text{var}}\ (\bar{y}_{PX3}) \doteq \frac{1}{M^2 n'(n'-1)}\sum^{n'}\left(\frac{M_i}{A_i}\right)^2 (\bar{Y}_i - \bar{y}_3)^2.$$

To estimate Eq. 8.52 we have

$$(8.53) \quad \widehat{\mathrm{var}}\left(\frac{\hat{Y}_{PX}}{\hat{Z}_{PX}}\right) = \left[\frac{\sum (Y_i/X_i)}{\sum (Z_i/X_i)}\right]^2 \left(\frac{1}{n'}\right)\left[\frac{\mathring{s}_{y\,|\,x}^2}{(\hat{Y}/X)^2} + \frac{\mathring{s}_{z|x}^2}{(\hat{Z}/X)^2} - 2\frac{\mathring{s}_{yz|x}^*}{(\hat{Y}X)\,(\hat{Z}/X)}\right],$$

and to estimate Eq. 8.52a we have

$$(8.53a) \quad\quad \mathrm{rel}\ \widehat{\mathrm{var}}\left(\frac{\hat{Y}_{PX}}{\hat{Z}_{PX}}\right) = \frac{1}{n'}\left(\mathring{v}_{y|x}^2 + \mathring{v}_{z|x}^2 - 2\mathring{v}_{yz|x}\right),$$

where

$$\mathring{S}^2_{|x} = \left(\frac{X^2}{N(N-1)}\right)\sum \frac{X_i}{X}\left(\frac{Y_i}{X_i} - \frac{Y}{X}\right)^2,$$

$$(8.54) \quad \mathring{S}^2_{z|x} = \left(\frac{X^2}{N(N-1)}\right)\sum \frac{X_i}{X}\left(\frac{Z_i}{X_i} - \frac{Z}{X}\right)^2,$$

$$\mathring{S}^2_{yz|x} = \left(\frac{X^2}{N(N-1)}\right)\sum \frac{X_i}{X}\left(\frac{Y_i}{X_i} - \frac{Y}{X}\right)\left(\frac{Z_i}{Z} - \frac{Z}{X}\right);$$

$$\mathring{s}^2_{y|x} = \left(\frac{\bar{X}^2}{n'-1}\right)\sum \left(\frac{Y_i}{X_i} - \frac{\hat{Y}}{X}\right)^2 = \frac{\bar{X}^2}{n'-1}\left\{\sum \left(\frac{Y_i}{X_i}\right)^2 - \frac{1}{n'}\left[\sum\left(\frac{Y_i}{X_i}\right)\right]^2\right\},$$

$$(8.54a) \quad \mathring{s}^2_{z|x} = \left(\frac{\bar{X}^2}{n'-1}\right)\sum \left(\frac{Z_i}{X_i} - \frac{\hat{Z}}{X}\right)^2 = \frac{\bar{X}^2}{n'-1}\left\{\sum \left(\frac{Z_i}{X_i}\right)^2 - \frac{1}{n'}\left[\sum\left(\frac{Z_i}{X_i}\right)\right]^2\right\},$$

$$\mathring{s}^2_{yz|x} = \left(\frac{\bar{X}^2}{n'-1}\right)\sum \left(\frac{Y_i}{X_i} - \frac{\hat{Y}}{X}\right)\left(\frac{Z_i}{X_i} - \frac{\hat{Z}}{X}\right)$$

$$= \frac{\bar{X}^2}{n'-1}\left\{\sum \left(\frac{Y_i}{X_i} \cdot \frac{Z_i}{X_i}\right) - \frac{1}{n'}\left[\sum\left(\frac{Y_i}{X_i}\right)\sum\left(\frac{Z_i}{X_i}\right)\right]\right\};$$

$$(8.54b) \quad \mathring{V}^2_{y|x} = \frac{\mathring{S}^2_{y|x}}{(Y/X)^2}, \quad \mathring{V}^2_{z|x} = \frac{\mathring{S}^2_{z|x}}{(Z/X)^2}, \quad \mathring{V}_{yz|x} = \frac{\mathring{S}_{yz|x}}{(Y/X)\,(Z/X)};$$

$$(8.54c) \quad \mathring{v}^2_{y|x} = \frac{\mathring{s}^2_{y|x}}{\bar{y}^2_{PX}} = \left(\frac{n'^2}{n'-1}\right)\frac{\sum (Y_i/X_i)^2 - (1/n')[\sum (Y_i/X_i)]^2}{[\sum (Y_i/X_i)]^2};$$

and similarly for $\mathring{v}^2$ and $\mathring{v}_{yz|x}$.

Estimating N, When Unknown.   Occasionally one may wish to estimate the number of SUs in the frame itself when selection probabilities, either M_i/M or

X_i/X, are known but N is not. In this case, if every SU has a positive M_i or X_i, we can estimate N by

$$(8.55) \qquad \hat{N}_{PX} = \left(\frac{1}{n'}\right) \sum^{n'} \left(\frac{1}{P_i}\right) = \frac{X}{n'} \sum^{n'} \left(\frac{1}{X_i}\right),$$

when X_i is the MOS. The estimator, although a mean of reciprocals of a random variable, is nevertheless unbiased. Its variance is given by

$$(8.56) \qquad \text{var } (\hat{N}_{PX}) = \left(\frac{X^2}{n'}\right) \sum \left(\frac{X_i}{X}\right) \left(\frac{1}{X_i} - \frac{N}{X}\right)^2 = \frac{1}{n'} \sum A_i \left(\frac{1}{A_i} - N\right)^2,$$

or, in terms of weighted S^2-variances, by

$$(8.56a) \qquad \text{var } (\hat{N}_{PX}) = N^2 \left(\frac{N-1}{Nn'}\right) \overset{*}{S}{}^2_{1/X} = \frac{N(N-1)}{n'} \overset{*}{S}{}^2_{1/X},$$

where

$$(8.57) \qquad \overset{*}{S}{}^2_{1/X} = \left(\frac{X^2}{N(N-1)}\right) \sum \left(\frac{X_i}{X}\right) \left(\frac{1}{X_i} - \frac{N}{X}\right)^2.$$

The variance can be estimated from the sample by

$$(8.58) \qquad \widehat{\text{var}} \, (\hat{N}_{PX}) = \left(\frac{1}{n'(n'-1)}\right) \sum^{n'} [(\hat{N})_i - \hat{N}]^2$$

$$(8.58a) \qquad = \frac{X^2}{n'(n'-1)} \left\{ \sum \left(\frac{1}{X_i}\right)^2 - \frac{1}{n'} \left[\sum \left(\frac{1}{X_i}\right) \right]^2 \right\}$$

$$(8.58b) \qquad = \left(\frac{N^2}{n'}\right) \overset{*}{s}{}^2_{1/X};$$

hence

$$(8.59) \qquad \overset{*}{s}{}^2_{1/X} = \left(\frac{\bar{X}^2}{n'-1}\right) \left\{ \sum \left(\frac{1}{X_i}\right)^2 - \frac{1}{n'} \left[\sum \left(\frac{1}{X_i}\right) \right]^2 \right\}$$

and

$$(8.59a) \qquad \overset{*}{v}{}^2_{1/X} = \frac{\overset{*}{s}{}^2_{1/X}}{\bar{X}^2 [(1/n) \sum (1/X_i)]^2}$$

$$= \left(\frac{n'^2}{n'-1}\right) \frac{\sum (1/X_i)^2 - (1/n') [\sum (1/X_i)]^2}{[\sum (1/X_i)]^2},$$

whence

$$(8.59b) \qquad \text{rel } \widehat{\text{var}} \, (\hat{N}_{PX}) = \frac{\overset{*}{v}{}^2_{1/X}}{n'}$$

Estimating $\bar{Y}$ When N Is Unknown.   To estimate $\bar{Y}$ we have the ratio of the two random variables $\hat{Y}$ and $\hat{N}$; thus from Eqs. 8.11 and 8.55, substituting X_i/X

In substance, this procedure amounts to drawing PPS (probabilities proportional to size) samples in steps with the measure of size modified following each draw, so that the remaining relative sizes always add to unity. If we use as relative sizes the P_i given in the example of Table 8.1, we have:

Element, i: 1   2   3   4

Relative size, A_i: 0.1   0.2   0.3   0.4

The probability of selecting element 1 followed by element 2 in RNR sampling is

$$P\{1, 2\} = P\{1\} \cdot P\{2|1\}$$

$$= \frac{1}{N} \cdot \frac{1}{N-1} = \frac{1}{4} \cdot \frac{1}{3} = \frac{1}{12},$$

and for PNR(1) sampling it is

$$P\{1, 2\} = (A_i)(A_2')$$

$$= (0.1)\left(\frac{A_2}{A_2 + A_3 + A_4}\right) = \frac{(.1)0.2}{0.2 + 0.3 + 0.4} = \frac{(.1)0.2}{0.9}$$

$$= .022.$$

The 12 possible ordered samples of $n = 2$ and the probabilities of selecting each of them by the two selection methods are presented in Table 8.7. Note that under the RNR method all samples of different ordering of the *same* elements are equally likely and, in fact, all samples of *different* elements are equally likely; whereas in the PNR(1) method, probabilities vary among samples of the same elements as well as those of different elements.

Inclusion Probability.   The probability that element i appears in a sample of n draws shall be called its inclusion probability, designated by the symbol P_i. When an element is drawn in any nonreplacement scheme, that event may have taken place in any one of the n draws. But once it is drawn, it cannot be drawn again. In RNR sampling, the probability that it is drawn is equal for each draw—that is, $1/N$. In PNR sampling, the probability that it is drawn is not necessarily equal for each draw. Note in Table 8.7 that under PNR(1) sampling, element 1 has a probability of .100 of being the first drawn and .135 of being second. Its *average probability* of being drawn on one of the two draws is .1175; i.e., $(.100) + .135)/2$. If we ignore (or do not know) the order in which a sample is drawn, this is the probability we would attach to the ith element's being drawn on each of the n draws. Let us denote this probability by $\bar{P}_i$, defined as

$$\bar{P}_i = P_i/n \tag{8.64}$$

for PNR or RNR sampling.

Consider the following reasoning: Using Eq. 8.1 for each draw, i, we can estimate the total y of a universe by simply dividing the observed Y_i by P_i, the probability of selecting the element on draw i; or for, say, n draws, let us use the simple mean of the n estimates as given in Eq. 8.11. This reasoning can be shown to be valid *whenever we know the appropriate probabilities of observing an event* (element), $\bar{P}_i$. For PNR sampling let us consider the estimator of Y,

$$\hat{Y}_{\text{PNR}} = \frac{1}{n}\left(\frac{Y_1}{\bar{P}_1} + \frac{Y_2}{\bar{P}_2} + \cdots, \frac{Y_n}{\bar{P}_n}\right) \tag{8.65}$$

$$= \frac{1}{n}\sum^n\left(\frac{Y_i}{\bar{P}_i}\right) \tag{8.65a}$$

or, since $P_i = n\bar{P}_i$,

$$= \sum^n\left(\frac{Y_i}{P_i}\right). \tag{8.65b}$$

Table 8.7. Probabilities of selecting ordered samples in RNR and PNR(1) sampling $(N = 4, n = 2)$

Sample no.	Elements and order in sample	Probability of tth sample, $P\{t\}$, if selection is	
(t)	(i, j)	RNR	PNR(1)
1	1, 2	$(1/4)(1/3) = .083$	$(.1)(.2/.9) = .022$
2	1, 3	$(1/4)(1/3) = .084$	$(.1)(.3/.9) = .034$
3	1, 4	$(1/4)(1/3) = .083$	$(.1)(.4/.9) = .044$
4	2, 1	$(1/4)(1/3) = .083$	$(.2)(.1/.8) = .025$
5	2, 3	$(1/4)(1/3) = .084$	$(.2)(.3/.8) = .075$
6	2, 4	$(1/4)(1/3) = .083$	$(.2)(.4/.8) = .100$
7	3, 1	$(1/4)(1/3) = .083$	$(.3)(.1/.7) = .043$
8	3, 2	$(1/4)(1/3) = .083$	$(.3)(.2\cdot.7) = .086$
9	3, 4	$(1/4)(1/3) = .084$	$(.3)(.4/.7) = .171$
10	4, 1	$(1/4)(1/3) = .083$	$(.4)(.1/.6) = .067$
11	4, 2	$(1/4)(1/3) = .084$	$(.4)(.2/.6) = .133$
12	4, 3	$(1/4)(1/3) = .083$	$(.4)(.3/.6) = .200$
		1.000	1.000

Note: PNR(1): where selection is PPS in both draws.

Equation 8.65a is an unbiased estimator of Y, provided the $\bar{P}_i$ for each element is greater than zero and the sampling scheme is such that

$$(8.66) \qquad\qquad \sum_i^n \bar{P}_i = 1.$$

To check its unbiasedness empirically, we shall continue with the universe of $N = 4$ and samples of $n = 2$ of Table 8.7. Let us again consider scheme PNR(1), where samples were selected with PPS—that is, probability proportional to a relative measure of size A_i on draw 1, to A'_j on draw 2, A'_k on draw 3, and so on. Table 8.8 shows the computations leading to the determination of the P_is and $\bar{P}_i$s. The $P\{t\}$s, the probabilities of selecting each of the possible samples, are taken from Table 8.7 and are now associated with each element appearing in that sample. Summing the $P\{t\}$s for each of the N elements separately provides the inclusion probabilities, P_i. The $\bar{P}_i$s are then obtained by dividing the P_is by n.

Having the $\bar{P}_i$s, we can compute the estimates of Y using Eq. 8.65a; these are shown in Table 8.9.

Table 8.8.   **Inclusion probabilities, P_i, and $\bar{P}_i$, for PNR(1)**
sampling $(N = 4, n = 2)$

Sample no. (t)	Elements (i, j)	Probability of sample (P{t})	Element (i) 1	2	3	4
1	1, 2	.022	.022	.022		
2	1, 3	.034	.034		.034	
3	1, 4	.044	.044			.044
4	2, 1	.025	.025	.025		
5	2, 3	.075		.075	.075	
6	2, 4	.100		.100		.100
7	3, 1	.043	.043		.043	
8	3, 2	.086		.086	.086	
9	3, 4	.171			.171	.171
10	4, 1	.067	.067			.067
11	4, 2	.133		.133		.133
12	4, 3	.200			.200	.200
$\sum$		1.000	.235	.441	.609	.715
$P_i: \sum P_i =$		2.000	.235	.441	.609	.715
$\bar{P}_i = P_i/n: \sum \bar{P}_i =$		1.000	.1175	.2205	.3045	.3575

When weighted by the probabilities of sample t selection, $P\{t\}$ (see Table 8.9), we obtain as the expectation of the estimates, 23.998, which is the true value, 24, except for rounding errors. This is not an isolated incident; the estimates of Y given by Eq. 8.65a are unbiased. To estimate $\bar{Y}$ where $\bar{Y} = Y/N$, we need only divide $\hat{Y}$ by N to obtain an unbiased estimate of $\bar{Y}$.

Table 8.9.   Estimates of Y and their expectation in PNR(1) sampling $(N = 4, n = 2)$

Sample[a] i, j	Y_i, Y_j	P_i	P_j	Y_i/P_i	Y_j/P_j	$\hat{Y}_t$	$P\{t\}$[b]	$P\{t\}\hat{Y}_t$
1, 2	3, 4	.235	.441	12.76	9.07	21.83	.047	1.026
1, 3	3, 7	.235	.609	12.76	11.49	24.25	.077	1.867
1, 4	3, 10	.235	.715	12.76	13.99	26.75	.111	2.969
2, 3	4, 7	.441	.609	9.07	11.49	20.56	.161	3.310
2, 4	4, 10	.441	.715	9.07	13.99	23.06	.233	5.373
3, 4	7, 10	.609	.715	11.49	13.99	25.48	.371	9.453
							1.000	23.998

[a] Ignoring order—hence $ij = (i, j) + (j, i)$.
[b] Sum over orders—hence $P(ij) = P(i, j) + P(j, i)$.

Variance of $\hat{Y}$.   The variance of PNR sampling regardless of the selection method followed, as long as $0 < P_i \leq 1$ and $\sum^N P_i = n$, is given by

$$(8.67) \quad \text{var}(\hat{Y}_{PNR}) = \sum^N \left(\frac{Y_i^2}{P_i}\right)(1 - P_i) + 2\sum_i^N \sum_{j>i}^N \left(\frac{Y_i}{P_i}\frac{Y_j}{P_j}\right)(P_i P_j - P_{ij})$$

or, in a more compact form,

$$(8.67a) \quad \text{var}(\hat{Y}_{PNR}) = \sum_i^N \sum_{j>i}^N (P_i P_j - P_{ij})\left(\frac{Y_i}{P_i} - \frac{Y_j}{P_j}\right)^2.$$

Equation 8.67 is due to Narain (1951) and Horvitz and Thompson (1952), and where P_{ij} is the inclusion probability of both elements i and j. Equation 8.67a is due to Yates and Grundy (1953).

It can be seen that the variance of $\hat{Y}$ under PNR sampling is affected by the P_is and P_{ij}s obtained. Since these depend on the selection scheme adopted (see next section), it is of interest to examine some of the properties of the variance formulas, Eqs. 8.67 and 8.67a.

It can be seen in Eq. 8.67a that if the P_is are proportional to the Y_is, the squared term will be zero and hence var $(\hat{Y}) = 0$. Since the Y_i are unknown and usually the object of a sampling investigation anyway, it will not be possible to fully utilize this principle of optimization. However, let us assume that we know a set of A_i, which is in some degree proportional to Y_i. We might then expect some reduction in var $(\hat{Y})$ if the P_i can be made proportional to the A_i. The effectiveness will, of course, depend on the nature of the relationship between P_i and A_i. A further study of Eq. 8.67a may strongly suggest that if the P_is were fixed, it may be possible to alter the P_{ij}s so that var $(\hat{Y})$ is reduced. It may be noted that considerable flexibility exists in the values the P_i and P_{ij} may take on. It is necessary, however, that

$$(8.68) \qquad \sum_i^N P_i = n,$$

$$(8.69) \qquad \sum_i^N \sum_{j>i}^N P_{ij} = \frac{P_i(n-1)}{2},$$

and that $P_i > 0$ in order that unbiased estimates of Y can be obtained.

Computation of var $(\hat{Y})$; an Example. To compute the value of var $(\hat{Y})$ for our microuniverse using PNR(1) sampling, we shall use Eq. 8.67a. The P_is are obtained from Table 8.8; the P_{ij}s are compiled from Table 8.10 by aggregating

Table 8.10.   *Inclusion probabilities of element pairs,*
P_{ij}, in PNR(1) sampling (N = 4, n = 2)

Sample (t)	Elements (i, j)	Probability of sample ($P\{t\}$)	Elements (i, j) 12	13	14	23	24	34
1	1, 2	.022	.022					
2	1, 3	.034		.034				
3	1, 4	.044			.044			
4	2, 1	.025	.025					
5	2, 3	.075				.075		
6	2, 4	.100					.100	
7	3, 1	.043		.043				
8	3, 2	.086				.086		
9	3, 4	.171						.171
10	4, 1	.067		.067				
11	4, 2	.133					.133	
12	4, 3	.200						.200
$\sum$		1.000	.047	.077	.111	.161	.233	.371
	$P_{ij}: \sum^N P_{ij} = 1.000$		.047	.077	.111	.161	.233	.371

the probabilities of those samples in which elements i and j occur together. The computations of var $(\hat{Y})$ are shown in Table 8.11, where the contribution from each of the six possible samples is provided. In this case, var $(\hat{Y}) = 3.989$, with the sample comprising elements 2 and 4 being the largest "contributor." See below for the real contributor.

The variance of $\hat{Y}$ can also be obtained by evaluating $E\{(\hat{Y} - Y)^2\}$. The results of this procedure are presented in Table 8.12. Note that in this case the sample comprising elements 2 and 3 is the largest contributor to variance—it is the most divergent from the parameter and has a reasonably large probability of occurrence.

Estimating var $(\hat{Y})$ from a PNR Sample. An unbiased estimate of var (y) for PNR sampling, based only on the sample, is provided by the estimator

$$(8.70) \qquad \widehat{\text{var}} \, (\hat{Y}_{\text{PNR}}) = \sum_{i}^{n} \sum_{j>i}^{n} \frac{P_i P_j - P_{ij}}{P_{ij}} \left(\frac{Y_i}{P_i} - \frac{Y_j}{P_j} \right)^2 .$$

Table 8.11. Computations of var $(\hat{Y})$

Element		Y_i	Y_i	P_i	P_j	$\dfrac{Y_i}{P_i}$	$\dfrac{Y_j}{P_j}$	$\dfrac{Y_i}{P_i} - \dfrac{Y_j}{P_j}$	$\left(\dfrac{Y_i}{P_i} - \dfrac{Y_j}{P_j} \right)^2$
i	j								
1	2	3	4	.235	.441	12.76	9.07	+3.69	13.6161
	3	3	7	.235	.609	12.76	11.49	+1.27	1.6129
	4	3	10	.235	.715	12.76	13.99	−1.23	1.5129
2	3	4	7	.441	.609	9.07	11.49	−2.42	5.8564
	4	4	10	.441	.715	9.07	13.99	−4.92	24.2064
3	4	7	10	.609	.715	11.49	13.99	−2.50	6.2500

Element		$P_i P_j$	P_{ij}	$P_i P_j - P_{ij}$	$\left(P_i P_j - P_{ij} \right)\left(\dfrac{Y_i}{P_{ij}} - \dfrac{Y_i}{P_j} \right)^2$
i	j				
1	2	0.103,635	.047	0.056,635	0.771,148
	3	0.143,115	.077	0.066,115	0.106,637
	4	0.168,025	.111	0.057,025	0.086,273
2	3	0.268,569	.161	0.107,569	0.629,967
	4	0.315,315	.233	0.082,315	1.992,550
3	4	0.435,435	.371	0.064,435	0.402,719
		1.434,094	1.000	0.434,094	3.989,294

Table 8.12.   Determination of var($\hat{Y}$) PNR(1)
sampling directly ($N = 4$, $n = 2$)

Sample (t)	Elements (ij)	$\hat{Y}_t$	$\hat{Y} - Y$	$(\hat{Y} - Y)^2$	$P\{t\}$	$P\{t\}(\hat{Y} - Y)^2$
1	12	21.83	-2.17	4.7089	.047	0.2213
2	13	24.25	-0.25	0.0625	.077	0.0048
3	14	26.75	$+2.75$	7.5625	.111	0.8394
4	23	20.56	-3.44	11.8336	.161	1.9052
5	24	23.06	-0.94	0.8836	.233	0.2059
6	34	25.48	$+1.48$	2.1904	.371	0.8126
						3.9892

Table 8.13.   Computation of $\widehat{var}$ ($\hat{Y}$)s for $n = 2$

Sample no.	Elements ij	$W_{ij}D_{ij}^2$	P_{ij}	$\left(\dfrac{W_{ij}}{P_{ij}}\right)D_{ij}^2$
1	12	0.771,148	0.047	16.407
2	13	0.106,637	0.077	1.385
3	14	0.086,273	0.111	0.777
4	23	0.629,967	0.161	3.913
5	24	1.992,550	0.233	8.552
6	34	0.402,719	0.371	1.085

This formula, due to Yates and Grundy (1953), requires that P_i and P_{ij} be greater than zero for all i and ij. It may result in negative estimates in some cases. This event can be avoided, however, by the proper choice of sampling scheme, where P_{ij} are made proportionate to $P_i P_j$.

By letting

$$(8.71) \qquad P_i P_j - P_{ij} = W_{ij}$$

and

$$(8.72) \qquad \frac{Y_i}{P_i} - \frac{Y_j}{P_j} = D_{ij},$$

we can put Eqs. 8.67a and 8.70 into the more compact forms:

$$(8.73) \qquad \text{var}(\hat{Y}_{\text{PNR}}) = \sum_{i}^{N} \sum_{j>i}^{N} W_{ij}\, D_{ij}^2$$

and

$$(8.74) \qquad \widehat{\text{var}}(\hat{Y}_{\text{PNR}}) = \sum_{i}^{n} \sum_{j>i}^{n} \frac{W_{ij}}{P_{ij}}\, D_{ij}^2,$$

respectively. It is quite apparent now that variance is the sum of the weighted differences of all possible pairs of weighted observations.

It may be of interest to estimate the variance of $\hat{Y}_{PNR}$ from each of the possible samples in our example. The computations are displayed in Table 8.13. The estimates in this case range from 0.777 to 16.407 (where, it may be recalled, the true variance of $\hat{Y}$ is 3.989).

8.9.   Estimating $\bar{Y}$, Y/Z, and N with PNR Sampling

When sampling with PNR the estimators of $\bar{Y}$, Y/Z, and N, for both PPM_i and PPX_i given for PR sampling, can be used without alteration (see Sections 8.5–8.7). The variance of these estimators, however, will be different, depending on the method of PNR employed. Since these are matters somewhat more sophisticated than the scope of this book, we shall not include them. Some references at the end of the chapter will give some guidance to those interested.

8.10.   Some Methods of Selecting PNR Samples

In the previous section we considered the elements of PNR sampling for a particular method of selection, namely, PPS at each step. However, the theory presented was quite general—allowing adaptability to any set of P_i and P_{ij} within some rather modest restraints. We shall now consider other methods of selection and examine their properties.

PNR(2): Or, the Midzuno-Lahiri Scheme.   In the first step an element is drawn with PPS, in the second and subsequent steps with equal probability from the remaining elements. In this case it can be shown that the probability of selecting the set of n elements is proportional to the aggregate size of the set [see Midzuno (1950) and Lahiri (1951)].

PNR(3): Or, the Random-Systematic Scheme.   The elements are put into a random order from 1 to N, and the cumulative totals of sizes are calculated [see Hartley and Rao, (1962)]. Thus we have:

$$
\begin{aligned}
&\text{Elements:} &&1, 2, \ldots, i, \ldots, N \\
&\text{Relative size:} &&A_1, A_2, \ldots, A_i, \ldots, A_N \\
&\text{Cumulative relative size:} &&A_1, A_1 + A_2, \ldots, \sum_{i=1}^{i} A_i, \ldots, \sum_{i=1}^{N} A_i = 1
\end{aligned}
$$

The sampling interval k is determined by

$$
k = \frac{A}{n} = \frac{1}{n}
$$

and a random number, RN, is drawn in the interval $0 < RN < 1/n$ for the first draw, $RN + 1/n$ for the second, $RN + 2/n$ for the third, and so on. The elements whose cumulative sizes include the RNs are the selected elements.

A number of other methods that have been proposed, and to a a certain extent evaluated [see Jessen (1969)] will be briefly outlined later in this section. A scheme that may have some merit, particularly in dealing with more complex controls, will be presented in the next section.

A Comparison of the Three Schemes.   It may be of interest to examine the two additional PNR schemes and to compare them with PNR(1), with PR, and with the two more familiar equal-probability schemes, RNR and RR. The essential statistical properties of the various schemes are to some extent revealed by the P_i and P_{ij} each possesses, since it is on these and their relationships with the A_i and Y_i that sampling variance, var $(\hat{Y})$, depends. Table 8.14 presents the data for the same four-element universe used previously. Note that although the PNR schemes favor the selection of the "larger" (i.e., larger A_i) elements, they differ somewhat in the amount of probability assigned to each. This is true for pairs as well as individual elements (i.e., P_{ij} as well as P_i). The resulting variances and efficiencies relative to RNR sampling are as follows:

Table 8.14.   Comparison of six sampling schemes with regard to inclusion probabilities, P_i and P_{ij}

	Sampling scheme					
P_i or P_{ij}	RR	RNR	PR	PNR(1)	PNR(2)	PNR(3)
P_1	.500[a]	.500	.200[a]	.235	.400	.200
P_2	.500[a]	.500	.400[a]	.441	.467	.400
P_3	.500[a]	.500	.600[a]	.609	.533	.600
P_4	.500[a]	.500	.800[a]	.715	.600	.800
P_{12}	.1250	.166	.040	.047	.100	.066
P_{13}	.1250	.167	.060	.077	.134	.067
P_{14}	.1250	.166	.080	.111	.166	.067
P_{23}	.1250	.167	.120	.161	.166	.066
P_{24}	.1250	.167	.160	.233	.201	.267
P_{34}	.1250	.167	.240	.371	.233	.467
P_{11}	.0625	—	.010	—	—	—
P_{22}	.0625	—	.040	—	—	—
P_{33}	.0625	—	.090	—	—	—
P_{44}	.0625	—	.160	—	—	—

[a] Inclusion probability in this case being defined as $P_i = \sum_{j \neq i} P_{ij} + 2P_{ij}$.

Sampling method		var $(\hat{Y})$	Relative efficiency
RR:	Random replacement	60.00	67
RNR:	Random nonreplacement	40.00	100
PR:	Probability replacement	3.67	1090
PNR:	Probability nonreplacement		
	(1) PPS each step	3.99	1005
	(2) PPS first step, equal thereafter	16.82	238
	(3) Random-systematic	2.34	1715

In the mini-universe dealt with, rather large gains in efficiency were achieved by some of the schemes—particularly the PNR scheme using random-systematic selection. Note that two PNR schemes are inferior to simple PR—a situation that would not occur in equal probability sampling. It appears that PNR sampling is very sensitive to the particular method of selection one adopts.

Outlines of Other Schemes.   A number of procedures for selecting PNR samples have been proposed, but at the time of this writing no one of them seems to have the sort of general acceptance that the RNR (random nonreplacement) schemes have. This may be due, in part, to the fact that in the case of small universes (perhaps the more relevant group for this method of sampling) the variance of estimates of interest are quite sensitive to the method of selection as well as to the nature of the variable being dealt with.

It may be convenient to group the procedures described in the literature into six groups as follows:

GROUP 1. PPS.   Procedures whose essential characteristic seems to be that the samples of size n are drawn in n successive steps and on step 1 the probability of selecting the element is PPS, that is, A_i. On the second and subsequent steps the basis for selecting elements may follow any of several rules. Included in this group are schemes suggested by Midzuno (1950) and Lahiri (1951), where at the second and subsequent steps elements are drawn with equal probability; and Yates and Grundy (1953), where probabilities for these steps are based on PPS of the *remaining* sizes.

GROUP 2. RANDOMIZED-SYSTEMATIC.   The N elements are put into random order, the A_i are cumulated, and a systematic sample of n elements is selected. Early description given by Goodman and Kish (1950) and its properties examined by Hartley and Rao (1962), Hartley (1966), and Connor (1966).

GROUP 3. WORKING PROBABILITIES.   Like those in group 1, the schemes in this group draw samples in n successive steps, but a set of calculated "working probabilities," or A_i', are determined to satisfy the condition that, using these A_i', after n successive steps $P_i = nA_i$. Contributors to this group are Narain (1951), Horvitz and Thompson (1952), Yates and Grundy (1953), Brewer and Undy (1962), and others. One problem of interest is that of determining a practicable procedure for calculating the A_i', particularly for $n > 2$.

GROUP 4. REJECTIVE. The distinguishing characteristic of this group is that although acceptable samples contain n distinct elements, they are visualized as drawn with the *replacement* of elements. If any element appears more than once, the whole sample is "rejected" and a new trial is undertaken. The scheme, using probabilities based on A_i, is presented by Yates and Grundy (1953) and Durbin (1953). Rao (1963) and Hajek (1964) present the scheme suggesting the use of working probabilities. A_i', in order to achieve $P_i = nA_i$.

GROUP 5. "PREFERRED" P_{ij}. A scheme proposed by Raj (1956) is put here. Dealing with the $n = 2$ case, Raj was concerned with a scheme for minimizing var $(\hat{Y})$ through the choice of P_{ij}, when it is assumed that $Y_i = \alpha + \beta A_i$ and required that $P_i = nA_i$.

GROUP 6. RANDOMIZED GROUPINGS. The N elements are randomized into n groups. Two subgroups can be distinguished: (i) Rao, Hartley and Cochran (1962) propose the groups be equal as possible in number of elements, and (ii) Cochran (1963) suggests the groups be as equal as possible in aggregate measure of size. Having the n randomized groupings we obtain the sample by selecting one element with PPS from each of the n groups.

8.11. Additional Methods of Selecting PNR Samples

Two selection schemes that provide $P_i = nA_i$ and in some practical cases a rather simple method of draw will be presented from Jessen (1969). One scheme can deal with any n with little difficulty; the other has some properties of an optimal nature but, as presented here, is limited to $n = 2$.

The Decremental Scheme. We shall introduce the method by an example. Suppose we have a universe of four elements, $N = 4$, where A_i, $i = 1, \ldots, 4$, is the relative measure of size, and we wish to select samples of two distinct elements, $n = 2$, from the four such that $P_i = nA_i$—that is, the probability that element i is included in the sample is nA_i. Suppose we establish a sampling tableau and designate a set of samples, U_m, as shown in Table 8.15.

We have 10 samples of $n = 2$, which, taken as a set, have some useful properties. For example, if each of the 10 samples is assigned a probability of selection of $1/10$ ($= D_m$) it will be noted that $P_i = nA_i$, since the ith element appears in $10A_i$ of the samples. The set of U_m can be regarded as a "feasible set" of samples that satisfy the requirement that $P_i = nA_i$ provided each sample in the set is given the probability of selection, D_m.

Table 8.15.   Tableau for decremental method ($N = 4$, $n = 2$)

| | | i: 1 | 2 | 3 | 4 | |
| | | A_i: .1 | .2 | .3 | .4 | |

(U_m)	Sample elements, nA_i:	.2	.4	.6	.8	Decrement (D_m)
U_1	3, 4	$\frac{-}{.2}$	$\frac{-}{.4}$	$\frac{.1}{.5}$	$\frac{.1}{.7}$	.1
U_2	3, 4	$\frac{-}{.2}$	$\frac{-}{.4}$	$\frac{.1}{.4}$	$\frac{.1}{.6}$	.1
U_3	3, 4	$\frac{-}{.2}$	$\frac{-}{.4}$	$\frac{.1}{.3}$	$\frac{.1}{.5}$	.1
U_4	2, 4	$\frac{-}{.2}$	$\frac{.1}{.3}$	$\frac{-}{.3}$	$\frac{.1}{.4}$	.1
U_5	3, 4	$\frac{-}{.2}$	$\frac{-}{.3}$	$\frac{.1}{.2}$	$\frac{.1}{.3}$	.1
U_6	2, 4	$\frac{-}{.2}$	$\frac{.1}{.2}$	$\frac{-}{.2}$	$\frac{.1}{.2}$	.1
U_7	1, 2	$\frac{.1}{.1}$	$\frac{.1}{.1}$	$\frac{-}{.2}$	$\frac{-}{.2}$	.1
U_8	3, 4	$\frac{-}{.1}$	$\frac{-}{.1}$	$\frac{.1}{.1}$	$\frac{.1}{.1}$	.1
U_9	2, 3	$\frac{-}{.1}$	$\frac{.1}{0}$	$\frac{.1}{0}$	$\frac{-}{.1}$	.1
U_{10}	1, 4	$\frac{.1}{0}$	$\frac{-}{0}$	$\frac{-}{0}$	$\frac{.1}{0}$	$\frac{.1}{1.0}$

The procedure followed here was that of designating for each U_m those elements having the largest amount of residual nA_i, where nA_i is looked upon as a sort of supply of stock and that part of it is lost each time the element appears in a sample. When residual stocks of each element are the same, any arbitrary selection is taken. A solution will always result if these rules are followed.

If duplications—that is, those containing the same elements—are combined, the set of 10 samples can be reduced to five distinct ones, as follows.

Distinct samples (U_t)	Feasible samples (U_m)	Elements (ij)	D_t
1	7	12	.1
2	10	14	.1
3	9	23	.1
4	4, 6	24	.2
5	1, 2, 3, 5, 8	34	.5
			—
			1.0

The procedure generalizes easily to any n. With some modification, the number of steps can be reduced by determining the maximum "bite"—that is, D_m—that can be taken for each sample. If at each step those elements possessing the largest values of nA_i (or their subsequent residuals) are designated for a feasible sample, and if a decrement, Δ, as large as the smallest designated nA_i is taken, then a set of feasible samples will usually be generated with a minimum number of sets. The Δs obtained will serve as the appropriate probabilities to be associated with them. For example, applying this method to the above universe of four elements, we have:

Element, i: 1   2   3   4
   nA_i: .2   .4   .6   .8

Sample, 1:   .2   .4   ⑥   ⑧ $\Delta_1 = .6$
     2:   .2   ④ —   ② $\Delta_2 = .2$
     3:   ②   ② —   — $\Delta_3 = .2$

Hence these three samples, with probabilities equaling their Δs provide P_is that are proportional to A_is. However, it may be noted that some P_{ij}s are zero. We shall refer to this selection procedure as decremental method 2. For details see Jessen (1969).

◇   ◇   ◇

8.12.   Discussion and Summary

Sampling with unequal probabilities adds a powerful tool to the list of the sample designer. It provides another method by which prior information on a universe can be utilized to increase the precision of sampling, which to a certain extent is in competition with sampling-unit equalization, estimation, and stratification.

There are situations when probability sampling should be seriously considered for adoption in sampling design:

1. When the measure of size X_i is proportionately related to Y_i—that is, the relationship is linear and goes through the origin.
2. When an unbiased ratio estimate of y to x is required and sample sizes are small. (This is at present the only method of obtaining unbiased estimates of such ratios.)
3. When map frames are used and point sampling provides a simple and inexpensive method of selecting area elements.
4. When dealing with cases where unequal probability samples arise naturally or are easily available.

Like stratification and sampling-unit equalization, probability sampling is most effective when only one parameter is to be estimated. With several unrelated parameters the problem of finding an overall optimum arises, and then serious consideration should be given to the use of better estimators or the combination of selection probabilities and estimators.

PNR sampling, as it is now understood, provides some opportunities for gain over PR sampling, perhaps about what one would expect from RNR over RR sampling. But there is no certainty in this gain, and there can even be losses. With our present understanding, it may be wise to avoid this type of sampling unless one feels confident he can take care of himself in this somewhat treacherous area.

8.13. Review Illustrations

1. A method developed to estimate the total number of fruits on a tree uses the following procedure. Starting from the trunk of a tree, one chooses which of several branches to follow up a tree by a probability scheme where the branch at each forking is chosen with probability proportional to its cross-sectional area. The procedure is carried through the forkings until one arrives at a branch small enough so that the fruits on it can be easily and accurately counted. Suppose two branches were selected at the end of a series of forkings with the following results:

Sample 1: $P_1 = .50, P_2 = .12, P_3 = .70, P_4 = .10, Y_1 = 20.$
Sample 2: $P_1 = .10, P_2 = .20, P_3 = .40, Y_2 = 10.$

(a) What is the estimated total fruit on the tree?
(b) What is its estimated relative standard error?

Solution

(a) Probability of selecting branch is

$$P_1 = (.50)\,(.12)\,(.70)\,(.10) = .0042,$$
$$P_2 = (.10)\,(.20)\,(.40) = .0080;$$
$$(\hat{Y})_1 = Y_1/P_1 = 20/(.0042) = 4762,$$
$$(\hat{Y})_2 = Y_2/P_2 = 10/(.0080) = 1250.$$

Hence $\hat{Y} = (4762 + 1250)/2 = 3006.$

(b)
$$\widehat{\text{var}}\,(\hat{Y}_{PR}) = [1/n\,(n-1)]\left[\sum(\hat{Y})_i^2 - n\hat{Y}_{PR}^2\right]$$
$$= (1/n)\left[(\hat{Y}_1 - \hat{Y}_2)^2/2\right]$$
$$= (1/2)\left[12,\,334,\,144/2\right]$$
$$= 3{,}083{,}536.$$
$$\widehat{\text{RSE}} = \sqrt{\text{var}/\hat{Y}} = (3{,}083{,}536)^{1/2}/3066 = 0.584 \quad \text{or}$$
$$58\%.$$

2. It is desired to determine the average cost of producing peaches in a certain peach production area. Records are available from each grower (through the growers' association) on total tons produced last year, X_i, but detailed cost records for each activity (pruning, spraying, picking, etc.) are kept only on a small sample of growers for a year. Suppose average cost per ton for the industry is defined as

$$\theta = \frac{\sum_{i=1}^{N} Y_i}{\sum_{i=1}^{N} X_i},$$

where Y_i is total production costs of the *i*th grower.

(a) If sampling is PR on X_i, what would be the appropriate estimator?

(b) In your judgment, how would (a) compare in efficiency with RR sampling using the estimator:

$$\hat{\theta}_{RR} = \frac{\sum_{i=1}^{n.} Y_i}{\sum_{i=1}^{n'} X_i}.$$

Solution

(a)
$$\hat{\theta}_{PR} = \frac{(1/n)\sum Y_i/(X_i/X)}{(1/n)\sum X_i/(X_i/X)}$$

$$= \frac{1}{n}\sum Y_i/X_i$$

or, if $X_i = M_i$,

$$\hat{\theta}_{PR} = \frac{1}{n}\sum \bar{Y}_i.$$

(b)   Probably fairly good if Y_i/X_i is most variable for large X_i; however, simple RNR sampling with ratio estimator is likely to be nearly as good. (An additional reason for PR here was to permit more stratification, such as lattice sampling, presented in Chapter 11.)

3.   In a study of new methods for estimating the production of corn in the United States some data were collected that will help illustrate some of the uses of probability sampling. For a portion of the overall study we have, in Table 8.16, the data collected on 36 farms in southcentral Iowa. The farms were selected by putting down at random 36 points onto maps of the 11-county area; hence, farms were selected with probabilities proportional to their overall sizes. (Although the sample was actually stratified by area, let us ignore this fact in order to keep the description simple.) The farms on which the points fell were located in the field, and from each farm information was obtained on X_i (the area of the farm, in acres), Y_i (the total quantity of corn produced during the crop year, in bushels), and Z_i (the area from which corn was harvested, in acres).

We wish to estimate the following parameters and the relative standard errors of our estimates:

(a)   Total production of corn, Y.
(b)   Total area of corn harvested, Z.
(c)   Total number of farms in the universe, N
(d)   Production of corn per acre harvested, Y/Z.
(e)   Fraction of farm area in corn, Z/X.
(f)   Production of corn per farm, Y/N.
(g)   Average size of farm, X/N.
(h)   Production of corn per acre of farm, Y/X.

The basic computations are given in Table 8.16. The total area of the population X is given by the Iowa Annual Farm Census as $X = 3,471,367$ acres. (This census does not give the total number of farms.)

Solution

(a)   Total production of corn: Y. We have from Eq. 8.23

$$\hat{Y}_{PR} = (X/n') \sum (Y_i/X_i)$$

$$= (3,471,367)\,(211.917)/36 = 20,434,000 \text{ bushels.}$$

Table 8.16. Basic data and computations for the 36 farms

i Sample farm no. (1)	X_i Area in farm (2)	Y_i Production of corn (bushels) (3)	Z_i Area in corn (acres) (4)	i Sample farm no. (1)	X_i Area in farm (2)	Y_i Production of corn (bushels) (3)	Z_i Area in corn (acres) (4)
1	519	5350	107	19	240	1960	49
2	303	960	32	20	10	240	4
3	320	416	13	21	54	975	15
4	160	1840	46	22	680	2250	150
5	560	3160	79	23	86	0	0
6	347	1540	44	24	120	1600	40
7	40	210	7	25	510	5439	111
8	320	1824	57	26	359	2940	98
9	450	1760	88	27	160	600	20
10	440	1500	60	28	180	240	12
11	578	1224	51	29	80	0	0
12	150	440	22	30	320	2430	81
13	120	945	27	31	200	1650	33
14	110	1176	21	32	780	0	0
15	238	1540	44	33	255	1400	70
16	250	0	0	34	120	0	0
17	440	1260	42	35	187	640	16
18	320	1500	50	36	520	2128	56
Total					10,526	51,137	1545

Summary of computations

$\sum X_i = 10{,}526$ $\sum X_i^2 = 4{,}355{,}294$ $\sum 1/X_i = 0.293{,}415$ $\sum 1/X_i^2 = 0.011{,}926{,}361$

$\sum Y_i = 51{,}137$ $\sum Y_i^2 = 130{,}997{,}239$ $\sum Y_i/X_i = 211.917$ $\sum Y_i^2/X_i^2 = 2{,}188.60$

$\sum Z_i = 1545$ $\sum Z_i^2 = 113{,}185$ $\sum Z_i/X_i = 5.6844$ $\sum Z_i^2/X_i^2 = 1.243{,}383$

                                       $\sum Y_i/Z_i = 1091$ $\sum Y_i^2/Z_i^2 = 42{,}855$

                                                             $\sum Y_i Z_i/X_i^2 = 49.0696$

                                                                        $\sum Y_i/X_i^2 = 3.600646$

An estimate of the relative variance is given by

$$\text{rel } \widehat{\text{var}}\, (\hat{Y}) = \frac{\widehat{\text{var}}\, (\hat{Y})}{\hat{Y}^2} = \frac{\hat{v}_{y|x}^{*2}}{n'}$$

$$= \frac{(n')^2}{n'-1} \; \frac{\sum (Y_i/X_i)^2 - (1/n')[\sum Y_i/X_i]^2}{n'[\sum (Y_i/X_i)]^2} \qquad \text{from Eq. 8.54c}$$

$$= \frac{36}{35} \cdot \frac{2188.60 - (211.917)^2/36}{(211.917)^2}$$

$$= \frac{36}{35} \cdot \frac{941.13}{44{,}908.8} = 0.0216;$$

hence $\widehat{RSE}\,(\hat{Y}) = (0.0216)^{1/2} = 0.147$ or 14.7%.

(b)   Total area of corn harvested: Z. We have

$$\hat{Z}_{PR} = (X/n')\sum(Z_i/X_i)$$

$$= (3,471,367)\,(5.6844)/36 = 548,129 \text{ acres.}$$

As in (a), we have

$$\text{rel }\widehat{var}\,(\hat{Z}) = \frac{\widehat{var}\,(Z)}{\hat{Z}^2} = \frac{\overset{*2}{v}_{z|x}}{n'}$$

$$= \frac{(n')^2}{n'-1}\cdot\frac{\sum(Z_i/X_i)^2-(1/n')[\sum(Z_i/X_i)]^2}{n'[\sum(Z_i/X_i)]^2} \qquad \text{from Eq. 8.54c}$$

$$= \frac{36}{35}\cdot\frac{1.243,383-(5.6844)^2/36}{(5.6844)^2}$$

$$= \frac{36}{35}\cdot\frac{0.3458}{32.312} = 0.01101;$$

hence RSE $(\hat{Z}) = (0.01101)^{1/2} = 0.105$ or 10.5%.

(c)   The total number of farms: N. We have from Eq. 8.55

$$\hat{N} = \frac{X}{n'}\sum\frac{1}{X_i} = \frac{3,471,367}{36}\,(0.2934) = 28,300 \text{ farms.}$$

Its estimated relative variance is given by Eqs. 8.59a and 8.59b.

$$\text{rel }\widehat{var}\,(\hat{N}) = \left(\frac{1}{n'}\right)\overset{*2}{v}_{1/x} = \left(\frac{1}{n'}\right)\cdot\frac{(n')^2}{n'-1}\cdot\frac{\sum(1/X_i)^2-(1/n')[\sum(1/X_i)]^2}{[\sum 1/X_i]^2}$$

$$= \frac{36}{35}\cdot\frac{0.011,926-(0.2934)^2/36}{(0.2934)^2} = 0.1108.$$

And $\widehat{RSE} = (0.1108)^{1/2} = 0.333$ or 33.3% (indicating this is not a very efficient way to estimate the number of farms).

(d)   Production of corn per acre harvested: Y/Z. We have from Eq. 8.51

$$\hat{Y}/\hat{Z} = \frac{\sum Y_i/X_i}{\sum Z_i/X_i} = \frac{211.917}{5.6844} = 37.3 \text{ bushels per acre.}$$

To estimate its relative standard error we have from Eq. 8.53a

$$\widehat{RSE}\,(\hat{Y}/\hat{Z}) = \left[\left(\frac{1}{n'}\right)\left(\overset{*2}{v}_{y|x}+\overset{*2}{v}_{z|x}-2\overset{*}{v}_{yz|x}\right)\right]^{1/2},$$

where

$$
\begin{aligned}
\overset{*2}{v}_{y|x} &= \frac{(n')^2}{n'-1} \cdot \frac{\sum (Y_i/X_i)^2 - (1/n')[\sum Y_i/X_i]^2}{[\sum Y_i/X_i]^2} \\
&= \frac{(36)^2}{35} \cdot \frac{2,188.60 - (211.917)^2/36}{(211.917)^2} = 0.7760,
\end{aligned}
$$

$$
\overset{*2}{v}_{z|x} = \frac{(36)^2}{35} \cdot \frac{1.243,383 - (5.6844)^2/36}{(5.6844)^2} = 0.3963,
$$

$$
\overset{*}{v}_{yz|x} = \frac{(36)^2}{35} \cdot \frac{49.0696 - (211.917)(5.6844)/36}{(211.917)(5.6844)} = 0.4798.
$$

Therefore,

$$
\widehat{\text{RSE}}\,(\hat{Y}/\hat{Z}) = [(1/36)(0.7760 + 0.3963 - 2(0.4798))]^{1/2}
$$
$$
= 0.0769 \text{ or } 7.7\%.
$$

(e)   Fraction of farm area in corn: Z/X. We have $\hat{Z}$ from part (b); hence,

$$
\frac{\hat{Z}}{X} = \frac{548,129}{3,471,367} = 0.1579 \text{ corn acres per farm acre.}
$$

Its estimated variance is the same as that for $\hat{Z}$ except for the factor $1/X^2$, and the RSE will be identical to that for Z. Hence, from (b),

$$
\text{RSE}\,(\hat{Z}/X) = \text{RSE}\,(\hat{Z}) = 0.105 \text{ or } 10.5\%.
$$

(f)   Production of corn per farm: Y/N. We have Eq. 8.60:

$$
\frac{\hat{Y}}{\hat{X}} = \frac{\sum (Y_i/X_i)}{\sum (1/X_i)} = \frac{211.917}{0.293415} = 722.24 \text{ bushels per farm.}
$$

Its estimated relative variance is given by Eq. 8.62a, whence

$$
\widehat{\text{RSE}}\,(\hat{Y}/\hat{N}) = \left[\left(\frac{1}{n'}\right) \left(\overset{*2}{v}_{y\cdot x} + \overset{*2}{v}_{1/x} - 2\overset{*}{v}_{y(1/X)} \right) \right]^{1/2},
$$

where

$$
\overset{*2}{v}_{y|x} = 0.7760 \qquad \text{from part (d),}
$$

$$
\overset{*2}{v}_{1/x} = 36\,(0.1170) = 4.212 \qquad \text{from Eq. 8.59a,}
$$

$$
\overset{*}{v}_{y(1/X)} = \frac{(n')^2}{n'-1} \cdot \frac{\sum [(Y_i/X_i)(1/X_i)] - (1/n')(\sum Y_i/X_i)(\sum 1/X_i)}{(\sum Y_i/X_i)(\sum 1/X_i)}
$$

from Eq. 8.63c,

$$= \frac{(36)^2}{35} \cdot \frac{3.600,646 - (211.917)\,(0.293,415)/36}{(211.917)\,(0.293,415)}$$

$$= 1.1156,$$

$$\text{RSE}\,(\hat{Y}/\hat{N}) = \left[\left(\frac{1}{n'} \right) \left(\overset{*}{v}^2_{y|x} + \overset{*}{v}^2_{1/x} - 2\overset{*}{v}_{y(1/X)} \right) \right]^{1/2} \qquad \text{from Eq. 8.62a,}$$

$$= \{ \tfrac{1}{36}\, [0.7760 + 3.9888 - 2(1.1156)] \}^{1/2}$$

$$= (2.5336/36) = (0.704)^{1/2} = 0.265 \text{ or } 27\%.$$

(g)   Average size of farm: X/N. To estimate X/N we have the estimator:

$$\frac{X}{\hat{N}} = \frac{n'}{1/X_i} = \frac{36}{0.2934} = 122.7 \text{ acres per farm,}$$

and its estimated relative variance is given by Eq. 8.62a, wherein $\overset{*}{v}^2_{x|x}$ $= \overset{*}{v}_{x(1/X)|x} = 0$. Thus

$$\text{rel } \widehat{\text{var}}\,(X/\hat{N}) = \left(\frac{1}{n'} \right) \overset{*}{v}^2_{1/x} = 0.1108 \qquad \text{from part } (c).$$

Hence,

$$\text{RSE}\,(X/\hat{N}) = (0.1108)^{1/2} = 0.333 \text{ or } 33.3\%.$$

(h)   Production of corn per acre of farm: Y/X. Obtaining $\hat{Y}$ from (a),

$$\frac{\hat{Y}}{X} = \frac{20,434,000}{3,471,367}$$

$$= 5.89 \text{ bushels per farm acre.}$$

Since Y/X is of the same form as Z/X—that is, a variable divided by the measure of size—we find that the estimator or RSE is that of the variable; hence

$$\text{RSE}\,(\hat{Y}/X) = \text{RSE}\,(\hat{Y}) = 14.7\% \qquad \text{from } (a).$$

4.   Consider the problem of estimating the total value of M numbers located on a wheel. Suppose that when the wheel is spun, all points on the periphery are equally likely to stop at the arrow, A. Two observations can be made: (1) the distance (say clockwise) to the last Y_i that passed point A and (2) the value of that Y_i. The Y_i are arbitrarily distributed on the wheel. What is an appropriate estimator of Y?

$$Y = \sum Y_i, \quad i = 1, 2, \ldots, M$$
$$X = \sum X_i$$
$$P_3 - P_4 = X_3$$
$$P_3 - P_A = x_3$$

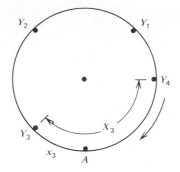

Solution

$\hat{Y} = Y_i/(X_i/X) = X(Y_i/X_i)$, but we don't know X_i here. An unbiased estimate of X_i is given by $\hat{X}_i = 2x_i$, where x_i is the portion of X_i that can be observed. Hence a possible estimator is

$$\hat{Y} = X(Y_i/\hat{X}_i) = \frac{X Y_i}{2x_i}.$$

However, this is not unbiased.

5. Here is a universe of five elements with relative sizes, A_i. Select a probability sample of $n = 2$ from this universe such that $P_i \propto A_i$, using the systematic method, and letting RF (random fraction) = 0.4.

i: 1 2 3 4 5
A_i: .1 .1 .2 .3 .3

Solution

The tabular format is as follows:

i	A_i	Cumulative A_i	Sample no.
1	.1	.1	
2	.1	.2	.2
3	.2	.4	
4	.3	.7	.7
5	.3	1.0	

Since $n = 2$, the sampling interval, M/m, the total number of selection numbers in cumulative A_i, divided by total elements desired in the sample. Hence M/m

$= 1.0/2 = 0.5$. Since $RF = 0.4$, $RN =$ the range $0.5 \times RF = (0.5)(0.4) = 0.2$. Hence our selection numbers are 0.2, 0.7 ($= 0.2 + 0.5$). Therefore elements 2 and 4 are selected.

6. Here is the same universe presented in problem 5 above. Now select a probability sample with the requirement that $P_i \propto A_i$ but use the decremental method 2 of Section 8.11. Again, take $RF = 0.4$.

i:	1	2	3	4	5
A_i:	.1	.1	.2	.3	.3

Solution

The format by step follows:

i:	1	2	3	4	5		Cumulative	Sample no.
nA_i:	.2	.2	.4	.6	.6			
	.2	.2	(.4)	.6	(.6)	$\Delta = .4$	.4	.4
	.2	.2	—	(.6)	(.2)	$\Delta = .2$	.6	
	.2	(.2)	—	(.4)	—	$\Delta = .2$	.8	
	(.2)	—	—	(.2)	—	$\Delta = .2$	1.0	

$$\Sigma = 1.0$$

Here the $RN = RF$, since M, the range for RF, is the same. Hence the sample selected consists of elements 3 and 5. (Note that RF should not be known to the person doing the ordering in selection.)

8.14. References

Brewer, K. R. W. Samples of Two Units Drawn with Unequal Probabilities Without
Undy, G. C.   Replacement. *Australian Journal of Statistics*, 4(3): 89–100.
1962

Cochran, W. G. *Sampling Techniques*. (2nd ed.) New York: John Wiley & Sons.
1953, 1963

Connor, W. S.   An Exact Formula for the Probability That Two Specified Sampling Units
1966   Will Occur in a Sample Drawn with Unequal Probabilities and Without
                Replacement. *Journal of the American Statistical Association*, **61**: 384–390.

Durbin, J.   Some Results in Sampling Theory When Units Are Selected with Unequal
1953   Probabilities. *Journal of the Royal Statistical Society*, Series B, **15**(2): 262–269.

Goodman, J. R. Controlled Selection—A Technique in Probability Sampling. *Journal of the*
Kish, L.   *American Statistical Association*, **45**: 300–372.
1950

Hajek, J.   Asymptotic Theory of Rejective Sampling with Varying Probabilities from a
1964   Finite Population. *Annals of Mathematical Statistics*, **35**(4): 1491–1523.

Hansen, M. H.   On the Theory of Sampling from Finite Populations. *Annals of Mathematical
Hurwitz, W. N.   Statistics*, **14**(4): 333–362.
1943

Hansen, M. H.   *Sample Survey Methods and Theory*. Vols. 1 and 2. New York: John Wiley &
Hurwitz, W. N.   Sons.
Madow, W. G.
1953

Hartley, H. O.   Systematic Sampling with Unequal Probability and Without Replacement.
1966   *Journal of the American Statistical Association*, **61**: 739–748.

Hartley, H. O.   Sampling with Unequal Probabilities and Without Replacement. *Annals of
Rao, J. N. K.   Mathematical Statistics* **32**(2): 350–374.
1962

Horvitz, D. G.   A Generalization of Sampling Without Replacement from a Finite Universe.
Thompson, D. J.   *Journal of the American Statistical Association*, **47**: 663–685.
1952

Jessen, R. J.   Encuesta por Muestreo de las Fincas en la Provincia de Buenos Aires,
Thompson, D. J.   Argentina. *Estadistica: Journal of the Inter-American Statistical Institute*
1958   (December). [Example of use of unequal probabilities in selecting farms for a
                   cluster (area) sampling unit.]

Jessen, R. J.   Some Methods of Probability Non-Replacement Sampling. *Journal of the
1969   American Statistical Association*, **64**(1): 175–193.

Lahiri, D. B.   A Method of Sample Selection Providing Unbiased Ratio Estimates. *Bulletin
1951   of the International Statistical Institute*, **33**(2): 133–140.

Midzuno, H.   An Outline of the Theory of Sampling Systems. *Annals of the Institute of
1950   Statistical Mathematics* (Japan), **1**: 149–156.

Midzuno, H.   On the Sampling System with Probability Proportional to Sum of Sizes.
1952   *Annals of the Institute of Statistical Mathematics*, **3**: 99–108.

Narain, R. D.   On Sampling Without Replacement with Varying Probabilities. *Journal of the
1951   Indian Society of Agricultural Statistics*, **3**(2): 169–175.

Raj, Des   A Note on the Determination of Optimum Probabilities in Sampling Without
1956   Replacement. *Sankhyā*, **17**: 197–200.

Rao, J. N. K.   A Simple Procedure of Unequal Probability Sampling Without Replacement.
Hartley, H. O.   *Journal of the Royal Statistical Society*, Series B, **24**(2): 482–491.
Cochran, W. G.
1962

Rao, J. N. K.   On Three Procedures of Uhequal Probability Sampling Without
1963   Replacement. *Journal of the American Statistical Association*, **58**: 202–215.

Sukhatme, P. V.   *Sampling Theory of Surveys with Applications*. Ames, Iowa: Iowa State
1954   College Press.

Thompson, D. J.   *A Theory of Sampling Finite Universes with Unequal Probabilities*. Ph.D.
1952   Dissertation, Iowa State College.

Yates, F.   Selection Without Replacement from Within Strata and with Probability
Grundy, P. M.   Proportional to Size. *Journal of the Royal Statistical Society*, Series B, **15**:
1953   253–261.

8.15   Mathematical Notes

Notation

Frame unit, i: $1, 2, \ldots, n, \ldots, N$
y-Population, Y_i: $Y_1, Y_2, \ldots, Y_N$.
x-Population, X_i: $X_1, X_2, \ldots, X_N$
No. of elements, M_i: $M_1, M_2, \ldots, M_N$

whence $Y = \sum Y_i, X = \sum X_i, M = \sum M_i, \bar{Y} = Y/N, \bar{X} = X/N, \bar{M} = M/N$.

If a sample of n' is drawn with P_i for each trial, consider the estimator of Y,

$$\hat{Y}_{PR} = \frac{1}{n'} \sum^{n'} \frac{Y_i}{P_i},$$

and the estimator of $\bar{Y}$,

$$\bar{y}_{PR} = \frac{1}{N} \hat{Y}_{PR}.$$

PR Sampling; $E[\hat{Y}_{PR}]$ **and var** $(\hat{Y}_{PR})$; $\overset{*}{S}{}^2$.   Regard n' trials as n' replicates of samples of size $n = 1$. Draw a sample of one with probability proportional to X_i, or $P_i = X_i/X$.

$$\hat{Y}_{PR} = \frac{1}{n'} \sum^{n'} \frac{Y_i}{P_i},$$

$$E[\hat{Y}_{PR}] = \frac{1}{n'} \sum^{n'} \left[\frac{Y_i}{P_i} \right]$$

$$= \frac{n'Y}{n'} = Y.$$

$$\text{var}\,(\hat{Y}_{PR}) = E[(\hat{Y}_{PR} - Y)^2]$$

$$= \frac{1}{(n')^2} \sum^{n'} E[Y_i^2/P_i^2] - Y^2$$

$$= \frac{n'}{(n')^2} [N(N-1)\overset{*}{S}{}^2 + Y^2] - Y^2$$

$$= \frac{N(N-1)\overset{*}{S}{}^2}{n'},$$

where

$$\overset{*}{S}{}^2 \equiv \frac{\sum^N Y_i^2/P_i - Y^2}{N(N-1)}$$

$E[\hat{Y}_{PR}^2]$; $E[\overset{*}{s}_0^2]$; var $(\hat{Y}_{PR})$

$$E[\hat{Y}_{PR}^2] = E\left[\left(\frac{\sum Y_i/P_i}{n'}\right)^2\right]$$

$$= \frac{1}{(n')^2} E\left[\sum^{n'} \frac{Y_i^2}{P_i^2} + \sum_{j \neq i} \left(\frac{Y_i}{P_i}\right)\left(\frac{Y_j}{P_j}\right)\right]$$

$$= \frac{1}{(n')^2} \left[n'N(N-1)\overset{*}{S}{}^2 + Y^2 + n'(n'-1)Y^2\right]$$

$$= \frac{N(N-1)\overset{*}{S}{}^2}{n'} + Y^2.$$

$$E[\overset{*}{s}_0^2] = E\left[\frac{1}{N^2(n'-1)} \sum^{n'}\left(\frac{Y_i}{P_i} - \hat{Y}_{PR}\right)^2\right]$$

$$= \frac{1}{N^2(n'-1)} \left[n'N(N-1)\overset{*}{S}{}^2 + n'Y^2 - n'N(N-1)\frac{\overset{*}{S}{}^2}{n'} - n'Y^2\right]$$

$$= \left(\frac{N-1}{N}\right)\overset{*}{S}{}^2.$$

$$E[\widehat{\text{var}}\,(\hat{Y}_{PR})] = E\left[\sum^{n'}\left(\frac{Y_i}{P_i} - \hat{Y}_{PR}\right)^2\right]$$

$$= E\left[\frac{N^2}{n'}\overset{*}{s}_0^2\right]$$

$$= \frac{N(N-1)}{n'}\overset{*}{S}{}^2.$$

Optimal P_is. For samples of size $n = 1$,

$$\text{var}\,(\hat{Y}) = \sum \frac{Y_i}{P_i} - Y^2.$$

To minimize $\text{var}(\hat{Y})$ where $\sum P_i = 1$ we have

$$F(P_i, \lambda) = \sum \frac{Y_i^2}{P_i} - Y^2 + \lambda(\sum P_i - 1),$$

$$\frac{\partial F}{\partial P_i} = 0 = -\frac{Y_i^2}{P_i^2} + \lambda, \cdot \quad \frac{\partial F}{\partial \lambda} = 0 = \sum P_i - 1 = 0, \qquad \text{since } \sum P_i = 1$$

$$\lambda = \frac{Y_i^2}{P_i^2}, \qquad \text{whence } \lambda = Y^2, \quad \text{since } \sum P_i = 1.$$

But

$$P_i^2 = \frac{Y_i^2}{\lambda}, \quad P_i = \left|\frac{Y_i}{Y}\right|, \qquad \text{since } P_i > 0.$$

This will be true for $n > 1$ for PR sampling, since each trial is a replicate of the $n = 1$ case.

$E[\hat{Y}_{\text{PNR}}]$. Since $\hat{Y}_{\text{PNR}} = (1/n) \sum^n Y_i/\bar{P}_i$, where $\bar{P}_i$ is P_i/n—that is, the probability element i is selected in a particular position among the possible n—then

$$E[\hat{Y}_{\text{PNR}}] = \frac{1}{n} E \sum^N \frac{Y_i I_i}{\bar{P}_i},$$

where I_i is 1 if element i is selected in a sample of n and 0 otherwise.

$$E[\hat{Y}_{\text{PNR}}] = \frac{1}{n} \sum^N \frac{Y_i E[I_i]}{\bar{P}_i}$$

$$= \frac{1}{n} \sum^N \frac{Y_i(n\bar{P}_i)}{\bar{P}_i} = Y,$$

since $E[I_i] = P_i = n\bar{P}_i$.

var $(\hat{Y}_{\text{PNR}})$

$$\text{var}(\hat{Y}_{\text{PNR}}) = E[(\hat{Y}_{\text{PNR}} - Y)^2]$$

$$= E\left[\left(\sum^N \frac{Y_i I_i}{P_i} - Y\right)^2\right]$$

$$= E\left[\sum^N \frac{Y_i^2 I_i^2}{P_i^2} + \sum_i^N \sum_{j \neq i}^N \frac{Y_i Y_j I_i I_j}{P_i P_j}\right] - Y^2$$

$$= \sum^N Y_i^2 \left(\frac{1 - P_i}{P_i}\right) + \sum_i^N \sum_{j \neq i}^N Y_i Y_j \left(\frac{P_{ij} - P_i P_j}{P_i P_j}\right),$$

since $E[I_i^2] = P_i$, $E[I_i I_j] = P_{ij}$, and

$$Y^2 = \left(\sum^N Y_i\right)^2 = \sum^N Y_i^2 + \sum^N \sum^N Y_i Y_j.$$

$\widehat{var} \ (\hat{Y}_{PNR})$

$$var \ (\hat{Y}_{PNR}) = \sum^N Y_i^2 \left(\frac{1-P_i}{P_i}\right) + \sum_{j \neq i}^N \sum^N Y_i Y_j \left(\frac{P_{ij} - P_i P_j}{P_i P_j}\right)$$

To estimate $\widehat{var} \ (\hat{Y}_{PNR})$ we observe that

$$\sum^N \left(\frac{Y_i^2}{P_i}\right) \quad \text{unbiasedly estimates} \quad \sum Y_i^2$$

and

$$\frac{1}{n} \sum_i^n \sum_{j \neq i}^n \left(\frac{Y_i Y_j}{P_{ij}}\right) \quad \text{unbiasedly estimates} \quad \sum^N \sum^N Y_i Y_j;$$

hence

$$\widehat{var} \ (\hat{Y}_{PNR}) = \sum^n \frac{Y_i^2}{P_i^2} (1 - P_i) + \sum_i^n \sum_{j \neq i}^n Y_i Y_j \left(\frac{P_{ij} - P_i P_j}{P_i P_j P_{ij}}\right)$$

is an unbiased estimator of $var \ (\hat{Y}_{PNR})$.

8.16. Exercises

1. A study of cotton buyers in a certain state is being considered. It is learned that no list of such buyers is available but a list of all cotton farmers (sellers) is. Suppose a random sample m of the M cotton sellers is taken to determine the names of buyers of their cotton. From the list of sellers (unduplicated) so obtained, a random sample of n is selected for the survey.

(a) If Y_i is the total number of the ith buyer, what is an appropriate estimator of $\bar{Y}$, the mean number of employees per buyer, in the state? What additional information, if any, is required? (Assume each farmer sells to one and only one buyer.)

(b) Suppose some farmers sell to more than one buyer. Would this fact affect your choice of estimator for $\bar{Y}$? What would it affect, if anything?

2. A scheme for determining the nature and extent of error in a company's books is as follows. Let entries in the books be numbered serially from 1 to N. Denote the "book" value and the "true" value (obtained by audit) of entry i by

X_i and Y_i, respectively. Denote the largest X_i by X_L. Now choose a pair of random numbers, $\overset{*}{N}$ and $\overset{*}{X}$, where $1 \le \overset{*}{N} \le N$ and $1 \le \overset{*}{X} \le X_L$. Entry i is regarded as chosen for audit if $\overset{*}{N} = i$ and $\overset{*}{X} \le X_i$ for that i. Otherwise that pair of $\overset{*}{N}$ and $\overset{*}{X}$ are rejected and a new pair is drawn by the same procedure. The process is repeated until n' entries are drawn, with replacement. The audit is carried out on the n' entries.

The proposed estimator of Y/X, the ratio of correct to book value over the total of N entries, is given by

$$\frac{\hat{Y}}{X} = \frac{1}{n'} \sum_{i=1}^{n'} \frac{Y_i}{X_i}.$$

(a) Evaluate the scheme for bias.

(b) What is the expression for the variance of this estimator?

(c) Compare this scheme with an RR sample of the same size using a ratio estimator. Is it likely to be more or less accurate?

3. Here is a small ($N = 3$) universe with y and relative measure of size populations as follows:

			Σ
U_i:	1	2	3	
Y_i:	1.6	3.5	5.4	10.5
A_i:	.20	.35	.45	1.0

(a) For PNR sampling ($n = 2$) using the decremental method as a selection procedure, determine $P(s)$ for all possible samples.

(b) Determine the P_i.

(c) Determine the P_{ij}.

(d) Compute the $\hat{Y}_s$.

(e) Compute $E\{(\hat{Y} - Y)^2\}$—that is, var $(\hat{Y})$.

4. A sample of farms is obtained by putting down six points at random on a map of the universe being dealt with. Suppose the data obtained are:

Random point no.	Area of farms (acres)	Tenure of operator	Area in soybeans (acres)
1	160	Owner	20
2	1600	Owner	100
3	40	Renter	10
4	-------- (Same farm as no. 2) --------------
5	80	Renter	10
6	20	Renter	0

(a)   Estimate the mean acreage of soybeans per farm in the universe.

(b)   Estimate the fraction of farms that are owned by the operator.

(c)   Suppose we know that the total area in farms in this universe is 100,000 acres. Estimate the number of farms.

(d)   What is the standard error of (c)?

5.   For a number of the N blocks of a city the population in 1960 is known, and for a sample of n blocks $(n = N/20)$ the population in 1966 is to be counted for the purpose of estimating the 1966 population (Y) of all N blocks. Make suggestions for at least two suitable designs and resulting estimators of Y, giving reasons for their relative merits under certain assumptions about the city's population. Statements on relative precision of estimators should be supported by quotations of variance formulas.

CHAPTER 9

Subsampling (Multistage)

9.1.   Introduction

The technique of subsampling or sampling at two or more stages is widely used in surveys of various universes. Often the universe with which we are concerned comprises groups of elements on which information is desired; for example, persons are grouped according to state, county, or city in which they reside; farms are grouped also by geographic area; production of some products may be grouped by the output of a day or a shift; and so on. In these cases it may be convenient in sampling to first select a sample of these *groups* and then, in order to conserve sampling resources, to sample only a portion of the elements in each selected *group*. In cluster sampling it may be apparent that the elements in a cluster are so much alike that it is a waste of time to obtain costly information on all the elements when it seems a sample may do. In these and related cases the technique of subsampling is applicable, and, as we shall see, many of the ideas and principles discussed thus far become merely special cases of this rather flexible and useful design.

9.2.   Simple Subsampling; ANOVA

Suppose a universe consists of N groups of M_0 elements each. As in simple cluster sampling, we select at random n of the N groups for our sample of groups. But instead of observing all M_0 elements in each sample group, we select at random m_0 of them. Hence we sample in two stages, n of the N groups or *primaries* and m_0 of the M_0 elements or *secondaries* from each primary.

A simple subsample is represented by Fig. 9.1 for a universe consisting of

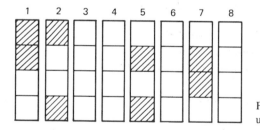

Fig. 9.1.   Two-stage sample of a small universe.

eight primaries of four secondaries each. The sample consists of four primaries of two secondaries each. In generalizing the procedure, we shall use the same notation as for cluster sampling, with some additions:

N, n = number of primaries in universe and in sample respectively,

M_0, m_0 = number of secondaries in each primary in universe and in sample respectively,

M, m = number of secondaries in the universe and in the sample respectively, $M = NM_0$; $m = nm_0$,

Y_{ij} = the value of the characteristic y for the jth secondary in the ith primary,

Y_i = total y in the ith primary, $Y_i = \sum Y_{ij}$ $(j = 1, 2, \ldots, M_0)$,

$Y, \bar{Y}, \bar{\bar{Y}}$ = total y, mean y per primary and mean y per secondary, respectively, in the universe; $Y = \sum Y_i$, $\bar{Y} = Y/N$, $\bar{\bar{Y}} = Y/M$,

$\bar{y}_i$ = sample mean of y per secondary in the ith primary, $\bar{y}_i$
$= \sum Y_{ij}/m_0$ $(j = 1, 2, \ldots, m_0)$,

$\bar{\bar{y}}$ = sample mean of y per secondary in n primaries in the sample; $\bar{\bar{y}}$
$= \sum \bar{y}_i/n$ $(i = 1, 2, \ldots, n)$.

To define the variances to be used with subsampling it is convenient to use an analysis of variance table for the universe such as that of Table 9.1.

Table 9.1. *Analysis of variance of a universe (on a per secondary basis)*

Source of variation	Degree of freedom	Mean square	Mean square defined as	
			Component of variance	Intraclass correlation[a]
Total	$NM_0 - 1$	MS (T)	S^2	S^2
Primaries	$N - 1$	MS (B)	$S_w^2 + M_0 S_B^2$	$S^2[1 + (M_0 - 1)\delta]$
Secondaries/ primaries	$N(M_0 - 1)$	MS (W)	S_w^2	$S^2[1 - \delta]$

[a]Approximate. Exact is given in Section 4.14.

The ANOVA (analysis-of-variance) table serves as a useful device for defining certain quantities and to show their relationship with each other. Also it indicates the computational procedure required to obtain the defined parameters. In Table 9.1 notice that the total mean square, MS (T), is as before (in cluster sampling, Section 4.5) the variance of elements (or what we are now calling secondaries) over the whole universe. The between mean square is defined as the sum of two terms (as in Sections 4.5 and 7.5). The between primary component of variance S_B^2 is obtained from the relationship

$$(9.1) \qquad S_B^2 = \frac{MS(B) - MS(W)}{M_0},$$

which we shall allow to be either positive or negative. When negative, we shall regard it as a "correction term", rather than a component of variance, owing to the presence of negative intraclass correlation.

The variance of a mean per secondary of a two-stage sample comprised of n randomly selected primaries and from each of these m_0 randomly selected secondaries is given in terms of mean squares by

$$(9.2) \qquad \text{var}(\bar{\bar{y}}) \doteq \left(\frac{N-n}{N}\right)\frac{MS(B)}{nM_0} + \left(\frac{M_0 - m_0}{M_0}\right)\frac{MS(W)}{nm_0},$$

or, in terms of variance components, by

$$(9.2a) \qquad \text{var}\,(\bar{\bar{y}}) \doteq \left(\frac{N-n}{N}\right)\frac{S_B^2}{n} + \left(\frac{NM_0 - nm_0}{NM_0}\right)\frac{S_W^2}{nm_0},$$

or, in terms of intraclass correlation (with N large so $N - 1 \doteq N$),

$$(9.2b) \qquad \text{var}\,(\bar{\bar{y}}) \doteq S^2 \left\{\left(\frac{N-n}{N}\right)\frac{1+\delta(M_0-1)}{nM_0} + \left(\frac{M_0-m_0}{M_0}\right)\frac{1-\delta}{nm_0}\right\},$$

which, if FPCs can be ignored and M_0 is large, simplifies to

$$(9.2c) \qquad \text{var}\,(\bar{\bar{y}}) \doteq \frac{S^2}{nm_0}[1+\delta(m_0-1)]\,.$$

In two-stage sampling, the variance of the sample mean is affected by variances from two sources: (i) the variation among primaries and (ii) the variation among secondaries within primaries. In the case where FPCs can be ignored, Eq. 9.2a simplifies to

$$(9.2d) \qquad \text{var}\,(\bar{\bar{y}}) = \frac{S_B^2}{n} + \frac{S_W^2}{nm_0} = \frac{1}{n}\left(S_B^2 + \frac{S_W^2}{m_0}\right).$$

Increasing the number of secondaries taken reduces only the "within-variance" term, S_W^2/nm_0, whereas increasing the number of primaries taken and keeping m_0 constant reduces both terms.

To estimate the variance of the sample mean we can use Eq. 9.2a by substituting estimates for S_B^2 and S_W^2. These can be obtained from an ANOVA of the sample results, as shown in Table 9.2.

Table 9.2. Analysis of variance of a sample (on a per secondary basis)

Source of variation	Degrees of freedom	Mean square	Defined mean square
Total	$nm_0 - 1$	MS (t)	
Primaries	$n - 1$	MS (b)	$s_W^2 + m_0 s_B^2$
Sections/primaries	$n(m_0 - 1)$	MS (w)	s_W^2

To estimate the variance of the mean per secondary in two-stage sampling we have from Eq. 9.2a and the ANOVA for the sample the unbiased estimator

$$(9.3) \qquad \widehat{\text{var}}\,(\bar{\bar{y}}) = \left(\frac{N-n}{N}\right)\frac{s_B^2}{n} + \left(\frac{NM_0 - nm_0}{NM_0}\right)\frac{s_W^2}{nm_0},$$

where

(9.4)
$$s_B^2 = \frac{MS(b) - MS(w)}{m_0}.$$

Example 9.1. An experimental survey was carried out in Iowa in 1940 to study methods of estimating areas in crops by measuring acreages on area sampling units. The section (1 square mile) was the sampling unit and the township comprising 36 sections was the stratum, of which there were 1584 in the state. In the experimental survey, two sections were drawn at random from each township. An ANOVA of acreage in corn is presented in Table 9.3. What would be the efficiency of a sample of one-fourth of the townships and four times as many sections in each compared with the sample as taken?

Table 9.3.   ANOVA of acres in corn in the sample (on a per section basis)

Source of variation	Degrees of freedom	Mean square	Components of variance
Total	3167	4273	
Townships	1583	6592	$s_W^2 + 2s_B^2$
Sections/ townships	1584	1954	s_W^2

From the ANOVA we obtain $s_W^2 = MS(w) = 1954$; and $s_B^2 = [MS(b) - MS(w)]/2 = (6592 - 1954) \div 2$ or 2319. To solve our problem, we must (a) obtain $\widehat{var}(\bar{y}_{strat})$ for the sample as it was taken—that is, a stratified sample with 1584 strata—and then (b) obtain $\widehat{var}(\bar{y})$ for the proposed subsample. Since in both cases the number of sections in both samples is the same—that is, $n = 1584 \times 2 = 3168 = nm_0 = (\frac{1}{4})(1584) \times (2)(4) = 3168$—the estimated efficiency of subsampling relative to stratified sampling is given by

$$\widehat{RE}(sub/strat) = \frac{\widehat{var}(strat)}{\widehat{var}(sub)},$$

$$\widehat{var}(strat) = \left(\frac{N-n}{N}\right)\frac{MS(w)}{n} = \left(\frac{57,024 - 3168}{57,024}\right)\frac{1954}{3168}$$

$$= 0.583,$$

where $N = (1584)(36) = 57,024$ sections in the state, $n = (1584)(2) = 3168$, and MS $(w) = s_W^2 = 1954$.

$$\widehat{\text{var}} \text{ (sub)} = \left(\frac{N-n}{N}\right) \frac{s_B^2}{n} + \left(\frac{NM_0 - nm_0}{NM_0}\right) \left(\frac{s_W^2}{nm_0}\right),$$

where $N = 1584$ townships (primaries) in the state, $n = 1584/4 = 396$ (given), $M_0 = 36$ sections per township, $m_0 = 4 \times 2 = 8$ (given), $s_W^2 = 1954$, and $s_B^2 = 2319$. Hence

$$\widehat{\text{var}} \text{ (sub)} = \frac{1584 - 396}{1584} \left(\frac{2319}{396}\right) + \frac{1584 \times 36 - 396 \times 8}{1584 \times 36} \left(\frac{1954}{396 \times 8}\right)$$

$$= (0.750)(5.856) + (0.944)(0.617)$$

$$= 4.392 + 0.582 = 4.974$$

and

$$\widehat{\text{RE}} \text{ (sub/strat)} = \frac{\widehat{\text{var}} \text{ (strat)}}{\widehat{\text{var}} \text{ (sub)}} = \frac{0.583}{4.974} = 0.117,$$

hence this simple subsampling scheme is only 12 % as efficient as the simple stratified random scheme.

Cluster Sampling a Special Case.   Two designs previously dealt with, cluster sampling and stratification, can be regarded as special cases of subsampling. Returning to Eq. 9.2a, we have as the variance of the mean per secondary, in subsampling,

(9.2a) $$\text{var} (\bar{\bar{y}}) = \left(\frac{N-n}{N}\right) \frac{S_B^2}{n} + \left(\frac{NM_0 - nm_0}{NM_0}\right) \frac{S_W^2}{nm_0}.$$

Suppose we take a 100 % sample of the secondaries in each sample primary— that is, $m_0 = M_0$; then

$$\text{var} (\bar{\bar{y}}) = \left(\frac{N-n}{N}\right) \frac{S_B^2}{n} + \left(\frac{NM_0 - nm_0}{NM_0}\right) \frac{S_W^2}{nm_0}$$

(9.5) $$= \left(\frac{N-n}{Nn}\right) \left(S_B^2 + \frac{S_W^2}{M_0}\right),$$

and, multiplying and dividing by M_0,

$$= \left(\frac{N-n}{NnM_0}\right) (M_0 S_B^2 + S_W^2),$$

but according to the ANOVA for the universe, Table 9.1,

(9.6) $$M_0 S_B^2 + S_W^2 = \text{MS} (B);$$

therefore Eq. 9.5 can be written

$$(9.5a) \qquad \text{var}(\bar{\bar{y}}) = \left(\frac{N-n}{N}\right)\frac{\text{MS}(B)}{nM_0},$$

which is identical to that of cluster sampling, Eq. 4.9, when clusters are all the same size, M_0. Hence cluster sampling can be regarded as subsampling where primaries are subsampled at a 100% rate (see Fig. 9.2).

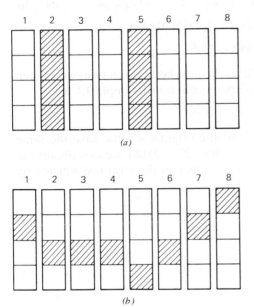

Fig. 9.2.   Special cases of subsampling a small universe. *(a)* Cluster sampling where the sample primaries are subsampled at a 100% rate ($n = 2$, $m_0 = 4$). *(b)* Stratified samplings where primaries are not subsampled at a 100% rate but all primaries are sample ($n = 8$, $M^0 = 1$).

Stratified Sampling a Special Case.   Suppose now we subsample at a rate less than 100% but sample primaries at 100%, that is, we have $n = N$. Substituting N for n in Eq. 9.2a, we have

$$\text{var}(\bar{\bar{y}}) = \left(\frac{N-n}{N}\right)\frac{S_B^2}{n} + \left(\frac{NM_0 - Nm_0}{NM_0}\right)\frac{S_W^2}{Nm_0}$$

$$(9.7) \qquad\qquad = \left(\frac{M_0 - m_0}{M_0}\right)\frac{S_W^2}{Nm_0},$$

which is equivalent to Eq. 7.17a for stratified sampling, where instead of L strata we now have N primaries, and instead of n_h elements per stratum we now have m_0 secondaries per primary (see Fig. 9.2).

Example 9.2. Suppose in the corn acreage study of Example 9.1 we choose a sample consisting of 88 townships subsampled 100%. How would the efficiency of this cluster sample compare with original stratified sample in efficiency?

The variance is given by Eq. 9.2, where the total number of townships is $N = 1584$; $n = 88$; the total sections per township is $M_0 = m_0 = 36$; $s_B^2 = 2319$; $s_W^2 = 1954$.

$$\widehat{\text{var}}\,(\bar{y}) = \left(\frac{1584 - 88}{1584}\right)\frac{2319}{88} + \left(\frac{1584 \times 36 - 88 \times 36}{1584 \times 36}\right)\frac{1954}{88 \times 36}$$

$$= (0.945)\,(26.4) + (0.945)\,(0.617)$$

$$= 24.9 + 0.583 = 25.5.$$

The variance of the original sampling scheme, where two sections were taken from each of the 1584 strata (townships), is given in Example 9.1 as

$$\widehat{\text{var}}\,(\text{strat}) = 0.583,$$

and, since both our cluster scheme and the original scheme have the same number of sections in them ($1584 \times 2 = 88 \times 36 = 3168$), we can obtain the efficiency of the cluster scheme relative to that of the stratified scheme by simply calculating the ratio

$$\widehat{\text{RE}}\,(\text{cluster/strat}) = \frac{\widehat{\text{var}}\,(\text{strat})}{\widehat{\text{var}}\,\text{cluster})} = \frac{0.583}{25.5}$$

$$= 0.0228 \text{ or } 2.28\%.$$

The sheer inefficiency of using large clusters and no stratification is clear in this example.

9.3. Optimal Choice of n and m_0

Since the variance of the mean of a two-stage sample depends, apart from the underlying variances, on both the number of primaries and the number of secondaries drawn for the sample, it is of interest to know the best values of n and m_0 to choose. By examining Eq. 9.2a we can see that by adding to m_0 the number of secondaries in the sample, we shall decrease the "within" portion of the variance, that arising from S_W^2, and not affect the "between" portion, that arising from S_B^2. If we increase n, the number of primaries in the sample, then both components are decreased. Since S_B^2 will almost always be positive, it will therefore almost always be desirable to add primaries to the sample, thereby *decreasing both* components. When carried to its logical conclusion, the best sample allocation for a fixed sample of $m = m_0 n$ secondaries, whenever S_B^2 is positive, is one where m_0 is 1 and n is therefore equal to m. Thus we should take as many primaries as we can with only a secondary sample of one each.

On the other hand, if S_B^2 is negative, then all sample secondaries should be taken from one sample primary. If one prefers, however—as we can see by looking at Eq. 9.2c, where var ($\bar{\bar{y}}$) is given in terms of intraclass correlation— if δ is zero, the choice n and m_0 for a given m_0 can be arbitrary, since the secondaries are in effect grouped at random into primaries. However, if δ is positive, the usual case, the best choice of m_0 is unity, hence imposing a maximum spread of the sample over primaries. To obtain practicable answers to the problem of allocation, it will be helpful to consider an important constraint—costs.

In ordinary two-stage sampling, certain costs are incurred in selecting a primary and carrying out certain operations on it preliminary to observing the secondaries; and certain costs are incurred for selecting a secondary and obtaining the required observations on it. Usually these unit costs are not independent of n and m_0, therefore they should be determined for the values of n and m_0 that are likely to be used. If the cost per primary is denoted by c_p and the cost per secondary is c_s, then the total cost, aside from overhead, is given by

$$(9.8) \qquad C = nc_p + nm_0c_s.$$

For a fixed total cost, C, the variance of the subsample mean, var ($\bar{\bar{y}}$), becomes a minimum when

$$(9.9) \qquad \text{opt} (m_0) = \sqrt{(S_W^2/S_B^2) \cdot (c_p/c_s)},$$

or, approximately, in terms of intraclass correlation,

$$(9.9a) \qquad \text{opt} (m_0) \doteq \sqrt{(1-\delta)/\delta) \cdot (c_p/c_s)},$$

and

$$(9.10) \qquad \text{opt} (n) = \frac{C}{c_p + m_0c_s}.$$

When the within-primary variance, S_W^2, is large relative to the between component, we should take a *larger* number of elements in each sample primary. If the costs of observing elements are high relative to the costs of primaries, we should take *fewer* elements per primary and more primaries. If the costs associated with a primary are very low or negligible, which is sometimes the case where primaries are conveniently available file cases or where primary costs are simply map costs and maps may be available without charge, the best allocation is that where opt (m_0) is 1, its smallest practical value.

Example 9.3. A study of costs and variances for sampling crops in Brookings County, S.D., using a plot one mile square (section) for measuring actual area of land in various crops (with combined use of aerial photographs and an actual visit to the site), revealed the following facts for acres in barley:

Mean square of square-mile plots within townships, 1504.
Mean square of township (6×6 miles), 6901.

Assume that:

Cost associated with a township is \$2.20.
Cost to examine a square-mile plot is \$3.80.

A township contains 36 of the square-mile plots and may be a useful primary unit. In a two-stage sampling design for a barley survey of the county, how many square mile plots should be taken per township?

The appropriate formula is Eq. 9.6. To obtain S_B^2 we have:

Mean square for townships $= 6901 = S_W^2 + 36S_B^2$,

Mean square, plots within townships $= 1504 = S_W^2$,

and

$$36S_B^2 = 6901 - 1504 = 5397,$$

$$S_B^2 = 149.9.$$

In summary we have

$$S_W^2 = 1504, \quad S_B^2 = 149.9, \quad c_p = 2.20, \quad c_s = 3.80.$$

To determine optimum size of sample with primaries, we have for Eq. 9.9

$$m_{opt} = \sqrt{(S_W^2/S_B^2) \cdot (c_p/c_s)}$$

$$= \sqrt{(1504/149.9)(2.20/3.80)} = \sqrt{5.81} = 2.4.$$

Hence we should have either two or three plots per township. A more precise answer can be obtained by determining n (from Eq. 9.10) for both values of m_0 and solving Eq. 9.2 or its variants 9.2a–d for var $(\bar{y})$, obtaining and choosing that which yields the lower var $(\bar{y})$. Generally var $(\bar{y})$ does not change drastically for unit changes in m_0.

With More Complex Costs. The simple function for costs used above. Eq. 9.8, where c_p and c_s are fixed unit costs of primaries and secondaries, respectively, it is not very realistic for many practical situations. Where observing or dealing with a sample unit involves travel, the costs per unit may decline as the number of units in the sample is increased. Some illustrations of the nature of this relationship were presented in Section 4.7. The following cost function (Eq. 4.28a), taking into account a "square-root" component for travel, was given (except we now write m_0 for M_0 its equivalent in cluster sampling):

$$(9.11) \qquad C = c_1 n + c_2 n m_0 + c_3 \sqrt{n}.$$

This function for cluster sampling will be a suitable base for a fairly general function for two-stage sampling. By adding two components to cover costs associated with the establishment of the sampling frame, we obtain for total costs of both sampling and surveying in two-stage designs:

$$(9.12) \qquad C_T = c_0 N + c_1 M_0 n + c_2 n m_0 + c_3 \sqrt{n} + c_4 n \sqrt{m_0 M_0},$$

where the coefficients c_1, c_2, and c_3 play the same general role as before in Eq. 4.28a, except c_1 is replaced by $c_1 M_0$; c_4 is the unit cost of dealing with a secondary in the sample of primaries (such as the cost of "listing" dwelling units), and c_0 is the unit cost of establishing the frame of N primary units.

Our problem now is to determine the optimal values of m_0 and n when we have a cost function of the form of Eq. 9.12. Writing Eq. 9.12 as a quadratic in $\sqrt{n}$, we have

$$n(c_1 M_0 + c_2 m_0 + c_4 \sqrt{m_0 M_0}) + c_3 \sqrt{n} + c_0 N - C_T = 0.$$

If we choose certain values for m_0, then for a fixed budget, C_T, we can solve for n, where

$$n = \left(\frac{-c_3 + \sqrt{c_3^2 + 4aC}}{2a} \right)^2,$$

in which

$$a = c_1 M_0 + c_2 m_0 + c_4 \sqrt{m_0 M_0}$$

and

$$C = C_T - c_0 N.$$

In the case of simple costs, Eq. 9.8, the optimal value of m_0 for a given budget is given by Eq. 9.9,

$$\text{opt } (m_0) = \sqrt{(S_W^2 / S_B^2)(c_p / c_s)},$$

which, in terms of intraclass correlation, becomes

$$\text{opt } (m_0) = \sqrt{[(1 - \delta)/\delta](c_p / c_s)}.$$

We can utilize this relationship by expressing the complex function in terms of the simple one by letting

$$c_1 M_0 + \frac{c_3}{\sqrt{n}} = c_p$$

and

$$c_2 + c_4\sqrt{M_0 m_0} = c_s;$$

then

(9.13)
$$\frac{c_p}{c_s} = \frac{c_1 M + c_3\sqrt{n}}{c_2 + c_4\sqrt{m_0 M_0}}.$$

We cannot solve for opt (m_0) without knowing the cost ratio c_p/c_s, which depends on m_0. We can solve for opt (m_0) iteratively by guessing at opt (m_0), determining c_p/c_s for that value, Eq. 9.13, and thence to the resulting value of opt (m_0), Eq. 9.9. Subsequent guesses in opt (m_0) should bring one near enough to the true value fairly quickly.

9.4. Subsampling with Stratification of Primaries; ANOVA

When stratification is combined with simple subsampling and each stratum is subsampled independently as in Section 9.3, then an unbiased estimate of $\bar{Y}$, the mean per secondary in the universe, is given by the weighted mean

(9.14)
$$\bar{y} = \sum_{}^{L} \left(\frac{M_h}{M}\right) \bar{\bar{y}}_h,$$

where $\bar{\bar{y}}_h$ = the mean per secondary in the hth stratum,

M_h = the number of secondaries in the hth stratum,

$$M = \sum_{}^{L} M_h,$$

L = total number of strata.

The variance of the weighted mean (Eq. 9.14) is given by

(9.15)
$$\text{var}(\bar{y}) = \sum_{}^{L} \left(\frac{M_h^2}{M^2}\right) \text{var}(\bar{\bar{y}}_h),$$

where $\text{var}(\bar{\bar{y}}_h)$ is the variance of the sample mean in the hth stratum.

To estimate $\text{var}(\bar{y})$ we can substitute estimates of $\text{var}(\bar{\bar{y}}_h)$ in Eq. 9.15, using Eq. 9.3.

To achieve optimum allocation of sampling resources among strata on one hand and between stages on the other, where cost functions are the simple linear types, we can use a procedure as follows: Let

C = total funds available for survey (except for overhead),

C_h = the amount of C to allocate to the hth stratum,

$\text{var}(\bar{\bar{y}}_h)$ = variance of the sample mean in the hth stratum for $n_h = 1$.

First determine the optimum values of $(m_o)_h$, using Eq. 9.9. Then from Eq. 9.8 we find that the cost per unit of n_h in the hth stratum is given by

$$(9.16) \qquad C_h = (C_p)_h + (m_0)_h + (m_0)_h (C_s)_h.$$

We can now substitute this value of cost per "sampling unit" in stratum h in Eq. 7.10, and we have for the optimum number of primaries to be taken from the hth stratum

$$(9.17) \qquad \text{opt } (n_h) \doteq n \, \frac{[N_h \sqrt{\text{var }(\bar{y}_h)} / \sqrt{c_h}]}{\sum [N_h \sqrt{\text{var }(\bar{y}_h)} / \sqrt{c_h}]},$$

where c_h and var $(\bar{y}_h)$ are defined above and N_h is the number of primaries in the hth stratum. Equation 9.17 is approximate because we are ignoring the finite population.

When it is desirable to have a self-weighting sample using stratification, the allocation of the sample among the strata should be proportional to the number of secondaries, M_h, when $\bar{Y}$ is being estimated, and proportional to the number of primaries, N_h, when the mean per primary, $\bar{Y}$, is being estimated. (This latter case is relevant when primaries are *not* of *equal* size.) In this case of simple allocation in stratified subsampling—that is, $n_h \propto N_h$ and $(m_0)_h \propto (M_0)_h$—the sample mean becomes

$$(9.18) \qquad \bar{\bar{y}} = \frac{\sum_{h=1}^{L} \sum_{i=1}^{n_h} \sum_{j=1}^{(m_0)_h} Y_{hij}}{m},$$

which is simply the sample total of y divided by the total sample size. Also, the variance of this mean can be expressed as a version of Eq. 9.2a,

$$(9.19) \qquad \text{var } (\bar{\bar{y}}) = \left(1 - \frac{n}{N}\right) \frac{\bar{S}_B^2}{n} + \left(1 - \frac{m}{M}\right) \frac{\bar{S}_W^2}{m},$$

where $\bar{S}_B^2$ and $\bar{S}_w^2$ are the average within-stratum components of variance that can be identified in an analysis of variance.

If all strata have N_0 primaries each and each primary has M_0 secondaries, an analysis of variance can be written as shown in Table 9.4. Likewise if a sample is

Table 9.4.   ANOVA of a two-stage stratified universe (on a per secondary basis)

Source of variation	Degrees of freedom	Mean square defined as
Total	$NM_0 - 1$	
Strata	$L - 1$	
Primaries/strata	$N_0 - L$	$\bar{S}_w^2 + M_0 \bar{S}_B^2$
Secondaries/primaries	$N_0(M_0 - 1)$	$\bar{S}_w^2$

allocated with $n_0 \propto N_0$ and $m_0 \propto M_0$ for all L strata, then estimates of $\bar{S}_w^2$ and $\bar{S}_B^2$ can be made unbiasedly from the corresponding statistics, s_w^2 and s_B^2, from samples. However, in a sample where m_0 secondaries are taken from each, the component-of-variance mean square above will be replaced by $s_w^2 + m_0 s_B^2$.

9.5.　Simple Three-Stage Sampling; ANOVA

We can easily extend two-stage sampling to three stages by further sampling the secondaries instead of examining each of the elements of which they are composed. For example, a three-stage sample of the United States may consist first of counties (primaries), second of listing units (such as blocks) within counties (secondaries), and third of households within listing units (tertiaries).

Assuming, as in the two-stage case, that we have N primaries containing M_0 secondaries each, and supposing that each secondary contains K_0 tertiaries, we have the necessary parameters described in Table 9.5.

Table 9.5.　ANOVA of a three-stage universe (on a per tertiary basis)

Source of variation	df	Mean square defined as
Total	$NM_0K_0 - 1$	
Primaries	$N - 1$	$S_3^2 + K_0 S_2^2 + M_0 K_0 S_1^2$
Secondaries/primaries	$N(M_0 - 1)$	$S_3^2 + K_0 S_2^2$
Tertiaries/secondaries	$NM_0(K_0 - 1)$	S_3^2

If we draw a random (RNR) sample of n primaries, and from each of these an RNR sample of m_0 secondaries, and from each of these an RNR sample of k_0 tertiaries, the simple sample mean, $\bar{\bar{y}}$, will be an unbiased estimate of $\bar{\bar{Y}}$, the population mean of y per tertiary, and its sampling variance will be given by

$$(9.20) \qquad \mathrm{var}\,(\bar{\bar{y}}) = \left(\frac{N-n}{N}\right)\frac{S_1^2}{n} + \left(\frac{NM_0 - nm_0}{NM_0}\right)\frac{S_2^2}{nm_0}$$

$$+ \left(\frac{NM_0K_0 - nm_0k_0}{NM_0K_0}\right)\left(\frac{S_3^2}{nm_0k_0}\right),$$

where the S^2s are defined by Table 9.5. Note the structural relationship to simple two-stage sampling, Eq. 9.2a, and the suggested form that sampling at still higher stages might take. When sampling fractions are small, the FPCs can be ignored and hence further simplicity is achieved.

Optimum allocation for three stages is also analogous to that of two stages. If we denote the cost per unit of observing a tertiary as c_t, then our total budget will be spent among the three stages thus:

$$(9.21) \qquad C = c_p n + c_s n m_0 + c_t n m_0 k_0,$$

and sampling variance, Eq. 9.20, will be a minimum if

$$(9.22) \qquad k_0 = \sqrt{(S_3^2/S_2^2)(c_s/c_t)}$$

and

$$(9.23) \qquad m_0 = \sqrt{(S_2^2/S_1^2)(c_p/c_s)}$$

Note that opt (k_0) can be found without knowledge of costs or variances of primaries and that, because of the analogy of Eq. 9.20 to Eq. 9.2a, these allocation formulas can be extended to higher stages rather easily.

9.6. Unequal-Sized Primaries with PE

The theory and practice of two-stage sampling dealt with so far in this chapter was confined to the case where all primaries contained the same number of secondaries, M_0, and in the case of three stages the secondaries contained a constant number of tertiaries, K_0. However, a much more accommodating theory and practice are required if we are to deal with unequal sizes in primaries, which are very common in many fields. To do this we shall employ the principles of two design factors already introduced and discussed: estimation (Chapter 5) and selection probabilities (Chapter 8). Most of the situations commonly encountered in two-stage statistical surveys can be met adequately by four types of estimators and three types of selection probabilities. In this section we shall consider the use of three of these estimators, using equal probabilities of selecting primaries (that is, RNR sampling at the first stage). In Section 9.7 we shall consider selecting primaries with PPS where size is measured by number of secondaries, M_i, and in Section 9.8 by PPS where size is measured arbitrarily by X_i. The fourth estimator, a ratio of two variables, is considered in Section 9.10, along with an estimator of Y. It should be noted that the average of M_i is denoted by $\bar{M}$, which in the equal-sized primary case was M_0.

Some Feasible Estimators. Suppose a sample of n primaries has been selected and an arbitrary number of secondaries, m_i, has been selected in each. Several estimators appear to be feasible for estimating $\bar{Y}$, such as (i) a *simple mean* of the sample means per secondary for the n primaries, (ii) the same as (i) except that the primary means are *weighted* by primary sizes, and (iii) a *ratio* of the estimated totals in sample primaries to their total sizes. Specifically we have

$$(9.24) \qquad \bar{\bar{y}}_{\text{PE1}} = \frac{1}{n}\sum_{i}^{n} \bar{y}_i,$$

$$(9.25) \qquad \bar{\bar{y}}_{\mathrm{PE2}} = \frac{1}{n\bar{M}} \sum_{i}^{n} M_i \bar{y}_i,$$

$$(9.26) \qquad \bar{\bar{y}}_{\mathrm{PE3}} = \frac{\sum M_i \bar{y}_i}{\sum M_i},$$

where

$$(9.27) \qquad \bar{y}_i = \frac{1}{m_i} = \frac{1}{m_i} \sum^{m_i} Y_{ij},$$

the sample mean per secondary in the ith primary.

In order to determine the performance of these estimators we shall examine their (i) biasedness and (ii) accuracy as measured by MSE. The first and third, Eqs. 9.24 and 9.26, are biased in general but the second, Eq. 9.25, is unbiased. (The subscript indicates "probabilities equal, number 1, . . . ," and so on.) The MSE for Eq. 9.24 is given by

$$(9.28) \quad \mathrm{MSE}\,(\bar{\bar{y}}_{\mathrm{PE1}}) = \left(\frac{N-n}{Nn}\right) \underbrace{\sum^{N} \frac{(\bar{Y}_i - \bar{\bar{Y}}_{\mathrm{PE1}})^2}{N-1}}_{\substack{\text{between} \\ \text{primaries}}} + \underbrace{\frac{1}{Nn}\sum^{N}\left(\frac{M_i - m_i}{M_i}\right)\frac{S_i^2}{m_i}}_{\substack{\text{within} \\ \text{primaries}}}$$

$$+ \underbrace{(\bar{\bar{Y}}_{\mathrm{PE1}} - \bar{Y})^2,}_{\text{bias}}$$

for Eq. 9.25 by

$$(9.29) \quad \mathrm{MSE}\,(\bar{\bar{y}}_{\mathrm{PE2}}) = \left(\frac{N-n}{Nn}\right)\underbrace{\sum^{N}\frac{(M_i \bar{Y}_i/\bar{M} - \bar{Y})^2}{N-1}}_{\substack{\text{between} \\ \text{primaries}}} + \underbrace{\frac{1}{Nn}\sum^{N}\left(\frac{M_i - m_i}{M_i}\right)\left(\frac{M_i}{\bar{M}}\right)^2 \frac{S_i^2}{m_i}}_{\substack{\text{within} \\ \text{primaries}}}$$

$$+ \underbrace{0,}_{\substack{\text{no} \\ \text{bias}}}$$

and for Eq. 9.26 approximately by

$$(9.30) \quad \mathrm{MSE}\,(\bar{\bar{y}}_{\mathrm{PE3}}) \doteq \left(\frac{N-n}{Nn}\right)\underbrace{\sum^{N}\left(\frac{M_i}{\bar{M}}\right)^2\frac{(\bar{Y}_i - \bar{Y})^2}{N-1}}_{\substack{\text{between} \\ \text{primaries}}} + \underbrace{\frac{1}{Nn}\sum^{N}\left(\frac{M_i - m_i}{M_i}\right)\left(\frac{M_i}{\bar{M}}\right)^2\frac{S_i^2}{m_i}}_{\substack{\text{within} \\ \text{primaries}}}$$

$$+ \underbrace{(Y_{\mathrm{PE3}|n} - \bar{Y})^2}_{\text{bias}}$$

In Eq. 9.28 the symbol $\bar{\bar{Y}}_{PE1}$ denotes the expected value of $\bar{y}_{PE1}$, which is

$$(9.31) \qquad \bar{\bar{Y}}_{PE1} = \frac{1}{N} \sum_{i}^{N} \bar{Y}_i$$

and unaffected by sample size n. On the other hand, $\bar{\bar{Y}}_{PE3|n}$ in the bias term of Eq. 9.30 is also dependent on n and is negligible if n is large enough (see Section 5.3).

There are circumstances under which any one of these estimators would be more accurate than the others, but none is to be preferred generally. It may be noted at this point, however, that the first estimator, $\bar{y}_{PE1}$, requires no extrasample information whatever, whereas $\bar{y}_{PE2}$ requires that M_i for primaries in the sample and the universe value, $\bar{M}$, be known. The third, $\bar{y}_{PE3}$, requires M_i for the primaries in the sample but not $\bar{M}$.

To get a closer look at the properties of these estimators, let us first try to determine a good way to allocate M_i.

Allocation Policy: Self-Weighting and Efficiency. A simple policy for allocating will make the estimator self-weighting. If m_i is made proportional to M_i, say $m_i = f_2 M_i$, estimators $\bar{y}_{PE2}$ and $\bar{y}_{PE3}$ are self-weighting ($\bar{y}_{PE1}$ is "self-weighting" only if $M_i = \bar{M} = M_0$), whence Eq. 9.25 becomes

$$(9.25a) \qquad \bar{y}_{PE2} = \frac{1}{n\bar{M}} \sum_{i}^{n} M_i \sum_{j}^{m_i} \frac{Y_{ij}}{m_i} = \frac{1}{n\bar{M}f_2} \sum_{i}^{n} \sum_{j}^{m_i} Y_{ij},$$

where $n\bar{M}f_2$ is a constant, and Eq. 9.26 becomes

$$(9.26a) \qquad \bar{y}_{PE3} = \frac{\sum_{i}^{n} M_i \sum_{j}^{m_i} Y_{ij}/m_i}{\sum_{i}^{n} M_i} = \frac{\sum_{i}^{n} \sum_{j}^{m_i} Y_{ij}}{\sum_{i}^{n} m_i},$$

where $\sum_{i}^{n} m_i$ is a random variable.

The MSEs for these two cases become

$$(9.29a) \quad \text{MSE}(\bar{y}_{PE2}) = \left(\frac{N-n}{Nn}\right) \sum^{N} \frac{[(M_i/\bar{M})\,\bar{Y}_i - \bar{\bar{Y}}]^2}{N-1} + \left(\frac{1-f_2}{n\bar{m}}\right) \sum \left(\frac{M_i}{\bar{M}}\right) S_i^2$$

and

$$(9.30a) \quad \text{MSE}(\bar{y}_{PE3}) = \left(\frac{N-n}{Nn}\right) \sum \left(\frac{M_i}{\bar{M}}\right)^2 \left[\frac{(\bar{Y}_i - \bar{\bar{Y}})^2}{N-1}\right] + \left(\frac{1-f_2}{n\bar{m}}\right) \sum \left(\frac{M_i}{\bar{M}}\right) S_i^2,$$

which may be compared with the equal-sized case where Eq. 9.2 becomes, when written in terms of sums of squares,

$$(9.32) \qquad \text{var}(\bar{y}) = \left(\frac{N-n}{Nn}\right) \sum \left[\frac{(\bar{Y}_i - \bar{\bar{Y}})^2}{N-1}\right] + \left(\frac{1-f_2}{n\bar{m}}\right) \sum \left(\frac{1}{N}\right) S_i^2,$$

where $f_2 = m_i/M_i = \text{constant} = \bar{m}/\bar{M}$. It may be noted that $\bar{y}$, $\bar{y}_{PE2}$, and $\bar{y}_{PE3}$, when n is constant and $m_i \propto M_i$, differ in accuracy (apart from bias) by the effects of different weighting of the between and within sources of variance. Some comments on this will be made in Section 9.9.

If costs are taken into account in determining allocation policy, we find that the within primary contributor to MSE for estimators $\bar{y}$, y_{PE2}, and $\bar{y}_{PE3}$ can be minimized somewhat for a given expected $m = \sum_{m_i}^{n}$, if

(9.33) $$\text{opt } (m_i) \propto \frac{M_i S_i}{\sqrt{c_i}},$$

since, for a given set of primaries selected for the sample, the allocation of secondaries to primaries should be analogous to that for stratified sampling, where we now have n strata.

To determine the allocation of resources between n primaries and m secondaries, we can use the equations given for the equal-sized primary case by substituting estimated values for the between and within sources of variance in the appropriate MSE equation. For example, to determine the $E[\text{opt } (\bar{m})]$ for $\bar{y}_{PE2}$ we can use as a rough approximation of Eq. 9.9,

(9.34) $$E[\text{opt } (\bar{m})] = \sqrt{[\text{var} (w)/\text{var} (b)] (c_p/c_s)},$$

where var (b) and var (w) are the variances in the first and second terms of Eq. 9.29a, excluding FPCs. Having obtained this value of $\bar{m}$, one can then calculate the m_i/M_i, which will provide

(9.35) $$\frac{m_i}{M_i} = \frac{E[\text{opt } (\bar{m})]}{\bar{M}}.$$

For more sophisticated and accurate procedures, see the references at the end of the chapter.

Estimation of MSEs from a Sample.   The variance portions of the MSEs for the foregoing estimators can be estimated from samples by replacing parametric values by their estimates. Thus we have

(9.36) $$\widehat{\text{var}} (\bar{y}_{PE1}) = \left(\frac{N-n}{Nn}\right) \sum^{n} \frac{(\bar{y}_i - \bar{y}_{PE1})^2}{n-1} + \frac{f_1}{n^2} \sum^{n} \left(\frac{M_i}{\bar{M}}\right)^2 \frac{(1-f_i)s_i^2}{m_i},$$

(9.37) $$\widehat{\text{var}} (\bar{y}_{PE2}) = \left(\frac{N-n}{Nn}\right) \sum^{n} \left(\frac{M_i}{\bar{M}}\right)^2 \frac{(\bar{y}_i - \bar{y}_{PE2})^2}{n-1} + \frac{f_1}{n^2} \sum^{n} \left(\frac{M_i}{\bar{M}}\right)^2 \frac{(1-f_i)s_i^2}{m_i},$$

(9.38) $$\widehat{\text{var}} (\bar{y}_{PE3}) = \left(\frac{N-n}{Nn}\right) \sum^{n} \left(\frac{M_i}{\bar{M}}\right)^2 \frac{(\bar{y}_i - \bar{y}_{PE3})^2}{n-1} + \frac{f_1}{n^2} \sum^{n} \left(\frac{M_i}{\bar{M}}\right)^2 \frac{(1-f_i)s_i^2}{m_i}$$

as estimators of the variance portions of Eqs. 9.28, 9.29, and 9.30, respectively.

9.7. Unequal-Sized Primaries with PPM_i

In an attempt to increase the accuracy of two-stage sampling when primaries vary in size, Hansen and Hurwitz (1943) proposed that primaries be selected with probabilities proportional to the number of secondaries they contain. A welcome consequence of this procedure, when applied to many practical situations, is a reduction in the between primary portion of the MSE of $\bar{\bar{y}}$. The appropriate estimator of $\bar{\bar{y}}$ in this case is

$$(9.39) \qquad \bar{\bar{y}}_{PM1} = \frac{1}{n}\sum_{}^{n} \bar{y}_i,$$

which is identical in appearance to $\bar{\bar{y}}_{PE1}$. However, because of the manner in which the primaries are drawn, its properties have important differences. For example, if primaries are selected either singly or in multiples with replacement with probabilities equal to M_i/M, then $\bar{\bar{y}}_{PM1}$ is unbiased and has a variance

$$(9.40) \qquad \mathrm{var}\,(\bar{\bar{y}}_{PM1}) = \frac{1}{n}\sum_{}^{N} \left(\frac{M_i}{M}\right)(\bar{Y}_i - \bar{\bar{Y}})^2 + \frac{1}{n}\sum_{}^{N}\left(\frac{M_i - m_i}{M_i m_i}\right)\left(\frac{M_i}{M}\right)S_i^2 .$$

Whether this method is more accurate than any of the equal-probability estimators depends largely on the relationship between M_i and $\bar{Y}_i$—that is, between the number of secondaries in primaries and their average y values. Some empirical evidence in several fields indicates that circumstances frequently favor this method; see Hansen and Hurwitz (1943), Madow (1950), and Jebe (1952). Comparisions with other methods will be considered in Section 9.9.

Policy for Allocating m_i.   A rather useful feature of this method is that taking a fixed number·of secondaries, $m_i = m_0$, in each primary is a near-optimal allocation policy under rather general practical situations and that this also makes the estimator $\bar{\bar{y}}_{PM1}$ self-weighting. In this case, $f = nm_0/M = $ constant.

In case the primaries are selected with PNR rather than PR, but where M_i remains the measure of size, the estimator remains unbiased and the variance will be modified by replacing the first term by its appropriate variance (see Chapter 8).

Estimation of MSEs from a Sample.   To estimate var $(\bar{\bar{y}}_{PM1})$ from a sample we can use the appropriate sample estimates; hence if the n primaries are drawn with PR,

$$(9.41) \qquad \widehat{\mathrm{var}}\,(\bar{\bar{y}}_{PM1}) = \frac{1}{n(n-1)}\sum_{}^{n}(\bar{y}_i - \bar{\bar{y}})^2,$$

which, if the usual policy of $m_i = m_0$ is used, and if FPC can be ignored, simplifies to

(9.42)
$$\hat{\text{var}}\,(\bar{\bar{y}}_{PM1}) = \frac{1}{m_0^2 n(n-1)} \sum_{i}^{n} (m_0 \bar{y}_i - \bar{y})^2$$

(9.42a)
$$\doteq \frac{MS\,(b)}{nm_0}$$

from a simple ANOVA wherein the MS (b) is appropriately self-weighted.

9.8.   Unequal-Sized Primaries with PPX_i

In many practical situations we do not know the M_i but are willing to estimate them by some presumed related characteristic, say X_i. Is it still possible to gain some accuracy over equal selection probabilities but using probabilities proportional to X_i? In this case we have the usual three types of estimators of $\bar{\bar{Y}}$, the simple unweighted, the weighted, and the ratio type:

(9.43)
$$\bar{\bar{y}}_{PX1} = \frac{1}{n} \sum^{n} \bar{y}_i,$$

(9.44)
$$\bar{\bar{y}}_{PX2} = \frac{1}{nM} \sum \frac{M_i \bar{y}_i}{A_i},$$

and

(9.45)
$$\bar{\bar{y}}_{PX3} = \frac{\sum M_i \bar{y}_i / A_i}{\sum M_i / A_i},$$

where $A_i = X_i / X$. Of these, $\bar{\bar{y}}_{PX2}$, the weighted estimator, is unbiased; the others are biased. The bias in $\bar{\bar{y}}_{PX1}$ depends on the relationship between X_i and M_i; if $X \propto M_i$, the bias is zero. The bias in $\bar{\bar{y}}_{PX3}$ depends on n, hence should decrease rapidly with larger samples of primaries.

The accuracies of these estimators in terms of MSEs are

(9.46) $\text{MSE}\,(\bar{\bar{y}}_{PX1}) = \underbrace{\frac{1}{n} \sum^{N} A_i (\bar{Y}_i - \bar{\bar{Y}}_{PX1})^2}_{\text{between}} + \underbrace{\frac{1}{n} \sum^{N} \left(\frac{M_i - m_i}{M_i m_i} \right) A_i S_i^2}_{\text{within}} + \underbrace{(\bar{\bar{Y}}_{PX1} - \bar{\bar{Y}})^2}_{\text{bias}},$

(9.47) $\text{MSE}\,(\bar{\bar{y}}_{PX2}) = \underbrace{\frac{1}{n} \sum A_i \left(\frac{M_i}{M} \right)^2 \left(\frac{\bar{Y}_i}{A_i} - \bar{\bar{Y}} \right)^2}_{\text{between}} + \underbrace{\frac{1}{n} \sum \left(\frac{M_i}{M} \right)^2 \left(\frac{M_i - m_i}{M_i m_i} \right) \frac{S_i^2}{A_i}}_{\text{within}},$

and

(9.48) $\text{MSE}\,(\bar{\bar{y}}_{PX3}) \doteq \underbrace{\frac{1}{n} \sum \left(\frac{1}{A_i} \right) \left(\frac{M_i}{M} \right)^2 (\bar{Y}_i - \bar{\bar{Y}})^2}_{\text{between}} + \underbrace{\frac{1}{n} \sum \left(\frac{M_i}{M} \right)^2 \left(\frac{M_i - m_i}{M_i m_i} \right) \frac{S_i^2}{A_i}}_{\text{within}}$

$$+ \,(\text{bias})^2,$$

where

(9.49) $$\bar{\bar{Y}}_{PX1} = \frac{1}{N} \sum_{i}^{N} A_i \bar{Y}_i.$$

Policy for Allocating m_i.   Again it may be noted that if $A_i \propto M_i/M$, then the within primary source of MSE is identical for all three estimators for a given m_i policy, but the between primary source can differ among methods. Hence the policy of allocating m_i is usually not an important consideration unless the within source is the important source of MSE. In practice it is common to apply the self-weighting criterion, which results in requiring that

(9.50) $$m_i = M_i f_i = \frac{M_i f}{n A_i}.$$

This m_i policy will make both $\bar{y}_{PX2}$ and $\bar{y}_{PX3}$ always self-weighting and will be self-weighting for $\bar{y}_{PX1}$ when $A_i \doteq M_i/M$.

Estimation of MSEs from a Sample.   We can estimate the MSEs for each estimator from a sample by the following formula, omitting the case of $\bar{y}_{PX1}$. Again we assume primaries are selected with PR.

(9.51) $$\widehat{\text{var}} \, (\bar{y}_{PX2}) = \frac{1}{n(n-1)} \sum_{i}^{n} \left(\frac{M_i \bar{y}_i}{M A_i} - \bar{y}_{PX2} \right)^2,$$

which, if $m_i = M_i f/n A_i$—that is, the sample is self-weighted—becomes

(9.51a) $$\widehat{\text{var}} \, (\bar{y}_{PX2}) = \frac{n}{f^2 M^2(n-1)} \sum_{i}^{n} (m_i \bar{y}_i - \bar{y})^2,$$

where $y_i = \sum_{j=1}^{m_i} Y_{ij}$ and $\bar{y} = \sum_{i=1}^{n} y_i/n$. And for $\bar{y}_{PX3}$ we have

(9.52) $$\widehat{\text{var}} \, (\bar{y}_{PX3}) = \frac{1}{n(n-1)} \sum_{i}^{n} \left(\frac{M_i}{M A_i} \right)^2 (\bar{y}_i - \bar{y}_{PX3})^2.$$

9.9.   Estimating $\bar{\bar{Y}}$; Comparison of Alternative Methods

To compare in detail the several methods presented in Sections 9.6–9.8 for accuracy and general practical desirability depends on an analysis and discussion beyond the scope of this book. We will note, however, that each method has its own data requirements (e.g., some require that M be known, some that all M_i be known, others only that M_i in the sample be known, etc.) Also, in order that the PPX methods be effective in estimating $\bar{\bar{Y}}$, it is required that X_i be related in some manner to the M_i. Moreover, in order to determine an efficient policy toward m_i, it is necessary to know the costs of observing in each primary and the values of M_i and S_i^2. When this information is missing, reasonable estimates are helpful.

Table 9.6. Summary of estimators

	Selection Probabilities		
Estimation type	Equal, PE	Proportional to M_i, PPM_i	Proportional to X_i, PPX_i
Simple mean of means	$\bar{y}_{PE1}^{*} = \dfrac{1}{n}\sum \bar{y}_i$	$\bar{y}_{PM1} = \dfrac{1}{n}\sum \bar{y}_i$	$\bar{y}_{PX1}^{*} = \dfrac{1}{n}\sum \bar{y}_i$
Weighted mean of means	$\bar{y}_{PE2} = \dfrac{1}{n\bar{M}}\sum M_i\bar{y}_i$	a^{*}	$\bar{y}_{PX2} = \dfrac{1}{n\bar{M}}\sum \dfrac{M_i\bar{y}_i}{A_i}$
Ratio on M_is	$\bar{y}_{PE3}^{*} = \dfrac{\sum M_i\bar{y}_i}{\sum M_i}$	b	$\bar{y}_{PX3}^{*} = \dfrac{\sum M_i\bar{y}_i/A_i}{\sum M_i/A_i}$

But the choice of method often depends on empirical experience, however meager, and the circumstances of the particular investigation at hand. To gain some empirical experience, we shall apply the seven methods described above to a synthetic universe. Two cases will be considered: (i) a sample of $n = 1$ with $m_i = 2$ and $m_i = M_i/2$ and (ii) $n = 2$ with $m_i = 2$.

The estimators for the seven methods are presented in Table 9.6. Those with stars are biased. The logical estimator for cell a has no intuitive appeal, since it applies weights where selection probabilities should suffice, so it is omitted. The logical estimator for cell b degenerates to $\bar{y}_{PM1}$, hence is redundant.

The data for the test, taken from Cochran (1953, p. 296) are presented in Table 9.7. It is a microuniverse of three primaries with a total of 12 secondaries. Basic parametric values are given.

Table 9.7. A synthetic universe of three primaries and 12 secondaries

Primary (i)	Observation Y_{ij}	Y_i	M_i	S_i^2	$\bar{Y}_i$	$A_i = X_i/X$
1	0, 1	1	2	0.500	0.5	0.2
2	1, 2, 2, 3	8	4	0.667	2.0	0.2
3	3, 3, 4, 4, 5, 5	24	6	0.800	4.0	0.4
$\sum$		33	12	1.967	6.5	1.0

$Y = \sum Y_i = 33$, $M = \sum M_i = 12$, $N = 3$, $\bar{Y} = Y/M = 33/12 = 2.75$, $\bar{M} = M/N = 12/3 = 4$.

Results for the case of samples of $n = 1$ and $m_i = 2$ and $m_i = M_i/2$ are presented in Table 9.8. In this comparison the two methods using ratio type estimators were omitted, because when $n = 1$ they are degenerate. Note that most of the MSE is due to the between primary variance and for the weighted unbiased estimator, $\bar{y}_{PE2}$, it is very large, making it the least accurate method. Choosing primaries with PPM has reduced this variance markedly, see $\bar{y}_{PM1}$, although basing probabilities on X_i has done as well. The allocation of m_i, whether $m_i = 2$ or $m_i = M_i/2$, has had a minor effect on overall MSE.

The results are fairly representative of widely varying fields of application. The superiority of PPM selection over PE is quite common. The sometimes small loss due to the use of X_is to estimate M_i is also not uncommon if the X_is happen to be reasonable choices.

Table 9.8. Comparisons of MSEs of five methods on a microuniverse ($n = 1$, $m_i = 2$, and $m_i = M_i/2$)

	Portion of MSE due to				
	Between primaries		Within		
Method and m_i policy	Variance	Bias	primaries variance	Total MSE	Accuracy rank
$\bar{y}_{PE1}$: $m_i = 2$	2.056	0.340	0.145	2.541	3
$m_i = M_i/2$	2.056	0.340	0.183	2.579	
$\bar{y}_{PE2}$: $m_i = 2$	5.792	0	0.256	6.048	5
$m_i = M_i/2$	5.792	0	0.176	5.968	
$\bar{y}_{PM1}$, $m_i = 2$	1.813	0	0.189	2.002	1
$\bar{y}_{PX1}$, $m_i = 2$	1.800	0.062	0.173	2.035	2
$\bar{y}_{PX2}$, $m_i = 2$	3.583	0	0.213	3.796	4

In order to examine the behavior of the methods involving a ratio type estimator, we shall now let $n = 2$ and hold $m_i = 2$. To do this for the PPS cases, we shall choose the primaries with PNR, following a procedure similar to that described in Section 8.11. (In this case, for the PM samples, primaries 1,3 and 2,3 were selected with 2/3 and 1/2 probabilities, respectively; and for PX samples, primaries 1,2 and 1,3 and 2,3 with .2, .2, and .6 probabilities, respectively.) The results are presented in Table 9.9.

Table 9.9. *Comparison of MSEs of seven methods on a microuniverse ($n=2$, $m_i = 2$)*

Method	Portion of MSE due to		Within primaries variance	Total MSE	Accuracy rank
	Between primaries				
	Variance	Bias			
$\bar{y}_{PE1}$	0.514	0.314	0.072	0.926	6
$\bar{y}_{PE2}$	1.448	0	0.128	1.576	7
$\bar{y}_{PE3}$, theory[a]	0.446[a]	—	0.128[a]	0.574[a]	
empirical	0.615[b]	0.020[b]	0.128[a]	0.763[b]	4
$\bar{y}_{PM1}$	0.125	0	0.095	0.220	1
$\bar{y}_{PX1}$	0.475	0.063	0.087	0.625	2
$\bar{y}_{PX2}$	0.787	0	0.106	0.893	5
$\bar{y}_{PX3}$, theory[c]	0.915	—	0.106[c]	1.121[c]	
empirical	0.576	0.004	0.106[c]	0.686[b]	3

[a]Approximate theoretical values. See Eq. 9.30.
[b]Empirical values.
[c]Approximate theoretical values. See Eq. 9.48.

It may be noted here that the ratio type estimators of $\bar{y}_{PE3}$ and $\bar{y}_{PX3}$ are not very good for small samples such as this. The superiority of $\bar{y}_{PM1}$ over its competitors is much greater than the case of $n = 1$. (This may be ascribed in part to the fact that with the type of PNR sampling adopted here primary 3 became in a sense a stratum and therefore the rather "bad" sample of primaries 1 with 2 was excluded.) The good showing of $\bar{y}_{PX3}$ is heartening, since this is a rather accommodating method, requiring no knowledge of M_i (except those chosen in the sample) or M and accepting X_i as estimates of M_i. However, it does not have an exact expression for MSE, and good sample estimates of the MSE may be a bit troublesome in the case of small samples.

9.10. Estimating $\bar{Y}/\bar{Z}$, $\bar{Y}/\bar{X}$, and $\hat{Y}$

The foregoing estimators were estimating $\bar{Y}$, the mean of characteristic y per secondary unit, where M, the total number of secondaries in the universe, may or may not be known and M_i and X_i may be the measure of size. The type 3 estimators were of the ratio type, since the denominator contained M_is or X_is, which were random variables. We shall now consider a slightly more general estimator for estimating the ratio of two random variables where the measure of size is either M_i or X_i, an extension of Sections 8.6 and 8.7 to the two-stage case.

The estimator of $\bar{Y}/\bar{Z}$ when selection is PPX$_i$ is given by

(9.53)
$$w_{rPX} = \frac{\sum^n M_i \bar{y}_i / A_i}{\sum^n M_i \bar{z}_i / A_i},$$

which, if the sample is self-weighted—that is $f_i = f/nA_i$—is

(9.53a)
$$w_{rPX} = \frac{\sum\sum Y_{ij}}{\sum\sum Z_{ij}},$$

both of which are generally biased like the usual estimator of ratios. Again $A_i = X_i/X$. If primaries are selected with replacement, the MSE of w_{rPX} is given by

(9.54)
$$\text{MSE}\,(w_{rPX}) \doteq \frac{1}{nZ^2}\sum_{}^{N}\frac{1}{A_i}\left[Y_i - \left(\frac{\bar{Y}}{\bar{Z}}\right)Z_i\right]^2 + \frac{1}{nZ^2}\sum\frac{M_i^2}{A_i}\left(\frac{M_i - m_i}{M_i m_i}\right)S_{di}^2$$

where

(9.55)
$$S_{di}^2 = \sum_{j=1}^{M_i}\frac{\left[Y_{ij} - (Y/Z)Z_{ij}\right]^2}{M_i - 1}.$$

Equation 9.54 can be estimated from a sample by

(9.56)
$$\widehat{\text{var}}\,(w_{rPX}) \doteq \frac{1}{n(n-1)Z^2}\sum_{}^{n}\left[\frac{M_i\bar{y}_i}{A_i} - (w_{rPX})\frac{M_i\bar{z}_i}{A_i}\right]^2,$$

which becomes, assuming Z is unknown and the sample is self-weighted,

(9.56a)
$$\widehat{\text{var}}\,(w_{rPX}) = \frac{n}{\left(\sum m_i\bar{z}_i\right)^2}\left[\frac{\sum^n(m_i\bar{y}_i - w_{rPX}\,m_i\bar{z}_i)^2}{n-1}\right].$$

It may be noted that by altering the selecting schemes to PPM and PE and by considering also the estimation of $\bar{Y}/\bar{X}$, we can easily derive from Eq. 9.53 the estimators given in Table 9.10.

Table 9.10. **Summary of ratio estimators with different selection probabilities**

Estimation type	Selection probabilities		
	PE	PM	PX
Ratio on Z	$w_{rPE}{}^* = \dfrac{\sum M_i\bar{y}_i}{\sum M_i\bar{z}_i}$	$w_{rPM}{}^* = \dfrac{\sum \bar{y}_i}{\sum \bar{z}_i}$	$w_{rPX}{}^* = \dfrac{\sum M_i\bar{y}_i/A_i}{\sum M_i\bar{z}_i/A_i}$
Ratio on X $(A_i = X_i/X)$	$w_{rPE}{}^* = \dfrac{\sum M_i\bar{y}_i}{\sum M_i\bar{x}_i}$	$w_{rPM}{}^* = \dfrac{\sum \bar{y}_i}{\sum \bar{x}_i}$	$w_{rPX} = \dfrac{\sum M_i\bar{y}_i/A_i}{\sum M_i\bar{x}_i/A_i}$

* Biased.

Also by making appropriate modifications, we can adapt Eqs. 9.54 and 9.56 to each estimator in Table 9.10. Moreover, if sampling is PNR on the primaries, the between primary variance terms will need to be appropriately adopted. The problem of estimating $\bar{Y}/\bar{X}$ is identical to that for $\bar{Y}/\bar{Z}$ except when X_i is used as a measure of size. In that case the estimator becomes unbiased.

Estimating $\hat{Y}$.   In order to estimate Y or the related $\bar{Y}$, it may not be necessary or even advisable to estimate $\bar{Y}$ and then multiply by M. It may be that M is unknown. For example, in the common case where we wish to make the sample as self-weighted as possible by letting $m_i = M_i f/nA_i$, then if we choose to fix the overall sampling rate, f, at some preferred level, we have

$$(9.57) \qquad\qquad f = nA_i \frac{m_i}{M_i}$$

for the case where selection of primaries is PPX (that is, PPA_i). For PPM, $f = nm_i/M$, and for PE, $f = nm_i/NM_i$.

In any of these cases, when f is fixed for all ij observations, the sample is self-weighted and hence

$$(9.58) \qquad\qquad \hat{Y} = \frac{1}{f} \sum_{i=1}^{n} \sum_{j=1}^{m_i} Y_{ij}$$

is an unbiased estimator of Y, with the variance estimator

$$(9.59) \qquad\qquad \widehat{\text{var}} \, (\hat{Y}) = \frac{n}{f^2} \frac{\sum^n (y_i - \bar{y})^2}{n-1},$$

where $y_i = \sum_{j=1}^{m_i} Y_{ij}$ and $\bar{y} = (1/n) \sum^n y_i$. Note the relationship to Eq. 9.51a, the variance estimator of $\bar{y}_{PX2}$.

$$\diamond \qquad \diamond \qquad \diamond$$

9.11.   Some Special Techniques

In designing multistage samples a number of techniques have been developed to simplify the procedure, to increase efficiency, or to deal with certain technical problems that always seem to be arising. We shall briefly mention some of these techniques here and make a few comments on them. This discussion is by no means an adequate coverage of all such techniques, but it may indicate the current flexibility in survey design and the great amount of useful innovation that has taken place.

Systematic Selection of Primaries with PPS.   Because of feared difficulties in dealing with the selection of more than one event with PPS without replacement from finite universes, most sample designers appear to choose one of three alternatives: (i) use PR sampling, (ii) if a sample is to contain n primaries, create n strata and draw a single primary with PPS from each, and (iii) draw a systematic sample with PPS. Procedure (ii) will be discussed below under "Collapsed Strata." Procedure (iii), a systematic scheme, is usually carried out by listing the primaries in some order that puts the "alike" units together as neighbors as much as possible, as one would in forming strata, and then computing the cumulative totals of the measure of size being used. A "sampling interval" is calculated by $SI = X/n$, selecting a random number between 1 and SI as a starting number and thence every SIth number throughout the listing. The primaries containing one of the designated numbers are regarded as selected. For example, let us use Table 9.7, where the universe consisted of three primaries. Following the order given, we have the following listing together with our measure of size, say M_i, and cumulative M_i.

Primary	M_i	Cumulative M_i	Sample no.
1	2	2	2
2	4	6	
3	6	12	8

We wish to choose $n = 2$ with PPM. Then $SI = \frac{12}{2} = 6$. Choosing an RN (random number), where $1 \leq RN \leq 6$, we obtain, say, 2; our sample numbers are 2 and 8 (or $2 + 6$). Hence primaries 1 and 3 are selected. Note that primaries 1 and 3 have probabilities 2/6 and 6/6 of selection, respectively. Hence primary 3 comes in the sample with certainty.

Unless the listing order is randomized, it is not possible to obtain unbiased or even good estimates of between primary variance. One approximation is given below under "Collapsed Strata." The objection to randomizing the listing order is that stratification effect is lost—unless the method is applied to stratified groupings of the units. It appears that most users of the systematic selection procedure do not wish to obtain good estimates of between primary variance from the sample anyway; they depend on experience for the information, or are otherwise uninterested.

The use of systematic selection within primaries is also common practice, but here its use does not inhibit obtaining good estimates of sampling variance, since the within primary variance need not be obtained separately in the variance estimators for useful approximations.

Stratification. Stratification for two-stage sampling with unequal primaries follows the principles presented for equal primaries (Section 9.4). In the more general case, the variance among primaries, whether equal or unequal, to be minimized by grouping, should depend on the combination of estimator type and selection probability scheme as well as the characteristic under measurement. The estimation procedure is made slightly more complex because strata weights must be added when aggregating estimates from individual strata. Hence

(9.60)
$$\bar{\bar{y}}_{strat} = \frac{1}{M} \sum_{h=1}^{L} M_h \bar{\bar{y}}_{wh}$$

and

(9.61)
$$\widehat{var} \, (\bar{\bar{y}}_{strat}) = \frac{1}{M^2} \sum_{h=1}^{L} M_h^2 \, \widehat{var} \, (\bar{\bar{y}}_{wh}),$$

where $\bar{\bar{y}}_{wh} =$ the estimator, whatever its form, of $\bar{\bar{Y}}_h$,

$\bar{\bar{Y}}_h =$ the mean per secondary in stratum h,

$M_h =$ the total number of secondaries in stratum h,

$M =$ the total over all strata.

"Certainty Primaries."   This expression refers to primaries that are so large in measure of size that they are to be treated as strata. Sometimes these are called "self-representing" primaries. When systematic PPS selection is used, any primary with a measure of size equal to or exceeding the sampling interval will automatically become a "stratum," since it has a probability 1 of being selected. Some PNR schemes of selection also have this automatic feature for identifying such strata (see Chapter 8). Obviously such primaries when so classified and treated do not contribute to the between primary portion of MSE.

"Collapsed Strata."   This term is sometimes used for a method of obtaining an estimate of between primary variance when the sample consisted of one primary selected from each of n strata. The method involves the pairing of strata such that the "most-alike" strata are put together. In the analysis of variation, one regards the two original strata as "collapsed" into a single new stratum with two primaries. The between primary variance in these strata is used as the estimate of what it is in the original strata. For example, if $2L$ strata are collapsed into L strata, where $M_I, M_{II}, \bar{\bar{y}}_I,$ and $\bar{\bar{y}}_{II}$ are numbers of secondaries and sample mean per secondary for the I and II original strata in each pair, then the mean over the L collapsed strata is given by

(9.62)
$$\bar{\bar{y}}_{col} = \frac{1}{M^2} \sum^{L} (M_I \bar{\bar{y}}_I + M_{II} \bar{\bar{y}}_{II})_h$$

and its variance is estimated by

(9.63)
$$\widehat{var} \, (\bar{\bar{y}}_{col}) = \frac{1}{2} \sum \frac{(M_I \bar{\bar{y}}_I - M_{II} \bar{\bar{y}}_{II})^2}{M_h^2} \, .$$

The estimate, Eq. 9.62, is almost always an overestimate, since it includes a component of variance, *between* original strata within collapsed strata, in addition to the desired variance of primaries *within* original strata. Almost always this bias is positive, the amount depending on how different the means per secondary, $\bar{Y}_I$ and $\bar{Y}_{II}$, of the original strata within the collapsed strata turn out to be. If the original stratification is rather fine—that is, strata are small and numerous—and if the original strata chosen for collapsing are those most nearly alike, then the bias in this method of estimating variance will generally be small and hence may be inconsequential. If, however, the original strata are quite different in regard to each other, so that "homogeneous pairs" do not logically exist, and an accurate measure of sampling variance is required, it would be better to start with the "collapsed" strata in the first place and therein draw two primaries with PNR or PR, preferably the former if strata are small and efficiency is important.

9.12. Two-Stage Frame Design

Although in many situations the matter of what grouping to use for primaries is obvious, often it is up to the sample designer to decide among alternatives. For example, suppose one wishes to obtain a sample of 400 grade 1 public school classes in the United States for detailed study. A listing of all such classes for the United States does not exist; even if it did, would it be desirable to choose the 400 classes directly from the list (of, say, some 150,000 such classes in the universe)? The state, the county, and the school district are possible alternative primaries for a multistage frame. Since a listing of classes may not be available at a grouping level higher than a school district, M_is would not be available for all counties or states; hence an X_i of some sort may have to be considered, based possibly on enrollment figures or numbers of teachers reported by counties and thence by states—perhaps only for the previous year! Primary and measure of size must generally be chosen with some consideration for costs—if listings have to be made in order to assure an accurate frame for randomization within the selected primaries—and variances. Frequently these considerations are based on judgments, vague anticipations, speculation, and experience with hopefully similar situations. Nevertheless, decisions are made and the operations are carried through. How well we guessed is all too often never looked into; consequently, when the next investigation comes up for design, we find ourselves facing the same kinds of things we were ignorant of before.

A great variety of possible alternatives come to mind when one is designing a multistage frame. Many of them are ideas and practices that occur to the designer as useful if they exist. If they do not exist, it may be feasible to compile certain lists and measures of size that he believes will provide efficiency worth more than they cost. But considerations of this sort soon go beyond the scope of this book. We shall give a few principles here for guidance that seem to serve well in a number of situations.

Type of Primary. Suppose we wish to obtain a sample of "admits" to a large hospital during a year, and we are told that they are classified by (1) date, or (2) doctor. Suppose we narrow our choices down to the 52 weeks or the 50 doctors under which the year's admits are classified. Which is the better choice of primary? In this case the average size of primary, $\bar{M}$, is about the same. Assuming the listing within an attending doctor is equal in quality to that within a week, the choice is narrowed to that having the least between primary variance. Presumably, the M_i will be more variable from doctor to doctor than from week to week. Also, since doctors are likely to attract patients who have somewhat similar complaints, we would expect the S_i^2 for many patient characteristics to be less for doctors than for weeks, where only seasonal factors are likely to be influential. Hence we should favor the week as a primary, since the between primary variance will likely be smaller, and hence for a fixed m the MSE of $\bar{y}$ will be smaller.

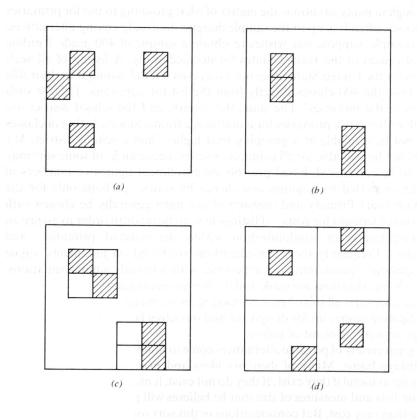

Fig. 9.3.   Notmal and extreme cases of subsampling resulting from varying the size of primaries. (*a*) Simple random. (*b*) Cluster. (*c*) Normal two-stage. (*d*) Stratified random.

Size of Primary.   The effect of size of primary on sampling efficiency depends on what happens to costs of listing and finding the secondaries as well as the between primary variance. In general, if within costs are not affected by increasing the size of primary, then it is advantageous to let the primary be as large as possible (in this case, the limit will be where $M/\bar{M} = N = n$, hence where primaries become strata.) Normally costs are such that the optimal sized primary will fall somewhere between a single secondary and a stratum. Figure 9.3 shows a universe of 36 secondaries with primaries varying from $M_0 = 1$ to $M_0 = 9$. If the costs within and the variance between primaries follow simple mathematical functions as size increases (see Sections 4.7 and 4.8), it would be possible to determine an optimal size. However, since most surveys deal with a number of observed characteristics rather than one, and each is likely to have its own optimum, the practicality of such a determination does not look very attractive for multi-item surveys.

9.13. Remarks and Summary

In the usual situation, two-stage samples are not as efficient for a given m (where $m = m_0 n$) as single-stage samples unless $m_0 = 1$. If framing costs (such as c_0 and part of c_1) and observing costs on primaries (such as part of c_1 and c_3) are high relative to observing costs on secondaries (such as c_2 and c_4) two-stage may be more efficient than one-stage sampling. The situations when this is true include those where the universe of elements is unlisted and where travel is a big part of the cost of data collection. Ordinarily mail and telephone inquiries have less need for multistages than interview surveys. It also appears that more efficient framing techniques (area frames, more economical listing procedures, and so on) are making smaller m_0 values more efficient in two-stage designs.

Increasing the number of sampling stages from one to two or more has increased considerably the complexity of the theory and practice of sampling. Although this makes it more difficult to determine the "best" sampling procedure for a given situation, this complexity offers a great deal of opportunity and challenges the sample designer. He must consider a number of alternative estimators, alternative selection probabilities, and the availability and value of different measures of size, and he must determine whether it is worthwhile to explore for more and better information in order to make his design the best that it can be for the problem at hand. To the student of sampling this state of affairs may appear messy and unnecessary. However, if we are satisfied with something less than maximum efficiency, then some basic principles—as presented in this chapter—are useful for guidance and assurance.

9.14. Review Illustrations

1. Here is a synthetic microuniverse of four groups of four elements each,

with 16 observations:

$$\frac{1}{9,\ 8,\ 6,\ 7},\ \frac{2}{9,\ 5,\ 7,\ 3},\ \frac{3}{10,\ 1,\ 7,\ 4},\ \frac{4}{14,\ 6,\ 10,\ 18}.$$

(a) Summarize the relevant computations in an ANOVA table.
(b) Regarding the groups as primaries, compute var $(\bar{y})$, where $n = 2$ and $m_0 = 2$.
(c) Regarding the groups as clusters, compute var $(\bar{y})$, where $n = 1$ and $M_0 = 4$.
(d) Regarding the 16 observations as serially numbered from 1 to 16, compute var $(\bar{y})$ for systematic samples of 1 in 4 with random start points.
(e) Compare the results of (a)–(d).

Solution

(a)

Source	df	SS	MS	MS defined as
Gross	16	1216		
Mean	1	961		
Total	15	255	17.00	S^2
Groups (rows)	3	105	35.00	$S_W^2 + 4S_B^2$, or $S_G^2/4$
Within Gs	12	150	12.50	S_W^2
Sys (cols)	3	244	81.33	$S_W^2 + 4S_S^2$, or $S_{sys}^2/4$

(b) $$\mathrm{var}\,(\bar{y}_{\mathrm{sub}}) = \left(\frac{N-n}{N}\right)\frac{S_B^2}{n} + \left(\frac{NM_0 - nm_0}{NM_0}\right)\frac{S_W^2}{nm_0},$$

where $N = 4,\ n = 2,\ M_0 = 4,\ m_0 = 2,$

$$S_B^2 = \frac{35.00 - 12.50}{4} = \frac{22.5}{4} = 5.625$$

$$S_W^2 = 12.50$$

$$= \frac{4-2}{2} \cdot \frac{5.625}{2} + \left(\frac{16-4}{16}\right)\frac{12.50}{4} = 1.41 + 2.34$$

$$\underline{= 3.75.}$$

(c) $$\text{var}\,(\bar{\bar{y}}_{\text{clust}}) = (1-f)\,\frac{\text{MS\,(clust)}}{nM_0} \quad \text{or} \quad (1-f)\,\frac{S_G^2}{n},$$

where
$$f = \tfrac{1}{4},\ \text{MS}\,(C) = 35.00,\ S_G^2 = M_0\,\text{MS}\,(C) = 140.0,$$
$$n = 1,\ M_0 = 4,$$

$$= \left(\frac{1-1}{4}\right)\frac{35.00}{(1)\,(4)} = \frac{3}{4}\cdot\frac{35.00}{4} = \frac{105.0}{16} = \underline{6.56}.$$

(d) $$\text{var}\,(\bar{\bar{y}}_{\text{sys}}) = (1-f)\,\frac{\text{MS\,(sys)}}{nM_0}$$

where
$$f = \tfrac{1}{4},\ \text{MS}\,(\text{sys}) = 81.33,\ n = 1,\ M_0 = 4,$$

$$= \left(\frac{1-1}{4}\right)\frac{81.33}{(1)\,(4)} = \frac{3}{4}\cdot\frac{81.33}{4} = \frac{243.99}{16} = \underline{15.25}.$$

(e)

Scheme	var $(\bar{y})$	RE
(b) sub	3.75	85.1
(c) clust	6.56	49.0
(d) sys	15.25	21.0
(—) ran	3.19	100.0

where

$$\text{var}\,(\bar{\bar{y}}_{\text{ran}}) = (1-f)\,\frac{S^2}{m} = \left(\frac{1-1}{4}\right)\frac{17.00}{4} = \frac{3}{4}\cdot\frac{17.00}{4} = 3.19.$$

In these cases, simple random sampling is the most efficient.

2. An examiner for the Food and Drug Administration has sampled a truckload of canned corn to determine the mean number of corn borer fragments per can. The truck contained 1000 cases, each case containing 50 cans. Ten cases were selected at random and from each, two cans were selected at random. The data for the sample are given below. The numbers were the number of corn borer fragments discovered in each of the selected cans.

Cases

		1	2	3	4	5	6	7	8	9	10	Total
Cans	1	4	2	9	8	8	5	0	4	1	7	48
	2	6	5	5	9	4	1	6	4	4	8	52
Total		10	7	14	17	12	6	6	8	5	15	100

ANOVA (on a mean per-can basis)

Source	df	SS	MS
Cases	9	82	9.11
Cans/cases	10	54	5.40

(a)   Estimate the mean number of fragments per can in the truck.

(b)   Estimate the variance of this estimate.

(c)   How many cases would it be necessary to sample to achieve the same accuracy, if five cans were drawn from each case instead of two?

(d)   If it requires two units of time to locate and open a random case and five units to locate and examine a random can in a case, what is the most efficient number of sample cans per case?

Solution

(a)   Estimate of number of fragments per can.

$$\bar{y} = \frac{\sum\sum Y_{ij}}{nm_0} = \frac{100}{10 \times 2} = \frac{100}{20} = 5.0.$$

(b) $$\text{var}(\bar{y}_2) = \left(\frac{N-n}{N}\right)\frac{s_B^2}{n} + \left(\frac{NM_0 - nm_0}{NM_0}\right)\frac{s_W^2}{nm_0}$$

$$\left[s_W^2 = 5.40,\ s_B^2 = (9.11 - 5.40)/2 = 1.86,\ n = 10,\ m_0 = 2 \right]$$

$$= \left(\frac{1000 - 10}{1000}\right)\frac{1.86}{10} + \left(\frac{(1000)(50) - (10)(2)}{(1000)(50)}\right)\frac{5.40}{(10)(2)}$$

$$= \frac{1.86}{10} + \frac{5.40}{20} = \frac{9.12}{20} = 0.456.$$

(c)   If $m_0 = 5$ rather than 2, how many n for the same $\widehat{\text{var}}(\bar{y})$?

$$\text{var}(\bar{y}_5) = \text{var}(\bar{y}_2) = 0.456 = \frac{s_B^2}{n} + \frac{s_W^2}{n(5)} = \frac{1.86}{n} + \frac{5.40}{5n},$$

$$(0.456)n = 1.86 + 1.08 = \frac{2.94}{0.456} = 6.45.$$

Hence an $n \doteq 7$ cases instead of 10.

(d) opt $(m) = \sqrt{(S_W^2/S_B^2)(C_P/C_S)} = \sqrt{(5.40/1.86) \cdot \frac{2}{5}} = 1.08$

or 1 can per case

3.   A survey of areas in various crops in Iowa was carried out in 1940 to study the most efficient sampling designs for crop acreage estimation. The trial design consisted of 3168 sections (square-mile "plots") obtained by selecting two at random from each of the 1584 townships in the state. Regard the frame as nine districts, each consisting of 11 counties, each consisting of 16 townships, each consisting of 36 sections. All land in a given crop lying within the boundaries of the section was measured by field workers' visiting the plot and identifying the crop in each field, and by the determination, by mapping and planimetering, of the areas in each crop. Below is an analysis of variance, on a section basis, of acreage in corn, where the sample mean per section was 161.5 acres.

ANOVA (on a section basis)

Source	DF	SS	MS
Total	3167	13,530,215	4273
District	8	4,309,240	538,655
Counties/districts	90	2,062,879	22,921
Townships/counties	1485	4,062,960	2736
Sections/townships	1584	3,095,136	1954

(a)   Estimate the variance of the sample mean per section.

(b)   Estimate the total acreage in corn in Iowa in 1940 and its standard error.

(c)   If one-half the townships in each county and twice as many sections in each township were selected, what would be the estimated var $(\bar{y})$ per section?

(d)   Regarding (c) as a stratified three-stage design with 100% sampling in the first stage, work out the estimated variance of var $(\bar{y})$ per section.

Solution

(a)   This is a stratified random design with simple allocation. Hence

$$\widehat{\text{var}}\,(\bar{y}) = \left(\frac{N_h - n_h}{N_h}\right)\frac{s_W^2}{n}$$

where s_W^2 is estimated by mean square within townships,

$$= \left(\frac{36-2}{36}\right)\frac{(1954)}{(2)\,(1485)} = \left(\frac{34}{36}\right)\frac{1954}{2970} = 0.5825.$$

(b) $\hat{Y} = N\bar{y} = (36)\,(1584)\,(161.5) = 9{,}209{,}000$ acres.

$$\widehat{\text{RSE}}\,(\hat{Y}) = \frac{1}{\bar{y}}\sqrt{\widehat{\text{var}}\,(\bar{y})} = \frac{(0.5825)^{1/2}}{161.5} = 0.004{,}73 \quad \text{or} \quad 0.47\%.$$

(c)   Regard as a stratified two-stage design, where

$L = 99$ strata (counties),

$N_0 = 16$ primaries (townships) per stratum, of which
$\quad n_0 = 8$ (instead of 16 originally),

$M_0 = 36$ secondaries (sections) per primary, of which
$\quad m_0 = 4$ (instead of 2 originally),

and where s_W^2 (for sections within townships) $= 1954$, s_B^2 (for township components) $= 391.0$ (this   comes   from   the   relationship: MS (townships/counties) $= 2736 = s_W^2 + 2s_B^2$).

Then

$$\text{var}\,(\bar{y}) = \left(1 - \frac{n_0}{N_0}\right)\frac{\bar{S}_B^2}{Ln_0} + \left(1 - \frac{n_0 m_0}{N_0 M_0}\right)\frac{\bar{S}_W^2}{Ln_0 m_0}$$

$$= \left(1 - \frac{8}{16}\right)\frac{391.0}{(99)\,(8)} + \left(1 - \frac{(8)\,(4)}{(16)\,(36)}\right)\frac{1954}{(99)\,(8)\,(4)}$$

$$= (0.50)\,(0.49) + (0.94)\,(0.62) = 0.83$$

compared to 0.58 for the allocation in the original survey.

(d)   Regarding (c) as a stratified three-stage design, then

$L = 11$ strata (districts),

$N_0 = 9$ primaries (counties) per district, of which $n_0 = 9$,

$M_0 = 16$ secondaries (townships) per county, of which $m_0 = 8$,

$K_0 = 36$ tertiaries (sections) per township, of which $k_0 = 4$,

and

$s_3^2 = 1924$, variance of sections within townships,

$s_2^2 = 391$, variance component of townships/counties,
   (from the relationship: MS (townships/counties) $= 2736 = s_3^2 + 2s_2^2$ in the ANOVA table),

$s_1^2 = 631$, variance component of counties/districts
   [from   the   relationship: MS   (counties/districts) $= 22{,}921 = s_3^2 + 2s_2^2 + (2)\,(16)s_1^2$]

Then

$$\text{var}\,(\bar{\bar{y}}) = \left(1 - \frac{n_0}{N_0}\right)\frac{S_1^2}{Ln_0} + \left(1 - \frac{n_0 m_0}{N_0 M_0}\right)\frac{S_2^2}{Ln_0 m_0} + \left(1 - \frac{n_0 m_0 k_0}{N_0 M_0 K_0}\right)\frac{S_3^2}{Ln_0 m_0 k_0}$$

$$= \left(1 - \frac{9}{9}\right) \frac{631}{(11)(9)} + \left[1 - \frac{(9)(8)}{(9)(16)}\right] \frac{391}{(11)(9)(8)}$$

$$+ \left[1 - \frac{(9)(8)(4)}{(9)(16)(36)}\right] \frac{1924}{(11)(9)(8)(4)}$$

$$= (0)6.37 + (0.50)(0.49) + (0.94)(0.62) = 0.83.$$

4. Suppose a large city consists of HHs containing either 1, 2, or 3 adults comprising 30%, 40%, and 30% of all HHs, respectively. In regard to some public issue the adults in favor have the following percentages by size of HH:

Size (number of adults):	1	2	3
Percent of adults in favor:	70%	60%	40%

Where Y_{ij} is 1 if the jth adult in the ith HH is in favor of the issue and 0 otherwise, calculations give the following statistical properties of this population.

$$\frac{\sum^N (\bar{Y}_i - \bar{\bar{Y}}_{PE1})^2}{N-1} = 0.161,345, \qquad \bar{\bar{Y}}_{PE1} = 0.630,$$

$$\frac{\sum^N (Y_i - \bar{Y})^2}{N-1} = 0.427,429, \qquad \bar{Y} = 0.485,$$

$$\frac{\sum^N (M_i/\bar{M})^2 (\bar{Y}_i - \bar{Y})^2}{N-1} = 0.128,237, \qquad \bar{M} = 2,$$

and S_i^2 is indeterminate for all $M_i = 1$, average 0.25 for all $M_i = 2$, and 0.24 for all $M_i = 3$.

Suppose an RNR sample of $n = 100$ HHs and $m_i = 1$ is taken.

(a) Calculate the MSEs of $\bar{y}_{PE1}$, $\bar{y}_{PE2}$, and $\bar{y}_{PE3}$ for the fraction of adults in favor of the issue.

(b) Calculate the variance of p for an RNR sample of 100 adults.

(c) Summarize the results of (a) and (b) and comment.

Solution

(a) Calculation of MSEs for $\bar{y}_{PE1}$, $\bar{y}_{PE2}$, and $\bar{y}_{PE3}$.

$$\text{MSE}(\bar{y}_{PE1}) = \left(\frac{N-n}{Nn}\right) \sum^N \frac{(\bar{Y}_i - \bar{\bar{Y}}_{PE1})^2}{N-1} + \frac{1}{Nn} \sum \left(\frac{M_i - m_i}{M_i}\right) \frac{S_i^2}{m_i} + (\bar{\bar{Y}}_{PE1} - \bar{Y})^2$$

$$= \frac{1}{100} \left\{ 0.161{,}135 + \left[(0.4) \left(\frac{2-1}{2} \right) \left(\frac{0.25}{1} \right) \right. \right.$$

$$\left. \left. + (0.3) \left(\frac{3-1}{3} \right) \left(\frac{0.24}{1} \right) \right] \right\} + (0.630 - 0.485)^2$$

$$= \tfrac{1}{100} (0.161 + 0.098) + 0.021{,}03$$

$$= 0.002{,}59 + 0.021{,}03 = 0.023{,}62.$$

From Eq. 9.29 (modified),

$$\text{MSE} (\bar{\bar{y}}_{PE2}) = \left(\frac{N-n}{Nn} \right) \sum \frac{(Y_i - \bar{Y})^2}{\bar{M}^2 (N-1)} + \frac{1}{Nn} \sum \left(\frac{M_i - m_i}{M_i} \right) \left(\frac{M_i}{\bar{M}} \right)^2 \frac{S_i^2}{m_i}$$

$$= \frac{1}{100} \left\{ \frac{0.427{,}429}{4} + \left[(0.4) \left(\frac{2-1}{2} \right) \left(\frac{2}{2} \right)^2 \left(\frac{0.25}{1} \right) \right. \right.$$

$$\left. \left. + (0.3) \left(\frac{3-1}{3} \right) \left(\frac{3}{2} \right)^2 \left(\frac{0.24}{1} \right) \right] \right\}$$

$$= \tfrac{1}{100} (0.107 + 0.158) = 0.002{,}65,$$

$$\text{MSE} (\bar{\bar{y}}_{PE3}) = \left(\frac{N-n}{Nn} \right) \sum \left(\frac{M_i}{\bar{M}} \right)^2 \frac{(\bar{Y}_i - \bar{Y})^2}{N-1}$$

$$+ \frac{1}{Nn} \sum \left(\frac{M_i - m_i}{M_i} \right) \left(\frac{M_i}{\bar{M}} \right)^2 \frac{S_i^2}{m_i} + (Y_{PE3|n} - \bar{Y})^2$$

$$= \frac{1}{100} \left\{ 0.128{,}237 + \left[(0.4) \left(\frac{2-1}{2} \right) \left(\frac{2}{2} \right)^2 \left(\frac{0.25}{1} \right) \right. \right.$$

$$\left. \left. + (0.3) \left(\frac{3-1}{3} \right) \left(\frac{3}{2} \right)^2 \left(\frac{0.24}{1} \right) \right] \right\} + 0 \text{ (approx)}$$

$$= \tfrac{1}{100} (0.128 + 0.158) = 0.002{,}86.$$

(b)   To compute var (p) for an RNR sample of 100 adults.

$$P = \bar{\bar{Y}} = 0.485, \qquad S^2 = \frac{NPQ}{N-1} = 0.25,$$

$$\text{var} (p) = \left(\frac{M-m}{M} \right) \frac{S^2}{n} = \frac{0.25}{100} = 0.002{,}50.$$

(*c*)   Summary:

Method	MSE	Comments
p	0.002,50	Most accurate but difficult to do.
$\bar{y}_{PE1}$	0.023,62	Least accurate, due to large bias.
$\bar{y}_{PE2}$	0.002,65 ⎱	More or less equal accuracy, but PE2 is slightly
$\bar{y}_{PE3}$	0.002,86 ⎰	better in this case.

5.   (Refer to Review Illustration 4, above.) If HHs are selected with PPM_i (with replacement) and one interview is taken from the M_i at random:

(*a*)   Calculate the variance of the estimator $\bar{\bar{y}} = \sum \bar{y}_i/n$, where $n = 100$. [Note: In this case $\sum^N (M_i/M) (\bar{Y}_i - \bar{Y})^2 = 0.153{,}148$].

(*b*)   How does the efficiency of PPM_i compare in efficiency with RNR on adults?

Solution

(*a*)

$$\text{var}(\bar{\bar{y}}_{PM1}) = \frac{1}{n}\sum\left(\frac{M_i}{M}\right)(\bar{Y}_i - \bar{Y})^2 + \frac{1}{n}\sum\left(\frac{M_i - m_i}{M_i}\right)\left(\frac{M_i}{M}\right)\frac{S_i^2}{m_i}$$

$$= \frac{1}{100}\left[0.153{,}148 + \left(\frac{2-1}{2}\right)(0.4)\left(\frac{0.25}{1}\right)\right.$$
$$\left. + \left(\frac{3-1}{3}\right)(0.3)\left(\frac{0.24}{1}\right)\right]$$

$$= \tfrac{1}{100}(0.153 + 0.098) = 0.002{,}51.$$

(*b*)   An RNR sample would have the variance

$$\text{var}(p) = \left(\frac{N-n}{Nn}\right)\frac{NPQ}{N-1} = \tfrac{1}{100}(0.485)(0.515) = 0.002{,}50.$$

Hence the efficiency is about the same.

6.   A 1% sample of students is to be taken from the following 10 universities. It appears that funds will permit investigators to visit only three of the campuses to carry out the survey. A PNR sample of the 10 schools resulted in the numbers 1, 6 and 9 being selected.

(*a*)   What are the appropriate f_is for the selected schools for a self-weighted sample?

(b)   How many students will be selected from each campus?

School	Students	School	Students
1	20,000	6	500
2	5,000	7	1,500
3	12,000	8	3,300
4	1,500	9	7,000
5	7,200	10	14,000
			72,000

Solution

(a) $f_i = fM_i/nM$;

$$f_1 = \left(\frac{0.01}{3}\right)\frac{72}{20} = \frac{12}{1000} \quad \text{or} \quad 0.0120,$$

$$f_6 = \left(\frac{0.01}{3}\right)\frac{72}{0.5} = \frac{144}{300} \quad \text{or} \quad 0.4800,$$

$$f_9 = \left(\frac{0.1}{3}\right)\frac{72}{7} = \frac{24}{700} \quad \text{or} \quad 0.0343.$$

(b) $m_i = M_i f_i$;

$$f_1 = \frac{20,000\,(12)}{1000} = 240$$

$$f_2 = \frac{500\,(144)}{300} = 240,$$

$$f_3 = \frac{7000\,(24)}{700} = 240.$$

7.   A sample of Los Angeles County consists of 32 census tracts with four segments in each, a total of 128 segments. The tracts were selected with probability proportional to size (number of segments). The numbers of tracts are proportional to the number of HHs in each of the nine regions into which the county has been divided. Four segments were drawn at random from each tract. Total occupied dwelling units are counted on each segment and compared with the number given (or imputed) to the segment in 1960 by the census. Hence, a survey on the 128 segments gave changes in numbers of

occupied dwelling units from April 1960 to July 1964. The ANOVA of the results are:

ANOVA (on a per segment basis)

Source	df	SS	MS
Total	127	17,210	135.5
Regions	8	2442	305.3
Tracts	31	4808	155.1
Tracts/regions	23	2366	102.8
Segments/tracts	96	12,403	129.2

The mean change in the sample is 0.9383 per segment, where the average segment size is 32.890 (1960 HHs).

(a) Write a brief report on sources of variation of population changes in Los Angeles County.

(b) Estimate the standard error of $\bar{\bar{d}}$ ($= 0.9383$).

(c) What loss in efficiency would result if stratification by region were abandoned?

(d) What loss in efficiency would result if the number of tracts were halved and the number of segments within each were doubled? (Assume no stratification).

(e) What is the optimum number of segments per tract if it costs $50 to add a tract and $50 to complete a segment? (Assume no stratification.)

Solution

(a) It appears that although regions differ in growth rates, there is little evidence that tracts differ among themselves within the same district. This can easily be attributed to sampling variation (note small F-value: $F = 1.25$ with 96 and 23 df), possibly due to a few extreme cases within tracts.

(b)
$$\widehat{\text{var}}\,(\bar{d}) = \frac{\text{mean square for tracts}}{\text{total segments}} = \frac{102.8}{128} = 0.805,$$

$$\text{SE}\,(\bar{d}) = \sqrt{0.805} = 0.897,$$

$$\bar{d} + \text{SE}\,(\bar{d}) = 0.9383 + 0.897.$$

(c) $\text{RE (strat/ran)} = \dfrac{\text{tract with region mean square}}{\text{tract mean square}} = \dfrac{155.1}{102.8} = 1.51.$

Loss would be $1 - 1/1.51 = 1 - 0.66$, or 34%.

(d) $\operatorname{var}(\bar{d}) = \dfrac{S_T^2}{n} + \dfrac{S_{S(T)}^2}{n\bar{m}} = \dfrac{6.5}{16} + \dfrac{129.2}{128} = 0.406 + 1.01 = 1.41,$

$$S_T^2 = \dfrac{155.1 - 129.2}{4} = 6.5.$$

Therefore, loss is given by $1 - 1.41/1.51 = 1 - 0.935$ or 6.5%.

(d) $\operatorname{opt}(m) = \sqrt{(S_W^2/S_B^2)\,(c_p/c_s)} = \sqrt{(129.2/6.5)\,\tfrac{50}{50}} = 4.45$

or about 4 segments each.

8. In a study of the incidence of certain health characteristics of a metropolitan area a sample of eight blocks was selected with PR using X_i, the published number of HHs per block as a measure of size (with replacements) from the 120 in one of the strata. Sampling fractions within blocks were set at $f_2 = X_f/X_i n$ with expected $f = 0.03$. The data on number of persons, Z_i, and those having "health conditions," Y_i, found in the sampled portion of each block i are:

Block, i:	1	2	3	4	5	6	7	8	Σ
"Conditions," y_i:	3	6	8	6	10	6	2	3	44
"Persons," z_i:	12	8	20	20	12	14	10	4	100
HHs, M_i:	4	5	6	8	4	6	4	3	40

(a) Estimate the mean "conditions" per HH in the stratum using an unbiased estimator. Assume $M = 1500$.

(b) Estimate the standard error of estimate in (a).

Solution

(a) In this case we wish to estimate $\bar{Y}$ by $\bar{y}_{PX2}$, where the sample is self weighted, hence we use Eq. 9.51a.

$$\bar{y}_{PX2} = \dfrac{1}{nM} \sum \dfrac{M_i \bar{y}_i}{A_i} = \dfrac{1}{Mf} \sum\sum Y_{ij} = \dfrac{44}{(1500)\,(0.03)} = 0.978.$$

(b) To estimate variance use Eq. 9.51a.

$$\widehat{\operatorname{var}}(\bar{y}_{PX2}) = \dfrac{n}{f^2 M^2 (n-1)} \sum^n (y_i - \bar{y})^2 = \dfrac{(8)\,(52)}{(0.03)^2 (1500)^2 (7)}$$
$$= 0.024{,}903.$$

Hence $\widehat{SE}(\bar{y}_{PX2}) = (0.024{,}903)^{1/2} = 0.157.$

9.15. References

Cochran, W. G.   *Sampling Techniques.* (1st ed., 1953) New York: John Wiley & Sons, Inc.
1953, 1963   (Chapters 10 and 11. Utilizes the ANOVA point of view.)

Hansen, M. H.   On the Theory of Sampling from Finite Populations. *Annals of*
Hurwitz, W. N.   *Mathematical Statistics,* **14**: 333–362. (Original presentation of the PPM$_i$
1943   and PPX$_i$ methods.)

Hansen, M. H.   *Sample Survey Methods and Theory.* Vol. I. New York: John Wiley & Sons,
Hurwitz, W. N.   Inc. (Chapters 6, 7, 8 and 9. Excellent for practical procedures and
Madow, W. G.   illustrations. Uses the intraclass correlation point of view.)
1953

Hansen, M. H.   *Sample Survey Methods and Theory.* Vol. II. New York: John Wiley & Sons,
Hurwitz, W. N.   Inc. (Theory for Vol. I.)
Madow, W. G.
1953

Hendricks, W. A.   *The Mathematical Theory of Sampling.* New Brunswick, N.J.: Scarecrow
1956   Press.

Jebe, E. H.   Estimation for Sub-Sampling Designs Employing the County as a Primary
1952   Sampling Unit. *Journal of the American Statistical Association,* **47**: 49–70.
                      (An empirical study on the efficiency of PPM versus PE selection
                      probabilities.)

Jessen, R. J., et al. On a Population Sample for Greece. *Journal of the American Statistical*
1947   *Association,* **42**: 357–384. (Example of using measure-of-size information in
                      various ways.)

Kish, L.   *Survey Sampling.* New York: John Wiley & Sons. (Chapters 5, 6, and 7.)
1965

Madow, L. H.   On the Use of the County as a Primary Sampling Unit for State Estimates.
1950   *Journal of the American Statistical Association,* **45**: 30–47.

Stephan, F. F.   *Sampling Opinions: An Analysis of Survey Procedure.* New York: John Wiley
McCarthy, P. C.   & Sons, Inc. (Chapter 20, pp. 403–424.)
1958

Sukhatme, P. V.   *Sampling Theory of Surveys with Applications.* Ames, Iowa: Iowa State
Sukhatme, B. V.   University Press. (Chapters 7 and 8.)
1954, 1970

Yates, F.   *Sampling Methods for Censuses and Surveys.* (2nd ed.) New York: Hafner
1953   Publishing Company.

Yates, F.   The Estimation of the Efficiency of Sampling with Special Reference to
Zacopanay, I.   Sampling in Cereal Experiments. *Journal of Agricultural Science,* **25**:
1935   545–577. (A pioneering study of subsampling theory and practice.)

9.16. Exercises

1. In 1937 the number of corn acres (also other crop acreages) in every section (a plot approximately one mile square) of Brookings County, S.D., was obtained by planimetering aerial photographs (see Example 9.3 on barley). From these data, the following mean squares, on a section basis, were computed:

ANOVA (on a per-section basis)

Source	df	MS
Total	719	3400
Townships (36 sections)	19	42,100
Sections with townships	700	2400
"Blocks" (four sections) in		
County	179	7400
Sections within blocks	540	2100

(The county is composed of 720 sections, 180 blocks, and 20 townships. Each township contains nine blocks, each block contains four sections.)

Considering the county as the universe, what is the variance of the estimate of the mean number of corn acres per section with each of the following sampling schemes?

(a)   Random sample of 20 sections from the whole county.
(b)   Random sample of five blocks (total of 20 sections).
(c)   Sample of 20 random sections stratified by township.
(d)   Sample taken by selecting 20 blocks at random from the total of 180 and selecting one section in each of the 20 blocks.

2.   Refer to Exercise 1 for basic data. What is the variance of a scheme wherein a block is selected at random from each of the 20 townships and one section selected at random from that block?

3.   A chain store operating 450 stores scattered around the western states desires a sample of its employees in order to study employee morale. The stores average 10 employees each. It was felt, when considering the between and within store variances, that a random sample of 100 stores should be taken and within each selected store a random sample of two-fifths of the employees should be selected and interviewed, thus yielding total of about 400 employees. The simple sample mean per employee is used as the estimator. Critically evaluate this scheme in regard to such matters as (a) unbiasedness and (b) efficiency. (c) What suggestions do you have that would improve the design? Explain.

4.   A sample of HHs is being selected from a stratum in Santa Monica which, according to the previous census, contained 1800 HHs (X) in 60 blocks. The sample is to consist of two blocks selected with PPX, and an overall sampling

fraction of $\frac{1}{100}$ is required. Determine the required sampling fractions, f_i, for the selected blocks, where their X_i are 20 and 60, respectively, if

(a) The blocks are selected with PR.

(b) The blocks are selected with PNR.

(c) The blocks were selected with RNR.

5. A city consists of a large number of HHs of two adults each. The universe of interest is adults. Frames (listings, say) of both individual adults and HHs are available for selecting samples.

(a) Suppose a sample of HHs is selected for the first stage and a random adult is selected at the second stage. How does this sample compare in efficiency with a simple random sample of adults if (i) $\delta = -1$? (ii) $\delta = +1$? (iii) $\delta = 0$?

(b) Same as (a), except that in the two-stage sample both adults are to be surveyed.

CHAPTER 10

Double Sampling (Multiphase)

10.1. Introduction

As subsampling can be regarded as a generalization of stratified and cluster sampling, so double sampling can be regarded as a generalization of the stratified, ratio, and regression estimators. It may be recalled that when stratification was dealt with, it was assumed that the universe number of elements in each stratum was known. Also, it may be recalled that when ratio or regression estimators were being discussed that the population values, X or $\bar{X}$, and M or $\bar{M}$, must be known in order that an improvement in precision through those estimators could be forthcoming. In the present chapter, methods will be dealt with that do not require that stratum sizes, or the population mean (or total) of an auxiliary variate (or total number of secondaries), be known. Double sampling comprises designs that have as one of their capabilities the estimation of these quantities from the sample itself.

10.2. Estimating the Ratio, $\bar{Y}/\bar{X}$, in Two Phases

In Section 5.3 we considered the sample ratio, $\bar{y}/\bar{x}$, as an estimator of $\bar{Y}/\bar{X}$ and discussed its properties, where $\bar{y}$ and $\bar{x}$ were the means of y and x observed on each of the sample's n elements. Let us now consider a sample of n' elements from a universe of N, on which observations are made on all n' of the X_i and only n'' of the Y_i, where n'' is a subsample of the n'. One reason for taking a smaller sample of Y_i than X_i may be that Y_i is very costly to observe relative to X_i. In this case an appropriate estimator is given by

$$(10.1) \qquad \frac{\bar{y}''}{\bar{x}'} = \frac{\sum^{n''} Y_i/n''}{\sum^{n'} X_i/n'},$$

where $\bar{x}'$ is the mean of x in the phase I sample and $\bar{y}''$ the mean of y in the phase II sample.

An approximate variance of Eq. 10.1 is given by

$$(10.2) \qquad \mathrm{var}\left(\frac{\bar{y}''}{\bar{x}'}\right) \doteq \frac{1}{\bar{X}^2}\left[\left(\frac{N-n'}{N}\right)\left(\frac{\bar{Y}}{\bar{X}}\right)^2\frac{S_x^2}{n'} + \left(\frac{N-n''}{N}\right)S_y^2 n'' - 2\left(\frac{N-n'}{N}\right)\left(\frac{\bar{Y}}{\bar{X}}\right)\frac{S_{xy}}{n''}\right]$$

$$(10.2a) \qquad \doteq \left(\frac{\bar{Y}}{\bar{X}}\right)^2\left(\frac{V_x^2}{n'} + \frac{V_y^2 - 2V_{xy}}{n''}\right)$$

if we ignore the FPCs and replace variances with coefficients of variation. Note that if n' is increased to a census, then Eq. 10.2 reduces to

$$(10.2b) \qquad \mathrm{var}\left(\frac{\bar{y}''}{\bar{x}'}\right) = \frac{1}{\bar{X}^2}\left(\frac{N-n''}{N}\right)\left(\frac{S_y^2}{n''}\right),$$

which is the variance of $\bar{y}/\bar{x}$, the case where the denominator is a known constant. Note also that when n'' is increased to n', Eq. 10.2 becomes the standard variance for $\bar{y}/\bar{x}$,

(10.2c) $$\text{var}\left(\frac{\bar{y}''}{\bar{x}'}\right) = \frac{1}{\bar{X}^2}\left(\frac{N-n'}{Nn'}\right)\left[\left(\frac{\bar{Y}}{\bar{X}}\right)^2 S_x^2 + S_y^2 - 2\left(\frac{\bar{Y}}{\bar{X}}\right)S_{xy}\right],$$

given by Eq. 5.8, since $2RS_xS_y = 2S_{xy}$. Also it may be noted that if n' and n'' are independent, the covariance term should be omitted.

To determine the situations in which it would be advisable to use double rather than single sampling (where $n'' = n'$), let us assume the simple cost function

(10.3) $$C = n'c' + n''c''$$

where c' and c'' are the unit costs of observing an X_i (in phase I) and a Y_i (in phase II), respectively, and C is the total budget. Also let us ignore the FPC on S_y^2 and write Eqs. 10.2 and 10.2a as

(10.4) $$\text{var}\left(\frac{\bar{y}''}{\bar{x}'}\right) = \frac{S_I^2}{n'} + \frac{S_{II}^2}{n''},$$

where

(10.5) $$S_I^2 = \left(\frac{\bar{Y}^2}{\bar{X}^4}\right)S_x^2 = \left(\frac{\bar{Y}}{\bar{X}}\right)^2 V_x^2$$

(10.6) $$S_{II}^2 = \left(\frac{1}{\bar{X}^2}\right)S_y^2 - 2\left(\frac{\bar{Y}}{\bar{X}^3}\right)S_{xy} = \left(\frac{\bar{Y}}{\bar{X}}\right)^2(V_y^2 - 2V_{xy}).$$

(Note that double sampling is feasible only when $V_y > 2RV_{x'}$) Then for a fixed budget, C, var $(\bar{y}''/\bar{x}')$ will be a minimum when

(10.7) $$\frac{n''}{n'} = \frac{1}{t} = \sqrt{(S_{II}^2/S_I^2)\,(c'/c'')}..$$

To solve for n' and n'' we have from Eq. 10.3

$$C = n'c' + \left(\frac{n'}{t}\right)c'',$$

(10.8) $$n' = \frac{C}{c' + c''/t},$$

and

(10.8a) $$n'' = \frac{n'}{t}.$$

Example 10.1. In an inquiry $V_x = 1.0$, $V_y = 2.8$, and $R = .9$ [hence V_{xy} $= (2.8)(1.0)(.9) = 2.52$]. The cost of observing $x = 1$ and of $y = 10$. What is the optimal size of n'' to n'? In this case

$$S_I^2 = \left(\frac{\bar{Y}}{\bar{X}}\right)^2 (V_x^2) = \left(\frac{\bar{Y}}{\bar{X}}\right)^2 (1.00)$$

and

$$S_{II}^2 = \left(\frac{\bar{Y}}{\bar{X}}\right)^2 [V_y^2 - 2V_{xy} = (2.8)^2 - 2(2.52)] = \left(\frac{\bar{Y}}{\bar{X}}\right)^2 (2.80).$$

Then

$$\frac{n''}{n'} = \frac{1}{t} = \sqrt{(2.80/1.00)\tfrac{1}{10}} = \sqrt{0.28} \doteq 0.5,$$

hence $t \doteq 2$. We should observe y on one-half of the elements on which x is observed.

Example 10.2. Calculate the efficiency of double relative to single sampling in Example 10.1.

Under single sampling the cost per element would be $c = c' + c''$, and since

$$C = nc = n'c' + \left(n''c'' = \frac{n'}{2c''}\right),$$

then $n = \tfrac{6}{11}n'$; that is, the budget would permit a single sample $\tfrac{6}{11}$ as large as phase I of a double sample. The variances are

$$\text{var }(s) = \left(\frac{\bar{Y}}{\bar{X}}\right)^2 \left[\frac{V_x^2 + V_y^2 - 2V_{xy}}{n}\right] = \left(\frac{\bar{Y}}{\bar{X}}\right)^2 \left[\frac{(1.0)^2 + (2.8)^2 - (2)(2.52)}{n}\right]$$

$$= \left(\frac{\bar{Y}}{\bar{X}}\right)^2 \frac{3.80}{n},$$

$$\text{var }(ds) = \left(\frac{\bar{Y}}{\bar{X}}\right)^2 \left[\frac{V_x^2}{n'} + \frac{V_y^2 - 2V_{xy}}{n''}\right] = \left(\frac{\bar{Y}}{\bar{X}}\right)^2 \left[\frac{1}{n'} + \frac{2.80}{n''}\right]$$

$$= \left(\frac{\bar{Y}}{\bar{X}}\right)^2 \left[\frac{1}{\tfrac{11}{6}n} + \frac{2.80}{\tfrac{11}{12}n}\right]$$

$$= \left(\frac{\bar{Y}}{\bar{X}}\right)^2 \frac{3.60}{n},$$

$$\text{RE }(d/s) = \frac{\text{var }(s)}{\text{var }(ds)} = \frac{3.80/n}{3.60/n} = 1.06 \quad \text{or } 106\%.$$

Example 10.3. Suppose in Example 10.1 $\bar{X}$ were known. Now the entire resources can be devoted to observing y. Calculate the efficiency of this case to that of single sampling for $\bar{y}/\bar{x}$.

The variance now will be

$$\text{var}\left(\frac{\bar{y}}{\bar{x}}\right) = \frac{1}{\bar{X}^2} \cdot \frac{S_x^2}{n_0} = \frac{V_x^2}{n_0} = \frac{(1.0)^2}{n_0} = \frac{1}{n_0}.$$

The cost now will be c'', since X_is are unnecessary. $C = nc = n_0 c''$, then $n_0 = nc/c''$, $n_0 = n(\frac{11}{10}) = 1.10n$,

$$\text{RE } (\bar{X} \text{ to } \bar{x} \text{ case}) = \frac{\text{var } (\bar{y}/\bar{x})}{\text{var } (\bar{y}/\bar{X})} = \frac{3.80/n}{1.10/n} = 3.45,$$

or over three times as efficient.

10.3.   Estimating the Product $\bar{Y}\bar{X}$, in Two Phases

Like the case of the ratio, we can extend the scope of the product estimator, $\bar{x}\bar{y}$, from the simple to double sampling. Hence we have the estimator,

$$(10.9) \qquad \bar{y}''\bar{x}' = \left(\sum^{n''} \frac{Y_i}{n''}\right)\left(\sum^{n'} \frac{X_i}{n'}\right),$$

where $\bar{x}'$ is the mean of x in the phase I sample and $\bar{y}''$ the mean of $\bar{y}$ in phase II. This estimator is appropriate whether $n'' \in n'$ or not.

An approximate variance of the above product, if $n'' \in n'$, is given by

$$(10.10) \ \text{var } (\bar{y}''\bar{x}') \doteq \bar{Y}^2\left(\frac{N-n'}{N}\right)\frac{S_x^2}{n'} + \bar{X}^2\left(\frac{N-n''}{N}\right)\frac{S_y^2}{n''} + 2\bar{X}\bar{Y}\left(\frac{N-n'}{N}\right)\frac{S_{xy}}{n'},$$

where all symbols have their usual meanings. If $n'' \bar{\in} n'$, the covariance term should be made null. For convenience let us ignore FPCs and adopt relative variances, where Eq. 10.10 becomes

$$(10.10a) \qquad \text{var } (\bar{y}''\bar{x}') \doteq (\bar{Y}\bar{X})^2\left(\frac{V_x^2}{n'} + \frac{V_y^2 + 2V_{xy}}{n''}\right),$$

which is analogous to the ratio case except for the sign on the covariance term. Using the same procedure presented in the previous section, we can determine the optimal choices of n' and n''. If we let

$$(10.11) \qquad\qquad S_I^2 = \bar{Y}^2 S_x^2 = \bar{Y}^2\bar{X}^2 V_x^2,$$

$$(10,11a) \qquad S_{II}^2 = \bar{X}^2 S_y^2 + 2\bar{X}\,\bar{Y}S_{xy} = \bar{Y}^2\bar{X}^2(V_y^2 + 2V_{xy}),$$

and assume the cost function, Eq. 10.3, the optimal distribution rate $1/t$ is given by Eq. 10.7.

An application of this scheme will be presented in Section 10.9.

10.4.   Estimating $\bar{Y}$ with the Help of x: With Ratio Estimator

We are interested in estimating the total or mean of some characteristic y that is expensive to observe or measure. A characteristic x, correlated with y, is much cheaper to observe. We are interested in improving our estimate of $\bar{Y}$ with the help of x by using a ratio type estimator. In Chapter 5, Eq. 5.19, we dealt with the ratio estimator for single-phase sampling,

$$(10.12) \qquad \bar{y}_{rat} = \bar{X}\left(\frac{\bar{y}}{\bar{x}}\right),$$

where $\bar{X}$, the population mean of x, is known. Suppose now we do not know $\bar{X}$ but estimate it from a two-phase sample. Suppose for the double sample we draw a random sample of n' elements from a universe of N, and on this sample only x is observed. From the n' elements of the first phase a random sample of n'' elements is drawn for the second phase in which both x and y are observed. Our estimator of $\bar{Y}$, the mean of y per element in the universe is given by

$$(10.13) \qquad \bar{y}_{d-rat} = \bar{x}'\left(\frac{\bar{y}''}{\bar{x}''}\right),$$

where $\bar{x}'$ = sample mean of x in phase I,
$\bar{y}''$, $\bar{x}''$ = sample mean of y and x in phase II.

This estimator, Eq. 10.13, like Eq. 10.12 is biased, but the bias is usually negligible if both n' and n'' are large.

The variance of Eq. 10.13 is given by

$$(10.14) \quad \text{var}\,(\bar{y}_{d-rat}) \doteq \left(\frac{\bar{Y}}{\bar{X}}\right)^2\left(\frac{n'-n''}{n'n''}\right)S_x^2 + \left(\frac{N-n''}{Nn''}\right)S_y^2 - 2\left(\frac{\bar{Y}}{\bar{X}}\right)\left(\frac{n'-n''}{n'n''}\right)S_{xy},$$

which becomes, if we group sources of variation by phase,

$$(10.14a)\; \text{var}\,(\bar{y}_{d-rat}) \doteq \frac{2(\bar{Y}/\bar{X})S_{xy} - (\bar{Y}/\bar{X})^2 S_x^2}{n'}$$

$$+ \frac{[(N-n'')/N]\, S_y^2 - 2(\bar{Y}/\bar{X})S_{xy} + (\bar{Y}/\bar{X})^2 S_x^2}{n''}.$$

If n' and n'' are drawn independently, then

$$(10.15)\; \text{var}\,(\bar{y}_{d-rat}) \doteq \frac{(\bar{Y}/\bar{X})^2 S_x^2}{n'} + \frac{[(N-n'')/N]\, S_y^2 + 2(\bar{Y}/\bar{X})S_{xy} + (\bar{Y}/\bar{X})^2 S_x^2}{n''}.$$

To estimate the above variances from a double sample, one can substitute sample estimates for the corresponding parameters.

To determine optimal choices of n' and n'' we use the procedures of Section

10.2. If we let

$$(10.16) \qquad S_I^2 = 2\left(\frac{\bar{Y}}{\bar{X}}\right)S_{xy} - \left(\frac{\bar{Y}}{\bar{X}}\right)^2 S_x^2$$

$$= \bar{Y}^2(2V_{xy} - V_x^2) = \bar{Y}^2 V_I^2,$$

$$(10.16a) \qquad S_{II}^2 = S_y^2 - 2\left(\frac{\bar{Y}}{\bar{X}}\right)S_{xy} + \left(\frac{\bar{Y}}{\bar{X}}\right)^2 S_x^2$$

$$= \bar{Y}^2(V_y^2 - 2V_{xy} + V_x^2) = \bar{Y}^2 V_{II}^2,$$

and assuming the cost function, Eq. 10.3, then the optimal fraction, $1/t$, is given by Eq. 10.7.

When optimal n' and n'' are used, the variance of the estimator is given by

$$(10.17) \qquad \text{var}\,(\bar{y}_{d-\text{rat}}) = \frac{(\sqrt{c'S_I^2} + \sqrt{c''S_{II}^2})^2}{C}.$$

Example 10.4. The problem is to estimate the total number of people in a town comprising 5000 blocks. It is proposed that a large sample of blocks selected at random be visited by two persons in a car (one driving, the other observing and recording) and a quick estimate of the numbers of households be made from the appearance of dwellings in the block. Next, a small sample of blocks will be drawn at random from the large sample and on these a detailed count of persons living there will be made. Assume that we have $100 and it costs $0.10 to "eye-cruise" a block and $1 to obtain an accurate count on a block, and that the correlation between the "cruise" and actual counts is .9. The basic data are:

N	C	c''	c'	$\bar{y}''$	$\bar{x}'$	$\bar{x}''$	V_x	V_y	V_{xy}
5000	100	1.00	0.10	80	22	20	0.924	1.000	0.832

(a) What is the best allocation for a double sample using a ratio estimator? Assuming our sampling fractions to be small, we shall proceed with Eq. 10.7,

$$\frac{n''}{n'} = \sqrt{(S_{II}^2/S_I^2)\,(c'/c'')};$$

since

$$S_{II}^2 = (V_y^2 + V_x^2 - 2V_{yx})\bar{Y}^2$$
$$= [(1.0)^2 + (0.924)^2 - 2(0.832)]\bar{Y}^2$$
$$= (0.190)\bar{Y}^2$$

and

$$S_I^2 = (2V_{xy} - V_x^2)\bar{Y}^2$$
$$= [2(0.832) - (0.924)^2]\bar{Y}^2$$
$$= (0.910)\bar{Y}^2.$$

then

$$\frac{n''}{n'} = \sqrt{(0.190/0.910)\,(0.10/1.00)}$$

$$= 0.145$$

and hence

$$n'' = 0.145n'.$$

Since from Eq. 10.3,

$$C = n'c' + n''c'',$$

$$C = n'c' + 0.145n'c'',$$

$$n' = \frac{C}{c' + (0.145)c''} = \frac{100}{0.10 + (0.145)\,(1.00)} = \frac{100}{0.245} = 408$$

and

$$n'' = \frac{C - n'c}{c''} = \frac{100 - (408)\,(0.10)}{1.00} = 59.$$

In this case neither n'/N ($= 8.2\%$) nor n''/n' ($= 14.5\%$) are negligible. Hence we have only an approximate optimum.

(b)   What is the variance of the double-sample estimator, using the ratio, and how does it compare with that of single sampling? From Eq. 12.17,

$$\text{opt var}\,(\bar{y}_{d-rat}) = \frac{(\sqrt{c'S_I^2} + \sqrt{c''S_{II}^2})^2}{C}.$$

Since $S_I^2 = \bar{Y}^2(0.910)$ and $S_{II}^2 = \bar{Y}^2(0.190)$,

$$\text{opt var} = \frac{\left[\sqrt{(0.10)\,(0.910)\bar{Y}^2} + \sqrt{(1.00)\,(0.190)\bar{Y}^2}\right]^2}{100}$$

$$= \bar{Y}^2(0.005,52).$$

For $\bar{Y}^2$ let us use $\bar{y}_{d-rat} = \bar{x}'(\bar{y}''/\bar{x}'') = (22.0)\,(\tfrac{80}{20}) = 88.0$ and hence opt var $= (88.0)^2(0.005,52) = 42.7$.

10.5. Estimating $\bar{Y}$ with the Help of x: With Regression Estimator

This type of double sampling, which is due to Cochran (see Watson, 1937), in its simplest form is based on a universe of N elements from which a relatively large sample, n', is drawn at random. On these n' elements we observe some characteristic x that is relatively easy or cheap to measure but that is correlated with y, the characteristic in which we are really interested. From the large sample n' we now draw at random a relatively small sample, n'', on which the expensive y measurements are taken.

Since we are interested in estimating $\bar{Y}$, we use as its estimator

$$(10.18) \qquad \bar{y}_{d-regr} = \bar{y}'' + b''(\bar{x}' - \bar{x}''),$$

where $\bar{y}'' = $ the mean of the ys in the small sample,

$\qquad \bar{x}'' = $ the mean of the xs in the small sample,

$\qquad \bar{x}' = $ the mean of the xs in the large sample,

$\qquad b'' = $ the regression coefficient of y on x, obtained from the small sample.

The estimator of the double sampling, $\bar{y}_{d-regr}$, is quite similar to that given as the "regression estimator," Eq. 5.23, except that the x-population mean $\bar{X}$ is replaced by $\bar{x}'$, its estimate from the large first-phase sample. Ordinarily, $\bar{y}_{d-regr}$ is a biased estimator of $\bar{Y}$ for the same reason the regression estimator $\bar{y}_{regr}$ is a biased estimator of $\bar{Y}$; that is, the regression of y on x will seldom be perfectly linear. However, in most practical applications the bias, which is approximately that given by Eq. 5.25, will be negligible except for very small values of n''.

The average variance of $\bar{y}_{d-regr}$—that is, the variance averaged over repeated selections of n' and n''—is given approximately by

$$(10.19) \quad \text{var}\,(\bar{y}_{d-regr}) \doteq \left(\frac{N-n''}{N}\right)\left[\frac{S_y^2(1-R^2)}{n''}\left(1+\frac{n'-n''}{n'}\cdot\frac{1}{n''-3}\right)+\frac{R^2 S_y^2}{n'}\right],$$

which is due to Cochran [see Jessen (1942)], and where FPCs and terms in $1/n'(n''-3)$ can be ignored,

$$(10.19a) \qquad \text{var}\,(\bar{y}_{d-regr}) \doteq \frac{R^2 S_y^2}{n'} + \frac{(1-R^2)S_y^2}{n''}$$

$$(10.19b) \qquad\qquad\qquad \doteq \frac{S_I^2}{n'} + \frac{S_{II}^2}{n''},$$

that is, the sum of variances contributed by the two phases, where

$\qquad S_y^2 = $ the variance of the y-population,

$\qquad R = $ the coefficient of correlation of y on x in a finite universe,

N = the universe number of elements,
n' = size of sample in phase I,
n'' = size of sample in phase II.

Equations 10.19 and 10.19a express the average or expected variance of $\bar{y}_{d-regr}$ that would result if the double-sampling method were used. However, the appropriate variance for a given sample will depend on the particular set of xs that happen to be drawn. This is given by

$$(10.20) \quad \text{var } (\bar{y}_{d-regr}) \doteq \left(\frac{N-n''}{N}\right)\left[S_y^2(1-R^2)\left[\frac{1}{n''}+\frac{\bar{x}'-\bar{x}''}{\sum(X_i-\bar{x}'')^2}\right]+\frac{R^2 S_y^2}{n'}\right],$$

where $\sum(X_i-\bar{x}'')^2$ is summed over n'', the phase II sample. Equation 10.20 is appropriate for estimating the precision of a *given* selected double sample, whereas Eq. 10.19 is appropriate for comparing the variance of the double-sampling design with *other* designs.

It may be of interest to see what happens to this double-sampling scheme if certain extremes are considered. First, let us increase the size of the large sample (phase I) to the point where it includes the universe—that is, $n' = N$. In this case, since $\bar{x}' = \bar{X}$, Eq. 10.18 becomes

$$(10.21) \qquad\qquad \bar{y}_{d-regr} = \bar{y}'' + b''(\bar{X}-\bar{x}''),$$

which is identical to the regression estimator given by Eq. 5.23 and its variance from Eq. 10.19 and, regarding $(n''-3) \doteq n''$, reduces to

$$(10.22) \qquad \text{var } (\bar{y}_{d-regr}) \doteq \left(\frac{N-n''}{N}\right)\frac{S_y^2(1-R^2)}{n''}\left(1+\frac{1}{n''}\right),$$

which is identical to that of the regression estimator, Eq. 5.30, and, where terms in $1/(n'')^2$ can be ignored, is identical to Eq. 5.30a; that is,

$$(10.22a) \qquad \text{var } (\bar{y}_{d-regr}) \doteq \left(\frac{N-n''}{N}\right)\frac{S_y^2(1-R^2)}{n''}.$$

The second case of interest is when the small sample is increased to the size of the large one and they become identical, then $n'' = n'$ and, from Eq. 10.18,

$$(10.23) \qquad\qquad \bar{y}_{d-regr} = \bar{y}'',$$

and its variance, from Eq. 10.22a, reduces to

$$(10.24) \qquad \text{var } (\bar{y}_{d-regr}) \doteq \left(\frac{N-n''}{N}\right)\frac{S_y^2}{n''},$$

which, of course, is the case of simple "single" sampling, since we are obtaining no additional information on $\bar{X}$ that can improve the estimate of $\bar{Y}$.

It is possible to see that some gain may be made by double over single sampling of (a) the cost of measuring or observing x is quite cheap compared to the cost of measuring y, and (b) the correlation between y and x is high.

If there were no cost of measuring x, then we would probably obtain x observations on all N elements in the universe (that is, $n' \to N$), and as our variance of the adjusted mean we would have Eq. 10.22. However, the more general situation is that in which the x information must be obtained as part of the survey, and the question arises: how much of the sampling resources should be put to measuring x and how much to measuring y?

For optimal choice of n' and n'' we again use the results of Section 10.2; hence var $(\bar{y}_{d-regr})$ will be a minimum approximately when

$$(10.25) \qquad \frac{n''}{n'} = \sqrt{(S_{II}^2/S_I^2)\,(c'/c'')} = \sqrt{[(1-R^2)/R^2]\,(c'/c'')},$$

where S_I^2 and S_{II}^2 are defined in Eqs. 10.19a and b.

The efficiency of double relative to single sampling when we have a fixed budget C and where optimum allocation of resources to the two phases (Eq. 10.8) is attained is given by

$$(10.26) \qquad \text{NRE (double/single)} = \frac{\text{var }(\bar{y}_{single})}{\text{var }(y_{d\ regr})}.$$

Since for the single sample

$$(10.27) \qquad C = nc'',$$

if the FPC is ignored,

$$(10.28) \qquad \text{var }(\bar{y}_{single}) = \frac{S_y^2}{C/c''}.$$

The var $(\bar{y}_{d-regr})$ is given by Eq. 10.19 with the condition of Eq. 10.25, whence

$$(10.29) \qquad \text{NRE (double/single)} = \frac{c''}{(\sqrt{(1-R^2)c''} + R\sqrt{c'})^2}.$$

In general, double will be more efficient than single sampling when

$$(10.30) \qquad \frac{c''}{c'} > \frac{R^2}{(1-\sqrt{1-R^2})^2}$$

or when

$$(10.30a) \qquad R^2 > \frac{4c''c'}{(c''+c')^2}.$$

In terms of cost ratios, the net relative efficiency of double to single sampling can be derived from Eq. 10.29 and expressed by

(10.31)
$$\frac{c''}{c'} = \frac{R^2}{(1/\sqrt{\text{NRE}} - \sqrt{1-R^2})^2}.$$

In Table 10.1 the cost ratios required to achieve a given *net relative efficiency*, Eq. 10.31, for various values of R are given for the NRE levels of 1.00, 1.25, 1.50, 2.00, 3.00, and 4.00. These values are also plotted in Figure 10.1.

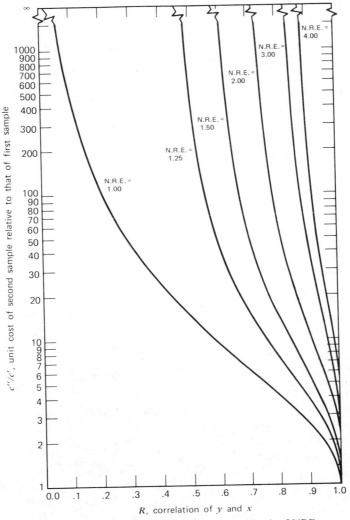

Fig. 10.1.   Plot of c''/c' against R for several levels of NRE.

Table 10.1. *Net relative efficiencies of double sampling compared with single sampling for various correlation and cost ratio levels (regression estimation)*

Correlation of y and x (R)	Unit cost of second sample relative to unit cost of first, c''/c'					
	Net relative efficiencies NRE (d/s)					
	1.00	1.25	1.50	2.00	3.00	4.00
0.0	∞					
.1	398.40					
.2	98.03					
.3	42.42					
.4	22.96					
.447		∞				
.5	13.93	309.96				
.550	11.14	86.23				
.577			∞			
.6	9.00	40.37	1322.31			
.625	8.12	30.16	303.49			
650	7.33	23.36	132.04			
.7	6.00	15.07	46.77			
.707				∞		
.750	4.91	10.36	23.40			
.8	4.00	7.38	13.65	55.79		
.817					∞	
( )						∞
.850	3.23	5.35	8.61	22.22	282.52	
.875	2.88	4.55	6.93	15.40	88.09	1978.00
.9	2.55	3.85	5.59	11.01	40.48	197.1
.925	2.23	3.23	4.49	8.00	21.96	59.4
.950	1.91	2.66	3.55	5.79	12.84	25.6
.975	1.57	2.10	2.69	4.05	7.55	12.3
1.0	1.00	1.25	1.50	2.00	3.00	4.00

Note that only when both the x, y correlation and the cost ratio are high can we expect a great gain in use of double over single sampling. When these conditions exist, substantial gains can be achieved. Some examples will be given later.

Example 10.5. For the problem of estimating the number of people in a town (Example 10.4), use the regression rather than the ratio estimator. The relevant data are:

N	C	c''	c'	R	b''
5000	100	1.00	0.10	0.9	3.9

(a) What is the best allocation of a double sample?
(b) Is double sampling in this case better than single sampling?
(c) How much better or worse?
(d) If $\bar{y}'' = 80.0$, $\bar{x}'' = 20.0$, $\bar{x}' = 22.0$, and $b'' = 3.9$, what is $\bar{y}_{d-regr}$?

To determine n'' we employ Eq. 10.25,

$$n'' = n' \sqrt{[(1-R^2)/R^2] (c'/c'')} = n' \sqrt{\{[1-(.9)^2]/(.9)^2\} (0.10/1.00)}$$
$$= 0.153n'.$$

Since, $C = c'n' + c''n'' = c'n' + c''(0.153)n'$,

$$n' = \frac{C}{c' + c''(0.153)} = \frac{100}{0.10 + (1.00)(0.153)} = 395$$

and

$$n'' = 0.153n' = (0.153)(395) = 60,$$

which means that the best allocation of the sampling between the two phases is 395 blocks for eye cruising and 60 blocks for detailed counting.

In answering (b) to determine whether double or single sampling is better in this case, we use Eq. 10.30a:

$$R^2 > \frac{4c''c'}{(c''+c')^2},$$

$$(.9)^2 > \frac{4(1.00)(0.10)}{(1.00+0.10)^2},$$

$$.81 > .33.$$

Therefore, since this condition is satisfied, double is more efficient than single sampling.

In answer to (c), which asks how much better or worse double sampling is in this case, we use Eq. 10.29:

$$\text{NRE (double/single)} = \frac{c''}{(\sqrt{(1-R^2)c''} + R\sqrt{c'})^2}$$

$$= \frac{1.00}{[\sqrt{(1-.81)(1.0)} + (.9)\sqrt{.10}]^2}$$

$$= \frac{1.00}{(.435+.284)^2} = \frac{1.00}{(.719)^2} = 1.94,$$

which means that in this case double sampling will give approximately twice as much information for the $100 as would single sampling.

To answer (d), estimating $\bar{y}_{(d-regr)}$, we use Eq. 10.18:

$$\bar{y}_{(d-regr)} = \bar{y}'' + b''(\bar{x}' - \bar{x}'')$$

$$= 80.0 + 3.9(22.0 - 20.0)$$

$$= 80.0 + 7.8 = 87.8.$$

The foregoing problem could be worked out graphically, as shown in Fig. 10.2. The equation for the regression line may be written

$$\overset{*}{y} - \bar{y}'' = b''(\overset{*}{x} - \bar{x}'')$$

or

$$\overset{*}{y} - 80.0 = 3.9(\overset{*}{x} - 20.0),$$

where $\overset{*}{y}$ and $\overset{*}{x}$ are the coordinates of any arbitrary point on the line. Since we want the y-intercept, $\overset{*}{x}$ is set equal to zero. Solving for $\overset{*}{y}$, we obtain

$$\overset{*}{y} = 80.0 - (3.9)(20.0) = 2.0.$$

Clearly, if $\overset{*}{y} = \bar{y}$ and $\overset{*}{x} = \bar{x}$, the equation for the line is satisfied. The regression line may therefore be established by the points (0, 2.0) and (20.0, 80.0). Since the large sample has a larger mean, 22.0 instead of 20.0, we expect its mean, which we are estimating by $\bar{y}_{(d-regr)}$, to be larger than 80.0, the y mean of the small sample. The $\bar{y}_{(d-regr)}$ value can be regarded as an extension of the point $x = 22.0$ vertically to the regression line, thence horizontally to the ordinate, where the value 87.8 is obtained.

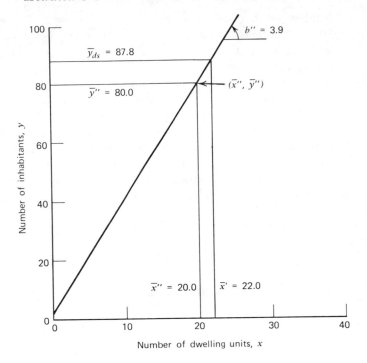

Fig. 10.2.   Graphic representation of double sampling using regression.

10.6.   Estimating $\bar{Y}$ With the Help of x: Two-Strata Case

A more general case of double sampling with stratification will be given in the next section. Here we shall be concerned with a simple case, so that we can introduce the basic principles without the clutter of the lengthy and somewhat complex formulas that the more general case entails.

Suppose we draw a sample in two steps or phases as in the previous section, but all we do in the first step is to decide whether a particular element belongs in stratum 1 or stratum 2; for example if $X_i = 10$ or more, it belongs to stratum 1; if X_i is less than 10, it belongs in stratum 2. All sample elements in stratum 1 are then to be examined for y, and only one in t_2 of stratum 2 elements are to be so examined. The estimator of $\bar{Y}$ will be

(10.32) $$\bar{y}_{ds} = w_1' \bar{y}_1' + w_2' \bar{y}_2'',$$

and its variance is given by

(10.33) $$\text{var} \, (\bar{y}_{ds}) = \left(\frac{N-n'}{N} \right) \frac{S^2}{n'} + \left(\frac{t_2-1}{n'} \right) W_2 S_2^2,$$

where N = total elements in the universe,

$\quad n'$ = size of first-phase sample,

$\quad S^2$ = variance of y in the universe,

$\quad S_2^2$ = variance of y in stratum 2,

$\quad W_2 = N_2/N$ = fraction of elements in stratum 2,

$\quad t_2$ = the inverse of the sampling fraction used in drawing the second-phase sample in stratum 2; $t_2 = n_2'/n_2''$.

In this design we are using sampling not only to estimate the stratum means, but also to estimate the relative sizes of the strata, which are presumed to be unknown. The variance consists of a component (the first term) that represents the variance that would arise if no subsampling were done, and a component representing the additional variance stemming from the subsampling in stratum 2.

To compare this scheme with that of simple one-phase sampling, we should introduce some costs. Let

$$(10.34) \qquad\qquad C = c'n' + c_1''n_1'' + c_2''n_2''$$

be the cost function for the two-phase design, and

$$(10.35) \qquad\qquad C = c_1 n_1 + c_2 n_2$$

be the cost function for the one-phase design, where

$\qquad\qquad n$ = total size of a one-phase RNR sample,

$\quad n_1, n_2$ = size of sample, n, falling into strata 1 and 2,

$\qquad\quad n'$ = total size of phase I sample in two-phase design,

$\quad n_1', n_2'$ = size of sample n' falling into strata 1 and 2,

$\quad n_1'', n_2''$ = size of phase II samples in strata 1 and 2,

$\qquad\quad C$ = total costs (or total available budget),

$\quad c_1, c_2$ = unit costs of observing Y_i in strata 1 and 2 for a one-phase design,

$\quad c_1'', c_2''$ = unit costs of observing Y_i in strata 1 and 2 for a two-phase design (assuming one knows what stratum one is in).

Ordinarily, $c_1 = c_1''$ or $c_1'' = c'$, depending on circumstances; likewise for c_2.

The efficiency of this method of sampling compared to single sampling can be estimated by obtaining an expression for the "net relative efficiency" of the two—that is, the relative variances where budgets are held constant. Ignoring FPCs, the precision of double sampling relative to single sampling is given by

$$(10.36) \qquad\qquad \text{RP } (ds/s) \doteq \frac{n'/n}{1 + (t_2 - 1)W_s S_2^2/S^2} \, ,$$

where n'/n is the size of the first phase of the double sample relative to the size of a single RNR sample. When a budget is fixed and costs are given by Eqs. 10.34

and 10.35, the value of the ratio n'/n is given by

(10.37)
$$\frac{n'}{n} = \frac{W_1 c_1 + W_2 c_2}{c' + W_1 c_1'' + W_2 c_2''/t_2},$$

and the net relative efficiency of double to single sampling is given by

(10.38)
$$\text{NRE } (ds/s) \doteq \frac{n'/n}{1 + (t_2 - 1)W_2 S_2^2/S^2}.$$

This kind of double sampling can be either better or worse than single sampling, depending on the situation and the choice of t_2, the rate of sampling in phase II. The optimal value of t_2 is given approximately by

(10.39)
$$t_{opt} = \frac{W_1 S_1 \sqrt{c_1''} + W_2 S_2 \sqrt{c_2''}}{W_2 S_2 \sqrt{c_2''}},$$

which follows from the principle of optimal allocation in stratification.

Example 10.6.   Suppose we have a universe in which the y-population is concentrated quite heavily in a relatively small number of elements. Examples of such populations might be cattle on farms, gross sales, and of manufacturers, amount of life insurance held by families. Suppose the concentration is such that the largest 10% of the units contain 80% of total y. Assuming $\bar{Y} = 10$ and $V_1^2 = V_2^2 = 4$, then

W_1	W_2	$\bar{Y}$	$\bar{Y}_1$	$\bar{Y}_2$	S_1^2	S_2^2	S^2
0.10	0.90	10	80	2.22	25,600	19.8	3.423

Assume the costs:

c_1	c_2	c'	c_1''	c_2''
4.5	3.5	0.50	4	3

In this case, where no prior information is available on which element belongs to the class of "big" elements (and it costs 0.50 units to find out), which will be more efficient, a single- or double-sampling scheme? We compute:

$$t_{opt} = \frac{(0.10)\sqrt{(25,600)\,(4)} + (0.90)\sqrt{(19.8)\,(3)}}{(0.90)\sqrt{(19.8)\,(3)}}$$

$$= \frac{32.0 + 7.02}{7.02} = 6,$$

$$\frac{n'}{n} = \frac{(0.10)\ (4.5) + (0.90)\ (3.5)}{0.50 + (0.10)\ (4) + (0.90)\ (3)/6} = \frac{3.60}{1.35} = 2.67,$$

and

$$\text{NRE } (ds/s) = \frac{2.67}{1 + (6-1)\ (0.90)\ (19.8)/(3423)} = \frac{2.67}{1.026}$$

$$= 2.60 \quad \text{or} \quad 260\%,$$

which means that a double-sampling design is far superior to a simple random sample for this situation.

Suppose the budget for the survey is $5000; what sizes of first- and second-phase samples are required?

Since

$$\frac{n'}{n} = 2.67$$

and

$$C = c_1 n_1 + c_2 n_2 = n(c_1 W_1 + c_2 W_2) = 5000,$$

then

$$n = \frac{5000}{(4.5)\ (0.10) + (3.5)\ (0.90)} = \frac{5000}{3.60} = 1388,$$

the size of the single-phase sample, and

$$n' = \left(\frac{n'}{n}\right) n = (2.67)\ (1388) = 3706,$$

the size of the phase I sample. For the second phase, we have for the expected size of the big element sample

$$n_1'' = n' W_1 = (3706) \cdot (0.10) = 372,$$

and for stratum 2,

$$n_2'' = \frac{(3706)\ (0.90)}{6} = 558.$$

In this case, although the double-sampling design requires the observation of only 929 elements rather than the single sampling's 1388, yet it yields more than $2\frac{1}{2}$ times the precision.

10.7. Estimating $\bar{Y}$ with the Help of x: L-Strata Case

Neyman (1938) presented this type of double sampling for a survey of families to obtain information on their household purchases. A relatively large sample of n' elements is selected at random from a universe of N. On these elements a characteristic x is observed and, on the basis of the x information, L strata are formed with n'_h elements in the hth stratum. A random sample of n''_h elements is drawn from each stratum, and on these elements the y characteristic is measured. An unbiased estimate of $\bar{Y}$ is given by the estimator

$$(10.40) \qquad \bar{y}_{d-strat} = \sum_{}^{L} w_h \bar{y}''_h,$$

where $\bar{y}''_h$ is the sample mean of y in the hth stratum, and w'_h is the fraction of the n' elements occurring in the hth stratum (i.e., $w'_h = n'_h/n'$).

In the general case—that is, for any allocation of the n'_h to the strata—the variance of the estimated mean, $\bar{y}_{d-strat}$, is given by

$$(10.41) \quad var\,(\bar{y}_{d-strat}) \doteq \sum_{}^{L} \left\{ \left[W_h^2 + \frac{W_h(1-W_h)}{n'} \right] \frac{S_h^2}{n''_h} + \frac{W_h(\bar{Y}_h - \bar{Y})^2}{n''} \right\}$$

if n'_h/N_h and n'/N can be ignored, where

N_h = universe number of elements in stratum h,
$W_h = N_h/N$ = fraction of the universe elements in the hth stratum,
$\bar{Y}$ = actual mean of the y-population in the universe,
$\bar{Y}_h$ = actual mean of the y-population in the hth stratum,
S_h^2 = variance of y in the hth stratum,
L = number of strata.

When Equation 10.41 is compared with that for ordinary stratification (Eq. 7.4a), it can be seen that estimating the stratum weights, W_h, from a double sample adds two terms to the variance expression (those terms with n' as a divisor). One of these is the expected between-strata component, the other is the rather unexpected addition to the within-stratum component.

In order to utilize this type of sampling most effectively, it is necessary to decide for given costs and a fixed budget: (a) How large should be the sample, n', for measuring x? (b) How large should be the sample, n'', for measuring y? (c) How should we allocate the n'' among the L strata?

Consider a cost function

$$(10.42) \qquad\qquad C = c'n' + c''n'',$$

where c' and c'' are the unit costs of obtaining observations in phase I and II samples, respectively. Let us allocate the phase II sample to strata according to the principle of optimum allocation for ordinary stratification, ignoring differential costs in strata; that is,

$$(10.43) \qquad n_h'' = n'' \frac{W_h S_h}{\sum^L W_h S_h}.$$

Then we obtain as the variance of $\bar{y}_{d-strat}$ under this approximately optimal allocation scheme

$$(10.44) \qquad \text{var}\,(\bar{y}_{optd-strat}) = \frac{(\sum^L W_h S_h)^2}{n''} + \left(\frac{N-n'}{N}\right)\frac{\sum^L W_h(\bar{Y}_h - \bar{Y})^2}{n'}$$

$$-\frac{1}{N}\sum_{h}^{L} W_h^2 S_h^2,$$

or, in a shortened form (and ignoring the last term, which is ordinarily small),

$$(10.44a) \qquad \text{var}\,(\bar{y}_{optd-strat}) = \frac{S_I^2}{n''} + \frac{S_{II}^2}{n'}.$$

When this variance is minimized for a fixed expenditure, Eq. 10.42, we obtain

$$(10.45) \qquad \frac{n''}{n'} \doteq \sqrt{(S_I^2/S_{II}^2)\,(c'/c'')}.$$

Note that Eq. 10.44, when the large-sample phase is extended to the universe—that is, $n' = N$—reduces to

$$(10.46) \qquad \text{var}\,(\bar{y}_{optd-strat}) = \frac{(\sum^L W_h S_h)^2}{n''} - \frac{1}{N}\sum_{h}^{L} W_h^2 S_h^2,$$

which is identical to Eq. 7.14, since $W_h = N_h/N$, the equivalent case dealt with in ordinary stratification with optimum allocation.

To compare optimum double with single sampling we have, ignoring FPCs,

$$(10.47) \qquad \text{var}\,(\bar{y}_{optd-strat}) = \left(\frac{1}{n} - \frac{1}{n''}\right)\left(\sum^L W_h S_h\right)^2 + \left(\frac{1}{n} - \frac{1}{n'}\right)\sum^L W_h(\bar{Y}_h - \bar{Y})^2$$

$$+\frac{1}{n}\sum^{L} W_h(S_h - \bar{S})^2,$$

where n is single sample size and $\bar{S}$ is $\sum^h W_h S_h$, the average within stratum standard deviation. In the case where the S_hs are equal, and hence the n_h'' are allocated proportionally to the strata sizes, it can be shown that no appropriate loss can result from the use of this type of double over single sampling. Hence this design can be a fairly safe one, if allocation of n'' among the L strata is proportional to the estimated strata sizes, w_h (where $w_h = n_h'/n'$). However, this "safeness" has a high price, since the potential gain from the simple proportional allocation case of double sampling over single sampling is usually very low. The big gains with this scheme usually come when variances within strata are so greatly different that the allocation of Eq. 10.43 is appropriate.

If a double sample with stratification is drawn and the value of the double-sample estimate of the population mean, $\bar{Y}$, is determined, how is its variance estimated from the sample? In the general case, the variance of $\bar{y}_{d-\text{strat}}$, given by Eq. 10.41, can be estimated unbiasedly, if n_h''/n' and n'/N are negligible, by

$$(10.48) \qquad \widehat{\text{var}}\,(\bar{y}_{d-\text{strat}}) = \frac{n'}{n'-1} \sum^L \left[\left(w_h^2 - \frac{w_h}{n'} \right) \frac{s_h^2}{n_h''} + \frac{w_h(\bar{y}_h - \bar{y}_{d-\text{strat}})^2}{n'} \right],$$

which, when w_h/n' is small relative to w_h^2, becomes

$$(10.48a) \qquad \widehat{\text{var}}\,(\bar{y}_{d-\text{strat}}) = \sum^L \left[\frac{w_h^2 s_h^2}{n_h''} + \frac{w_h(\bar{y}_h - \bar{y}_{d-\text{strat}})^2}{n'} \right].$$

Example 10.7. The general problem to be considered is that of estimating by means of a survey the total number of persons living within a given area, who watched 22 specified TV programs on specified dates. The survey was necessarily confined to one week, and information was to be obtained by a single interview with individual persons. Because of the possible difficulty people would have in recalling whether they did or did not watch a given program that occurred some days past, it was decided to confine the period of recall to the previous day only and to attempt by double sampling to get the best possible precision under this restriction. For simplicity we shall regard samples drawn for each phase as random.

The first-phase information was obtained by asking persons in the sample if they "usually" or "seldom" watched each of the 22 TV programs. In the second phase they were asked if they watched each of those programs that were telecast "yesterday," the day prior to that of the interview.

In this case, the first-phase sample consisted of 1048 persons. Since these persons were randomly allocated to seven groups of approximately 150 each, and each of the seven groups was assigned at random to each of the seven days in the survey week, we can regard each set of 150 persons as a randomly selected second-phase sample. Here N is the total number of persons in the area having a TV set in the home, and Y of those N watched program k on a given day. We are interested in estimating $\bar{Y}$, the fraction watching. The numerical results for the weekly program "Farm Facts" are presented below.

TV Program: "Farm Facts"   Time: 8:30–9:00 p.m.   Day: Friday

Phase	Stratum 1: "seldom" watch[a]	Stratum 2: "usually" watch[a]	Total phase
Large sample, n'_h	779	269	1048
Small sample, n''_h	102	33	135
Watched[b]	18	22	40
Did not watch[b]	84	11	95

[a] "Now we are interested in your *general* televiewing habits. Would you say that *you* USUALLY or SELDOM watch (name of program)?"

[b] "Here is a listing of yesterday's television programs on the Ames station. Which of these programs did *you* watch?"

To estimate the fraction of persons viewing the program by both simple and double sampling, we have

$$w_1 = \tfrac{779}{1048} = 0.743, \quad w_2 = \tfrac{269}{1048} = 0.257,$$

$$\bar{y}''_1 = \tfrac{18}{102} = 0.176, \quad \bar{y}''_2 = \tfrac{22}{33} = 0.667.$$

Hence,

$$\bar{y}'' = \tfrac{40}{135} = 0.297$$

and

$$\bar{y}_{d-\text{strat}} = (0.743)(0.176) + (0.257)(0.667) = 0.302.$$

To compute the estimated variances of the two estimates in order to compare their efficiencies, we have, since $\hat{S}^2 = npq/(n-1)$, and when FPCs are ignored,

$$\widehat{\text{var}}\,(\bar{y}'') = \frac{s_y^2}{n_s} = \frac{\frac{135}{134}\left(\frac{40}{135}\cdot\frac{95}{135}\right)}{135} = \frac{0.210}{135} = 0.001{,}556,$$

and for computing $\widehat{\text{var}}\,(\bar{y}_{d-\text{strat}})$ we obtain $n' = 1048$, $n'' = 135$,

$$\begin{aligned}
n''_1 &= 102, & n''_2 &= 33, \\
w_1 &= 0.743, & w_2 &= 0.257, \\
\bar{y}''_1 &= 0.176, & \bar{y}''_2 &= 0.667, & \bar{y}_{d-\text{strat}} &= 0.302, \\
s_1^2 &= 0.146, & s_2^2 &= 0.229,
\end{aligned}$$

where

$$s_1^2 = \frac{n_1''}{n_1'' - 1}, \qquad s_2^2 = \frac{n_2''}{n_2'' - 1},$$

$$\left(\frac{n_1''}{n_1'' - 1}\right)(\bar{y}_1'')(1 - \bar{y}_1'') = \frac{102}{101}(0.176)(0.824) = 0.146,$$

$$\left(\frac{n_2''}{n_2'' - 1}\right)(\bar{y}_2'')(1 - \bar{y}_2'') = \frac{33}{32}(0.667)(0.333) = 0.229.$$

Evaulating Eq. 10.48, we obtain

$$\widehat{\text{var}}\,(\bar{y}_{d-\text{strat}}) = \frac{1048}{1047}\left\{\left[(0.743)^2 - \frac{0.743}{1048}\right]\frac{0.146}{102} + \frac{0.743(0.176 - 0.302)^2}{1048}\right.$$

$$\left. + \left[(0.257)^2 - \frac{0.257}{1048}\right]\frac{0.229}{33} + \frac{0.257(0.667 - 0.302)^2}{1048}\right.$$

$$\doteq 0.000,789 + 0.000,011 + 0.000,457 + 0.000,033$$

$$= 0.001,290.$$

When this estimator is compared with the simple estimator of single sampling (where only $\bar{y}$ of the small sample is used), the estimated relative efficiency is given by

$$\text{RE (double/single)} = \frac{\widehat{\text{var}}\,(\bar{y}_s)}{\widehat{\text{var}}\,(\bar{y}_{d-\text{strat}})} = \frac{0.001,556}{0.001,290} = 1.21 \quad \text{or } 121\%.$$

Hence in this case an estimated 21 % gain was made by the use of the double-sample estimator. Note that in this problem no attempt was made to try to achieve an optimum allocation. The allocation, which was proportional, was a logical result of the nature of the survey. Considering the approximate nature of the formulas dealt with, it is doubtful whether an actual gain was made in this case. On some of the other programs, however, a substantial gain was made by this method of estimation.

10.8.   Estimating Y/M (or $\bar{\bar{Y}}$) and Y/Z, M Unknown; One-Stage

To estimate $\bar{\bar{Y}}$ in double sampling when M is unknown we may use

(10.49)
$$\frac{\hat{Y}}{\hat{M}} = \frac{N\bar{m}'(\bar{y}''/\bar{m}'')}{N\bar{m}'} = \left(\frac{\bar{y}''}{\bar{m}''}\right),$$

which is simply the ratio based on the phase II sample. In a similar manner, it can be shown that the estimator for Y/Z should be $\bar{y}''/\bar{z}''$. Any resources spent on phase I would be wasted.

10.9. Two-Stage, Two-Phase Design; Phase II Primaries Selected RNR

Suppose we carry out a two-phase design in two stages. For example, we have a universe of M secondaries comprising N primaries of M_i secondaries each. Y_{ij} is the observed y on the jth secondary in the ith primary. We do not know M, and hence M_i, but we wish to devote phase I to estimating $\bar{M}$ by means of an RNR sample of n' primaries and phase II to the estimation of $\bar{Y}$ by RNR samples of M_i from n'' primaries. Our estimator of $\bar{Y}$ is therefore

$$(10.50) \qquad \bar{y}_{d-PE2} = \bar{m}' \left(\frac{\bar{y}''}{\bar{m}''} \right) = \bar{m}' \bar{y}',$$

where

$$\bar{m}' = \frac{\sum^{n'} M_i}{n'}, \quad \bar{m}'' = \frac{\sum^{n''} M_i}{n''}, \quad \bar{y}'' = \frac{\sum^{n''} M_i \bar{y}''_i}{n''}, \quad y''_i = \frac{\sum^{m''_i} Y_{ij}}{m''_i},$$

and m''_i is the number of secondaries in ith phase II primary.

The appropriate variance of Eq. 10.50 is given by

$$(10.51) \qquad \text{var} \left[\bar{m}' \left(\frac{\bar{y}''}{\bar{m}''} \right) \right] \doteq \left(\frac{n'-n''}{n'n''} \right) \left[\left(\frac{\bar{Y}}{\bar{M}} \right)^2 S_m^2 - 2 \left(\frac{\bar{Y}}{\bar{M}} \right) S_{my} \right]$$

$$+ \frac{1}{Nn''} \sum \left(\frac{M_i - m_i}{M_i m_i} \right) M_i^2 S_{(y)i}^2,$$

where

$$S_{(y)i}^2 = \frac{\sum_{j=1}^{M_i} (Y_{ij} - \bar{Y}_i)^2}{M_i - 1}$$

and symbols have their usual meaning.

It may be noted that when $n' = N$, Eqs. 10.50 and 10.51 reduce to the case of a one-phase ratio estimator with subsampling (hence Eq. 10.14 with subsampling variance added). When $n'' = n'$, they reduce to the case of two-stage sampling for $\bar{Y}$; when $m_i = M_i$, they reduce to one-stage sampling with two phases for estimating $\bar{Y}$ with the help of x using the ratio (replacing x with

m). If $M_i = M_0$, then the case reduces to simple subsampling with equal-sized primaries with var $[\bar{M}\bar{y}]$ (see Eq. 9.2, where $S_y^2 = M_0[MS(B)]$). Moreover, if $n'' = n' = N$, the primaries will become strata, and the case reduces to stratified sampling, wherein the estimator, Eq. 10.56, becomes

$$\bar{y}'' = \frac{1}{N} \sum_{}^{N} M_i \bar{y}_i,$$

which is equivalent to $\bar{M}\bar{y}_{strat}$ (see Eq. 7.3), where M_0 is N/L, and the variance can be shown to be equivalent to Eq. 7.4a multiplied by $(N/L)^2$ to put it on a mean per stratum basis.

Optimal Allocation of n', n'', m'', ***and*** m_i''. Collecting terms in Eq. 10.51 by dependency on phase and stage, we have

(10.51a) $$\mathrm{var}\left[\bar{m}'\left(\frac{\bar{y}''}{\bar{m}''}\right)\right] = \frac{S_I^2}{n'} + \left(\frac{S_{II}^2}{n''} = \frac{S_1^2}{n''} + \frac{S_2^2}{n''\bar{m}''}\right),$$

where

$$S_I^2 = 2\left(\frac{\bar{Y}}{\bar{M}}\right) S_{my} - \left(\frac{\bar{Y}}{\bar{M}}\right)^2 S_m^2,$$

$$S_1^2 = \left(\frac{\bar{Y}}{\bar{M}}\right)^2 S_m^2 - 2\left(\frac{\bar{Y}}{\bar{M}}\right) S_{my} + \left(\frac{N-n'}{N}\right) S_y^2,$$

(10.52)

$$S_2^2 = \bar{S}_{(y)}^2 = \frac{1}{N} \sum M_i^2 S_{(y)i}^2,$$

$$\bar{m} = \frac{m''}{n''},$$

or, in terms of relative variances,

(10.51b) $$\mathrm{rel\ var}\left[\bar{m}'\left(\frac{\bar{y}''}{\bar{m}''}\right)\right] = \frac{V_I^2}{n'} + \left(\frac{V_{II}^2}{n''} = \frac{V_1^2}{n''} + \frac{V_2^2}{n''\bar{m}''}\right),$$

where

$$V_I^2 = 2V_{my} - V_m^2,$$

(10.52a) $$V_1^2 = V_m^2 - 2V_{my} - \left(\frac{N-n'}{N}\right) V_y^2,$$

$$V_2^2 = S_{(y)}^2 \bar{Y}2 = \bar{V}_{(y)}^2.$$

Since the phase II variance comprises two additive terms identical to that for two-stage sampling, we can use the same methods to determine an optimal value for $\bar{m}''$. Since $\bar{m}''$ is a random variable, the best expected value of $\bar{m}$ is given by

$$(10.53) \qquad \bar{m}''_{\text{opt}} = \sqrt{(S_2^2/S_1^2)\, c_1/c_2''} \quad \text{or} \quad \sqrt{(V_2^2/V_1^2)\, (c_1/c_2'')},$$

assuming that the unit cost of adding a primary (that is, the cost of observing an M_i) is $c' = c''$, the unit cost of adding a secondary (of observing Y_{ij}) is c_2'', and hence the cost of adding a primary in phase II is $c'' = \bar{m}'' c_2''$. Total sampling cost is therefore

$$(10.54) \qquad\qquad C = n'c' + (n''c'' = n''\bar{m}''c_2'').$$

Having determined $\bar{m}''$, we obtain the optimal ratio of n' to n'' by the relationship

$$(10.55) \qquad \frac{n'}{n''} = \sqrt{(S_1^2/S_{II}^2)\, (c''/c')} \quad \text{or} \quad \sqrt{(V_1^2/V_{II}^2)\, (c''/c')} = \phi$$

and, from Eqs. 10.54 and 10.55,

$$(10.55a) \qquad\qquad n'' = \frac{C}{\phi c' + c''} \quad \text{and} \quad n' = \phi n''$$

Example 10.8.   Suppose we wish to estimate the total inhabitants in a city by selecting an RNR sample of blocks on which a count (or simply an eyeball estimate) is made of the number of housing units, M_i, situated on each, and on a portion of these a subsample of housing units is selected to determine the average inhabitants per housing unit. The estimator to be used is $\hat{Y} = N\bar{m}'(\bar{y}''/\bar{m}'')$. The available facts (synthetic) are

$$N = 50{,}000, \qquad V_y = 0.6,$$
$$C = 10{,}000, \qquad V_m = 0.4,$$
$$c' = 0.25, \qquad \bar{V}_y = 2.0,$$
$$c_2'' = 2.00, \qquad 2V_{my} = 2RV_mV_y = 0.384.$$

(a)   Determine the optimal values of $\bar{m}$, n', and n''.
(b)   What is the expected relative standard error of $\hat{Y}$?

Solution

(a)   To determine $\bar{m}''_{\text{opt}}$ we use Eq. 10.53, for which $V_1^2 = V_m^2 - 2V_{my} + V_y^2 = 0.16 - 0.38 + 0.36 = 0.14$, $V_2^2 = 4,00$, $c_1 = c' = 0.25$, $c_2'' = 2.00$. Hence

$$\bar{m}''_{\text{opt}} = \sqrt{(V_2^2/V_1^2)\,(c_1/c_2'')} = \sqrt{(4.00/0.14)\,(0.25/2.00)} = \sqrt{3.6} \doteq 2.$$

With this result we obtain $c'' = \bar{m}''c_2'' = (2)(2.00) = 4.00$, and since $V_1^2 = 2V_{my}$ $- V_m^2 = 0.38 - 0.16 = 0.22$, $V_{II}^2 = V_1^2 + V_2^2/\bar{m} = 0.14 + 4.00/2 = 2.14$. From Eq. 10.55

$$\frac{n'}{n''} = \sqrt{(V_1^2/V_{II}^2)\,(c''/c')} = \sqrt{(0.22/2.14)\,(4.00/0.25)} = \sqrt{1.66} \doteq 2$$

and

$$n'' = \frac{C}{\phi c' + c''} = \frac{10{,}000}{(2)(0.25) + 4.00} = \frac{10{,}000}{4.50} = 2220$$

blocks, on which an average of two housing units will be selected for phase II.

$$n' = \phi n'' = (2)(2220) = 4440 \quad \text{blocks for phase I.}$$

(b)   To obtain RSE we have Eq. 10.51b,

$$\text{rel var}\left[\bar{m}'\left(\frac{\bar{y}''}{\bar{m}''}\right)\right] = \frac{V_1^2}{n'} + \frac{V_{II}^2}{n''} = \frac{0.22}{4440} + \frac{2.14}{2220}$$

$$= 0.000{,}050 + 0.000{,}965 = 0.001{,}015,$$

and
$$\text{RSE}\,[\hat{Y}] = \sqrt{0.001{,}015} = 0.032 \quad \text{or} \quad 3.2\%.$$

10.10. Estimating $\bar{Y}$ with the Help of M_i; PPS in Phase II

There are occasions when circumstances are such that one would like to use PPS sampling but a suitable MOS (measure of size) is lacking. Consider the case where we wish to estimate $\bar{Y}$ with a PPM_i sample but the M_i are lacking. We shall draw an RNR sample of n' of N clusters and observe the M_i on each. A cumulative total of the M_i is constructed and a sample of n'' are selected for phase II by some PPS scheme. On these, the Y_is are observed. An estimator of $\bar{Y}$ is

(10.56) $$\bar{y}_{d-PM1} = \bar{m}'\bar{y}'',$$

where

$$\bar{m}' = \frac{\sum^{n'} M_i}{n'}, \qquad \bar{y}'' = \frac{\sum^{n''} \bar{Y}_i}{n''}, \qquad \bar{Y}_i = \frac{Y_i}{M_i}.$$

If the n'' clusters in phase II are selected PR or RNR, then the product estimator, Eq. 10.9, will become $\bar{m}' \sum \bar{Y}_i/n''$, and its variance in general will be

Eq. 10.10, modified to

$$(10.57) \quad \text{var}\left(\bar{m}\frac{\sum \bar{Y}_i}{n''}\right) \doteq \bar{Y}^2\left(\frac{N-n'}{Nn'}\right)S_m^2 + 2\bar{M}\,\bar{Y}\left(\frac{N-n'}{Nn'}\right)S_{my} + \bar{M}^2[\text{var}\,(\bar{y}'')].$$

A simple case is where clusters in phase II are slected with PR. Then

$$(10.58) \qquad\qquad \text{var}\,(\bar{y}'') \doteq \left(\frac{1}{M}\right)^2 \frac{\sum^N (Y_i/P_i - \bar{Y})^2}{n'}$$

where $P_i = M_i/M$. Since M and P_i are not known, they may be estimated by

$$(10.59) \qquad \hat{M} = \bar{m}'N \qquad \text{and} \qquad \hat{P}_i = \frac{n'}{N}\cdot\frac{M_i}{\sum^{n'} M_i} \doteq \frac{M_i}{\hat{M}},$$

respectively.

10.11. Two-Stage, Two-Phase Design; Phase II Primaries Selected PR

Suppose a sample of n' primaries are selected RNR from a universe of N on which M_i is measured. For phase II a sample of m'' secondaries is selected by, say, selecting a primary from the n' in phase I with $P_i = M_i/\sum^{n'} M_i$, taking $\bar{m}_0''$ secondaries from it, and repeating this procedure n'' times with replacement. These secondaries are observed for Y_{ij}. To estimate $\bar{Y}$ (whence $Y = N\bar{Y}$), an estimator of interest has a form similar to that of Eq. 10.50,

$$(10.60) \qquad\qquad \bar{y}_{d-PM1} = \bar{m}'\bar{\bar{y}},$$

but now $\bar{m}' = \sum^{n'} M_i/n'$, $m'' = \sum^{n'} m_i$,

$$\bar{\bar{y}}'' = \frac{\sum^{n''}\sum^{\bar{m}_0''} Y_{ij}}{m''} \quad\text{or}\quad \frac{\sum^{n''} \bar{y}_i}{n''}, \qquad \bar{y}_i'' = \frac{\sum_{j=1}^{m_0''} Y_{ij}}{\bar{m}_0''},$$

and its variance is given approximately by

$$(10.61) \quad \text{var}\,[\bar{m}'\bar{\bar{y}}] \doteq \bar{Y}^2\left(\frac{N-n'}{Nn'}\right)S_m^2 + 2\bar{M}\,\bar{Y}\left(\frac{N-n'}{Nn'}\right)\overset{*}{S}_{m\bar{Y}_i} + \bar{M}^2\,\text{var}\,(\bar{y}_{PM1}).$$

It may be noted that Eq. 10.60 is a more general form of Eq. 9.39, to which this degenerates if phase I is extended to a complete census (i.e., $n' = N$). The approximation to the $\bar{y}_{PM1}$ estimator is based on the observation that under this type of sampling within the selected primaries, the probability of selecting any one of the M secondaries on a single draw of a secondary is given by

$$(10.62) \qquad \frac{n'}{N}\cdot\frac{M_i}{\sum^{n'} M_i}\cdot\frac{\bar{m}}{M_i} = \frac{1}{N}\frac{1}{\sum^{n'} M_i/n'} = \frac{1}{N\bar{M}} \doteq \frac{1}{M},$$

hence it is similar to that of drawing a single secondary from a primary drawn with $P_i = M_i/M$. For n' draws, the expected probability of a secondary appearing in the sample is m''/M, approximately, and hence we can regard the sampling for $\bar{y}$ as approximately that of PPM_i.

Viewed this way, we can regard this scheme of sampling as one in which information obtained from phase I is used to provide measures of size for the primaries to be used to determine probabilities for their selection to observe a limited number of Y_{ij}s in phase II.

$$\diamond \qquad \diamond \qquad \diamond$$

10.12. Discussion and Summary

In this chapter we have confined the concepts and illustrations of multiphase samples to the simpler cases. We have, however, considered two-stage, two-phase sampling with equal and with arbitrary selection probabilities. The extensions to more complex cases can and should be done if some practical problems are to be properly dealt with. An example follows.

An estimate of the total tonnage of oranges hanging on trees in Florida is desired as a basis for forecasting the crop to be harvested. A plan devised and carried out in 1953–54 [see Jessen (1972)] consisted of a six-phase design with one or more stages in each phase; it had a structure as presented in Table 10.2.

The estimator for total tonnage of fruit standing on trees at the time of the survey is

$$\hat{W} = N\hat{A}\hat{P}_A\hat{T}_P\hat{F}_T\hat{V}_F W_V,$$

where N = total number of square-mile areas (U.S. Public Land Survey Section) in the universe,

$\hat{A}$ = mean number of eyeball-estimated acres in citrus per section (as can be observed by scanning aerial photographs),

$\hat{P}_A$ = number of planting positions per "eyeball acre,"

$\hat{T}_P$ = proportion of positions occupied by bearing orange trees,

$\hat{F}_T$ = number of fruits per tree,

$\hat{V}_F$ = mean volume of fruits,

$\hat{W}_V$ = fruit weight per unit volume.

Phase I, making eyeball estimates for each section, was carried out on a 100% sample of sections. The remaining phases and stages were carried out on samples. The sizes of the samples were arrived at by considered judgements, since actual data on variation were generally nonexistent at the time.

In this chapter we have generally considered double sampling as a scheme wherein the estimator consists of a function of several characteristics, and where one of the important problems is that of deciding how best to sample each in order to minimize overall variance. Most often the sampling for each is not independent—that is, each sample is "nested" within another and hence covariances must be taken into account.

Table 10.2. *Structure of a survey to estimate total orange crops in Florida*

Parameter	Phase	Stage	Survey unit	Selection probability	MOS	Observation or operational task
A	1	1	County	Certainty	None	Listing of each square-mile section in each citrus county and an eyeball estimate of "area in citrus" on each
P_A	2	1	Square mile	PPS	"Area in citrus"	Listing of groves and the area of each
		2	Grove	PPS	Area in grove	Listing of rows in each sample grove
		3	Row	PE	None	Number of positions
T_P	3	1	Position	PE	None	Identification of "orange trees"
F_T	4	1	"Tree"	PE	None	Listing of limbs at a forking and the size of each
		...	Limb	PPS	Limb size	Listing of sublimbs at a forking and size of each
		K	Sublimb	PPS	Limb size	Number of fruits
V_F	5	1	Fruit	PE	None	Volume of each fruit
W_L	6	1	Fruit	PE	None	Weight of each fruit

The opportunity for improvisation is abundant in these situations. The material presented here suggests how they might be dealt with.

10.13. Review Illustrations

1. (Refer to the television audience problem of Example 10.7.) Now consider a ratio estimator for estimating the fraction of hours watching, where Y_i is 1 if a given program is watched and 0 if not and X_i is 1 and 0 if usually watched or not, respectively. Here $n' \doteq 6n''$, since X_i is asked of interviewees each of the six days, whereas Y_i is asked only of those interviewees contacted the day following the program presentation. The relevant data for the program "Farm Facts" are

$$n' = 779, \qquad\qquad n'' = 135,$$

$$\bar{x}' = 0.2567, \qquad\qquad \bar{y}'' = 0.2963, \qquad\qquad \bar{x}'' = 0.2444,$$

$$(\bar{x}')^2 = 0.065,895, \qquad (\bar{y}'')^2 = 0.087,794$$

$$s_x^2 = 0.1861, \qquad\qquad s_y^2 = 0.2101, \qquad\qquad s_{y/x}^2 = 0.1679,$$

$$v_x^2 = 2.824,190, \qquad\quad v_y^2 = 2.393,102,$$

$$v_x = 1.6805, \qquad\qquad v_y = 1.5469, \qquad\qquad v_{xy} = 1.2743,$$

$$r_{xy} = 0.461, \qquad\qquad b'' = 0.4902.$$

(a) Estimate the fraction of homes watching "Farm Facts."
(b) Estimate the variance of (a).
(c) Estimate the standard error of (a).

Solution

(a) $\quad \bar{y}_{\text{d-rat}} = \bar{x}' \left(\dfrac{\bar{y}''}{\bar{x}''} \right) = (0.2567) \left(\dfrac{0.2963}{0.2444} \right) = 0.3112.$

(b) $\quad \widehat{\text{var}} \, (\bar{y}_{\text{d-rat}}) \doteq \bar{y}^2 \left[\left(\dfrac{n'-n''}{n'n''} \right) s_x^2 + \dfrac{v_y^2}{n''} + 2 \left(\dfrac{n'-n''}{n'n''} \right) v_{xy} \right]$

$$\doteq (0.3112)^2 \left[\left(\frac{779-135}{779} \right) \left(\frac{2.8242+(2)\,(1.2743)}{135} \right) \right.$$

$$\left. + \frac{2.3931}{135} \right]$$

$$\doteq (0.096,845) \left[\left(\frac{644}{799} \right) \frac{5.3728}{135} + \frac{2.3931}{135} \right]$$

$$\doteq (0.096,845) \, [(0.806) \, (0.039,799) + (0.017,727)]$$
$$\doteq (0.096,845) \, (0.049,805) = 0.004,823.$$

(c) $\widehat{SE} \, (\bar{y}_{d-rat}) = \sqrt{var} = (0.004,823)^{1/2} = 0.0695.$

$$\widehat{RSE} = \frac{0.0695}{0.3112} = 0.22 \quad \text{or} \quad 22\%.$$

2.   (Refer to the television audience problem of Example 10.7 and Review Illustration 1.) By appropriate coding of the data we can consider the use of the regression estimator for estimating the fraction of persons viewing a specified program. In the case of the "Farm Facts" program we have

		102	33	135
y	1	18	22	40
	0	84	11	95
		0	1	
			x	

Code	y	x
0	"did not watch"	"seldom watch"
1	"watched"	"usually watch"

In this case

$$\bar{y}'' = 0.2963, \qquad n'' = 135, \qquad n' = 779,$$
$$\bar{x}'' = 0.2444, \qquad \bar{x}' = 0.2567,$$
$$s_x^2 = 0.1861, \qquad s_y^2 = 0.2101, \qquad s_{y/x}^2 = 0.1679,$$
$$r = 0.461 \, , \qquad b'' = 0.4902.$$

(a)   Compute the estimated fraction of viewers of the "Farm Facts" program, using double sampling with regression.

(b)   Compute the estimated variance of (a).

(c)   Using the results of Example 10.7 and those of (a) and (b), compare the simple, double-regression, and double-stratified estimators and their estimated variances. Comment on results.

Solution

(a)   To make an estimate of $\bar{Y}$ we use Eq. 10.18:

$$\bar{y}_{d-regr} = \bar{y}'' + b''(\bar{x}' - \bar{x}'')$$
$$= 0.2963 + (0.4902) \, (0.2567 - 0.2444)$$
$$= 0.2963 + 0.0060$$
$$= 0.3023.$$

(b) Its estimated average variance is given by Eq. 10.19:

$$\widehat{\text{var}}\,(\bar{y}_{d-\text{regr}}) \doteq \frac{S_y^2(1-R^2)}{n''}\left(1 + \frac{n'-n''}{n'}\cdot\frac{1}{n''-3}\right) + \frac{R^2 S_y^2}{n'}$$

$$\doteq \frac{(0.2101)(1-0.461^2)}{135}\left(1 + \frac{779-135}{779}\cdot\frac{1}{135-3}\right)$$
$$+ \frac{(0.461)^2(0.2101)}{779}$$

$$\doteq \frac{(2101)(0.7870)}{135} + \frac{0.0447}{779}$$

$$= 0.001,234 + 0.000,057 = 0.001,291.$$

(c) Comparing with the simple and stratified double-sample estimator (of Example 10.7), we have

Estimator	Estimate	Estimated variance	Relative efficiency (%)
simp	0.297	0.001,556	100
d-strat	0.302	0.001,290	121
d-regr	0.302	0.001,291	121

Regression seems to be no better than the stratified estimator here.

3. (Refer to the television audience problem of Example 10.7 and Review Illustrations 1 and 2.) Summarize the estimates of "Farm Facts" viewership, their estimated variances, and compute the efficiency relative to simple sampling for single sampling and for double sampling using the stratified, ratio, and regression estimators. Comment on the results.

Solution

Source	Estimator	Estimate	Estimated variance	Relative efficiency
Example 10.7	Simple	0.297	0.001,556	100
Example 10.7	d-strat	0.302	0.001.290	120
Rev. Illus. 1	d-rat	0.311	0.004,823	32
Rev. Illus. 2	d-regr	0.302	0.001,291	121

Apparently regression gives a better fit here than the ratio.

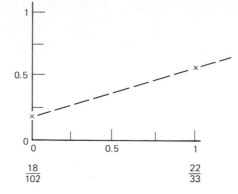

$$\frac{18}{102} \qquad\qquad\qquad \frac{22}{33}$$

4. Show that Eq. 10.41, when allocation of n_h'' is proportional to n_h' say n_h'' $= n_h'/t$, and n' moderately large, reduces to

$$\text{var}\,(\bar{y}_{d-\text{strat}}) \doteq \frac{1}{n'}\left[t\bar{S}_W^2 + \sum_{}^{L} W_h(\bar{Y}_h - \bar{Y})^2\right].$$

Solution

Equation 10.41 is

$$\text{var}\,(\bar{y}_{d-\text{strat}}) = \sum^{L}\left\{\left[W_h^2 + \frac{W_h(1-W_h)}{n'}\right]\frac{S_h^2}{n_h''} + \frac{W_h(\bar{Y}_h - \bar{Y})^2}{n'}\right\}$$

and we are given $n_h'' \propto n_h'$, hence $n_h'' = n_h'/t = n'W_h/t$; then

$$\text{var}\,(\bar{y}_{d-\text{strat}}) = \frac{1}{n'}\sum^{L}\left\{\left[\frac{1+W_h(n'-1)}{n'}\right]tS_h^2 + W_h\,\frac{(\bar{Y}_h - \bar{Y})^2}{n'}\right\}$$

$$= \frac{t\sum S_h^2}{(n')^2} + \frac{t\bar{S}_W^2}{n'} + \frac{\sum W_h(\bar{Y}_h - \bar{Y})^2}{n'}\,.$$

If n' is large enough, the first term will be negligible and can be ignored; therefore

$$\text{var}\,(\bar{y}_{d-\text{strat}}) = \frac{1}{n'}\left[t\bar{S}_W^2 + \sum W_h(\bar{Y}_h - \bar{Y})^2\right],$$

which, since $t/n' = n''$, is identical to stratified sampling, $\bar{S}_W^2/n''$, with a contribution from stratum differences in $\bar{Y}_h$.

10.14. References

Cochran, W. G. *Sampling Techniques.* New York: John Wiley & Sons, Inc. (Chapter 12.)
1953, 1963

Hansen, M. H. *Sample Survey Methods and Theory.* New York: John Wiley & Sons, Inc.
Hurwitz, W. N. (Pages 464–476.)
Madow, W. G.
1953

Jessen, R. J. Statistical Investigation of a Sample Survey for Obtaining Farm Facts.
1942 *Iowa Agricultural Research Bulletin* 304. (For an early presentation of the variance of the regression type due to W. G. Cochran.)

Jessen, R. J. On an Experiment in Forecasting a Tree Crop By Counts and
1972 Measurements. Chapter in *Statistical Papers in Honor of George W. Snedecor*, T. A. Bancroft, ed. (Example of multistage, multiphase sampling.)

Jessen, R. J. Statistical Investigation of Farm Sample Surveys Taken in Iowa, Florida,
Houseman, E. E. and California. *Iowa Agricultural Experiment Station Research Bulletin*
1944 329. Ames, Iowa: Iowa State College. (Example of double sampling with phase I to detect "zero" SUs..)

Neyman, J. Contributions to the Theory of Sampling Human Populations. *Journal of*
1938 *the American Statistical Association*, **33**: 101–116. (Perhaps the first development of the stratified type of double sampling.)

Sukhatme, P. V. *Sampling Theory of Surveys with Applications.* Ames, Iowa: Iowa State
1953 College Press. (Pages 112–120 and 223–231.)

Watson, D. F. The Estimation of Leaf Areas. *Journal of Agricultural Science*, **27**:
1937 474–483. (In this paper, leaf areas were measured in phase I and leaf weights in phase II. A regression estimator, with curvilinearity, is used. Cochran is credited as having "worked out the mathematical basis of the method.")

WOI-TV "The WOI-TV Audience Survey." Ames, Iowa: Iowa State College. (125
1952 pp., mimeographed.)

10.15. Exercises

1. Suppose the cost of measuring y is 20 times that for measuring x, and the correlation of x and y is .7.

(*a*) What is the best size of the phase II sample relative to phase I in this case? (Assume a regression estimator.)

(*b*) What is the NRE (net relative efficiency) of double to simple sampling if allocation is according to (*a*)?

(*c*) Check your answer to (*b*) against Fig. 10.1. Do they agree?

CHAPTER 11

Lattice Sampling (Multistratification)

11.1. Introduction

In practice it often happens that one has a number of factors (variables) available for use in stratification, but, because of size of sample and other design considerations, the number of strata that can be accommodated is so limited that one's options are restricted. Under these circumstances one either restricts himself to the "most effective" factor and abandons the others, or chooses the "most effective" strata from various combinations of two or more factors. In either case the resulting strata will be more heterogeneous than one would like, but, since the number of strata is limited, this is the price to be paid.

The idea of going beyond these restrictions in stratification has been considered and experimented with for some time under such labels as "deep stratification," "controlled selection," "lattice sampling," and "two-way stratification." Let us first consider a very simple case.

The Accommodation Problem. Suppose we have a universe of 16 elements that can be classified into four groups of four elements each according to some factor A, and into another four groups of four elements each according to another factor B. For simplicity, assume that when the elements are grouped on both factors simultaneously they will be arranged neatly into four rows of four columns each. If a sample of four elements is to be drawn, one could be tempted to adopt four equal-sized strata of four elements each and draw an element at random from each stratum. Suppose both factors A and B appear to be suitable candidates for strata, then a matter of choice comes up. If A is a better candidate than B, then choose A. If equally good, then one is indifferent. But in either case, one may feel a loss has been suffered and perhaps one that was somehow unnecessary. In an attempt to avoid this loss, one may try to devise some combination of factors A and B that generates strata embracing both factors but with fewer factor categories—for example, two instead of four, yielding four strata as a 2×2. Now the loss is in the coarseness with which the factors have been categorized.

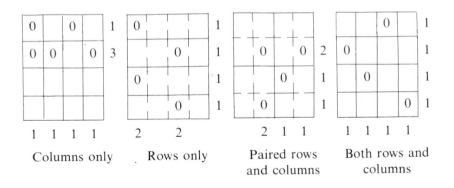

Fig. 11.1. Samples of four in a 4×4 frame

Two-Way Control. An intuitively appealing scheme is to hold onto both factors A and B and the original four levels of categorization. In this scheme the sample is drawn such that representation is balanced over both sets of candidate strata simultaneously. A graphic representation of these four schemes is shown in Fig. 11.1.

The two-way scheme can be randomly generated by any of several methods described in the next section, where its properties will be discussed.

11.2. Simple Random Lattices

Selection Methods for Two-Way Control. Patterson (1954) has distinguished four methods for selecting samples with control on both sets of strata in a two-way classification of a universe. The essentials can be illustrated fairly simply by using a 4×4 frame, as shown in Table 11.1. In the case where two elements are taken from each row and each column, the eight elements can be split into two sets of four, where each set of four will also have an element in each row and column. Hence each set of four can be regarded as a lattice sample, and since each has identical properties, in the sense of mathematical expectation, each is a replicate of the other. This property can be used to obtain an estimate of sampling variance; it will be discussed later.

A number of other procedures may be used to produce the types of randomizations shown in Table 11.1. For example, method 1 can be generated by selecting an element at random from the four in column 1, then one from the three rows in column two not occupied by the element chosen in column 1, then one from the two rows not already occupied by the elements chosen in columns 1 and 2, and so on. Likewise, choosing r elements at random from unoccupied rows in successive columns is equivalent to method 4. Moreover, methods 2 and 4 degenerate to method 1 when $r = 1$. All four methods provide equal selection probabilities for each element in the universe; therefore the sample mean, $\bar{y}$, will be an unbiased estimate of $\bar{Y}$ for each method. However, the four methods have distinguishing characteristics on sampling variance and in their ability to provide suitable estimates on sampling variance.

Method 2, the Latin lattice, can be adapted to the situation where the frame has an odd number of rows and columns. Following is an example for a 5×5.

5×5 Latin lattice

Setup	Randomization procedure	After randomization
(5) A B . . .	Columns	(4) . B . A .
(4) B A . . .	and rows	(2) B . . . A
(3) . . A . B		(5) . A . B .
(2) . . B A .		(1) . . A . B
(1) . . . B A		(3) A . B . .
(1) (2) (3) (4) (5)		(3) (1) (5) (2) (4)

Table 11.1.   Randomization schemes

Method	Setup	Randomization procedure	After randomization
1. The simple lattice. One replicate. $r = 1$, $n = 4$.	A . . . . A . . . . A . . . . A (1) (2) (3) (4)	Columns (or rows) only	. . A . . A . . . . . A A . . . (4) (2) (1) (3)
2. The latin lattice. Two replicates. $r = 2$, $n = 8$.	(4) A B . . (3) B A . . (2) . . A B (1) . . B A (1) (2) (3) (4)	Columns and rows	(3) . A B . (1) A . . B (4) . B A . (2) B . . A (4) (2) (1) (3)
3. The systematic lattice. Two replicates. $r = 2$, $n = 8$.	A B C D D A B C C D A B B C D A	Sets (two of four) at random	A B . . . A B . . . A B B . . A
4. The general lattice. Two replicates. $r = 2$, $n = 8$.	A B C D D A B C C D A B B C D A (1) (2) (3) (4)	Sets and columns (or rows)	. B A . . A . B B . . A A . B . (4) (2) (1) (3)

11.3.   Square Random Lattices

If we are willing to comply with rather limiting relationships among sample size, universe size, and number of strata, the square can serve as a fairly general model for a number of applications.

Notation.   Let a square consist of L rows by L columns, and

$N = L^2$ = total elements,

$n = rL$ = sample size,

r = number of replications in a row or column,

Y_{gh} = observed y in the hth column of the gth row,

$\bar{Y}$, $\bar{Y}_g$, $\bar{Y}_h$ = mean per element in the overall population, the gth row, and hth column, respectively,

$E_{gh} = Y_{gh} - \bar{Y}_g - \bar{Y}_h + \bar{Y}$ = the residual deviation of Y_{gh} after the removal of the general mean and the row and column deviations, $(\bar{Y}_g - \bar{Y})$ and $(\bar{Y} - \bar{Y})$,

$S_{RC}^2 = [1/(L-1)^2] \sum \sum E_{gh}^2$ = variance of the residual deviations.

If a lattice sample of $n = L$ is drawn from an $L \times L$ square, then a sample element must appear in each row and column. If $n = 2L$, then two sample elements appear in each row and column. Hence r is analogous to n_h in simple stratification, where n_h is a constant.

The variance of $\bar{y}$, the lattice sample mean, if drawn by methods 1, 2, or 4, is given by

$$(11.1) \qquad \qquad \text{var}(\bar{y}) = (1-f)\frac{S_{RC}^2}{n}.$$

Hence S_{RC}^2 replaces S^2 for random sampling and S_w^2 for simple stratified sampling.

Example 11.1.   Here is a 4×4 population with numbers easy to compute.

$$
\begin{array}{rrrr}
-5 & -6 & -4 & -1 \\
-3 & -2 & -2 & -1 \\
-1 & 2 & 4 & 3 \\
1 & 2 & 6 & 7
\end{array}
$$

Calculate S^2 and S_{RC}^2 and compare the efficiency of a lattice sample of $n = 4$ with that of a simple random sample.

Here $\bar{Y} = 0$, $\bar{Y}_g = [\bar{Y}_1 = -4, \bar{Y}_2 = -2, \bar{Y}_3 = 2, \bar{Y}_4 = 4]$, and $\bar{Y}_h = [\bar{Y}_1 = -2, \bar{Y}_2 = -1, \bar{Y}_3 = 1, \bar{Y}_4 = 2]$, and the E_{gh}s are

$$
\begin{array}{rrrr}
1 & -1 & -1 & 1 \\
1 & 1 & -1 & -1 \\
-1 & 1 & 1 & -1 \\
-1 & -1 & 1 & 1
\end{array}
$$

whence $\sum\sum E_{gh}^2 = 16$ and $S_{RC}^2 = 16/(4-1)^2 = \frac{16}{9} = 1.78$; S^2 works out to be $\frac{216}{15} = 14.40$. Hence in this case lattice sampling would be $14.40/1.78$, or 8.1 times as efficient as random sampling.

11.4.   ANOVA; Efficiency

A compact arrangement of concepts and definitions, useful to relate lattice sampling with random and stratified-random sampling, is that of the ANOVA table—for example, Table 11.2.

It may be noted that the $R \times C$ (row by column) interaction mean square is identical to S_{RC}^2 defined before; hence it can be calculated as a residual sum of squares rather than directly on the residuals as was done in Example 11.1.

Table 11.2.   **ANOVA Table for an $L \times L$ Square**

Source	df	SS	MS defined as
Gross	L^2	$\sum\sum Y_{gh}^2$	
Mean	1	$(\sum\sum Y_{gh})^2/L^2$	
Total	L^2-1	$\sum\sum (Y_{gh}-\bar{Y})^2$	MS(T) or S^2
Rows	$L-1$	$\sum\sum (\bar{Y}_g-\bar{Y})^2$	MS(R) or $S_{RC}^2+LS_R^2$
Columns	$L-1$	$\sum\sum (\bar{Y}_h-\bar{Y})^2$	MS(C) or $S_{RC}^2+LS_C^2$
$R\times C$	$(L-1)^2$	$\sum\sum (Y_{gh}-\bar{Y}_g-\bar{Y}_h+\bar{Y})^2$	MS$(R\times C)$ or S_{RC}^2
. . .			
Elements/rows	$L(L-1)$	$\sum\sum (Y_{gh}-\bar{Y}_g)^2$	MS$(W)_R$ or $S_{W(R)}^2$
Elements/columns	$L(L-1)$	$\sum\sum (Y_{gh}-\bar{Y}_h)^2$	MS$(W)_C$ or $S_{W(C)}^2$
Elements/RC	$(L-1)^2$	$\sum\sum (Y_{gh}-\bar{Y}_g-\bar{Y}_h+\bar{Y})^2$	$S_{W(RC)}^2$

Example 11.2.   Calculations on the beetle data of Fig. 4.1 (Chapter 4), which can be regarded as a 48×48 frame, can be arranged into an ANOVA table form such as Table 11.3. If a sample of 48 segments were selected from this field using a lattice design on a 48×48 frame, the variance of the sample mean would be

$$\text{var}(\bar{y}) = (1-f)\frac{S_{RC}^2}{n} = \left(1-\frac{48}{2304}\right)\frac{12.97}{48} \doteq 0.27.$$

Table 11.3.   **ANOVA of beetle data (on a per-segment basis)**

Source	df	SS	MS
Gross	2304	86,239	
Mean	1	51,690	
Total	2303	34,549	15.00
Rows	47	5289	112.53
Columns	47	2387	50.79
$R\times C$	2209	26,873	12.17
. . .			
Segments/rows	2256	29,260	12.97
Segments/columns	2256	32,162	14.26

Efficiency.   Using the properties of the ANOVA table, we can write S^2_{RC}, $S^2_{W(R)}$, $S^2_{W(C)}$, and S^2 in terms of components of variance. Thus

$$S^2_{W(R)} = S^2_{RC} + S^2_C, \qquad S^2_{W(C)} = S^2_{RC} + S^2_C$$

and

$$S^2 = S^2_{RC} + \left(\frac{L}{L+1}\right)(S^2_R + S^2_C),$$

so if S^2_C and S^2_R are positive, as is usually the case, then

$$S^2_{RC} \le S^2_{W(R)} \quad \text{or} \quad S^2_{W(C)} \le S^2,$$

and therefore, for a given sample size n,

$$\text{var (lattice)} \le \text{var (strat)} \le \text{var (random)}$$

unless the population has some unusual negative intraclass correlations (i.e., S^2_R or $S^2_C < 0$).

In the beetle data,

$$12.17 < 12.97 \quad \text{or} \quad 14.26 < 15.00$$

or, in terms of relative efficiency,

$$100 > 0.94 \quad \text{or} \quad 0.85 > 0.81.$$

11.5.   Estimating var $(\bar{y})$

Unbiased estimates of var $(\bar{y})$, or its component S^2_{RC}, can be made from lattice samples, if provision is made for suitable replication. In the general lattice (method 2) with $r = 2$ the n elements comprise two sets A and B. Each set can be regarded as a single lattice with $r = 1$. The set means $\bar{y}_A$ and $\bar{y}_B$, are unbiased estimates of $\bar{Y}$, and their difference, $\bar{y}_A - \bar{y}_B$, can supply some information on S^2_{RC}.

Consider the more general case for any $r \ge 2$. Regard each set (all elements of the same letter in Table 11.1) as a cluster of $M_0 = L$ elements; there are L such clusters in the universe. The mean square among sets (clusters) is given for the population by

(11.2) $$\text{MS }(S) = \frac{L \sum_{s=1}^{L} (\bar{y}_s - \bar{Y})^2}{L-1},$$

and for a sample of r sets (clusters) by

(11.3) $$\text{MS }(s) = \frac{L \sum_{s=1}^{r} (\bar{y}_s - \bar{y})^2}{r-1},$$

which is an unbiased estimator of Eq. 11.2. Then an unbiased estimator of var $(\bar{y})$ is given by

$$\text{(11.4)} \qquad \widehat{\text{var}}\ (\bar{y}) = (1-f)\ \frac{\sum^r (\bar{y}_s - \bar{y})^2}{r(r-1)}$$

with $r-1$ degrees of freedom, and whence

$$\text{(11.5)} \qquad \hat{S}^2_{RC} = \frac{n \sum^r (\bar{y}_s - \bar{y})^2}{r(r-1)}\ .$$

With $r = 2$, a simpler version of Eq. 11.4 is given by

$$\text{(11.6)} \qquad \widehat{\text{var}}\ (\bar{y}) = (1-f)\ \frac{(\bar{y}_A - \bar{y}_B)^2}{4}\ .$$

Latin and Nonlatin Lattices. In the course of generating general lattices using method 4, or its equivalent, it will be found that some samples will have configurations identical to those generated by method 2—that is, comprising exploded sets of $r \times r$ latin squares. Since the two types have different properties with respect to some characteristics, it is suggested they be distinguished from those that do not contain latin squares. (For example, in the 4×4 frame where $n = 8$, of the 90 possible lattices, 18 will be latin and 72 nonlatin.) The latin lattice provides an alternative estimate of var $(\bar{y})$ with somewhat more precision.

Var $(\bar{y})$ for Latin Lattices. The method of using replicate sets, described above for estimating sampling variance, can be used for latin lattices, but it uses only part of the information available. One replicate set, in the $r = 2$ case, is provided by $\bar{y}_A$ and $\bar{y}_B$. Another set is provided by the $\bar{y}_k$, the mean of each $r \times r$ latin. More sets are derivable by various combinations of these two sources. Like the $\bar{y}_s$, the $\bar{y}_k$ are also unbiased estimators of $\bar{Y}$.

An efficient method of obtaining this information and utilizing it is to deal with each latin square separately. The sample of n where L is even consists of $p\,r \times r$ latins, where $p = L/r$. If, for example, $r = 3$, then each latin will be of the form

$$
\begin{array}{ccc}
A & B & C \\
C & A & B \\
B & C & A
\end{array}
$$

where three replicate means, $\bar{y}_s$, are $\bar{y}_A$, $\bar{y}_B$ and $\bar{y}_C$. Then let

$$\text{(11.7)} \qquad (s^2_{RC})_k = \frac{r \sum_{s=1}^r (\bar{y}_s - \bar{y})^2}{r-1},$$

which can be computed for each of the p latins. Each $(s_{RC}^2)_k$ is an unbiased estimator of S_{RC}^2, and the pooled variance

$$(11.8) \qquad s_{RC}^2 = \frac{\sum_{k=1}^{p} (s_{RC}^2)_k}{p}$$

is an unbiased estimator of S_{RC}^2 with $p(r-1)$ degrees of freedom. If $r = 2$, a simplified computational form is

$$(11.9) \qquad s_{RC}^2 = \frac{1}{4} \frac{\sum_{k=1}^{p} (A-B)_k^2}{p} \,,$$

where $(A-B)_k$ is the difference between the totals of y on the A and B replicates, respectively, for latin k.

The Case of L an Odd Integer.   Pictorially the odd latin for $r=2$ will appear as follows:

$$\begin{array}{ccc} A & . & B \\ B & A & . \\ . & B & A \end{array}$$

and

$$s_{RC}^2 = \frac{(A-B)^2}{6r} = \frac{(A-B)^2}{12}$$

with one degree of freedom.

Use of ANOVA to Estimate S_{RC}^2 for Latin Lattices.   For the case of a sample of $n = pr^2$, comprising p $r \times r$ latins, we can write the ANOVA structure as in Table 11.4 to display sources of variation. This is an equivalent alternative to the method of examining individual latins described above. It may be preferred if additional information on sources of variation is of interest.

     Example 11.3.   A latin sample of $n = 8$ was drawn from the 4×4 population of Example 11.1, with the following results:

A_1 -5			B_1 -1
	A_2 -2	B_2 -2	
B_1 -1			A 3
	B_2 2	A_2 6	

Table 11.4. *ANOVA for a latin lattice*

Source	df	MS defined as	Expected mean square
Gross	n		
Mean	1		
	—		
Total	$n-1$		
Replicates × latin subclasses	$rp-1$		
Latins $p-1$		s^2_{RC}	S^2_{RC}
Subclasses/latins $p(r-1)$			S^2_{RC}
Replicates $r-1$			S^2_{RC}
Replicates × latins $(r-1)(p-1)$			S^2_{RC}
Within subclasses	$n-rp$		

Calculate the mean and its standard error. The letters refer to replicates, the subscript numbers refer to latins. For latin 1, $A = -2, B = -2$, and for latin 2, $A = 4, B = 0$; hence

$$\bar{y} = \tfrac{1}{4}[(A+B)_1 + (A+B)_2]$$
$$= \tfrac{1}{4}[-4+4] = 0,$$

and

$$s^2_{RC} = \frac{1}{4p}\sum_{k=1}^{p}(A-B)^2_k$$

$$= \frac{1}{(4)(2)}[(-2+2)^2 + (4-0)^2] = \tfrac{16}{8} = 2,$$

hence

$$\widehat{SE}\,(\bar{y}) = \left[(1-f)\frac{s^2_{RC}}{n}\right]^{1/2} = [(1-\tfrac{8}{16})\tfrac{2}{8}]^{1/2} = \sqrt{0.125} = 0.35.$$

11.6. Rectangular Lattices

Lattice sampling can be easily adapted to rectangular frames if the legs bear a simple multiple relationship with each other. In this case randomization methods can be devised to accommodate rectangles. For example, latin type lattices can be generated for a 4×6 frame as follows:

Setup	Randomization procedure	After randomization
$A\ B\ C$ . . .	Columns and	. B_2 . . $C_2\ A_2$
$C\ A\ B$ . . .	rows	B_1 . $C_1\ A_1$ . .
. . . $A\ B\ C$		C_1 . $A_1\ B_1$ . .
. . . $C\ A\ B$		. C_2 . . $A_2\ B_2$

The mean $\bar{y}$ is an unbiased estimator of $\bar{Y}$, and the variance of $\bar{y}$ will be given by

$$(11.10) \qquad\qquad \text{var } (\bar{y}) = (1 - f)\,\frac{S_{RC}^2}{n},$$

where

$$(11.11) \qquad\qquad S_{RC}^2 = \frac{(Y_{gh} - \bar{Y}_g - \bar{Y}_h + \bar{Y})^2}{(L - 1)\,(L' - 1)},$$

which is the mean square of rows × columns in an ANOVA table for an $L \times L'$ population.

An estimate of S_{RC}^2 would be provided by obtaining the mean square of rows × columns for each latin rectangle and pooling them over the p rectangles. In all, there will be $p(r - 1)(r' - 1)$ degrees of freedom for the estimate.

11.7. Cubic Random Lattices

When a universe of N elements can be classified into a cubic frame of L rows, L columns, and L files, and a sample of n is to be drawn with fixed sampling fractions on each row, column, and file, the sample designer will find many alternative procedures available. Four schemes, to be described here, present some of the practical problems involved in the selection of such samples, their properties (such as efficiency and limitations), and some principles generally helpful in sampling theory and practice.

Methods of Selection. The four basic schemes of selection are presented in Fig. 11.2 using a $4 \times 4 \times 4$ cubic frame as an example. (Regard each position in a solid square as a file *plane*, e.g., upper left as "l," etc.) The entries indicate the location of sample elements prior to the randomization of rows, columns, and files. Scheme 1, the simplest, arranges a sample of $n = L$ elements along the points of a diagonal in the cube. This arrangement assures that each line— that is, a row, column, or file—contains one sample element. Scheme 2 is a nonreplacement replication of scheme 1, where A and B designate the two replicates. Scheme 3 is a latin type where each set with the same subscript number is a 2^3 latin cube and the A and B letters designate a replicate for that cube. In this case the two cubes will provide four sample elements in each line of the frame but they are really two clusters of two elements each. Scheme 4 is an extension of scheme 3 where all lines are grouped into 2×2 sets containing a 2^3 latin cube. As in scheme 3, the A and B designate a replicate within each latin cube identified with a numerical subscript.

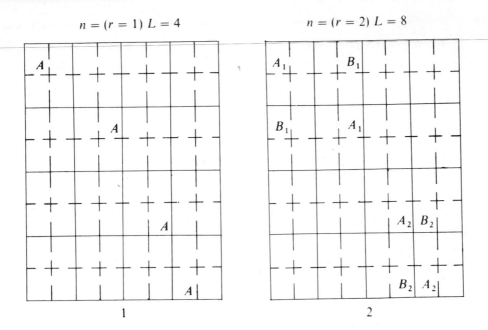

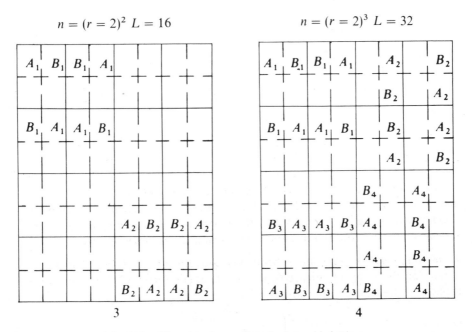

Fig. 11.2. Pictorial schemes for selecting cubic lattices.

Notation and ANOVA.   Let

$$Y_{ghf} = \text{observed } y \text{ in the } g\text{th row, } h\text{th column and } f\text{th file, } g = h = f$$
$$= 1, 2, \ldots, L,$$

$$\bar{Y}_g, \ \bar{Y}_h, \ \bar{Y}_f, \ \bar{Y} = \text{means per element for rows, columns, files, and overall, respectively,}$$

$$\bar{Y}_{gh}, \ \bar{Y}_{gf}, \ \bar{Y}_{hf} = \text{means per element for the combinations indicated.}$$

Other definitions and concepts are given in Table 11.5.

Since multiway stratification is an attempt to eliminate the effects of isolatable sources of variation in the population from samples drawn from it, we shall start with a standard procedure for isolating and evaluating sources of variation—the ANOVA table. Table 11.5 presents this structure for a cubic population.

Var $(\bar{y})$ for the Four Schemes.   Each of the four schemes provides a $\bar{y}$ that is an unbiased estimator of $\bar{Y}$. However, the precision of each depends on the variance structure of the population. The variances of each are

(11.12)   (1) $\text{var} (\bar{y}) = \left(1 - \dfrac{n}{L^2}\right) \left(\dfrac{S_{RC}^2 + S_{RF}^2 + S_{CF}^2}{n}\right) + \left(1 - \dfrac{n}{L^3}\right) \dfrac{S_{RCF}^2}{n}$

for $n = L$, no replicates;

(2) $\text{var} (\bar{y}) = \left(1 - \dfrac{n}{L^2}\right) \left(\dfrac{S_{RC}^2 + S_{RF}^2 + S_{CF}^2}{n}\right) + \left(1 - \dfrac{n}{L^3}\right) \dfrac{S_{RCL}^2}{n}$

for $n = rL$ and $r =$ number of replicate lines, replicate points ≤ 1;

(3) $\text{var} (\bar{y}) = \left(1 - \dfrac{2}{L}\right) \left(\dfrac{2(S_{RC}^2 + S_{RF}^2 + S_{CF}^2)}{4L}\right) + \left(1 - \dfrac{4}{L^2}\right) \dfrac{S_{RCF}^2}{4L}$

for $n = 4L = L/2$ $2 \times 2 \times 2$ cubes;

(4) $\text{var} (\bar{y}) = \left(1 - \dfrac{2}{L}\right) \left(\dfrac{S_{RCF}^2}{2L^2}\right)$

for $n = 2L^2 = L$ $2 \times 2 \times 2$ cubes.

In all cases S_{RCF}^2 is a common source of variation, and schemes 1, 2, and 3 also contain variation from the two-factor interactions. Scheme 1 is simply a special case of 2 when $r = 1$. When the two-factor interactions are zero, all schemes have essentially the same basic variance per element observed.

Estimating var $(\bar{y})$.   Scheme 1, being a single replicate, does not provide the means for obtaining an unbiased estimate of var $(\bar{y})$. Scheme 2, however, with r replicates, does permit an estimate. Thus

(11.13) $\text{var} (\bar{y}) = \left(1 - \dfrac{r}{L}\right) \dfrac{\sum_{s=1}^{r} (\bar{y}_s - \bar{y})^2}{r(r-1)},$

where $\bar{y}_s$ is the mean per element in replicate s.

Table 11.5. ANOVA for an $L \times L \times L$ population

Source	df	SS	MS defined as
Gross	L^3	$\sum\sum\sum Y^2_{ghf}$	
Mean	1	$(\sum\sum\sum Y_{ghf})^2/L^3$	S^2
Total			
Rows	$L-1$	$L^2\sum\sum\sum(\bar{Y}_g - \bar{Y})^2$	$S^2_{RCF} + L(S^2_{RC} + S^2_{RF}) + L^2 S^2_R$
Columns	$L-1$	$L^2\sum\sum\sum(\bar{Y}_h - \bar{Y})^2$	$S^2_{RCF} + L(S^2_{RC} + S^2_{CF}) + L^2 S^2_C$
Files	$L-1$	$L^2\sum\sum\sum(\bar{Y}_f - \bar{Y})^2$	$S^2_{RCF} + L(S^2_{RF} + S^2_{CF}) + L^2 S^2_F$
$R \times C$	$(L-1)^2$	$L\sum\sum\sum(\bar{Y}_{gh} - \bar{Y}_g - \bar{Y}_h + \bar{Y})^2$	$S^2_{RCF} + LS^2_{RC}$
$R \times F$	$(L-1)^2$	$L\sum\sum\sum(\bar{Y}_{hf} - \bar{Y}_h - \bar{Y}_f + \bar{Y})^2$	$S^2_{RCF} + LS^2_{RF}$
$C \times F$	$(L-1)^2$	$L\sum\sum\sum(\bar{Y}_{gf} - \bar{Y}_g - \bar{Y}_f + \bar{Y})^2$	$S^2_{RCF} + LS^2_{CF}$
$R \times C \times F$	$(L-1)^3$	$\sum\sum\sum(Y_{ghf} + 2M - I - \bar{Y})^{2\,a}$	S^2_{RCF}
...			
Elements/rows	$L(L^2-1)$	$L\sum\sum_{h\ f}\sum(Y_{ghf} - \bar{Y}_g)^2$	$S^2_{W(R)}$
Elements/columns	$L(L^2-1)$	$L\sum\sum_{g\ f}\sum(Y_{ghf} - \bar{Y}_h)^2$	$S^2_{W(C)}$
Elements/files	$L(L^2-1)$	$L\sum\sum_{g\ h}\sum(Y_{ghf} - \bar{Y}_f)^2$	$S^2_{W(F)}$
Elements/RCF	$(L-1)^3 + (L-1)^2$	$\sum\sum\sum(Y_{ghf} - \bar{Y}_g - \bar{Y}_h - \bar{Y}_f + 2\bar{Y})^2$	$S^2_{W(RCF)}$

$^a M = \bar{Y}_g + \bar{Y}_h + \bar{Y}_f;\ I = \bar{Y}_{gh} + \bar{Y}_{gf} + \bar{Y}_{hf}.$

371

Scheme 3 can provide an estimate, but it is quite wobbly; the two-factor interaction components involved are difficult to estimate with precision because they are obtained as differences between mean squares of the sample data [see Patterson (1954)]. However, if one wants to learn something about these components of variance, scheme 3 can provide at least crude estimates, something scheme 2 can hardly do at all. If good estimates of var ($\bar{y}$) are desired and a large sampling fraction is no problem, then scheme 4 should be used.

To estimate var ($\bar{y}$) with scheme 4, one needs an estimate only of S_{RCF}^2, which can be obtained from each $2 \times 2 \times 2$ cube. In terms of ANOVA this can be obtained from computations as indicated in Table 11.6.

Table 11.6.   *ANOVA of $2 \times 2 \times 2$ cube*

Source	df	MS defined as
Total	7	
Row	1	
Column	1	
File	1	
$R \times C$	1	$s_{RCF}^2 + 2s_{RC}^2$
$R \times F$	1	$s_{RCF}^2 + 2s_{RF}^2$
$C \times F$	1	$s_{RCF}^2 + 2s_{CF}^2$
$R \times C \times F$	1	s_{RCF}^2

In Table 11.6 the source of estimates of the two-factor interactions is also shown, in case they are of interest for investigating sources of variation. The required estimate of S_{RCF}^2 is the pooled s_{RCF}^2 over the p latin cubes.

11.8. Simple Probability Lattices

Up to now in this chapter the lattice scheme of sampling has been confined to those cases where a universe of elements can be classified in two or more dimensions and each intersection is occupied by one and only one element (or cluster). Such universes are not very common in many fields where surveys are employed—particularly in the social sciences. A more flexible scheme, one that can accommodate classifications in which cells may contain unequal numbers of elements, or none at all, is needed if the balancing feature of multiway stratification is to be applicable in many practical situations.

A class of problems in which two-way stratification may be desired, but unequal subclass numbers cause complications, is that of selecting committees or juries that "represent" their small constituencies on two factors, say profession and ethnicity, and yet provide equal probability of selection to each

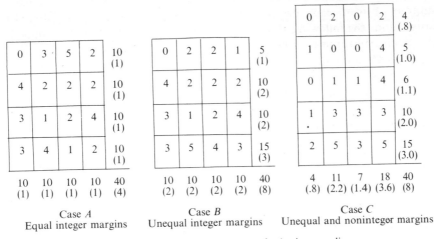

Fig. 11.3.   Some two-way frames for lattice sampling

member in the constituency. Suppose we have 40 members from which to select a committee of four and, although there are exactly 10 in each of four professions and 10 in each of four ethnic groups, the 16 profession-ethnic subclasses have numbers that vary from 0 to 5. It is desired to draw a sample of four such that it is stratified by profession and by ethnicity and each element has an equal chance of selection. This is shown as case A in Fig. 11.3. Also shown are the cases where marginal (row and column) requirements of the sample are unequal but integer (case B) and unequal and noninteger (case C).

A Simple Selection Method.   Let $M_i =$ number of elements in the ith subclass, $M =$ total elements in the universe, and $A_i = M_i/M$. Calculate nA_i, the number of elements expected in the sample, for each cell i and for their sums on each margin. Thus for case A we have the nA_is and their sums:

0	.3	.5	.2	1.0
.4	.2	.2	.2	1.0
.3	.1	.2	.4	1.0
.3	.4	.1	.2	1.0
1.0	1.0	1.0	1.0	4.0

With this format designate a feasible sample—one that meets the cells and marginal requirements (in this case, one in each row and column, none in cell 1, and a total $n = 4$). It is suggested that the largest nA_i be attempted first.

Subtract .1 from each designated nA_i and repeat the process. After 10 designations the following 10 feasible samples (those starred) were obtained. (A lot of arbitrariness is possible. Just make sure the cell and marginal requirements are met.)

	1					2					3					4					5		
0	3	5*	3		0	3	4*	3		0	3*	3	3		0	2	3*	3		0	2	2	2*
4*	2	2	2		3*	2	2	2		2	2	2*	2		2	2*	1	2		2*	1	1	2
3	1	2	4*		3	1	2	3*		3*	1	2	2		2	1	2	2*		2	1	2*	1
3	4*	1	2		3	3*	1	2		3	2	1	2*		3*	2	1	1		2	2*	1	1

	6					7					8					9					10		
0	2*	2	1		0	1	2*	1		0	1	1*	1		0	1	–	1*		0	1*	–	–
1	1	1	2*		1	1*	1	1		1*	–	1	1		–	–	1*	1		–	–	–	1*
2*	1	1	1		1	1	1	1*		1	1*	1	–		1*	–	1	–		–	–	1*	–
2	1	1*	1		2*	1	–	1		1	1	–	1*		1	1*	–	–		1*	–	–	–

This set of 10 feasible samples will provide a $P_i = nM_i/M$ for each cell in the frame if each is given a 1/10 probability of selection. As in two-stage sampling with primaries selected with PPS, where M_i is the measure of size, the jth element in the ith cell is selected at random from its M_i elements.

Cases B and C can be dealt with in a similar manner by designating feasible samples with 2 in each column and 1, 2 or 3 in the specified rows (case B), or (case C) designating 0 or 1 in row one, always 1 in column 2, and so on in the remaining rows and columns. The procedure just described might be called decremental method 1. Another method, also based on decrements, that tends to minimize the number of feasible samples required is given in the next section.

In the foregoing examples the cells were regarded as clusters of M_i elements and sampling was two-stage, with one element to be selected from each primary (cell). In this case the lattice sampling provided convenient probabilities for the selection of primaries; that is, $P_i = nM_i/M$. If cells are occupied by single elements with a measure of size, X_i, then $A_i = X_i/X$ instead of M_i/M and the procedure is unaltered. To estimate Y we use $\hat{Y} = (1/n)\sum_{i=1}^{n} Y_i/A_i$ as an unbiased estimator.

11.9. Square Probability Lattices

When the elements of a universe can be ranked on some factor, as for example census tracts can be ranked from low to high on median family income, percent black population, percent home owners, and so on, one has the opportunity of forming multiway classifications with equal "sizes" (e.g., sums of secondaries,

M_i, or some measure of size, X_i, in each margin). In this case the basic form is that of a square random lattice where $n = rL$, but now the cells are to be drawn with PPS.

Methods of Selection. The simple method given in the previous section may be adequate for many situations, especially for small frames. An algorithm for selecting probability lattices given by Jessen (1970) aims to minimize the number of feasible samples required to form a set. However, with a little practice the principles can be utilized quite effectively without such formality.

For example, we have a square population of ys and xs their measure of size, as follows:

i		Y_i					X_i		
1 2 3 4	0	1.05	2.25	0.80		0	3	5	2
5 6 7 8	0.60	0.70	0.65	0.40		4	2	2	2
9 10 11 12	0.60	0.20	0.50	0.60		3	1	2	4
13 14 15 16	0.30	0.80	0.25	0.30		3	4	1	2

A sample of $n = rL = (2)(4) = 8$ is required. Calculating $nA_i = 8X_i/X$, we obtain the basic selection tableau.

$$nA_i$$

0	.6	1.0	.4	2
.8	.4	.4	.4	2
.6	.2	.4	.8	2
.6	.8	.2	.4	2
2	2	2	2	8

Note that cell 3 must be selected with certainty, since $nA_3 = 1$. Designating the largest nA_1 (by starring), we obtain the first feasible sample.

1.   0   .6*   1.0*   .4   Since the smallest nA_i here is .4 (for $i = 7$), let us
     .8*   .4   .4*   .4   give this sample a $P_1 = .4$ and subtract .4 from
     .6*   .2   .4   .8*   each starred entry.
     .6   .8*   .2   .4*

2. | 0 | .2 | .6* | .4* | Again we star the largest nA_i; in this case the
    | .4* | .4* | — | .4 | smallest is .2, hence $P_2 = .2$
    | .2 | .2 | .4* | .4* |
    | .6* | .4* | .2* | — |

3. | 0 | .2* | .4* | .2 | The smallest nA_i here is again .2, hence $P_3 = .2$
    | .2* | .2 | — | .4* |
    | .2 | .2* | .2 | .2* |
    | .4* | .2 | .2* | — |

4. | 0 | — | .2* | .2* | The residual leaves one complete feasible sample
    | — | .2* | — | .2* | with $P_4 = .2$. The selection is completed.
    | .2* | — | .2* | — |
    | .2* | .2* | — | — |

The feasible samples consist of four with designated probabilities summing to 1.0. The final sample is that which is selected with the appropriate probability from this set.

Properties of Probability Lattices.   In the case where the cells in the two-way table are individual elements, the probability lattice sample can be regarded as simply a type of PNR sampling, described in Sections 8.8–8.11, with additional stratification control. Since the variance of samples of this sort depends on the selection method, and in this case the particular set of feasible samples finally generated, it is difficult to evaluate their effectiveness over alternatives by just examining population structure. As an indicator, however rough, one may hypothesize that if there are row and column "components" on the Y_i/M_i (or Y_i/X_i), then he may expect the lattice to provide less variance in estimating $\bar{\bar{Y}}$, or $\bar{Y}$, than one-way stratification, and one-way stratification less variance than simple PNR sampling. In the case where the cells are primaries, efficiency will depend on the relationship between Y_i and M_i (see Section 9.9). In the case where cells are elements and $\hat{Y}$ is being estimated, efficiency will depend also on the relationship of Y_i to X_i. (See Section 8.4.)

If calculations are carried out on the 4×4 population just described, we obtain the following results for $n = 8$.

Sampling design	var $(\hat{Y})$
RNR	4.14
Random lattice	4.14
PR	2.49
Probability lattice	0.24

For this rigged population the probability lattice has clearly outperformed unrestricted probability sampling, whereas the random lattice did no better than simple RNR sampling.

Estimating var $(\bar{y})$.   In the two-stage case $\bar{y}$ is calculated as discussed in Chapter 9 for PPS sampling of primaries. Lattice sampling merely provides an alternative method for selecting primaries with PPS. Hence unbiased estimates of $\bar{Y}$ are similarly available in lattice sampling. However, var $(\bar{y})$ is difficult to determine because of the dependence of the component of variance, due to primaries, on the selection method. Rough estimates of var $(\bar{y})$ can be obtained by the split-sample method described in Section 11.5. The method is not unbiased when sampling with PNR but may be useful. A PNR adaptation is given by

(11.14) $$\widehat{\text{var}}\ (\hat{Y}) = \left(1 - n\sum A_i^2\right) \frac{\sum_{t=1}(\hat{Y}_A - \hat{Y}_B)^2}{4}\ ,$$

where the term in brackets serves as an approximate FPC for PNR sampling (see Section 8.11) and A_i is the relative measure of size.

For example, the exact variance of $\hat{Y}$ where $\hat{Y} = \sum Y_i/nA_i$ for samples of eight in the feasible sample set of four in the 4×4 population above works out to be 0.244. The four sets of two splits (since feasible sample number 2 is a latin, an average of the two splits is taken here) yielded

$$(1-f) \frac{\sum_{t=1}^{4} P_t(\hat{Y}_A - \hat{Y}_B)_t^2}{4} = 0.183$$

and the individual one-degree-of-freedom estimates were 0.05, 0.58, 0.21, and 0.02, hence the estimates are biased and rough. This suggests that lattices of this sort are not suitable where accurate estimates of sampling error are desired from the sample.

11.10.   Irregular Two-Way Lattices

The methods described above can be adapted and extended to nonsquare lattices and to cases where marginal sample expectations are not integers— such as case C of Section 11.8. These will not be dealt with here, but the decremental method is discussed by Jessen (1970). A rectangular case, where the cells must be nonzero, is discussed by Bryant et al. (1960). Goodman and Kish (1950) also have dealt with this case.

Consider the committee selection problem, case C of Section 11.8, where committees of eight are to be formed from a constituency of 40 in a 4×5 grouping of irregular sizes. The data and the expected sample sizes, nA_i, are as follows:

	2 (.4)		2 (.4)	4 (.8)
1 (.2)			4 (.8)	5 (1.0)
	1 (.2)	1 (.2)	4 (.8)	6 (1.2)
2 (.4)	5* (1.0)	3 (.6)	5* (1.0)	15 (3.0)
4 (.8)	11 (2.2)	7 (1.4)	18 (3.6)	40 (8.0)

The starred cells containing five persons must be selected with certainty and therefore can be omitted in subsequent steps, which follow.

```
 .   4*   .   4  |  8
 2    .   .  8*  | 10(1)
 .   2    2  8*  | 12
 2   6   6*  6*  | 20(2)
 4*   .  6    .  | 10(1)
------------------------
 8  12  14  26   | 60(6)
```

Step 1. Designate a sample of 6 that meets marginal requirements. Row 1 can be either 0 or 1, rows 2 and 5 must be 1 only, row 3 can be either 1 or 2, and so on. "Must" values are shown in parentheses. Designating the largest values, we obtain the starred values. Maximum decrement here is $\Delta = 0.4$ for feasible sample 1.

```
 .    .   .   4  |  4
 2    .   .  4*  |  6(1)
 .   2   2*  4*  |  8
 2*  6*   2   2  | 12(2)
 .    .  6*   .  |  6(1)
------------------------
 4   8  10  14   | 36(6)
```

Step 2. Subtracting 4 from each unit in the previous sample yields the new tableau. The designated sample can yield a maximum decrement of $\Delta = 0.2$ for feasible sample 2.

```
 .    .   .  4*  |  4(1)
 2*   .   .   2  |  4(1)
 .   2    .  2*  |  4(1)
 .   4*   2  2*  |  8(2)
 .    .  4*   .  |  4(1)
------------------------
 2   6   6  10   | 24(6)
```

Step 3. The new tableau requires 1 unit from each row (except row 3, which requires 2) and variable numbers from each column. Maximum decrement here is $\Delta = 0.2$.

.	.	.	2*	2(1)	*Step* 4. The residual yields a sample that
.	.	.	2*	2(1)	satisfies all remaining requirements. Its
.	2*	.	.	2(1)	decrement clears the tableau with $\Delta = 0.2$.
.	2*	2*	.	4(2)	
.	.	2*	.	2(1)	
0	4	4	4	12(6)	
(0)	(2)	(2)	(2)		

The feasible set consists of four samples with the probabilities based on the indicated Δs. Selecting one of the feasible samples provides the first-stage sample. Selecting an individual at random from each cell in the first stage completes the selection of the committee.

In general, lattices that have noninteger marginal requirements will have additional components of variance attributable to the fact that rows and columns are not weighted equally from sample to sample. Hence it is advisable to avoid noninteger marginal requirements if possible in designing the frame or in choosing a sampling rate.

11.11. Cubic Probability Lattices

Although procedures become more complicated, the principles of probability lattices can be easily extended to three or more dimensions. If accuracy can be increased by extending stratification from one-way to two-way, then it appears that even further gains are possible by going on to three or even more dimensions—provided, of course, there are sensible factors to stratify by. There are a number of situations in which this appears to be the case. For example, when designing samples of households at the national, state, county, or city level, there are usually more factors for potential strata than one can accommodate with conventional single stratification designs. One usually chooses the "best" and sets aside, or ignores, those not-so-good. The possibility of utilizing other potential strata may simplify the choice problem and almost surely add to the effectiveness of the survey. The cost is that of obtaining and preparing the information needed and perhaps a loss in ability to obtain good estimates of sampling variance.

Methods of selecting cubic probability lattices are extensions of the same methods proposed for PNR sampling (one dimension) and square or other two-dimensional lattices described above, except for an increase in complexity. Details will not be given here, but some procedures and suggestion can be found in Jessen (1970 and 1975).

If properly replicated, an estimate of variance analogous to that for the square case, with similar limitations, is available. However, if it is important to have good estimates of sampling variance, one might wish to forego some of the controls in multistratified sampling.

11.12. Marginal Stratification

In Parts A and B of this chapter it was assumed that cell sizes were known. When they are not, but the margin sizes are, there may still be some incentive to attempt some sort of stratification, taking account of the information on margin sizes. This cases arises when we may be given information on a series of one-way classifications but none on two-way or crossed classifications. For example, we may know income distribution and ethnic distribution but not the distribution of income by ethnic classes.

Another example is that of students in a university for whom an alphabetical listing is readily obtainable but sex and class are recorded only in individual dossiers and are not generally made available, although a tabulation has been made of the total number by class and of totals by sex. The alphabetical list can be used to select individual students, and it is assumed that the class and sex information can be obtained from the dossiers for the selected students but that for a survey the additional information must be obtained from the students. Can this information be utilized to improve the sample?

A 2 × 2 Frame.   Suppose we can classify a universe into two rows and two columns with equal sizes, and we wish to select a sample of four. The marginal information suggests we select two from each row and at the same time two from each column. We have a simple listing for a sampling frame. How shall this be accomplished? Two cases can be distinguished: (1) controlling solely on marginal information and (2) utilizing sample information on cell sizes. Both are iterative.

In the first we draw a random sample of four elements and examine the marginal totals. If they are all twos, we can stop. If not, the excess should be omitted (at random) and replacements drawn. The process is repeated until a balance is achieved. Following are example iterations of two cases:

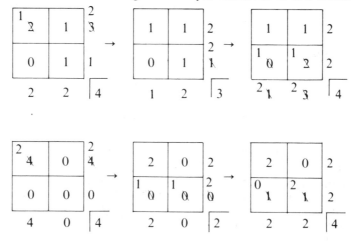

In the second case we try to utilize information on cell sizes. Let us draw an oversample of size $n' = 12$ and observe the distribution obtained. Here are some possible outcomes.

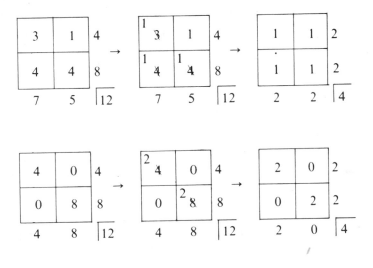

In the example above, where each cell is sampled, the presumption is that true cell sizes are essentially equal. In the second example we presume that only the diagonal cells are occupied and they are of equal size. A χ^2 test could be made to test the reasonableness of the configurations that turn up.

For any $R \times C$ Rectangle. For more realistic cases of $R \times C$ rectangles of varying marginal sizes, a method developed by Yates (1960) may be appropriate. The sampling variances for these cases apparently have not been explored, so their effectiveness is unknown.

◇   ◇   ◇

11.13. Constructing Frames for Lattices

The general principles for constructing frames for simple stratification apply to multiple stratification. However, in the former there was little incentive to try to equalize the sizes of strata. In multiple stratification, however, equal-sized strata, although not necessary, are certainly easier to deal with, and where flexibility exists equal sizes should be sought after. Where units can be ranked from high to low on the stratifying factor, this is usually no problem, except where unusually large elements exist. In this case the large ones can be made either individually, or in combination, into separate strata.

There is usually information somewhere around that can be used for stratification. A task of sample design is to sort them out and choose what appears most useful. For example, in household surveys, where census tracts are primary units, stratification factors might be median family income, percent owners, percent black population, percent with Spanish surname, and geographic location. These data can be obtained easily from United States Bureau of the Census publications. Where city blocks are being sampled, the stratification factors available in published form include similar information, except that income may have to be imputed from "home value" and/or "rent paid" and Spanish surname information is usually not available on a block basis.

Information such as this can be used to rank tracts or blocks from low to high. With number of households, or the equivalent, as a measure of size, fractiles of various levels and equality in size can be formed, as one might in simple stratification. With square or cubic lattices in prospect one might wish to equalize size to facilitate the selection, but it is not necessary.

To illustrate the many options in forming lattice frames, consider a household sample of the County of San Diego, Calif., where the tract is to be the first-stage frame unit. Let geography, family income, and ethnicity (combinations of black and Spanish surname population ratio) be the stratifying factors. The primary sample is to consist of 24 of the 316 tracts, hence up to 24 classes are available for each of the three factors. By constructing a 24^3 frame, one can decide later whether to re-form it into such alternatives as a 12^3 with two replications, either of two 12^2 frames with two replications and subsampling within the collapsed files, or perhaps any one of the three possible simple stratifications with 12 strata, or many possible schemes. The choice depends largely on how one comes out when he weighs the alternative prospective results with the purpose of the survey.

11.14. Discussion and Summary

In designing efficient samples, the sampler attempts to utilize all the information available to him. As we mentioned earlier, one place to utilize information is in stratification. But simple stratification, as discussed in Chapter 9, cannot always accommodate all the information that one may wish to use. Strata numbers cannot exceed sample size, and often many additional factors for stratification must be either ignored or only partially utilized in setting up strata. This chapter presented some schemes for going beyond usual simple stratification. The procedures are called multiple stratification, and the resulting samples are lattice samples.

For the cases where cell sizes are equal and equal selection probabilities can be effectively used, the mathematical theory has been worked out. However, where PPS sampling is used, such as in the cases where cells are unequal or even zeros, the theory on variance structure is not well developed. We have presented here the elements of the methods now available and only the rudiments of the theory. For further information the reader may wish to look into some of the references at the end of the chapter.

On the practical side there is strong evidence that lattices should be seriously considered whenever full utilization of prior knowledge in sampling design is believed important. More controls, if "sensible," are generally more effective than fewer controls if a single characteristic is being surveyed—and also if several characteristics are being surveyed, because then we have increased our chances that at least one of our several controls will be appropriate for each characteristic under study.

The lack of good estimates of sampling variance may not be much of a handicap where knowing that an estimate is more accurate than another is more important than knowing just how accurate it is. Usually, we are faced with a trade-off problem between these two goals. In the case where one does not wish to determine the magnitudes of individual sources of variance, although a good estimate of overall sampling variance is vital—such as in multistage sampling—replication within primaries, secondaries, and so on is generally not needed, since the between-primary mean square of the sample will include this component of variance with nearly its proper weight anyway. In these cases lattices should be looked into quite seriously.

11.15.   Review Illustrations

1.   Suppose we have a universe of people that can be divided into upper and lower halves according to age on one hand and income on the other. Suppose we are interested in their reaction to a new cola drink called Zooks and everyone is either for it (the 1s) or against it (the 0s), there being no in-betweens. Suppose also we know the preferences of everyone and they turn out to be as shown in the diagram; that is, everyone in the upper-income old folks are against Zooks, the lower-income old folks are all for it, and so on.

Income

		Upper	Lower
Age	Upper	A 0	B 1
	Lower	C 1	D 0

Consider the effects of using a quote or stratified sample of 100 persons in the following ways:

(1) 50 upper age and 50 lower age, no other control.
(2) 50 upper income and 50 lower income, no other control.
(3) 50 upper age and 50 lower age, *and* simultaneously 50 upper income and 50 lower income, no other control.
(4) 25 from each of the four cells.
(5) 100 at random from the whole universe.
(6) 50 from cell *A* and 50 from cell *B*.
(7) 50 from cell *B* and 50 from cell *C*.

(*a*) Which of the seven schemes will be the most precise? (There may be more than one.)
(*b*) Which will be the least?
(*c*) Which are the most accurate?
(*d*) What do you conclude about stratification and quota-setting policy?

Solution

We have the seven situations:

(1)
U | $p = .5$ | 50
L | $p = .5$ | 50

This is a one-way case, var $(\hat{p})$. $= 0.25/100$.

(2)

	U	L
	$p = .5$	$p = .5$
	50	50

This, too, is a one-way case, var $(\hat{p})$ $= 0.25/100$.

(3)

| | U | L |
| U | | | 50
| L | | | 50
| | 50 | 50 |

This is marginal stratification, but it appears there will be no stratification effect, hence var $(\hat{p}) = 0.25/100$.

(4)

| 25 | 25 |
| 25 | 25 |

This one-way stratification with four strata, but since s_W^2 is 0, var $(\hat{p}) = 0$.

(5)

| $p = .5$ |
| 100 |

This is a 0-way or simple random case, hence var $(\hat{p}) = 0.25/100$.

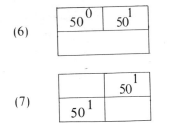

(6)

50 ^{0}	50 ^{1}

This is purposive selection, hence var $(\hat{p}) = 0$ and, fortunately, here bias = 0.

(7)

	50 ^{1}
50 ^{1}	

This is also a purposive selection; var $(\hat{p}) = 0$ but now bias = 0.5.

(a) Schemes (4), (6), and (7) are most precise (that is, have the least variance).

(b) Schemes (1), (2), (3), and (5).

(c) Schemes (4) and (6) are most accurate, since MSE $(\hat{p}) = 0$.

(d) One-way stratification on margins and marginal stratification were useless here. However, simple stratification by cells was very effective indeed. Quotas based on strata shown here would be equally effective with stratification.

2. The following data are from a latin lattice sample, $n = 12$, drawn from a 6 × 6 population.

(a) Estimate S^2_{RC} using the method of individual latins.

(b) Estimate S^2_{RC} using the ANOVA method. Confirm the results of (a).

(c) What is the estimate of S^2_{RC} if only A and B replicates were used?

	A_2				B_2
	13				25
A_1			B_1		
35			31		
		A_3		B_3	
		24		19	
B_1			A_1		
14			12		
	B_2				A_2
	7				20
		B_3		A_3	
		26		15	

Solution

(a) The $(A-B)$s for each of the three latins are $(47-45)$, $(33-32)$, $(39-45)$; the s^2_{RC}s are $(47-45)^2/4$, $(33-32)^2/4$, $(39-45)^2/4$, or 1, 0.25, 9; and pooled $s^2_{RC} = (1+0.25+9)/3 = 3.42$.

(b) The ANOVA layout is

Latin 1		Latin 2		Latin 3		$\sum$	
A	B	A	B	A	B	A	B
35	31	13	25	24	19		
12	14	20	7	15	26		
47	45	33	32	39	45	119	122
	92		65		84		241

Source	df	SS	MS
Gross	12	5607	
Mean	1	4840	
Total	11	767	
Subclasses	5	106.5	
Lattices	2	96.25	
Subclasses/latins	3	10.25	$3.42 = s_{RC}^2$
Replicates	1	0.83	
Replicates × latins	2	9.42	
Within subclasses	6	660.50	

The two methods produced the same $s_{RC}^2 = 3.42$.

(c) Using only the A and B replicates, the estimate would be the replicate mean square = 0.83.

3. A $4 \times 4 \times 4$ cubic population was constructed from the beetle data of Fig. 4.1 by splitting the field into quarters and then regarding the upper left quarter as file 1, upper right as file 2, lower right as file 3, and so on. Then each quarter (file) is regarded as consisting of four rows and columns (of 8×8 segments each). The data for the 65-cell cube are displayed below.

File 1				File 2			
231	295	328	280	131	116	113	78
277	324	326	397	224	157	140	99
175	198	187	182	306	274	230	162
157	128	112	154	188	201	212	154

File 3				File 4			
231	240	156	166	117	122	97	87
219	179	213	259	159	162	113	114
160	145	141	136	78	56	153	173
103	115	110	103	31	48	99	92

(a)   Calculate the ANOVA table.

(b)   Calculate var $(\bar{y})$ for an RNR sample of $n = 16$.

(c)   Calculate the var $(\bar{y})$ for a strat-random sample of $n = 16$, where strata are row-column combinations. Likewise for the cases where strata are row-file and column-file combinations.

(d)   Calculate the var $(\bar{y})$ for the square lattice $(L = 4)$, where the lattice structure is rows and columns, and eight row $\times$ column cells are selected as primaries with two clusters (6×6 segments) selected at random from each primary, hence $m = nm_0 = 16$. Likewise for the row $\times$ file and column $\times$ file lattices.

(e)   Calculate var $(\bar{y})$ for the cubic lattice $(L = 4)$, where $n = 16$ selected by schemes 2 (the nonlatin) and 3 (the latin case).

(f)   In the original two-dimensional form the variance S_{RC}^2 for an 8×8 lattice was 3403.7 on a 6×6 segment basis. Calculate the var $(\bar{y})$ for a lattice square with $n = 16$.

(g)   Summarize and comment.

Solution

(a)   The ANOVA table (on a per 6×6 cluster basis):

Source	df	SS	MS	MS defined as
Gross	64	2,226,911		
Mean	1	1,860,837		
Total	63	366,074	5811	S^2
Rows	3	57,886	19,295	
Columns	3	811	270	
Files	3	131,701	43.900	
$R \times C$	9	4906	545	$S_{RCF}^2 + 4S_{RC}^2$
$R \times F$	9	114,105	12,678	$S_{RCF}^2 + 4S_{RF}^2$
$C \times F$	9	22,516	2502	$S_{RCF}^2 + 4S_{CF}^2$
$R \times C \times F$	27	34,149	1265	S_{RCF}^2
...				
RC subclasses	15	63,603		
RF subclasses	15	303,692		
CF subclasses	15	155,027		
Within RC subclasses	48	302,471	6301	$S_{W(RC)}^2$
Within RF subclasses	48	62,382	1300	$S_{W(RF)}^2$
Within CF subclasses	48	211,047	4397	$S_{W(CF)}^2$
...				
Within $R \times C$s	54	361,168	6688	
Within $R \times F$s	54	251,969	4666	
Within $C \times F$s	54	343,558	6362	

(b)   RNR sampling:

$$\text{var} (\bar{y}) = \left(1 - \frac{n}{N}\right) \frac{S^2}{n} = \left(1 - \frac{16}{64}\right) \frac{5811}{16} = 272.4$$

(c)   Strat-random: strata are two-factor cells.

$$\text{var} (\bar{y}) = \left(1 - \frac{n}{N}\right) \frac{S_W^2}{n}$$

row-columns: $= \left(1 - \dfrac{16}{64}\right) \dfrac{6301}{16} = 295.4$

row-files: $= \left(1 = \dfrac{16}{64}\right) \dfrac{1300}{16} = 60.9$

column-files: $= \left(1 - \dfrac{16}{64}\right) \dfrac{4397}{16} = 206.1.$

(d)   Lattice squares with subsampling:

$$\text{var} (\bar{y}) = \left(1 - \frac{n}{N}\right) \frac{S_{RC}^2}{n} + \left(1 - \frac{nm_0}{NM_0}\right) \frac{S_{W(RC)}^2}{nm_0}$$

where $n = 8$, $m_0 = 2$; $S_{RC}^2 = [\text{MS} (R \times C) - S_{W(RC)}^2]/4 = (545 - 6301)/4,$

row-columns: $= \left(1 - \dfrac{8}{16}\right) \dfrac{-1439}{8} + \left(1 - \dfrac{16}{64}\right) \dfrac{6301}{16} = 205.5$

row-files: $= \left(1 - \dfrac{8}{16}\right) \dfrac{2845}{8} + \left(1 - \dfrac{16}{64}\right) \dfrac{1300}{16} = 238.7$

column-files: $= \left(1 - \dfrac{8}{16}\right) \dfrac{-474}{8} + \left(1 - \dfrac{16}{64}\right) \dfrac{4397}{16} = 176.5.$

(e)   Lattice cubes. For scheme 2:

$$\text{var} (\bar{y}) = \left(1 - \frac{n}{L^2}\right) \left[\frac{(S_{RC}^2 + S_{RF}^2 + S_{CF}^2)}{n}\right] + \left(1 - \frac{n}{L^3}\right) \frac{S_{RCL}^2}{n},$$

where $\quad S_{RC}^2 = (545.1 - 1265.8)/4 = -179.9,$

$\quad S_{RF}^2 = (12{,}678.3 - 1264.8)/4 = 2{,}853.4,$

$\quad S_{CF}^2 = (2501.8 - 1264.8)/4 = 309.3,$

$$= \left(1 - \frac{16}{16}\right)\left(\frac{2982.8}{16}\right) + \left(1 - \frac{16}{64}\right)\frac{1264.8}{16}$$

$$= 0 + 59.3 = 59.3.$$

For scheme 3:

$$\text{var}(\bar{y}) = \left(1 = \frac{2}{L}\right)\left[\frac{2(S_{RC}^2 + S_{RF}^2 + S_{CF}^2)}{4L}\right] + \left(1 - \frac{4}{L^2}\right)\frac{S_{RCF}}{4L}$$

$$= \left(1 - \frac{2}{4}\right)\left[\frac{2(2982.8)}{(4)\,(4)}\right] + \left(1 - \frac{4}{16}\right)\frac{1265}{(4)\,(4)}$$

$$= 11.7 + 59.3 = 71.0$$

$f)$ Lattice square on 8×8 frame:

$$\text{var}\,(\bar{y}) = \left(1 - \frac{n}{N}\right)\frac{S_{RC}^2}{n}, \qquad \text{where } S_{RC}^2 = 3403.7,$$

$$= \left(1 - \frac{16}{64}\right)\frac{3403.7}{16} = 159.6.$$

(g) In order of low to high on var $(\bar{y})$:

Design	var $(\bar{y})$
1. Lattice cube (no. 2)	59
2. Strat-random on RF subclasses	61
3. Lattice cube (no. 3)	71
4. Lattice square (8×8) on RC subclasses	160
5. Lattice square (4×4) on CF with subsampling	177
6. Lattice square (4×4) on RC with subsampling	206
7. Strat-random on CF subclasses	206
8. Lattice square (4×4) on RF with subsampling	239
9. RNR	273
10. Strat-random on RC subclasses	295

The var $(\bar{y})$s fall into three groups. The lowest group includes the cubic lattices and simple strat-random on both rows and files. The highest group includes square lattices on rows $\times$ files but with subsampling, RNR, and simple strat-random on rows and files—that is, no control, or control on the wrong factors or their combinations. The performance of the lattice square on RF is puzzling. Although rows and files were the right factor to control, apparently the $R \times F$ interaction was missed by this design, but caught by strat-random on the RF subclasses. The lattice squares were generally in the intermediate group. The important sources of variation were files, rows, and rows × files. The cubic lattices and the strat random on row-file classes did best in controlling these.

4. Here are some synthetic data in a 4×4 frame, where X_i and Y_i are the measures of size and observed values, respectively.

	i		
1	2	3	4
5	6	7	8
9	10	11	12
13	14	15	16

	X_i			Σ
0	3	5	2	10
5	0	2	3	10
3	2	1	4	10
2	5	2	1	10
Σ 10	10	10	10	$\overline{40}$

	Y_i			Σ
0	8	17	3	28
7	0	8	5	20
6	9	5	13	33
7	24	12	4	47
20	41	42	25	$\overline{128}$

(a) Compute a simple ANOVA table for this population of ys. Assuming that samples of $n = 4$ are selected with equal probabilities, compute var $(\hat{Y}_{ran})$, var $(\hat{Y}_{row})$, var $(\hat{Y}_{col})$, and var $(\hat{Y}_{lat})$.

(b) Consider sampling with PPS. For samples of $n = 4$, compute var $(\hat{Y}_{PR})$, var $(\hat{Y}_{PR-R})$, and var $(\hat{Y}_{PR-C})$, where PR − R and PR − C refer to stratification by row and column, respectively.

(c) Select a set of feasible probability lattices, $r = 1$, from the 4×4 frame. Compute var $(\hat{Y}_{P-lat})$ for this set.

(d) Summarize your results with regard to effectiveness of PPS versus PE and various degrees of stratification.

Solution

(a) The ANOVA table is as follows:

Source	df	SS	MS	MS defined as
Gross	16	1596.00		
Mean	1	1024.00		
Total	15	572.00	38.133	S^2
Rows	3	96.50	32.17	
Columns	3	93.50	31.17	
$R \times C$	9	382.00	42.44	S^2_{RC}
Elements/rows	12	475.50	39.625	$S^2_{W(R)}$
Elements/columns	12	478.50	39.875	$S^2_{W(C)}$

The variances are, since $N^2[(N-n)/Nn] = 48$,

$\text{var}(\hat{Y}_{ran}) = (48)S^2 = 1830.38$

$\text{var}(\hat{Y}_{col}) = (48)S^2_{W(C)} = 1914.00$

$\text{var}(\hat{Y}_{row}) = (48)S^2_{W(R)} = 1902.00$

$\text{var}(\hat{Y}_{lat}) = (48)S^2_{RC} = 2037.12$

(b)   When sampling is PPS, the analogous variances of the population and subpopulation involved are based on the relationship

$$\hat{S}^2 = \frac{\sum^N Y_i^2/P_i - Y^2}{N(N-1)} \qquad \text{(Eq. 8.6)}$$

In our case, $P_i = X_i/X$, and for the entire universe we have

		Y_i^2/X_i		$\sum$
—	$\dfrac{8^2}{3} = 21.33$	$\dfrac{17^2}{5} = 57.80$	$\dfrac{3^2}{2} = 4.50$	83.63
$\dfrac{7^2}{5} = 9.80$	—	$\dfrac{8^2}{2} = 32.00$	$\dfrac{5^2}{3} = 8.33$	50.13
$\dfrac{6^2}{3} = 12.00$	$\dfrac{9^2}{2} = 40.50$	$\dfrac{5^2}{1} = 25.00$	$\dfrac{13^2}{4} = 42.25$	119.75
$\dfrac{7^2}{2} = 24.50$	$\dfrac{24^2}{5} = 115.20$	$\dfrac{12^2}{2} = 72.00$	$\dfrac{4^2}{1} = 16.00$	227.70
$\sum$: 46.30	177.03	186.80	71.08	481.21

whence

$$\hat{S}^2 = \frac{40(\sum Y_i^2/X_i) - Y^2}{N(N-1)} = \frac{40(481.21) - (128)^2}{14(14-1)} = 15.738,$$

$$\hat{S}^2_w(\text{row } 1) = \frac{10(86.63) - (28)^2}{3(3-1)} = 8.72,$$

likewise

$$\check{S}_w^2(\text{row } 2) = 16.88, \quad \check{S}_W^2(\text{row } 3) = 9.04, \quad \check{S}_W^2(\text{row } 4) = 5.67;$$

$$S_W^2(\text{col } 1) = \frac{10(46.30 - (20)^2}{3(3-1)} = 10.50$$

and

$$S_W^2(\text{col } 2) = 14.88, \quad S_W^2(\text{column } 3) = 8.67, \quad S_W^2(\text{column } 4) = 7.15.$$

The required variance of $\hat{Y}$ with PR sampling of any n is given by Eq. 8.12b,

$$\text{var } (\hat{Y}_{PR}) = \frac{N(N-1)}{n} \overset{*}{S}{}^2 = \frac{14(14-1)}{4} (15.738) = 715.67,$$

and for stratifying by rows, var $(\hat{Y}_{PR})$ is the sum of the individual row variances, hence (not variations in strata sizes)

$$\text{var } (\hat{Y}_{PR-R}) = \frac{3(3-1)}{1} (8.72) + \frac{3(3-1)}{1} (16.88) + \frac{4(4-1)}{1} (9.04)$$

$$+ \frac{4(4-1)}{1} (5.67)$$

$$= 330.12.$$

Likewise for column strata,

var $(\hat{Y}_{PR-C}) = 342.12.$

(c)   Using the decremental method, we obtain

(1)

$$\begin{array}{cccc}
0 & 3 & ⑤ & 2 \\
⑤ & 0 & 2 & 3 \\
3 & 2 & 1 & ④ \\
2 & ⑤ & 2 & 1
\end{array}$$

max $\Delta_1 = 4$, $P_1 = \dfrac{4(4)}{40} = 0.4$

(2)

$$\begin{array}{cccc}
0 & ③ & 1 & 2 \\
1 & 0 & 2 & ③ \\
③ & 2 & 1 & — \\
2 & 1 & ② & 1
\end{array}$$

max $\Delta_2 = 2$, $P_2 = \dfrac{4(2)}{40} = 0.2$

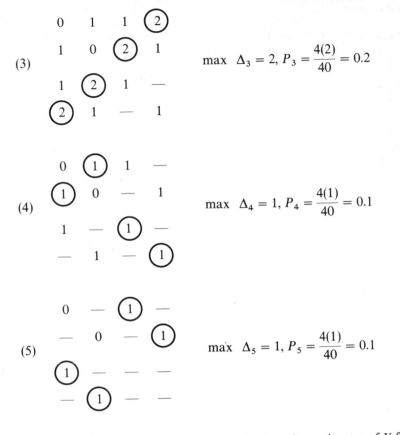

(3) $\text{max } \Delta_3 = 2, P_3 = \dfrac{4(2)}{40} = 0.2$

(4) $\text{max } \Delta_4 = 1, P_4 = \dfrac{4(1)}{40} = 0.1$

(5) $\text{max } \Delta_5 = 1, P_5 = \dfrac{4(1)}{40} = 0.1$

To compute var $(\hat{Y}_{\text{P–lat}})$ we first determine each of the five estimates of Y from the feasible set of five samples, where from Eq. 8.65b for **PNR** sampling,

$$\hat{Y} = \sum \frac{Y_i}{P_i} = \frac{X}{n}\sum \frac{Y_i}{X_i} = \frac{40}{4}\sum \frac{Y_i}{X_i} = 10\sum \frac{Y_i}{X_i},$$

since $P_i = nX_i/X$. The basic Y_i/X_is are:

Y_i/X_i

—	$\frac{8}{3}$	$\frac{17}{5}$	$\frac{3}{2}$
$\frac{7}{5}$	—	$\frac{8}{2}$	$\frac{5}{3}$
$\frac{6}{3}$	$\frac{9}{2}$	$\frac{5}{1}$	$\frac{13}{4}$
$\frac{7}{2}$	$\frac{24}{5}$	$\frac{12}{2}$	$\frac{4}{1}$

Feasible sample

t	Configuration	P_t	$10 \sum Y_i/X_i$	$\hat{Y}_t$	$P_t\hat{Y}_t$	$\hat{Y}-Y$	$(\hat{Y}-Y)^2$	$P_t(\hat{Y}-Y)^2$
1		.4	$10(\frac{17}{5}, \frac{7}{5}, \frac{13}{4}, \frac{24}{5})$	128.5	51.40	0.5	0.25	0.100
2		.2	$10(\frac{8}{3}, \frac{5}{3}, \frac{6}{3}, \frac{12}{2})$	123.3	24.66	-4.7	22.09	4.418
3		.2	$10(\frac{3}{2}, \frac{8}{2}, \frac{9}{2}, \frac{7}{2})$	135.0	27.00	$+7.0$	49.00	9.800
4		.1	$10(\frac{8}{3}, \frac{7}{5}, \frac{5}{1}, \frac{4}{1})$	130.7	13.07	$+2.7$	7.29	0.729
5		.1	$10(\frac{17}{5}, \frac{5}{3}, \frac{6}{3}, \frac{24}{5})$	118.7	11.87	-9.3	86.49	8.649
$\sum$		1.0			128.00			23.696

whence var $(\hat{Y}_{\text{P-lat}}) = \sum P_s(\hat{Y}-Y)^2 = 23.696$.

(d) To summarize these results, we have

	var $(\hat{Y})\|n = 4$	
Strata constraints	RNR	PR
No stratification	1830	716
Row stratification	1902	330
Column stratification	1914	340
Lattice	2037	23.7[a]

[a]In this case sampling is PNR.

It can be seen that for this population PR is much more efficient than RNR sampling regardless of stratification. Also, stratification, whether one-way or two-way, was not effective when selection probabilities were equal, but one-way stratification on either rows or columns reduces variance about one-half for PR sampling. The effectiveness of a probability lattice in this case is very large indeed. Since the data are synthetic, they are not to be regarded as typical of the real world. However, when it is observed that the Y_i/X_i have row and column "components," it may be concluded that probability lattice will be effective. In our case we have as simple averages of the Y_i/X_i by row and column:

Y_i/X_i				Σ	Means
(1.30)	2.67	3.40	1.50	8.87	2.22
1.40	(3.50)	4.00	1.67	10.57	2.64
2.00	4.50	5.00	3.25	14.75	3.69
3.50	4.80	6.00	4.00	18.30	4.56

Σ	8.20	15.49	18.40	10.42	52.49
Mean	2.05	3.87	4.60	2.60	

Note that there are prominent row and column components to the Y_i/X_i. The figures in parentheses are rough estimates of "missing values" for the empty cells, in order to obtain more valid row and column totals.

5. A hospital has 46 nurses, of whom 16 are to be selected for a sample. It is decided that there are four categories of class and type of nurse to be recognized and three categories of work shifts. A two-way arrangement of these categories gives the distribution as follows:

		Class				
		1	2	3	4	Σ
	1	13	4	3	1	21
Shift	2	10	2	3		15
	3	6		2	2	10
		29	6	8	3	46

(a) Set up the decremental methods tableau and designate a set of feasible lattice samples.

(b) Draw a feasible sample from the set in (a) and tell how it is to be employed here.

Solution

(a) The·table of nA_i is

13 (4.51)	4 (1.39)	3 (1.04)	1 (0.35)	21 (7.29)
10 (3.48)	2 (0.70)	3 (1.04)		15 (5.22)
6 (2.09)		2 (0.70)	2 (0.70)	10 (3.49)
29 (10.09)	6 (2.09)	8 (2.78)	3 (1.05)	46 (16.00)

Removing the certainty portions and proceeding with the standard tableau for decremental method 2, we have the following 10 feasible samples and their assigned probabilities of solution.

The decremental method:

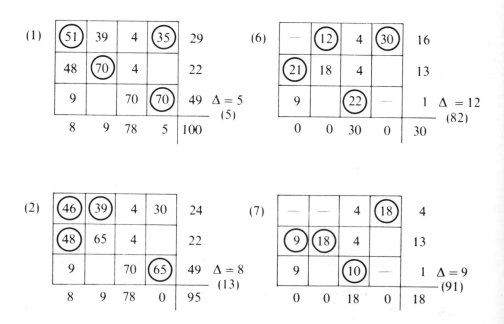

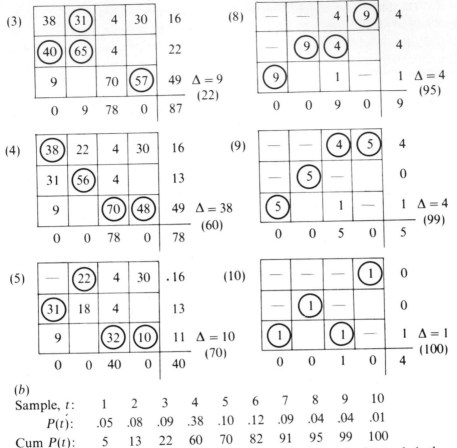

(b)

Sample, t: 1 2 3 4 5 6 7 8 9 10

$P(t)$: .05 .08 .09 .38 .10 .12 .09 .04 .04 .01

Cum $P(t)$: 5 13 22 60 70 82 91 95 99 100

Say $1 < RN < 100 = 46$; then $t = 4$ is designated. The selected sample is then

13* (5)	4 (1)	3 (1)	1 (0)	21 (7)
10 (3)	2* (1)	3 (1)		15 (5)
6 (2)		2* (1)	2* (1)	10 (4)
29 (10)	6 (2)	8 (3)	3 (1)	46 (16)

The starred cells are those receiving the "extra" allocation.

6. The following data represent the distribution of the number of households (1960) in the 12 tracts of Santa Monica, Calif., in a cubic frame of rough quartiles classifying the tracts on the following factors: median family income, percentage of home owners, and percentage of persons that are black. A sample of four tracts is desired with PPS (on HHs).

   (a) Using decremental method 2, designate as many feasible cubic lattice samples of 4 as you can. (The figures in parentheses are the calculated nA_is.)

   (b) Discuss why a complete set of cubic lattices cannot be found here.

Percent owners Σ

Income

					Σ
		2123 (0.24)			
2701 (0.31)	2706 (0.31)	2042 (0.23)			9572 (1.09)
	3330 (0.38)		3817 (0.44)		
					7147 (0.82)
		2423 (0.28)	4081 (0.47)		
				2963 (0.34)	9467 (1.08)
					2699 (0.31)
				3583 (0.41)	2602 (0.30)
					8884 (1.01)
8737 (1.00)	10,669 (1.22)	6780 (0.77)	8884 (1.01)		35,070 (4.00)

Percent black

8363 (0.95)	8767 (1.15)
8326 (0.95)	8271 (0.94)

6. *Solution* (*a*) A set of feasible samples follow:

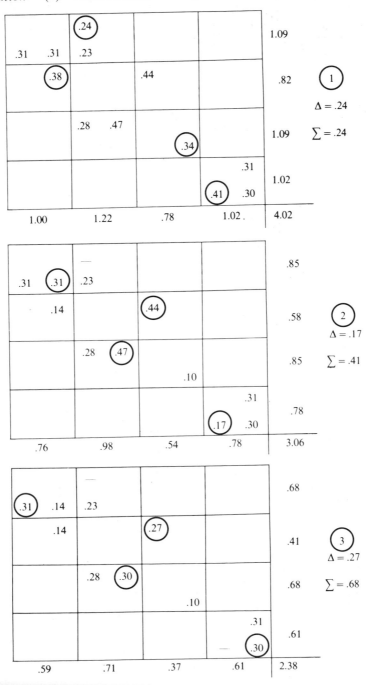

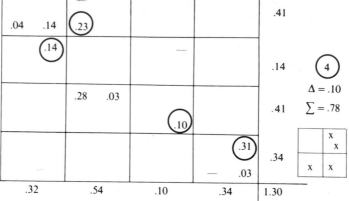

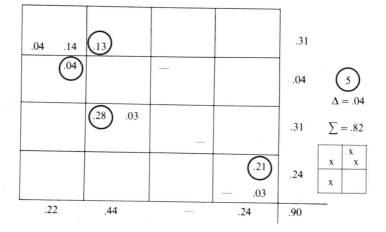

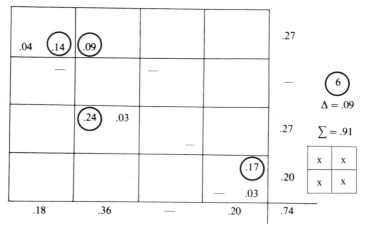

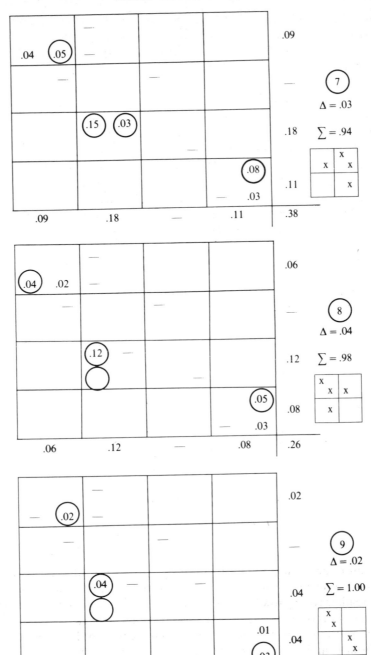

(*b*) Only the first three and 6 are cubic lattices. The others are off-cubes in one way or other. The reason is that we are not dealing with a perfect cube, as can be seen by examining the nonequality of the margins. However, 77 % of the probability in the feasible set is cubic.

7. In December 1969 the United States introduced a lottery for calling draftees for the military forces. The procedure adopted was essentially the random nonreplacement selection of the 366 days in a calendar year. Following the selection, a number of critics claimed the procedure did not produce a random order of the 366 days—for example, that early call-up dates were heavily concentrated in the first six calendar months.

(*a*) Devise a sampling scheme that should be free of such criticisms.

(*b*) Do persons born on February 29 have an equal chance of being drafted as others? State necessary or sufficient conditions.

Solution

(*a*) Set up a 12×31 rectangle and number as follows:

			Month			
Day	1	2	3	4	. . .	12
1	1	30	28	26	. . .	10
2	2	31	29	27		
3	3	1	30	28		
		2	31	29		
			1	30		
.	.	.		31		.
.	.	.		1		.
.	.	.				.
31	29	27				9

Randomize rows and columns. Regard the new order from top to bottom as the order group and the new column order as the ordering within order group. Then treading row by row from left to right will place every birthday in random order but with controls. Each 12 numbers will be in a different month, each 31 will be a different day of the month. Regard noncalendar days (e.g., 30 and 31 of February, and so on) as real days but with zero births!

(*b*) Yes.

11.16   References

Bryant, E. C.   Design and Estimation in Two-Way Stratification. *Journal of the American*
Hartley, H. O.   *Statistical Association.* **55**: 105–124.
Jessen, R. J.
1960

Dalenius, T.   Contributions to Statistics: Lattice Sampling by Means of Lahiri's
1963   Sampling Scheme. New York: Pergamon Press.

Deming, W. E.   On a Least Squares Adjustment of a Sampled Frequency Table When the
Stephan, F.   Expected Marginal Totals Are Known. *Annals of Mathematical Statistics.*
1940   **11**: 427–444.

Frankel, L. R.   On the Sample Survey of Unemployment. *Journal of the American*
Stock, J. S.   *Statistical Association*, **37**: 77–80.
1942

Goodman, J. R.   Controlled Selection—A Technique in Probability Sampling. *Journal of the*
Kish, L.   *American Statistical Association*, **45**: 350–372. (A pioneer paper on the
1950   selection of probability lattices.)

Hess, I.   *Probability Sampling of Hospitals and Patients.* (86 pp.) Ann Arbor:
Riedel, D. C.   University of Michigan.
Fitzpatrick, T. B.
1961

Jessen, R. J.   Probability Sampling with Marginal Constraints. *Journal of the American*
1970   *Statistical Association*, **63**: 776–796. (Suggests procedures for selecting
                       probability lattices.)

Jessen, R. J.   Square and Cubic Lattice Sampling. *Biometrics*, **31**: 449–471.
1975

Keyfitz, N.   The Canadian Sample for Labor Force and Population Data. *Population*
Robinson, H. L.   *Studies*, (2): 427–443.
1949

Maloney, C.   *Stratification in Survey Sampling.* Ph.D. Dissertation. (142 pp.) Ames, Iowa:
1948   Iowa State University Library.

Patterson, H. D.   The Errors of Lattice Sampling. *Journal of the Royal Statistical Society*,
1954   **B16**: 140–149. (A fundamental paper on PE lattice sampling.)

Sumner, G. C.   Examination of the Usefulness of Three-Way Probability Lattice Sampling
1973   for Household Populations. Ph.D. Dissertation. (168 pp.) University of
                       California, Los Angeles.

Tepping, B. J.   On the Efficiency of Deep Stratification in Block Sampling. *Journal of the*
Hurwitz, W. H.   *American Statistical Association*, **38**: 93–100.
Deming, W. E.
1943

Williams, W. H.   Sample Selection and the Choice of Estimator in Two-Way Stratified
1964   Populations. *Journal of the American Statistical Association*, **59**: 1054–1062.
                       (December).

Yates, F.   *Sampling Methods for Censuses and Surveys.* (3rd ed.) London: Charles
1953, 1960   Griffin & Co. (1st ed., 1949; 2nd, 1953; 3rd, 1960.)

11.17. Exercises

1. Refer to Section 11.2. For the data in the 4×4 square, determine the E_{gh} and show that $S^2_{RC} = 1.78$. Confirm for a single lattice ($r = 1$) that var $(\bar{y}_{lat})$ = 0.333.

2. Here is a synthetic microuniverse of 16 observations arranged in a 4×4 square:

		Column			
Row	1	2	3	4	Total
1	9	8	6	7	30
2	9	5	7	3	24
3	10	1	7	4	22
4	14	6	10	18	48
Total	42	20	30	32	124

(a) Summarize the relevant computations in an ANOVA table. Suppose samples of $n = 4$ are to be taken. Compare the efficiency of the following schemes relative to that of simple random (RNR) selection:
(b) Stratification by rows.
(c) Stratification by columns.
(d) Two-way stratification (lattice sampling).

3. A lattice sample of size 12 was randomly selected from a 6×6 square with the following results:

13		24			
	32				35
	20		27		
14				2	
20				7	
		26			27

(a) Estimate the population mean and its variance. For the population from which this sample was drawn, estimate the variance which would be obtained if sampling were:

(b) Simple stratified with columns as strata.

(c) Compare the estimated efficiency of the lattice relative to simple stratification on columns.

4. Here is a set of data in a 5×5 square. Using the data as a measure of size, draw a set of feasible probability lattices of $r = 2$ and determine their corresponding selection probabilities.

	1	2	3	4	5	Σ
1	2	2	4	6	6	20
2	4	2	5	2	7	20
3	5	5	2	3	5	20
4	0	3	6	9	2	20
5	9	8	3	0	0	20
Σ	20	20	20	20	20	100

5. Suppose we have a universe of people that when classified into high and low income and old and young age groups has the same number of persons in each group. Moreover we are interested in sampling this universe to determine the proportion, P, in favor of proposition X. The pollster, of course, does not know the true proportions for each group, but you do know, and they are as follows:

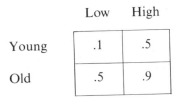

	Low	High
Young	.1	.5
Old	.5	.9

The pollster plans to draw a sample of 100 persons. What mean square error will he obtain if his sample is:

(a) A two-way stratified design on the margins?

(b) A one-way stratified design on age?

(c) A one-way stratified design on income?

(d) A complete stratified design on cells?

CHAPTER 12

Miscellaneous Survey Techniques

12.1. Introduction

In this chapter some schemes and techniques will be presented to deal with standard problems of statistical surveys in a somewhat more sophisticated and more efficient manner and, more important, to deal with a number of problems that arise because of unusual situations or uncommon objectives of the survey. Dealing with the former are more elaborate methods of systematic selection and the use of multiple frames. Special problems include unusual cost situations, multiple populations, multiple purposes of survey, and sampling populations through time. Moreover, the concept of the observation unit (OU) and its relationship to the survey unit will be looked into and some methods of dealing with incomplete OUs as a means of avoiding the full revelation of sensitive information will be considered. The problem of missing data in surveys—its consequences and what to do about it—will be briefly dealt with. In order to keep this chapter reasonably short, the techniques will be dealt with briefly and references will be given for further information.

12.2. Systematic Selection

Section 4.6 presented the nature and some of the properties of systematic selection. Here we shall briefly consider them in a little different way, extending the scheme of systematic selection to the case where probabilities of inclusion are desired to be proportional to some measure of size. (See Section 8.10.)

To apply systematic selection in its most common form, the N elements in the universe of interest are numbered serially from 1 to N. If a sampling fraction of 10% is desired, then some number between 1 and 10 is selected (preferably at random), say 3, whence elements 3, 13 ($= 10 + 3$), 23 ($= 2 \times 10 + 3$), 33 ($= 3 \times 10 + 3$), and so on through all N numbers constitute the sample. When the elements are unnumbered but are in a real or imagined linear order (such as file cards), the systematic selection can be carried out by simply counting. Three schemes are distinguished.

(a) PE Selection. Fixed k ; Intraclass Correlation.   In order to examine the nature of systematic selection, let us regard the N elements in the universe as arranged in a two-way frame consisting of k rows each with m elements (assuming for the moment that $N = mk$), where the original numbering: 1, 2, 3, ..., N is replaced by putting elements 1, 2, ..., k in column 1; elements $k + 1, k + 2, ..., 2k$ in column 2, and so on. Let Y_{ij} be the observed y in the ith row ($i = 1, 2, ..., k$) and jth column ($j = 1, 2, ..., m$). Tabularly we have

Column

Row	1	2	3	...	m	Σ
1	Y_{11}	Y_{12}	Y_{13}		Y_{1m}	Y_1
2	Y_{21}	Y_{22}	Y_{23}		Y_{2m}	Y_2
3	Y_{31}	Y_{32}	Y_{33}		Y_{3m}	Y_3
$\vdots$						
k	Y_{k1}	Y_{k2}	Y_{k3}		Y_{km}	$Y_{k.}$
Σ						$Y_{..}$

In systematic selection where the sampling fraction is $1/k$, we in effect select all the elements in one of the k rows—hence one of the k clusters that systematic selection generates.

If we define the sample mean as in cluster sampling, we have for any selected cluster, i,

$$(12.1) \qquad \bar{\bar{y}}_{\text{sys}} = \frac{Y_{i.}}{m},$$

the variance of which is given by

$$(12.2) \qquad \text{var}(\bar{\bar{y}}_{\text{sys}}) = \left(\frac{k-1}{k}\right)\frac{S_c^2}{m^2},$$

where

$$S_c^2 = \left(\frac{1}{k-1}\right)\sum_{i=1}^{k}(Y_i - \bar{Y})^2 \quad \text{and} \quad \bar{Y} = \frac{Y}{k}.$$

At times it is convenient to express Eq. (12.2) in terms of intraclass correlation, whence it becomes

$$(12.2a) \qquad \text{var}(\bar{\bar{y}}_{\text{sys}}) = \left(\frac{k-1}{km}\right)S^2[1 + (m-1)\delta],$$

where S^2 is the variance of the y-population over all $N = mk$ elements and, from section 4.6,

$$(12.3) \qquad \delta = \frac{[(N-1)/N]\,\text{MS}(B) - \text{MS}(W)}{mS^2}.$$

Since δ depends on the degree to which the y-values within the k clusters are "alike," its value will depend on the basic nature of the y-values relative to their ordering. If the order is random, δ will have an expected value of 0. If the y-values within each cluster are all the same, then $\delta = +1$ and Eq. 12.2a will be at its maximum. In general most ordered natural populations appear to have a $\delta < 0$, and therefore systematic selection produces more efficient samples than random samples (that is, Eq. 12.2a with $\delta = 0$).

It may be noted that if N/k is not an integer, the resulting samples will contain either n or $n + 1$ elements. In this case $\bar{y}_{sys}$ will not estimate $\bar{Y}$ unbiasedly, but generally the bias, if any, will be trivial.

(b) PE Selection. Fixed n.   In a number of situations one would prefer to fix n and let k be a noninteger, if it should work out that way, and yet assure each element an equal probability of inclusion. In this case we must know N (which is not necessary when we fix k). A suitable procedure is as follows.

1. Calculate $k = N/n$.
2. Choose random number RN, where $0 < RN \leq k$.
3. Calculate sample numbers SN, where
$$SN = RN + (t - 1)k, \qquad (t = 1, 2, ..., n).$$
4. Round the SNs of step 3 up to nearest whole number. These are the required elements in the sample.

Example 12.1.   Select a systematic sample of $n = 3$ from $N = 5$ elements. Using the foregoing steps, we have

1. $k = N/n = \frac{5}{3} = 1\frac{2}{3}$ (or 1.67 approximately).
2. $(0 < RN < 1\frac{2}{3}) = 0.35$, say.
3. For unrounded values we have $SN = RN + (t - 1)k$:

t	$(t-1)k$	RN, unrounded	RN, rounded
1	$0 = 0.00$	0.35	1
2	$1\frac{2}{3} = 1.67$	2·02	3
3	$3\frac{1}{3} = 3.33$	3.65	4

4. Hence the selected elements are 1, 3, and 4.

(c) PPS Selection.   Systematic selection procedures can be easily extended to case where elements are selected with PPS (probabilities proportional to "size"). In the more general case where k may or may not be an integer and X_i is the measure of size of element i (where $i = 1, 2, ..., N$) we now define k

$$k = \frac{X}{n},$$

where $X = \sum X_i$. Again let the random start number, RN, be $0 < RN < k$; then the sample numbers on x will be

$$(SN)_X = RN + (t - 1)k.$$

If the X_i are whole numbers, then the $(SN)_X$ may be rounded up to the nearest whole number. The $(SN)_X$ can be converted into the identification numbers of the N elements by the relationship

$$(SN)_N = i \quad \text{if cum } X_{i-1} \leq (SN)_N \leq \text{cum } X_i,$$

where cum $X_i = \sum_{i=1}^{i} X_i$.

Example 12.2. Select a systematic sample of $n = 3$ from the $N = 8$ with PPS. The data are:

Element, i:	1	2	3	4	5	6	7	8
X_i:	3	10	20	15	12	8	15	17

Here $X = 100$, $k = X/n = \frac{80}{3} = 26\frac{2}{3}$. Let $(0 < RN \leq 26\frac{2}{3}) = 8.35$. Then

t	$(t-1)k$	$(RN)_X$, unrounded	$(RN)_X$, rounded
1	0	8.35	9
2	$26\frac{2}{3}$	35.02	36
3	$53\frac{1}{3}$	61.68	62

To convert the $(RN)_X$ to $(RN)_N$ we calculate the cum X_i:

i	X_i	cum X_i	$(SN)_X$	$(SN)_N$
1	3	3		
2	10	13	9	2
3	20	33		
4	15	48	36	4
5	12	60		
6	8	68	62	6
7	15	83		
8	17	100		

For $(SN)_X = 9$, we note by scanning the cum X_i column that $(SN)_X$s 1, 2, and 3 designate element 1, that numbers $4, 5, ..., 13$ designate element 2, $14, 15, ..., 33$ designate 3, and so on. Hence we obtain elements 2, 4, and 6 as the selected elements.

Note that element 1 will be selected with all RNs from 0.01 to 3.00 out of a total set of 26.67. Hence $P_1 = 3/26.67 = X_1/k = nX_1/X$. Likewise $P_2 = (3.00 - 13.00)/26.67 = 10/26.67 = nX_2/X$. Hence $P_i = nX_i/X$, the necessary condition for PPS selection.

Estimating var $(\bar{\bar{y}}_{sys})$. Since systematic selection, as it is most commonly carried out, selects only one of k possible clusters, it is not possible to estimate sampling variance from the sample. Various techniques have been devised to provide acceptable approximations. Perhaps the most common one is to regard the systematic sample as comprising $k/2$ strata, where stratum 1 contains sample elements 1 and 2, stratum 2 contains elements 3 and 4, and so on. (If $k/2$ is odd, the singleton stratum can be ignored or combined with its neighbor to make a stratum of $n_h = 3$.)

Regarding the sample as stratified-random, $\widehat{\text{var}}(\bar{y}_{\text{strat}})$ is computed accordingly. Generally this estimator will overestimate, but not always, since in effect neighboring strata of $n_h = 2$, and hence differences between the substrata are included in the estimates. If one really wants a good estimate of $\text{var}(\bar{y})$ from his sample, he should avoid systematic selection.

An alternative procedure is to employ a k twice as large as standard and then select two random start numbers. An estimate of $\text{var}(\bar{\bar{y}}_{\text{sys}})$ is given by

$$(12.4) \qquad \widehat{\text{var}}(\bar{\bar{y}}_{\text{sys}}) = \left(\frac{k-2}{k}\right)\left(\frac{\bar{\bar{y}}_1 - \bar{\bar{y}}_2}{2}\right)^2,$$

where $\bar{\bar{y}}_1$ and $\bar{\bar{y}}_2$ are the means per element for starts 1 and 2, respectively. The extension to more than two start numbers should be fairly easy if one regards the sample as comprising a set of one or more clusters.

Replicated systematic sampling, such as that just described, is usually not as efficient as the single replicate or standard scheme, since the "quasi" strata on which it is based are larger and hence more variable. Moreover, one or two degrees of freedom on which to base the variance estimate may be inadequate for the purposes in mind. If this is so, stratified-random may be a more desirable alternative.

12.3. Cost-Oriented Schemes; Rare Elements

Some surveys are particularly irksome to design (or finance!) because of heavy costs for doing certain operations presumed to be required. These are essentially of three types: (i) those with heavy costs of seeking out and identifying the elements to be surveyed, sometimes called the rare-element case, (ii) those with heavy costs of making the observations themselves, and (iii) those for which the planner is quite uncertain what the costs of the various operations might be. Each case will be considered individually.

High Search Costs; Rare Elements.   Surveys of certain special groups may be particularly difficult to deal with because of high costs and other problems. A "special group" may be, for example, "homes with children under 5," "Mexican-American homes," "mentally retarded persons," "persons afflicted with gout-type diseases," "dairy farms," "dealers in farm machinery and equipment," "community leaders," or "hunters and fishermen." Generally a special group is a subgroup of some larger and generally more framable group, such as households in general, farms in general, and business establishments in general. The elements in the subgroup are not identified prior to the proposed survey, it being part of the burden of the survey to seek out and identify the subgroup membership. This search procedure, called "screening," "filtering," "sifting," or the like, if carefully carried out can be expensive and leave very little of the budget for obtaining observations on the subgroup itself.

To deal with the rare-element problem, practitioners generally do two kinds of things: (i) employ all the appropriate and generally known principles of sampling (such as choice of SU and its size, use of measure-of-size information, stratification, double sampling, and so on) and (ii) look for hitherto unknown information on lists, estimates on the likely distribution pattern of the rare elements, innovation of more efficient search techniques, and so on and put them to full use on the problem at hand. Techniques of the first type and their use can be summarized briefly as follows for different levels of rarity fractions and evenness of SU size in terms of the rare element. Let M_i, M_i', and $\hat{M}_i'$ be the number of gross elements (those in the frame being sampled), the rare elements, and the estimated (from MOS information) rare elements in the ith SU, respectively.

Rare-group fraction	Distribution of M_i'	
	Fairly equal	Quite equal
Large (say $\frac{1}{5}$ or more)	Equalize cluster size to $\hat{M}_i'$ to optimum: $\hat{M}_0'$. Stratify conventionally.	Try to obtain $\hat{M}_i'$; use to equalize SU size or as PPS. Stratify conventionally.
Small (say $\frac{1}{20}$ or less)	Same as above. Look for a very efficient search procedure.	Same as above except stratify by "density," $\hat{M}_i'/M_i$. Allocate $n_h \propto N_h/(c_h)^{1/2}$. Look for very efficient search procedure. Try multiple frames, multiple reporting, and double sampling.

Multiple frames and multiple reporting will be discussed below. The remaining techniques are applications of conventional techniques and principles to a case where costs of search are high and the need for some auxiliary information (such as an MOS) is unusually great.

High Determination Costs; "Composite" Sampling. In some investigations the cost of making the observation can be unusually high—for example, in chemically or biologically determining the vitamin content of foods, the chemical content of soils, or the nature and extent of pollutants in air, water, or food. In these cases it may be possible to aggregate the individual batches of material gathered from the various sample sites into one big batch, grind and mix it into a fairly homogenous slurry, and take perhaps a small portion for a single determination. Compositing the *n* elements of the sample this way to produce a single observation may be the most practical way to estimate the mean content. However, unless appropriate modifications are made, there is no valid estimate of sources of sampling variation.

Unknown Operational Costs; Two-Step Sampling. Costs of performing each of the operations in a survey—particularly the large-scale nonrepetitive ones—and of feasible alternative procedures are seldom known with suitable accuracy. Perhaps experience and shrewd guessing come close enough for most purposes. In cases where the planner feels he needs more information on costs or other matters, he will want a pilot survey. When properly designed and analyzed, the pilot survey can be of great help in reducing uncertainties on costs and other problems of data collection. Since pilot surveys cost money, take time, and therefore usually reduce the funds and time available for the survey proper, a certain amount of reasoning can be put to good use in their planning. For the situation where the pilot survey is viewed as the first step of a two-step survey, being used to determine costs which in turn are used to determine the optimal design for step two, Marshall (1956) has examined the problem of determining the size of the step-two sample that will have a calculated risk of exceeding a certain budget limit, or, alternatively, the size of budget needed for achieving a desired overall accuracy for an estimate. Because of learning effects and scale effects, the practical problem can be very complex.

12.4. Sampling Through Time; Panels

Many surveys conducted by agencies of the U.S. government require that estimates be made at various points in time, such as numbers of unemployed each month, retail sales each month, and numbers of livestock, by type, on farms each January 1. Likewise, opinion surveys are conducted at regular or irregular intervals during a political campaign, or opinions are desired on some rather continuing public issues (e.g., stand on death penalty, stand on confidence in the government) at least once each year, or more frequently. Because of the dynamic nature of these and many other matters that we wish to know about, a one-time determination is of limited value, even if ascertained very precisely, such as by a census. What we prefer is a series of investigations to determine and follow trends in certain characteristics through time.

When several investigations are to be made on the same universe through time, the survey planner has a number of new problems to consider. Should the samples be independent of each other? should one keep the same sample but make new observations on each occasion (sometimes called a panel)? or should there be some mixture of the two, where on each new occasion part of the sample is old and part new (hence some kind of changeover policy)?

Purposes of the Investigations. The purposes of sampling through time may be any one of three individual purposes or some combination of them. We may wish to estimate (1) the change in Y or $\bar{Y}$ from one time to another, (2) the general average of Y or $\bar{Y}$ over the whole period, or (3) the value of $\bar{Y}$ on each occasion.

In general, to achieve purpose (1) it would be best to use the same sample on both occasions—that is, a panel. For purpose (2) it would be best to use independently drawn samples each time. However, for (3) it would be best to use a combination of the two schemes. It is assumed here that repeated observations on the same unit have no effect on its subsequent behaviour nor on the accuracy of its measurement. If these assumptions do not hold, there still may be good reasons to prefer the use of one or the other of those sampling schemes. Moreover, costs are usually different for observing elements the first and subsequent times, and this has a bearing on choosing the most efficient sampling plan.

The Two-Occasion Case. Suppose a sample of n elements is taken on both occasions 1 and 2, but on occasion 2 the sample comprises m ("matched") elements from the previous sample and u ("unmatched") drawn independently such that $u + m = n$. We are observing some characteristic, y, and wish to estimate $\bar{Y}_1$ and $\bar{Y}_2$ for the two occasions, respectively. To estimate $\bar{Y}_1$ we have

$$\bar{y}_1 = \frac{\sum^n Y_i}{n}$$

and for $\bar{Y}_2$ we have two estimators, one for each of the matched and unmatched portions of that sample. Hence

(12.5)
$$\bar{y}_{2u} = \frac{\sum^u Y_i}{u}$$

with variance

$$\frac{S^2}{n} = \frac{1}{W_u}$$

and

(12.6) $$\bar{y}_{2m} = \bar{y}_{1m} + b(\bar{y}_1 - \bar{y}_{1m})$$

with variance

$$\frac{S^2(1-\rho^2)}{m} + \rho^2 \frac{S^2}{n} = \frac{1}{W_m},$$

where $\bar{y}_{2u}$ = the mean of the unmatched portion of the sample in time 2,
    $\bar{y}_{2m}$ = the mean of the matched portion of the sample in time 2,
    $\bar{y}_{1m}$ = the mean of the matched portion of the sample in time 1,
    S^2 = the variance of y, assumed the same on each occasion,
    ρ = the correlation of ys between the two occasions.
Note that Eq. 12.6 regards the two-occasion sample as a form of double sampling, where the first-phase sample is the entire sample of n in time 1 and the second phase is the m portion of time 2.

Combining the two estimates of $\bar{Y}_2$ provided by the matched and unmatched portions of the sample by weighting them inversely to their variances, we obtain

(12.7) $$\bar{y}_2 = \frac{W_u \bar{y}_{2u} + W_m \bar{y}_{2m}}{W_u + W_m}$$

with the variance

(12.8) $$\operatorname{var}(\bar{y}_2) = \frac{S^2}{n}\left(\frac{1 - (u/n)\rho^2}{1 - (u/n)^2 \rho^2}\right).$$

Suppose we wish our estimates of $\bar{Y}_1$ and $\bar{Y}_2$ to have equal accuracy. What fraction of the second-occasion sample should be matched? In this case, by minimizing Eq. 12.8 with respect to u, we obtain

(12.9) $$\frac{u}{n} = \frac{1}{1 + \sqrt{1 - \rho^2}},$$

$$\frac{m}{n} = \frac{\sqrt{1 - \rho^2}}{1 + \sqrt{1 - \rho^2}},$$

and the variance of $\bar{y}_2$, with optimal m for a fixed n, say $\bar{y}_2^*$, is given by

(12.10) $$\operatorname{var}(\bar{y}_2^*) = \frac{S^2}{2n}(1 + \sqrt{1 - \rho^2}).$$

The efficiency of the matched estimator, $\bar{y}_2$, relative to the unmatched estimator, $\bar{y}$, is given by

$$(12.11) \quad RE\,(\text{matched/unmatched}) = \frac{1 - (u/n)^2 \rho^2}{1 - (u/n)\rho^2} = \frac{1 - (1 - m/n)^2 \rho^2}{1 - (1 - m/n) \rho^2},$$

which, for optimal m, is

$$(12.11a) \qquad RE\,(\text{opt match/unmatched}) = \frac{2}{1 + \sqrt{1 - \rho^2}}.$$

The gain, or loss, in efficiency in percentages is obtained by $100(RE - 1)$. Calculations of gains in efficiency from matching several selected matching ratios, m/n, and levels of ρ are given in Table 12.1.

Table 12.1. Relative efficiency of matching

		Percent gain when			
ρ	m/n opt	m/n opt	m/n 1/2	m/n 1/3	m/n 1/4
0.5	0.46	7	7	7	6
0.6	0.44	11	11	11	9
0.7	0.42	17	16	17	15
0.8	0.38	25	24	25	23
0.9	0.30	39	34	39	39
0.95	0.24	52	41	50	52
1.0	0	100	50	67	75

It may be noted that with $\rho = 0$ it doesn't matter whether one matches or not. Moreover, it takes a sizable ρ, say 0.5, before the gains are very useful, and even if $\rho = 1$ the maximum possible gain is 100%. Note also that from 25% to 50% matching gives substantially the same efficiency for each ρ ranging from 0.5 to 0.95; hence a precisely optimal m/n is not very important, and other considerations may well dominate the decision—costs and quality of information, for example. Generally costs are lower for revisits, and quality of information may be better from revisits where cooperation and conceptual difficulties may be important.

The Multioccasion Case. The general theory of replacement policy in surveys through time has been studied. Since applications are rather limited, this theory will not be dealt with here. Those interested may refer to Patterson (1950) and Daly and Gurney (1965).

"Contamination" Effects of Panels. In dealing with human populations, each contact may affect the subsequent behavior or attitude of the person contacted, making the sample containing such individuals "unrepresentative." This is particularly likely for surveys involving a great deal of explanation in the measuring process (e.g., depth interviewing) to obtain the desired information. In this case a panel type survey will tend to lose in validity what it may gain in precision through matching.

"Births" and "Deaths" in Panels.   In some universes of interest the number of elements may be changing through time as well as the characteristics of the elements. Matching in this case may prove to be difficult or expensive, unless alternatives are found. For example, in dealing with changes in household characteristics through time, rather than trying to follow moving HHs it would be advisable to follow *housing units.* Families are usually replaced by similar families, and the correlation of characteristics through time will be only slightly less than for HHs. This is one way to avoid the problem of births and deaths of SUs. By the use of area SUs, even the creation of new HUs (housing units) and destruction of old HUs can be avoided entirely (assuming "areas" are not "created" and "destroyed").

12.5.   Multiple Populations

Up to this point we have been interested in estimating the characteristics of one population, y. Although other populations, such as x and z, have been considered from time to time, they have been of interest primarily as a source of information to improve the survey design. For example, they were used to improve the choice of SU, to form strata, to measure size for assigning probabilities of selection, or as part of the estimator. Many if not most surveys, however, are concerned with the estimation of parameters of more than one population. Since these populations may differ one from the other in their distributions over frame units, strata, and so on and in their functional relationships to each other, an optimal design based on one will not result in an optimal design for another. To deal with multiple populations generally leads one to the problem of compromises.

The problems of compromises in survey design will be sketched only briefly here. Although such problems have received some attention, it appears that a large gap exists between theories and practice. We shall discuss these problems as they affect some of the more important design determinations.

Effects on Determining Sample Size.   Suppose we have a universe of N elements, on each of which we observe a Y_i and Z_i. We wish to estimate the means of populations y and z with accuracies e_y and e_z, respectively, and it works out that the required sample sizes are n_y and n_z, respectively. If the es are inflexible and the cost of observing either Y_i or Z_i, having contacted the ith OU, is negligible, then the choice is fixed on the larger of the two ns. If the costs are not negligible, then one may omit observing the characteristic requiring the smaller sample, say z, on $n_y - n_z$ of the OUs. This may be done a number of ways—for example, by sampling items within the questionnaire (sometimes called a "split questionnaire").

Effects on Stratification.   Designing optimal strata and determining the optimal allocation of the sample among them becomes quite formidable when dealing with multiple populations. In designing strata, practitioners generally choose a "key" population—one that is "most important" or "most representative" of the set—and carry out the designing around it. After a tentative design has been established based on this "key" population, it is checked to see how well it meets the needs for estimating the characteristics of the remaining populations. If it falls short, then additions are made. These may consist of simply increasing the total sample size, n, or perhaps increasing the n_h in some strata. These measures are usually sufficient to achieve an approximate optimum design.

Attempts have been made to obtain methods of forming strata such that they possess some sort of generalized minimum variance over the several populations. Although such methods have a certain intellectual appeal, the problem of determining suitable measures of the "value of information" obtained on each population, together with the fact that stratification is a rather limited and usually somewhat weak method of improving the accuracy of sampling even in the single-population case, makes the prospects for practical success somewhat less than exciting.

Effects on Selecting Primaries; "Tethering."   In multistage designs dealing with multipopulations, it may be possible to achieve some gain in net efficiency by the manner in which primaries (or other-stage units) are selected. Consider a two-stage design with four populations, each with a measure of size, X^k, $k = 1, 2, 3, 4$. We wish to select primaries in the universe in Table 12.2 such that (i) each primary is selected with PPS; (ii) each population is represented by a single primary; and (iii) the total number of primaries selected is a minimum.

Constraints (i) and (ii) can be met by simply selecting a single primary independently with PPS for each of the four populations. However, to satisfy constraint (iii) we may adopt a procedure such as the following. Select a single random fraction and apply this to the cumulative MOS totals of all four populations. This procedure is indicated in Table 12.2 by converting the MOS to relative sizes, X_i^k/X^k. The RF (random fraction) assumed chosen is $RF = 0.50$. Note that primaries 3, 4, and 5 are selected, which doesn't appear to be much of a gain, if any, over independent selection. However, by selecting 20 trials for each case, we obtain the following results for the data in Table 12.2.

Table 12.2.   Relative MOSs for four populations

Primary (i)	X_i^1/X^1	X_i^2/X^2	X_i^3/X^3	X_i^4/X^4
1	0.10 (10)	0.13 (13)	0.16 (16)	0.03 (3)
2	0.22 (32)	0.10 (23)	0.17 (33)	0.10 (13)
3*	0.13 (45)	0.27 (50)*	0.17 (50)*	0.15 (28)
4*	0.34 (79)*	0.20 (70)	0.16 (66)	0.20 (48)
5*	0.06 (85)	0.07 (77)	0.17 (83)	0.22 (70)*
6	0.15 (100)	0.23 (100)	0.17 (100)	0.30 (100)
Total	1.00	1.00	1.00	1.00

Number of primaries	Independent selection		Constrained or tethered selection	
	Frequency	Relative frequency	Frequency	Relative frequency
1	3	15%	0	0%
2	13	65%	6	30%
3	4	20%	7	35%
4	0	0%	7	35%
Σ	20	100	20	100

Note that on the average 3.05 and 2.05 primaries are required for independent and tethered schemes, respectively, a welcome gain in overlap without additional cost in selection. If costs of travel between primaries is also a matter of concern, then it is suggested that the ordering of the primaries in the listing be such that primaries that are contiguous geographically appear "contiguously" in the list. A simple serpentine ordering may suffice.

Effect on Estimation.   Since each population can be dealt with individually to determine its own best estimator, a multiple-population survey brings forth no new problems—except possibly where one has oriented the design to enhance the effectiveness of a particular estimator scheme. But in general the flexibility of estimators in adapting to particular situations is an important property of this method of improving the accuracy of survey estimates.

12.6.   Two-Way Frames; Single Sampling; Multiple Reporting

Up to now we have generally regarded the universe of elements as a simple linear arrangement of individuals directly or in groups such that the procedure for selection could reasonably be carried out. In Chapter 11 we complicated this concept a bit by considering two-way and even multiway arrangements of the elements in order to exercise more control on selection procedure. We

Table 12.3.   *Notation for a two-way frame*

Frame A	Frame B 1	2	3	...	j	...	M	$\Sigma = I_{i\cdot}$
1	I_{11}	I_{12}	I_{13}		I_{1j}		I_{1M}	$I_{1\cdot}$
2								$I_{2\cdot}$
3								
⋮								
i	I_{i1}				I_{ij}		I_{iM}	$I_{i\cdot}$
⋮								
$N.$								
$\Sigma = I_{\cdot j}$	$I_{\cdot 1}$	$I_{\cdot 2}$	$I_{\cdot 3}$		$I_{\cdot j}$		$I_{\cdot M}$	$I_{\cdot\cdot}$

shall now reconsider the two-way frame and see if we can obtain even more sampling tricks from it and perhaps some help for solving new problems.

In this section we shall confine attention to those universes of elements wherein each element appears or is in some manner associated with at least one unit in each of two possible frames, A and B, into which the universe can be arranged either physically or conceptually. In the next section this requirement will be altered.

Six Basic Frame Types.   Suppose for a given universe of M elements, we construct two frames, A and B. Let frame A consist of units $1, 2, 3, ..., N$ and frame B of the elements $1, 2, 3, ..., M$. By selecting a sample of n of the N units of frame A we wish to (i) estimate M, the number of elements in B, and (ii) obtain a sample of the M elements that contain known properties. For example, frame B may be the individual households in a city and frame A the set of blocks in a map of the city. We wish to estimate the number of households in the city and also obtain a sample of households with known probabilities. The two-way frame is shown in Table 12.3.

Let I_{ij} be 1 if the ith unit of frame A "includes" element j of frame B and 0 otherwise. Six useful cases can be distinguished, depending on how we "associate" one frame with the other and what we may believe when we select a unit of frame A. The six cases follow.

1. $I_{\cdot j} = 1, I_{i\cdot} = 1$; *the simple case.*   This may be illustrated by the following picture:

Frame A	Frame B 1	2	3	4	5	$I_{i\cdot}$
1	1					1
2					1	1
3			1			1
4		1				1
5				1		1
$I_{\cdot j}$	1	1	1	1	1	$5 = I_{\cdot\cdot}$

This is the simple case of a one-to-one correspondence between the two frames. An example would be where frame A is a list of addresses of occupied housing units and frame B is the corresponding households. In this case $M = N$ and is presumed known. Hence

$$(12.12) \qquad \hat{M} = N,$$

and an estimate of $\tilde{Y} = Y/M$ is given by

$$(12.13) \qquad \tilde{y} = \frac{\hat{Y}}{\hat{M}} = \frac{N \sum_i^n Y_i/n}{N} = \bar{y}.$$

2. $I_{.j} = 1$, $I_{i.} \neq 1$; *simple cluster case.* This one can be represented as follows:

	Frame B						$I_{i.}$
Frame A	1	2	3	4	5	6	
1	1	1					2
2							0
3			1				1
4				1	1	1	3
$I_{.j}$	1	1	1	1	1	1	$6 = I_{..}$

This is the case where the A frame consists of simple clusters of B frame elements. An area frame is a simple example. In the case where farms are the units of observation, the location of "headquarters" designates the area unit to which the farm is associated, without regard to where the land may be. Similar rules of association can be constructed for other types of elements where unique associations are not obvious.

To estimate $\hat{M}$ we can use

$$(12.14) \qquad \hat{M} = \frac{N}{n} = \sum_i^n I_{i.} ,$$

and an estimator of $\tilde{Y}$, that is Y/M, is given by the ratio estimator

$$(12.15) \qquad \tilde{y} = \frac{\hat{Y}}{\hat{M}} = \frac{\sum_i^n \sum_j^{j \in i} Y_{ij}}{I_{i.}} .$$

3. $I_{.j} \neq 1$, $I_{i.} = 1$; $I_{.j}$s *determinable.* To illustrate, we have

	Frame B				$I_{i.}$
Frame A	1	2	3	4	
1	1				1
2	1				1
3		1			1
4			1		1
5			1		1
6				1	1
$I_{.j}$	2	1	2	1	$6 = I_{..}$

This case assumes that a particular B frame unit may intersect one or more A frame units but each A frame unit intersects one and only one B frame unit. It is also assumed that when a B frame unit is "hit," the total $I_{.j}$ is determinable. Then an unbiased estimator of M is given by

$$(12.16) \qquad \hat{M} = \frac{1}{m} \sum_{j}^{m} \frac{N}{I_{.j}},$$

where m is the number of different B frame units obtained in the n A frame units. Note that P_j, the probability of including element j in the sample when selecting an A unit at random, is $I_{.j}/N$. Hence Eq. 12.16 is analogous to Eq. 8.55. If a mean of some characteristic of B units is required, it is customary to use as the weighted estimator

$$(12.17) \qquad \bar{y} = \frac{\hat{Y}}{\hat{M}} = \frac{\sum_{j=1}^{m} Y_j/I_{.j}}{m}.$$

Sampling schemes that illustrate the use of this case are Birnbaum and Sirken (1965) and "point sampling" of farms (Review Illustration 3 of Chapter 8), where I_{ij}/N can be regarded as the fraction of total area in farm j.

4. $I_{.j} \neq 1$, $I_{i.} = 1$; $I_{.j}s$ not determinable. This differs from the case above in that we have no way to determine the number of times the jth B unit intersects the set of A units. This case presents somewhat difficult problems in estimating M unbiasedly or accurately. It may be illustrated with the problem of estimating the number of species (B frame) of some organism from samples of individuals taken completely at random (A frame). References [Mosteller (1949), Goodman (1949)] can be consulted for some of the theory and proposed estimators.

5. $I_{.j} \neq 1$, $I_{i.} \neq 1$; $I_{.j}s$ determinable. To illustrate this case we have

Frame A	Frame B						$I_{i.}$
	1	2	3	4	5	6	
1	1		1		1	1	4
2			1		1		2
3		1			1		2
4				1	1		2
5					1		1
$I_{.j}$	1	1	2	1	5	1	$11 = I_{..}$

Here the frames overlap each other. An example is where the A-units are areas and the B-units are farms, but unlike case 2, here we may associate a farm with all area units into which the farm extends. See Jessen and Thompson (1958).

Another example is a household survey wherein each HU reports for other

HHs (such as for "siblings") [Sirken (1970)], sometimes called "multiple reporting."

In this case M can be estimated unbiasedly by

(12.18)
$$\hat{M} = \frac{N}{n} \sum_{i}^{n} \sum_{j}^{j \in i} \frac{I_{ij}}{I_{\cdot j}},$$

and an unbiased estimator of $\tilde{Y}$ is given by

(12.19)
$$\tilde{y} = \frac{\hat{Y}}{\hat{M}} = \frac{\sum_{i}^{n} \sum_{j}^{j \in i} Y_i I_{ij}/I_{\cdot j}}{n}.$$

6. $I_{\cdot j} \neq 1$, $I_{i\cdot} \neq 1$; $I_{\cdot j}$s *not determinable*.   Here again each frame overlaps the other one or more times, but it is not possible to determine the $I_{\cdot j}$s. Some estimators have been proposed where the $I_{\cdot j}$s are estimated. Alternative estimators are available, but studies so far, Hsieh (1970), do not reveal one that is to be generally superior. Here is one, although biased, that may be useful:

(12.20)
$$\hat{M} = \frac{N}{n} \sum_{j} \frac{\sum_{i}^{n} I_{ij}}{1 + [(N-1)/(n-1)] \sum_{i}^{n} I'_{ij}},$$

where

$$I'_{ij} = \begin{cases} 0 & \text{if } ij \text{ is a reference union,} \\ I_{ij} & \text{otherwise.} \end{cases}$$

I_{ij} is regarded as being of two types. For example, the $I_{\cdot j}$ intersections of the ijth B-unit with all N A-units will have one designated as a "reference union" and all others as not. Hence, if farms are to be selected only if their headquarters fall within a sample area, then case 6 reduces to case 2, where the I_{ij} are the reference unions only.

12.7. Multiple Frames; Multiple Sampling

We now consider universes whose elements can be or are already arranged in two or more frames that may or may not be complete. Hence frames may consist of duds (nonexisting elements), or some elements may not appear on any of the frames. There are several reasons why it may be prudent to select samples from more than one frame: (1) a combination of frames may provide more coverage of elements than any one, (2) costs of surveying elements may be different from frame to frame, and therefore it may be more efficient to deal with two or more than any one, and (3) variability among units within frames may also vary from frame to frame, so that one would wish to avoid the variable frames if possible.

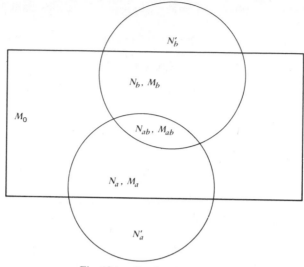

Fig. 12.1. Overlapping frames.

A General Structure of Frames. Consider the case where we have two frames, A and B, each with duds, each with exclusive members, an overlap group shared by both, and some elements completely missed. Diagrammatically, this case can be represented as in Fig. 12.1, where the Ns denote the number of frame units and the Ms the number of elements in each frame and class (or *domain*). N'_a and N'_b are the numbers of frame units in each frame that are duds. It is assumed that frame units need not be identical to elements. Hence frame A contains $N_A = N_a + N_{ab} + N'_a$ frame units and $M_a + M_{ab}$ elements. Note also that total elements $M = M_a + M_b + M_{ab} + M_0$. It is helpful to distinguish two cases: (i) the case where the M elements are included in the frames taken collectively, that is, $M_0 = 0$; and (ii) the case where M_0 is not 0.

M Covered by Frames (i.e., $M_0 = 0$). Hartley (1962) gives the general theory for this case under various assumptions about knowledge of the sizes of the domains involved. Four subcases are distinguished as displayed in Table 12.4. The relevant notation is:

Item	Frame		Domain		
	A	B	a	b	ab
Universe elements	N_A	N_B	N_a	N_b	N_{ab}
Sample elements	n_A	n_B	n_a	n_b	n'_{ab}, n''_{ab}
Population total	Y_A	Y_B	Y_a	Y_b	Y_{ab}
Population mean	$\bar{Y}_A$	$\bar{Y}_B$	$\bar{Y}_a$	$\bar{Y}_b$	$\bar{Y}_{ab}$
Sample mean	$\bar{y}_A$	$\bar{y}_B$	$\bar{y}_a$	$\bar{y}_b$	$\bar{y}'_{ab}, \bar{y}''_{ab}$
Cost per element	c_A	c_B			

We are here assuming that frame units and elements are identical: $N_A = M_A$, $N_B = M_B$, and so on. Note that n'_{ab} and n''_{ab} distinguish the sample elements coming from the A and B frames, respectively. It is understood that n_a, n_b, n'_{ab}, and n''_{ab} must be at least one or more before $\bar{y}_a$, $\bar{y}_b$, $\bar{y}'_{ab}$, and $\bar{y}''_{ab}$ can be computed.

Table 12.4.   Subcases of two-frame case of complete coverage

Subcase no.	Knowledge of domains	Sampling possibilities	Nature of domains
1	N_a, N_b, N_{ab}, etc. are known	Can allocate fixed sample sizes to domains	Domains are identical to strata
2	N_a, N_b, N_{ab}, etc. are known	Can allocate fixed sample sizes to frames only	Domains are identical to poststrata
3	N_a, N_b, N_{ab}, etc. are *not* known; N_A and N_B are known	Can allocate fixed sample sizes to frames only	Domains are treated as domains
4	N_a, N_b, N_{ab}, etc. are not known; N_A and N_B are known relatively only	Can allocate fixed sample sizes to frames only	Domains are treated as domains in universes of unknown sizes

Considering subcase 2, where N_a, N_b, N_{ab}, the sizes of the domains, are assumed known but n_A and n_B are simple random samples from frames A and B, respectively, we have the following estimator of Y:

$$(12.21) \qquad \hat{Y} = N_a \bar{y}_a + N_{ab}(p\bar{y}'_{ab} + q\bar{y}''_{ab}) + N_b \bar{y}_b,$$

where p and q are weights such that $p + q = 1$. It is presumed that, as in poststratification, $\bar{y}_a$ and $\bar{y}_{ab}$ are replaced by $\bar{y}_A$ if either $n_a = 0$ or $n'_{ab} = 0$ (and likewise for the B frame). Following the principles of poststratification, and ignoring FPCs, the variance of $\hat{Y}$ is given by

$$(12.22) \qquad \mathrm{var}(\hat{Y}) \doteq \frac{N_A^2}{n_A}\left[S_a^2\left(1 - \frac{N_{ab}}{N_A}\right) + p^2 S_{ab}^2\left(\frac{N_{ab}}{N_A}\right)\right]$$
$$+ \frac{N_B^2}{n_B}\left[S_b^2\left(1 - \frac{N_{ab}}{N_B}\right) + q^2 S_{ab}^2\left(\frac{N_{ab}}{N_B}\right)\right],$$

where S_a^2, S_b^2, and S_{ab}^2 are the population variances within the corresponding domains.

If we assume the usual cost function

$$(12.23) \qquad C = c_A n_A + c_B n_B,$$

where C is the budget allowed for nonoverhead expenditures, optimal values of p and q are obtained by solving the biquadratic,

$$(12.24) \qquad \frac{c_A p^2}{c_B q^2} = \frac{S_a^2 (1 - N_{ab}/N_A) + (N_{ab}/N_A) p^2 S_{ab}^2}{S_b^2 (1 - N_{ab}/N_B) + (N_{ab}/N_B) q^2 S_{ab}^2},$$

and for optimal p, optimal sampling fractions are given by

$$(12.25) \qquad \frac{n_A}{N_A} = C \left[\frac{S_a^2 (1 - N_{ab}/N_A) + (N_{ab}/N_B) p^2 S_{ab}^2}{c_A} \right]^{1/2},$$

$$\frac{n_B}{N_B} = C \left[\frac{S_b^2 (1 - N_{ab}/N_B) + (N_{ab}/N_B) q^2 S_{ab}^2}{c_B} \right]^{1/2}.$$

These formulas are considerably simplified if we restrict attention to the situation where the A frame has complete coverage (and the B frame remains incomplete). Then

$$(12.26) \qquad N_{ab} = N_B, \quad \frac{N_{ab}}{N_B} = 1, \quad S_B^2 = S_{ab}^2, \quad N_b = 0.$$

Then

$$(12.27) \qquad p^2 = \frac{S_a^2}{S_{ab}^2} \left(\frac{1 - N_{ab}/N_A}{c_A/c_B - N_B/N_A} \right)$$

and

$$(12.28) \qquad \hat{Y} = N_a \bar{y}_a + N_B (p \bar{y}_B' + q \bar{y}_B'')$$

with the variance

$$(12.29) \qquad \text{var}(\hat{Y}) = \frac{N_A^2}{n_A} \left[S_a^2 \left(1 - \frac{N_B}{N_A} \right) \right] + N_B S_B^2 \left[p^2 \left(\frac{N_A}{n_A} \right) + q^2 \left(\frac{N_B}{n_B} \right) \right].$$

The variance of the two-frame scheme where the A frame is complete, relative to that for a one-frame scheme where the poststratified estimator is used on domains a and $ab = B$, the estimator being $Y' = N_a \bar{y}_a + N_{ab} \bar{y}_{ab}$, is given by

$$(12.30) \qquad \frac{\text{var}(\hat{Y})}{\text{var}(Y')} = \frac{1 - (q/p)(N_B/N_A)(c_A/c_B)}{1 + [q(N_B/N_A)(1+p)]/[p^2(c_A/c_B)]}.$$

In general, the two-frame scheme is relatively more efficient than a one-frame scheme when

1. The cost of the incomplete B frame is cheaper than the complete A frame; $c_B < c_A$.
2. The variance of elements in the B frame is higher than in the A frame; $S_B^2 > S_A^2$.
3. The size of the B frame is near that of the A frame; $N_B \doteq N_A$.

Table 12.5 [from Hartley (1962)] will give the reader a feeling for the situations where two frames become practicable.

Table 12.5. Variance of two-frame relative to one-frame sampling

Sampling cost ratio c_B/c_A	N_B/N = proportion of universe in cheap frame						
	0.5	0.6	0.7	0.8	0.9	0.95	1.00
A: $S_B^2/S_a^2 = 16$							
0.01	0.096	0.076	0.059	0.045	0.031	0.024	0.010
0.05	0.154	0.134	0.118	0.102	0.086	0.075	0.050
0.10	0.206	0.188	0.174	0.160	0.143	0.131	0.100
0.20	0.288	0.278	0.269	0.261	0.248	0.237	0.200
0.30	0.359	0.356	0.355	0.353	0.347	0.338	0.300
0.40	0.423	0.428	0.435	0.440	0.441	0.436	0.400
0.50	0.483	0.496	0.510	0.524	0.533	0.532	0.500
1.00	0.735	0.784	0.836	0.889	0.944	0.972	1.000
B: $S_B^2/S_a^2 = 4$							
0.01	0.259	0.201	0.152	0.108	0.066	0.044	0.010
0.05	0.340	0.284	0.234	0.186	0.137	0.107	0.050
0.10	0.404	0.352	0.304	0.257	0.205	0.172	0.100
0.20	0.500	0.456	0.415	0.372	0.322	0.287	0.200
0.30	0.576	0.540	0.507	0.472	0.426	0.393	0.300
0.40	0.640	0.613	0.588	0.561	0.523	0.493	0.400
0.50	0.696	0.678	0.661	0.642	0.614	0.589	0.500
1.00	0.900	0.914	0.932	0.953	0.976	0.988	1.000
C: $S_B^2/S_a^2 = 1$							
0.01	0.571	0.477	0.379	0.276	0.164	0.101	0.010
0.05	0.656	0.573	0.482	0.381	0.260	0.186	0.050
0.10	0.718	0.645	0.562	0.465	0.344	0.263	0.100
0.20	0.800	0.752	0.674	0.589	0.475	0.392	0.200
0.30	0.857	0.812	0.757	0.686	0.582	0.503	0.300
0.40	0.900	0.866	0.824	0.765	0.676	0.604	0.400
0.50	0.933	0.909	0.877	0.832	0.759	0.695	0.500
1.00	1.000	1.000	1.000	1.000	1.000	1.000	1.000

M Not Covered Completely by Frames (i.e., $M_0 \neq 0$). This case is of some interest when one has a number of lists that presumably cover the elements of the universe of interest but that carry no assurance that they in fact do so. Deming and Glasser (1959) Goodman (1961) have presented schemes that attempt to deal with aspects of this problem and also to treat the general problem of estimating M and the extent of duplication by sampling from multiple frames.

12.8. Multipurpose Descriptive Surveys; Multiestimation

In surveys where it is of interest to estimate characteristics of various *domains* of study as well as characteristics of the universe as a whole, problems may arise such as (i) how to allocate samples among the strata, (ii) how to design the strata, and (iii) how to make estimates most efficiently. The problems arise primarily because of conflicts between the objectives of good domain estimates on the one hand and a good overall estimate on the other. An example might be a survey to obtain an accurate overall U.S. estimate, regional estimates of equal accuracy, and estimates for such zones as rural, suburban, and metropolitan, each with a precision not less than some specified level—this to be done with a minimum overall sample size.

Allocation.   In this case the problem of allocation is made complicated because we must try to satisfy several requirements, which may be in conflict. For the national estimate, the best allocation will be approximately proportional to region sizes, but to meet the specifications of equal accuracy for regions, they should receive about equal sample sizes. Moreover, the zones, which may vary in size from region to region, should also receive about equal sample sizes to meet their accuracy specifications. To solve a problem of this sort could probably be done fairly quickly by iteration, but a computerized linear program would be appreciated.

When domains are not identified in advance, and therefore sample sizes in them are random variables, one can first work out a solution on expected sample sizes and then increase them to whatever confidence level one wishes.

Estimation.   Where domains can be identified in advance of sampling, each can be assured at least one sample element, so that at least an unbiased estimate of its mean can be calculated. However, this estimate can be quite unsatisfactory if the sample is small; and if it is completely unrepresented by a sample, then an unbiased estimate is impossible.

Moreover, it may turn out that an aggregation of the separate domain estimates does not agree with the preferred overall estimate obtained by another estimator—that is, the whole does not agree with the sum of its parts. And it may be that the estimates do not agree with census or other extra-sample results. In these cases, if the differences are attributable to sampling errors, it may be simpler to accept them with their divergencies. However, readers of published data dislike such inconsistencies and may attribute them to mistakes or carelessness.

12.9. Analytic Surveys; Factorial Surveys

The essential difference between analytic and descriptive surveys is that in the former we are generally interested in relationships between characteristics (variables) and in the latter simply a set of measurements to determine the average level of one or more characteristics. Up to this point we have been concerned almost exclusively with the descriptive survey. Here we shall very briefly indicate some of the problems that seem peculiar to the analytic survey.

Section 7.7 described the concepts of "factors" and "levels" to help us discuss structure in the design of strata. These concepts will be helpful here, too.

One-Factor, Two-Level Survey.   The simplest analytical survey consists of a single factor at two "levels"—in other words, a survey involving two groups. An example would be a survey to determine if blood pressure differs between men and women. Here the factor is sex and its "levels" are male and female. If sex can be determined in advance of sampling, then an optimal allocation of observations between men and women will be given by $n_h \propto S_h$, or, if the S_h are regarded as equal, then n_h should be equal for both sexes in order to make $\mathrm{var}(\bar{y}_1 - \bar{y}_2)$ a minimum for a given $n = n_1 + n_2$, regardless of their actual numbers in the universe of interests.

One-Factor, k-Level Survey.   Suppose we still wish to study blood pressure, but now we would like to compare several ethnic groups—whites, blacks, chicanos, and asians. Again, on allocation we find that if we wish to compare all possible pairs of ethnics, we shall need equal accuracy for each group; hence, if the S_hs are equal, n_h should be equal for all groups despite their relative numbers in the universe. (To analyze the results efficiently, an analysis-of-variance procedure is suggested. See Section 14.6.

Two-Factor Surveys.   Now suppose we wish to combine the two foregoing examples and look into the matter of whether both sex and ethnicity are factors affecting blood pressure. This is regarded as a two-factor study with sex as one factor at two levels and ethnicity as the other with four levels. Without giving reasons, it will again be best (if factor interactions should be present) to allocate sampling resources equally to all eight ($= 2 \times 4$) subgroups. The analysis can now be extended to compare not only the sex effect over all ethnic groups and the ethnicity effect over both sex groups, but also whether these two factors "interact"—that is, whether the sex differences are the same over all ethnic groups. Again the analysis of such data can be done conveniently by analysis of variance if sampling is random in each of the eight subgroups. If the groups are not equal in size, the analysis will be somewhat more complicated.

General.   The number of factors and their respective levels can be increased to any number, but complexities, particularly if interactions exist, can make the analysis very complicated and (if one is not a careful worker) also hazardous. However, the design of surveys for analytical studies is somewhat different than for descriptive studies. The optimal allocations to the domains are different, and the analysis is more complicated.

If factors are *continuous* rather than *categorical* variables, then regression might be a preferred analytical procedure rather than analysis of variance. In this case, too, allocation principles are altered. They will not be considered here.

12.10. Hybrid Investigations

Many, if not most, investigations, inquiries, and studies seem to be involved not only with "finding the facts" but with looking for associations among the characteristics observed, inferring causative mechanisms, and—particularly in the social sciences—suggesting or demanding the adoption of "policies" that naturally follow from all this new knowledge. Most practical studies are hybrids of the pure simple types we suggested in Chapter 1. The principles learned in the simple types should be of help when we venture into the world of those more complex investigations.

Descriptive-Analytical Surveys. The conflict between descriptive and analytic type surveys in obtaining optimal allocation among domains can be a serious problem in the design of surveys that are dual-purpose. A workable solution can sometimes be achieved by simply deciding which half is more important, designing it as if it were a pure type, and, while the plan is still on paper, checking its suitability for achieving the lesser requirements. If the results are unsatisfactory, make the necessary alterations. Then check the altered plan for doing the main job. Iterate until all needs are met.

Survey-Experiments. The survey-experiment can be a very useful hybrid. In fact experiments imbedded within surveys can be very powerful investigative designs. If properly carried out, the experimental results will have wide applicability. Many problems in the social sciences seem to be applicable to survey-experiments, particularly if provision is made to detect and assess interactive factors.

12.11. The OU and Its Choice

The SU, the unit selected in the sampling process, and the OU, the unit actually measured or observed, are usually identical. However, in many situations they are not, and in those cases it is possible to consider alternative OUs for a given SU. These situations often arise when the purpose of the survey is to estimate a population total, or the ratio of two population totals, without regard to the unit used to obtain such estimates. For example, suppose the purpose of a survey is to obtain the total number of TV sets in the Los Angeles metropolitan area turned onto a particular program. The SU could be a segment of a block, and the OU could be each household, each individual person, or each TV set associated in some manner with that block segment. The choice of which OU to use should be based on some consideration of the usual matters as costs or accuracy or completeness of the data. In this section we shall consider a number of situations where choices are available and examine some of the principles that appear and some of the results of field experiences.

Linking OUs with SUs. Opportunities for alternative OUs and the ways in which they can be associated or linked with SUs are fairly rich for certain types of frames in certain situations. Area frames for agriculture are particularly rich, but alternative OUs and linkage rules seem to exist everywhere. Let us first consider an agricultural example, where the SU is an area grid and we wish to determine area in corn. Some OUs and linkage rules that have been used are presented in Table 12.6. Note that scheme 4 brings OUs into the sample with probabilities that depend on the number of different SUs a particular farm lies within. Alternative methods of estimation are available for obtaining unbiased estimates for the universe.

The best choice of OU and its linkage depends on the usual consideration of costs and variances and, sometimes quite importantly, on errors in measurement associated with them. For example, in Table 12.6 scheme 4 would cost more per SU than scheme 3 because it requires contacting more farmers. However, it was also found that the variance of scheme 4 for many

Table 12.6. Examples of some OU/SU linkages in agriculture. SU is an area

OU	OU/SU linkage	Method of measurement	Probability of selecting OU
(1) Fields or portions of fields	Take each field or portion thereof lying wholly within the SU grid	Obtain field areas by direct measurement (may use aerial photos)	Same as SU
(2) Whole fields	Take fields whose "NW corner"[a] is within SU grid	Direct measurement or by interview with farmer	Same as SU
(3) Whole farms	Take those farms whose "headquarters"[b] lie within the SU grid	Interview with farmer	Same as SU
(4) Whole farms	Take those farms, any part of which lies within the SU grid	Interview with farmer	$P\{SU\} N_j$, where N_j is the number of SUs in which the jth farm lies

[a] Can be more unambiguously defined as the westernmost point of the northernmost points.
[b] Can be defined as the "residence of the operator," "main operational buildings," or the like.

farm characteristics was less than that of scheme 3, and various errors of coverage and measurement were less [Jessen and Thompson (1958)].

Linkage problems when area SUs are used on surveys of people are usually fairly simple in structure, if not simple to apply. For example, a person who "lives here" (in this housing unit) is a member of the sample if the housing unit, HU, is within the area SU (or at the selected "address"). Persons who are "visiting," who are "temporarily" living here (students in dormitories, persons in motels, hotels, hospitals, and so on), persons who have more than one residence, or persons who have no "visible" residence must be accommodated by the survey with appropriate rules of inclusion or exclusion.

When SUs (in any stage) are street addresses, and OUs are HHs or individual persons, care must be taken in situations involving unusual living arrangements to set up rules of linkage that are appropriate for the survey objectives and yet can be satisfactorily carried out in the field. Business establishments can also present problems of linkage where complex arrangements exist and as a result the desired OUs may not be brought into the survey with their proper probabilities (zero or otherwise).

Some Unusual OU/SU Cases.   In order to suggest some of the problems that can be dealt with by innovative choices of OU/SU schemes, two examples will be described.

Example 12.3.   *The moving observer* [Yates (1949)].   Suppose we have a universe of elements continuously moving about in a specified area—for example, a crowd in a public square, on a beach, and so on. Devise a sampling scheme to estimate the total number of elements in the universe.

A pair of persons a fixed distance apart could walk along a transcept of the area (square, beach, or whatever) counting the people that pass between them, giving positive counts to those overtaken and negative counts to those overtaking them. See Fig. 12.2.

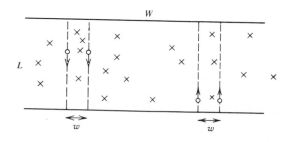

Fig. 12.2.   Moving observer scheme.

If x_i is the net count on the ith trip, w is the distance between the observers, and the transcepts are randomly selected, then $\bar{x}$ is the mean count per trip and

(12.31)
$$\hat{X} = \left(\frac{W}{w}\right)\bar{x}.$$

This is unbiased provided (a) persons do not move around in a manner such that they can be counted more than once, (b) the observers have no effect on the behavior of the crowd.

Example 12.4. A logging company is interested in a method of estimating the amount of wood left in an area recently cut in order to determine whether it is economically feasible to engage in a special operation to recover logs missed, rejected, or otherwise left in the worked area. Devise a scheme for estimating the volume of wood left.

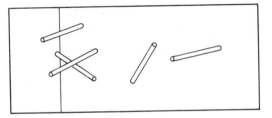

Fig. 12.3. Scrap log problem.

Suppose we represent the problem as shown in Fig. 12.3. A line running across the logged field is drawn at random. We measure the area of each log "cut" by the SU, in the direction of the SU, sum them, and divide by the length of the transcept to obtain the mean area per unit length. Denote this statistic for the ith transcept by a_i. If $V=$ volume of the logs in the area L by W, an estimate is obtained by

(12.32)
$$\hat{V}_i = WLa_i,$$

and for n transcepts

$$\frac{\sum a_i}{n} = \bar{a},$$

(12.32a)
$$\hat{V} = \frac{1}{n}\sum_{i=1}^{n} WLa_i = WL\bar{a}.$$

An alternative to the scrap log problem: Use logs as OUs and measure their lengths and diameters. Then the probability of "cutting" a log is given by the length displacement of the log perpendicular to the transcept. Call this

$$P_{ij} = \frac{L_{ij}}{L},$$

where L is the length of the "field."

(12.33)
$$\hat{V}_i = \sum_j^{m_i} \frac{V_{ij}}{L_{ij}},$$

where m_i is the number of logs "cut" by the ith line.

(12.33a)
$$\hat{V} = \frac{1}{n} \sum_i^n \sum_j^{m_i} \frac{V_{ij}}{L_{ij}}.$$

This scheme preserves the whole OU by appropriately modifying the probability of selection. The first method preserves equal probabilities but loses the individual OU characteristics.

Multiple OUs. Many surveys are concerned with more than a single OU. Farm surveys may be interested in characteristics of the farm as an economic unit, the operator's (if management is controlled by a person) characteristics, the characteristics of the operator's household, or perhaps the details of some event such as a sale of a herd of cattle or the purchase of a tractor. When this occurs it is well to recognize the fact and set up the reporting forms accordingly. Otherwise confusion may result.

12.12. Incomplete OUs

In the usual survey the observed element, whether it was called an OU, SU, or otherwise, was assumed to have some characteristic y (or x or z) that was observable without error. In Chapter 13 the matter of errors in these observations will be considered. Here we shall consider the situation where the observation may be an estimate of some sort, which may come about because (i) we wish to preserve privacy by obtaining only partial disclosure or (ii) a partial observation may be all that is available or practicable.

As an example of an incomplete observation, suppose we wish to determine the number of persons with a yearly calorie intake below some level k, and it is feasible to keep accurate records on the food consumption of a sample of people for, say, two weeks during the year. In this case we have observations on only two of the 52 required weeks. If the two weeks are randomly selected, we can obtain unbiased estimates of annual consumption of each person, but that does not lead directly to unbiased estimates of the number of $Y_i s < k$. Hence, to estimate total intake of calories during a year, we can easily obtain unbiased estimates by simply randomizing the selection of persons and weeks observed. The simple mean will suffice. But to determine the distribution of the annual calorie intakes of *persons* is somewhat more complicated if observations on OUs are "incomplete." The former is a simple case of sub-sampling; the latter is not, because the estimation problem is more complex.

In the case where $Y_i = 0$ or 1—that is, any qualitative characteristic—a scheme has been worked out that provides unbiased estimates of, say, P, the proportion of OUs that are 1s. The case where Y_i is a count or a continuous variate and the object is to estimate, say, certain percentiles of the distribution by OU, does not appear to be worked out. Some cases will be given where the OU is not the unit of analysis but where the incomplete observation is a rather clearer way to preserve privacy or improve accuracy of reporting.

Randomized Response. This scheme is due to Warner (1965). Suppose we have a universe of N persons of which N_1 would respond YES to a particular question and N_0 would respond NO. Let $N_1/N = P$, the true proportion of YESes in the population. Now suppose we use a device that is comprised of a dial and a spinner that can be held in one's hand. The dial is laid out in two sectors, region YES and region NO, where, say, θ = proportion of the dial in region YES. For example, the device may be represented as in Fig. 12.4. The spinner is handed the respondent and asked a question that requires YES or NO responses. "Now will you take this device and turn your back to me, spin the spinner, and tell me whether the spinner stops in the portion that indicates the correct answer, either 'Yes, it does' or 'No, it does not.' Then destroy the spinner location (by re-spinning if you wish) so I don't know where it stopped."

Let $\quad \theta$ = proportion of dial in YES region,

$$X_i = \begin{cases} 1 \text{ if } i\text{th OU says "Yes, it does,"} \\ 0 \text{ if } i\text{th OU says "No, it does not,"} \end{cases}$$

n = number of OUs in sample (assume N large),

n_1 = number of OUs reporting "Yes, it does" = $\sum^n X_i$,

$n - n_1$ = number of OUs reporting "No, it does not."

Then an unbiased estimate of P is given by

(12.34)
$$\hat{P} = \frac{1}{2\theta - 1}\left[(\theta - 1) + \frac{n_1}{n}\right]$$

with variance

(12.35)
$$\text{var}(\hat{P}) = \frac{1}{n}\left\{\left[\frac{1}{4} - \left(\hat{P} - \frac{1}{2}\right)^2\right] + \left[\frac{1}{16(\theta - \frac{1}{2})^2} - \frac{1}{4}\right]\right\}.$$

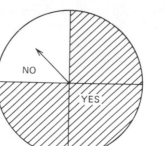

Fig. 12.4. Spinner.

It may be noted that when $\theta = 1$, $\hat{P} = n_1/n$; and when $\theta = 0$, $\hat{P} = (n - n_1)/n$. Hence the estimator reduces to the standard estimator, since the response is now complete. Note also that when $\theta = \frac{1}{2}$—that is, when the dial is divided equally into YES and NO regions—$\text{var}(\hat{P}) = \infty$, and when $\theta = 0$ or 1, $\text{var}(\hat{P}_i) = P(1 - P)/n$, which is the variance of the standard case. By setting θ at $\frac{1}{2}$ we reduce the information obtained from the informant to zero; by setting θ at 1 or 0 we obtain full information. Hence θ is a measure of the degree of information disclosed. The optimal level of θ would depend on the respondents' willingness to disclose the information desired. Noting that the first bracketed term can be written as $P(1 - P)$, we can express Eq. 12.35 in the two sources of variance: that due to sampling and that due to incompleteness of disclosure. Hence

$$(12.35a) \qquad \text{var}(\hat{P}) = \frac{P(1 - P)}{n} + \frac{1}{n}\left[\frac{1}{16(\theta - \frac{1}{2})^2} - \frac{1}{4}\right].$$

By substituting $\hat{P}$ for P in 12.35 or 12.35a we can obtain an estimate of $\text{var}(\hat{P})$.

Some of the properties of this technique can be seen in Fig. 12.5. The efficiency of the scheme depends on both the choice of θ and the value of P of the population. It is best relative to direct response (or the equivalent, when $\theta = 1$) when $P = 0.50$ and for high levels of θ. Rare populations, such as those of $P < 0.05$ (or their obverse, $P > 0.95$) do rather badly except for the high, and therefore unlikely to be practical, levels of θ.

Because incomplete disclosure may be free from bias (if respondents cooperate), it does cost in variance, and it may be that traditional methods of asking full disclosure and accepting a certain proportion of erroneous responses will have a smaller mean square error than $\hat{P}$. The best choice depends on circumstances.

Segmented OUs. A class of incomplete observations can be identified by regarding the OU as being made up of segments (in time or space), where for

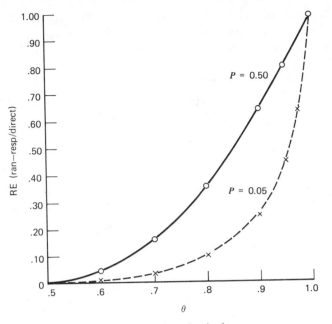

Fig. 12.5. Efficiency of randomized response.

one reason or another it is impossible to observe the whole OU but it is possible to observe the part or segment. Let us consider a simple model (see Review Illustration 8.4).

Regard the OU in Fig. 12.6 as consisting of four segments of sizes (lengths) X_1, X_2, X_3, and X_4, and y values Y_1, Y_2, Y_3, and Y_4, where $X = \sum X_i$ and $Y = \sum Y_i$ are the total size and total y value for the OU. For example, the Xs might represent the amount of time between purchases of a product, say

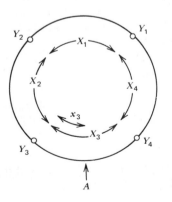

Fig. 12.6. Probability wheel.

liquor, that some people would regard as a private matter, and the Ys the amount purchased at those times. The desired observation for this OU is the total purchases of liquor during the year (the four occasions), Y. Suppose that the OU is visited at some instant during the year (some point on the periphery of the circle) and asked, "Have you ever purchased any liquor?" and if YES, "How long ago was that?" and "How much did you buy at that time?" Assuming that each answer is correct, an estimate of Y is given by (see Eq. 8.1 with $P_i = X_i/X$):

$$(12.36) \qquad\qquad \hat{Y} = X\frac{Y_i}{X_i},$$

provided one can determine the time that will elapse before the next purchase. Lacking this information, we can estimate X_i (unbiasedly if A is randomly located) by $X_i = 2x_i$ where x_i is the portion of time X_i already elapsed. Then we have

$$(12.36a) \qquad\qquad \hat{Y} = X\frac{Y_i}{2x_i}$$

as an estimate (but not unbiased) of Y. Unfortunately the properties of 12.36a do not appear to be examined, hence caution is advised.

In the calorie intake problem, the X_i are all $\frac{1}{52}$, and we have the simple case of random sampling to obtain $\hat{Y}$. The most common case, however, is when x_i is known but X_i is not.

Another variant is where the X_i are some measure of size. Suppose we wish to obtain an accurate unbiased estimate of the number of cattle on farms, even though farmers fail to tell the truth. But suppose a farmer has cattle in, say, four different fields. The interviewer asks the farmer to tell him the number of cattle he has in each field. The farmer provides answers, but they are suspect. The interviewer selects one of the fields (either with equal probability or with PPS on the X_i—the declared number of cattle on each field) and with some ruse tells the farmer he wishes to examine those cattle in field A to record their "condition," "types," "weights," and so on. The farmer agrees, and it is noted that there are Y_i rather than X_i cattle actually in the examined field. An estimate of Y is provided by the estimator

$$\hat{Y} = \frac{Y_i}{X_i/X} = \frac{X}{Y_i/X_i},$$

which is Eq. 12.36, or

$$\hat{Y} = \frac{Y_i}{1/N} = NY_i$$

if fields are selected with equal probability.

The problem of estimating percentiles of OUs on the true Y_i does not appear to be examined.

12.13. A Summing Up on Sampling Design

Since the next chapter will be devoted primarily to nonsampling problems, it may be helpful at this point to recapitulate briefly the material presented in the preceding 12 chapters—first a short history, then a look at some of the "laws" and principles that seem to underlie a lot of the procedures and methods used, and finally some remarks on seeking and utilizing "prior" information and on the creation of a "design."

Minihistory.   Most of the theory, principles, and procedures of sampling applied to statistical surveys have been developed in the present century. In the first decade, Student presented the *t*-distribution. In the 'teens, Bowley's survey of Reading, England, demonstrated the power of sampling in a household survey, although it was based on a quasi-random (essentially systematic) sample. In the twenties, stratified-random sampling was developed and shown to be generally more accurate than random (Bowley, Jensen, Tschuprov). The thirties brought forth the development of the areas as a sampling unit, optimal allocation in stratified sampling, the ratio estimator, multistage sampling, and the popularization of sampling in social surveys by the Gallup poll and the *Literary Digest* debacle. In the forties, Hansen and Hurwitz brought forth PPS sampling and its use in multistage sampling with unequal sizes of primaries. Also a number of studies were made on the nature and extent of errors in surveys. The fifties appear to have been a decade of improving, extending, and optimizing various techniques and principles and developing the idea of multiple frames. The sixties seem to have been almost entirely devoted to polish. Perhaps the big developments are over.

Principles.   Although sampling theory depends either explicitly or implicitly on randomization for achieving two of its usually accepted desired goals, unbiasedness and a measure of risk in the estimates made, yet most of the sample designer's attention is devoted to the imposition of various restrictions on randomization. These restrictions are carried out in either of two ways: (1) through allocation or (2) through estimation.

For optimal use of sampling resources, one cannot ignore costs. For proper consideration in design, costs must be known for each possible alternative for each survey operation. Hence the problem of survey design becomes one of constructing a scheme that provides the desired result with the least cost—a problem of production economics.

Several principles or "laws" seem to be present and useful, such as the "proximity principle," wherein elements located in close proximity (geographically, say) are more "alike" than those not, hence the intraclass correlation coefficient, δ, is generally positive. Another useful principle is that the average distance connecting random points in space (a travel route, say) is a square-root function of the number of points. An interesting principle is that larger clusters become more efficient relative to small clusters when stratification control is increased. Also within-stratum variance generally gets smaller when the number of strata are increased. Another is a "constancy rule" for call-backs—the completion rates should remain roughly constant for each call; if not, then look for inefficiency (or super efficiency if rates go up!) in the way calls are made.

Prior Information. The importance of information available to the sample designer prior to the design of the sample has been indicated a number of times in earlier chapters. This information may be used in the selection of sample elements by judgment, in forming cluster SUs, in forming strata, in assigning probabilities of selection, in choice of estimator, in deciding on stages and on whether multiphase sampling should be used. Rarely does this "prior" information seem to be readily at hand—it has to be sought out, usually following hunches. More often than not it is in an inappropriate form and therefore requires some modification, alteration, reorganization, and so on to put it into suitable form. But when it turns out to be effective, the designer can be very grateful indeed.

Ingenuity in Design. With the principles in mind and with prior information at hand, the problem now is to put something together that does a reasonably good job of meeting the specifications—if not the best. The options are many when the survey is expected to produce a variety of results and explores unexplored areas, whatever the subject matter. Although the principles and facts are helpful, the final determination of the "right" design seems to be unsystematic and hence is regarded more as an artistic rather than scientific achievement. I think it is both. And what is more scientific than good art!

◇　　◇　　◇

12.14. Review Illustrations

1. The following scheme is used for selecting a single individual at random from a household. A random fraction, $0 < RN < 1$, is assigned to each HH in advance to its visitation by the interviewer. At the visit the interviewer obtains a list of all occupants (previously unknown) by order of birthdate. By multiplying $N \times RF$ and rounding up, we obtain the sample number, SN.

(a)　If $N = 4$ and $RF = 0.277$, what is the SN?
(b)　If $N = 2$ and $RF = 0.277$, what is the SN?
(c)　Show that the procedure is unbiased.

Solution

(a) $N(\text{RF}) = (4)(0.277) - 1.108$ or $\text{SN} = 2$.
(b) $N(\text{RF}) = (2)(0.277) = 0.554$ or $\text{SN} = 1$.
(c)   Note that in (a) all RFs from 0.001 to 0.250 yield $\text{SN} = 1$ or $P_i = 0.250/1.000 = 1/4 = 1/N$.

2.   In the survey of farms in Iowa [Jessen (1942)] 452 were contacted January 1, 1939, and again on January 1, 1940. One item, number of swine on the farm January 1, increased from 14,583 to 19,903 in the 12-month period. The average s_W^2 (within-county strata) was 3913, and the correlation of swine numbers between the two years was .9.

(a)   Calculate the 95% confidence limits on the change in swine numbers between the two years. Comment.
(b)   If the samples were independent in the two years, would you likely come to the same conclusion in (a)?
(c)   Actually the 452 farms revisited in 1940 were half those in the 1939 survey. What should the optimal matching fraction be if the number of swine in 1940 is to be estimated most accurately and with 904 farms sampled each year?

Solution

(a) The variance of the difference of two sample means is given by Eq. 5.4,

$$\text{var}\,(\bar{y} - \bar{x}) = \left(\frac{N - n}{Nn}\right)(S_y^2 + S_x^2 - 2RS_y\,S_x).$$

In our case $s_y^2 = s_y^2 = s_W^2 = 3913$; N is large; $n = 452$, and $R = 0.9$.

$$\widehat{\text{var}}\,(\bar{y} - \bar{x}) = \frac{2s_W^2}{454}(1 - 0.9) = 1.724.$$

The 95% confidence limits are

$$(\bar{y} - \bar{x}) \pm 2\text{SE}\,(\bar{y} - \bar{x}) = \left(\frac{19{,}903}{452} - \frac{14{,}583}{452}\right) \pm 2(1.31)$$

$$= 11.77 \pm 2.62 \quad \text{or} \ (9.15, 14.39).$$

It can be concluded that the swine population in the state has increased at least 9.0 animals per farm with confidence of 95%.

(b) Now

$$\text{var}\,(\bar{y} - \bar{x}) = \left(\frac{N-n}{Nn}\right)(S_y^2 + S_x^2)$$

$$\doteq \frac{2s_W^2}{454} = 17.24,$$

and the 95% confidence limits are now 11.77 ± 8.30 or $(3.47, 20.07)$. We still can conclude that the swine population has increased, but we are much less confident on the magnitude of the increase.

(c) According to Table 12.1 the optimal match ratio should be 0.30, and the result gain in precision would be 39%.

3. The subscribers to magazines A and B are being studied by samples selected from subscription lists. Of particular interest is the group that subscribes to both A and B.

(a) What steps should be taken to assure that this group can be detected, assuming that the two lists cannot be collated for doing this?

(b) A ramdom sample of n_A and n_B names are selected from the N_A and N_B names from the A and B lists, respectively, where n_A/N_A is not necessarily equal to n_B/N_B. Determine a good estimator of N_{AB}, the number of subscribers of both magazines, and its variance.

(c) Determine a good estimator of $\overline{Y}_{AB}$, the mean of characteristic y for the dual group, and its variance.

Solution

(a) Ask each subscriber in the sample whether he is a subscriber to the other magazine or not.

(b)

$$N_A = N_a + (N_{AB} = N_{ab}),$$
$$N_B = N_b + (N_{AB} = N_{ab}),$$
$$n_A = n_a + n'_{ab},$$
$$n_B = n_b + n''_{ab}.$$

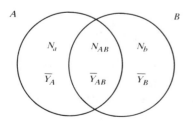

Two estimators of N_{AB} are immediately available; $\hat{N}'_{ab} = (N_A/n_A)\,n'_{ab}$ and $\hat{N}''_{ab} = (N_B/n_b)\,n''_{ab}$, one from each frame. Combined we have

$$\hat{N}_{ab} = W_A \left(\frac{N_A}{n_A}\right) n'_{ab} + (1 - W_A)\left(\frac{N_B}{n_B}\right) n''_{ab}.$$

To obtain a W_A that minimizes the variance of N_{ab} we first determine the variances of the two independent estimators.

$$\mathrm{var}\,(\hat{N}'_{ab}) = \left(\frac{N_A}{n_A}\right)^2 n_A \left[\left(\frac{N_{AB}}{N_A}\right)\left(\frac{1 - N_{AB}}{N_A}\right)\right],$$

$$\mathrm{var}\,(\hat{N}''_{ab}) = \left(\frac{N_B}{n_B}\right)^2 n_B \left[\left(\frac{N_{AB}}{N_B}\right)\left(\frac{1 - N_{AB}}{N_A}\right)\right],$$

if FPCs are ignored. If differences between the bracketed terms can be ignored, then an optimal W_A is given by

$$W_A = \frac{1/[(N_A/n_A)^2\, n_A]}{1/[(N_A/n_A)^2\, n_A + 1/[(N_B/n_B)n_B]}$$

or

$$W_A = \frac{f_A^2/n_A}{f_A^2/n_a + f_B^2/n_B}.$$

If the fs, the sampling fraction, are equal, then

$$W_A = \frac{n_B}{n_A + n_B} \quad \text{and} \quad 1 - W_A = \frac{n_A}{n_A + n_B}.$$

The variance of $\hat{N}_{ab}$, the combined estimator, is

$$\mathrm{var}\,(\hat{N}_{ab}) = W_A^2\,\mathrm{var}\,(\hat{N}'_{ab}) + (1 - W_A)^2\,\mathrm{var}\,(\hat{N}''_{ab}).$$

The variances of the individual estimates were given above.

(c) To estimate $\bar{Y}_{ab}$: $\bar{y}_{ab} = W_A \bar{y}'_{ab} + (1 - W_A)\bar{y}''_{ab}$. To obtain W_A, we note that the variances of $\bar{y}'_{ab}$ and $\bar{y}''_{ab}$ depend on a common S_{ab}^2 and the sample sizes. Ignoring FPC, $W_A = n_B/(n_A + n_B)$ and $1 - W_A = n_A/(n_A + n_B)$. Then,

$$\bar{y}_{ab} = \left(\frac{n_B}{n_A + n_B}\right)\bar{y}'_{ab} + \frac{n_A}{n_A + n_B}$$

and

$$\mathrm{var}\,(\bar{y}_{ab}) = S_{ab}^2 \left(\frac{W_A^2}{n'_{ab}} + \frac{(1 - W_A)^2}{n''_{ab}}\right).$$

4. Fisheries biologists are interested in a number of characteristics of persons fishing in a certain lake during the fishing season, where all persons fish from the shore. A sample was obtained by selecting a number of days at random

during the fishing season and, starting at random points on the shore of the lake, interviewing all persons found fishing. The shore of the lake was completely canvassed in this manner in a day. Data of the following type were obtained.

Day no.	Fisherman no.	Age	Sex	Length of time expected to be fishing today (hours)
1	1	20	M	$\frac{1}{2}$
	2	65	F	4
	3	16	M	2
2	1	40	M	8
	2	30	M	$\frac{1}{2}$
3	1	25	F	3
	2	30	M	1
	3	67	M	$3\frac{1}{2}$
	4	70	M	2

(a) Estimate the fraction of persons fishing in this lake that are men.

(b) Estimate the total hours of fishing effort (fisherman-hours) put on this lake (assume a season of 90 12-hour days).

Solution

Let X_i and Z_i be hours fished and "maleness" of a person i ($Z_i = 1$ if i is a male and 0 if female). Then P_i, the probability of selecting person i, is $P_i = kX_i$.

(a) Fraction of men:

$$\frac{\hat{Z}}{\hat{N}} = \frac{\sum Z_i/X_i}{\sum 1/X_i} = \frac{6.411}{6.994} = 9.18 \quad \text{or } 92\%.$$

(b)

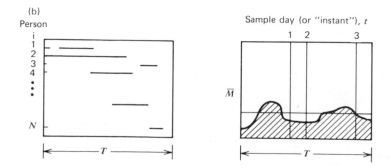

Person

i

1
2
3
4

N

$\longleftarrow T \longrightarrow$

Sample day (or "instant"), t

1 2   3

$\overline{M}$

$\longleftarrow T \longrightarrow$

(b)　We can regard the sample as one in which random "instants" are selected (except ours is "stretched-out") and the number of persons observed, M_j, is the number of fishermen fishing over the whole lake at instant t, where $t = 1, 2,$ and 3 in the sample. To estimate $\bar{M}$ we have three observations of M_t:

$$\bar{m} = \frac{3 + 2 + 4}{3} = \frac{9}{3} = 3.00 \text{ persons/"instant,"}$$

$$\bar{M} = T\bar{m} = (90)(12)(3.00) = 3240 \text{ man-hours.}$$

5.　In a health survey using the HH as a sampling unit the following problem arose. The respondent is asked whether he had any illness during, say, the past two weeks. If the answer is "yes," then he is asked when the "episode" of illness began and whether or not it is now ended. An objective of the survey is to estimate total "episodes" by kinds of illness involved and a number of characteristics of these episodes, such as length, type of person involved, days lost from work or school, and whether a doctor attended it or not. Suppose the households are selected at some random time during a year and Z is the total number of episodes occurring in the universe during the year, Y is the number of days lost from work or school from illness.

(a)　How can Y be estimated?
(b)　How can Z be estimated?

Solution

P, p = length of the total and sample period,
E_{ij}, e_{ij} = length of the jth episode and portion in sample period by the ith person.

To estimate Y_i, total days lost by the ith person,

$$\hat{Y}_i = \frac{Y_{ij}(e_{ij}/E_{ij})}{P/p} = \frac{Y_{ij}}{(E_{ij}/e_{ij})(p/P)}.$$

Hence the probability of picking up the jth episode is $(E_{ij}/e_{ij})(p/P)$.
(a) To estimate Y we have, for a sample of n from N,

$$\hat{Y} = \frac{N}{n}\sum_{i=1}^{n}\hat{Y}_i = \frac{NP}{np}\sum_{i=1}^{n}\frac{e_{ij}\,Y_{ij}}{E_{ij}}, \qquad j = 1.$$

[*Note*: We must sum over all persons in a household also.]

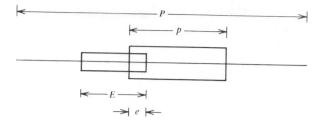

(b) To estimate Z, set

$$Z_{ij} = \begin{cases} 0 \text{ if no episode is "caught,"} \\ 1 \text{ if an episode is "caught";} \end{cases}$$

then

$$\hat{Z} = \frac{N}{n} \sum_{i=1}^{n} \hat{Z}_i.$$

6.   A life insurance company wishes to determine the proportion of its employees that are alcoholics. Presuming that alcoholics do in fact know they are alcoholics but are unwilling to disclose the fact, the survey is to be undertaken using the "randomized-response" method. A sample of 400 employees is selected at random. Of these 56 say "yes," the spinner agrees with his true status (alcoholic), where the "yes" portion of the spinner occupies 90% of the dial.

(a)   Estimate the true fraction of alcoholics in the firm.
(b)   Estimate the standard error of the estimate in (a).
(c)   Is it possible, despite the 56 yeses, that P is in fact zero? Comment.
(d)   Estimate the relative efficiency of the spinner technique here.

Solution

(a) $$\hat{P} = \frac{1}{2\theta - 1} \left[(\theta - 1) + \frac{n_1}{n} \right]$$

where $\theta = 0.9$, $n_1/n = \frac{56}{400} = 0.14$,

$$= \frac{1}{2(0.9) - 1} (0.9 - 1 + 0.14)$$

$$= \frac{1}{0.8} (-0.10 + 0.14) = \frac{0.04}{0.80} = 0.05 \quad \text{or } 5\%.$$

(b) $\qquad \hat{var}(\hat{P}) = \dfrac{\hat{P}(1-\hat{P})}{n} + \dfrac{1}{n}\left[\dfrac{1}{16(\theta-\frac{1}{2})^2} - \dfrac{1}{4}\right]$

$\qquad\qquad = \dfrac{(0.05)(0.95)}{400} + \dfrac{1}{400}\left[\dfrac{1}{16(0.9-0.5)^2} - \dfrac{1}{4}\right]$

$\qquad\qquad = 0.000{,}119 + 0.000{,}353 = 0.000{,}472$

And $\hat{SE}(\hat{P}) = \sqrt{0.000{,}472} = 0.022.$

(c) The 95% confidence limits are $0.050 \pm 2(0.022)$ or $(0.006, 0.094)$; therefore $P = 0$ lies outside the 95% confidence limits, which is unlikely to be the case here. The 56 yeses are not acknowledged alcoholics; they are simply cases where concordance has occurred. They may all be nonalcoholics where the spinner turned up with "no"—that is, nonconcordance.

(d) $RE(\text{spinner/direct}) = \dfrac{var(p)}{var(\hat{P})} = \dfrac{0.000{,}119}{0.000{,}472} = 0.25$

or a loss of about three-fourths of the potential information.

7. Considering budget, costs, and variances for a proposed survey of households in the state of California, you determine that optimum cluster size turns out to be four elements.

(a) Suppose the budget subsequently is doubled. What effect, if any, would this have on optimum cluster size?

(b) Suppose the interview is doubled in length; how is optimal cluster size affected?

(c) Suppose the initial sample design did not involve stratification. If stratification is imposed, what effect will this have on optimal cluster size?

Solution

(a) Optimal size becomes smaller (since a larger budget will permit a larger sample, economies of scale, such as on travel, will favor a smaller cluster).

(b) Optimal size becomes smaller, owing to larger C_W relative to C_B, and according to Eq. 4.37 or 4.38 $M(\text{opt})$ will decrease.

(c) Without stratification

$$RE(M_0/1) = \frac{MS(T)}{MS(B)} \doteq \frac{S_B^2 + S_W^2}{M_0 S_B^2 + S_W^2}.$$

With simple stratification

$$RE(M_0/1)_{\text{strat}} = \frac{MS(T)}{\sum W_h [MS(B)]_h} \doteq \frac{S_B^2 + S_W^2}{\sum W_h [M_0(S_B^2)_h + (S_W^2)_h]},$$

but $\sum W_h(S_W^2)_h = S_W^2$, and, if stratification is effective, $\sum W_h(S_B^2)_h < S_B^2$, and therefore $\mathrm{RE}(M_0/1)_{\mathrm{strat}} > \mathrm{RE}(M_0/1)$.

12.15. References

Birnbaum, Z. W.
Sirken, M. G.
1965
    Design of Sample Surveys to Estimate the Prevalence of Rare Diseases: Three Unbiased Estimates. *Vital and Health Statistics*, PHS Publication No. 1000, Series 2, No. 11 (October). Washington, D.C.: National Center for Health Statistics.

Chandrasekar, C.
Deming, W. E.
1949
    On a Method of Estimating Birth and Death Rates and the Extent of Registration. *Journal of the American Statistical Association*, **44**: 101–115 (March).

Cochran, W. G.
1946
    Relative Accuracy of Systematic and Stratified Random Samples for a Certain Class of Populations. *Annals of Mathematical Statistics*, **17**: 164–177.

Cochran, W. G.
1963
    Sampling Techniques. (2nd ed.) New York: John Wiley & Sons, Inc.

Dalenius, T.
1957
    Sampling in Sweden. Stockholm: Almquist and Wiksell.

Daly, J.
Gurney, M.
1965
    A Multivariate Approach to Estimation in Periodic Sample Surveys. (46 pp.) U.S. Bureau of the Census. Paper presented at the Annual Meeting of the ASA, September 1965, Philadelphia.

Deming, W. E.
Glasser, G. J.
1959
    On the Problem of Matching Lists by Sampling. *Journal of the American Statistical Association*, **54**: 403–415 (June).

Folsom, R. E.
Greenberg, B. G.
Horvitz, D. G.
Abernathy, J. R.
1973
    The Two Alternate Questions Randomized Response Model for Human Surveys. *Journal of the American Statistical Association*, **68**: 525–530 (September).

Frank, O.
1968
    Sampling from Overlapping Populations. Research report, University of Stockholm.

Goodman, L. A.
1949
    On the Estimation of Number of Classes in a Population. *Annals of Mathematical Statistics*, **20**: 572–579.

Goodman, L. A.
1961
    Snowball Sampling. *Annals of Mathematical Statistics*, **32**: 148–170.

Hansen, M. H.
Hurwitz, W. N.
Madow, W. G.
1953
    Sample Survey Methods and Theory, Vol. I: *Methods and Applications*. New York: John Wiley & Sons, Inc. (Good suggestions for dealing with "rare" items.)

Hartley, H. O.
1962
    Multiple Frame Surveys. *Proceedings of the Social Statistics Section of the American Statistical Association*, pp. 203–206.

Hsieh, Nan-Chang
1970
    Some Estimation Techniques for Utilizing Information from Elements Not in the Sample. Ph.D. Dissertation, University of California, Los Angeles.

Jessen, R. J.
1942
    Statistical Investigation of a Sample Survey for Obtaining Farm Facts. *Iowa Agricultural Experiment Station Research Bulletin* 304. Ames, Iowa: Iowa State University.

Jessen, R. J.
1962
On the Adequacy of Statistics for Economic Planning and Development in Costa Rica. Report to the U.S. Agency for International Development and the Banco Central de Costa Rica. (48 pp., mimeographed.)

Jessen, R. J.
1969
Some "Master" Sampling Frames for Social and Statistical Surveys in California, in N. L. Johnson and H. Smith, Jr. (eds.), *New Developments in Survey Sampling.* New York: John Wiley & Sons, Inc. (Pages 457–481. An example of a three-way frame.)

Jessen, R. J.
Houseman, E. E.
1944
Statistical Investigation of Farm Sample Surveys Taken in Iowa, Florida, and California. *Iowa Agricultural Experimental Station Bulletin* 329. Ames, Iowa: Iowa State College. (An example of use of aerial photos to detect "0" SUs.)

Jessen, R. J. et al.
1951
Accessibility and Availability of Farm Machinery in Iowa. V. I. Case Co (128 pp.) (An example of a search technique where one member of a "rare" frame is asked to identify others—in this case, competitors.)

Jessen, R. J.
Thompson, D. J.
1958
Encuesta por Mustreo de las fincas en la provincia de Buenos Aires, Argentina. *Estadistica: Journal of Inter-American Statistical Institute,* **16** (61): 464–504 (December). (An example of selecting farms by including any with portions lying within the SU.)

Kish, L.
1961
Efficient Allocation for a Multi-purpose Sample. *Econometrica,* **29**: 363–385.

Madow, L. H.
1946
Systematic Sampling and Its Relation to Other Sampling Designs. *Journal of the American Statistical Association,* **41**, 207–214.

Madow, W. G.
Madow, L. H.
1944
On the Theory of Systematic Sampling. *Annals of Mathematical Statistics,* **15**: 1–24.

Marshall, C. E.
1956
Cost Control of Sample Survey by Two-Step Designs. Ph.D. Dissertation, Iowa State University, Ames, Iowa.

Mosteller, F.
1949
Questions and Answers. *American Statistician,* **3**: 12–13 (June–July). (An early presentation of the problem of estimating the number of classes from a sample.)

Patterson, H. D.
1950.
Sampling on Successive Occasions with Partial Replacement of Units. *Journal of the Royal Statistical Society,* Series B, **12**: 241–255.

Peters, J. H.
Bucher, M. L.
ca. 1942
The 1940 Section Sample Survey of Crop Acreages in Indiana and Iowa. (44 pp., mimeographed.) U.S. Department of Agriculture, Bureau of Agricultural Economics.

Sirken, M. G.
1970
Household Surveys with Multiplicity. *Journal of the American Statistical Association,* **65**: 257–266 (March).

Snedecor, G. W.
1939
Design of Sampling Experiments in the Social Sciences. *Journal of Farm Economics,* **21**: 846–855 (November).

Stephen, F.
1948
History of the Uses of Modern Sampling Methods. *Journal of the American Statistical Association,* **43**: 12–39.

Sudman, S.
1972
On Sampling of Very Rare Human Populations. *Journal of the American Statistical Association,* **67**: 335–346 (June).

Warner, S. L.
1965
Randomized Response: A Survey Technique for Eliminating Evasive Answer Bias. *Journal of the American Statistical Association,* **60**: 63–69.

Woodruff, R. S.
1963
The Use of Rotating Samples in the Census Bureau's Monthly Surveys. *Journal of the American Statistical Association,* **58**: 454–467 (June).

Yates, F.
1949
Sampling Methods for Censuses and Surveys. (1st ed.) London: Charles Griffin & Sons.

Zarcovich, S. S.
1956
Notes on the History of Sampling in Russia. *Journal of the Royal Statistical Society,* **119**: 336–338.

12.16   Exercises

1.   Give an example of a problem for which you would almost certainly not want to use systematic sampling.

2.   A "panel" is a sample from which measurements are made at different time periods. For example, a sample of voters might be interviewed about voting intentions several times during the course of a political campaign. Using a political opinion poll as an example:

(*a*) What is an important advantage of the panel design?
(*b*) What potential disadvantages can you think of?

3.   A study is to be made of the subscribers to magazines A, B, and C. Some people subscribe to one, others to two or more. Suppose it is impossible to check subscriber lists of the three magazines to determine the "overlap" subscribers. Suggest a procedure by which unbiased estimates of those subscribing to one or more of the three magazines can be made of independent samples of the three magazine lists.

4.   Examine the following rules for associating observation units with sampling units and state whether each rule permits each observation unit to be selected with known probability.

(*a*) Farms are associated with SU containing largest portion of farm area.
(*b*) Farms are associated with SU containing residence of its operator.
(*c*) Fields are associated with SUs containing portions of a field by a random choice of the several SUs involved.
(*d*) Farm owners are associated with SU containing the headquarters of farm owned.
(*e*) Drainage areas are associated with the SU containing the outlet.

5.   The problem is to estimate certain characteristics of fishing in a given lake. Suppose fishing takes place from the shore and between the hours of 6 A.M. and 6 P.M. A sample of fishermen is obtained by first choosing a random point on the shore perimeter and then selecting all persons met while proceeding along the shore for a fixed distance p of a total P. Data obtained from each intercepted fisherman include (1) time spent and expected to be spent fishing

that day, (2) age, (3) sex, (4) total catch to the time of contact. Devise appropriate estimators for

(a)   Total man-hours of fishing done on the lake.
(b)   Average age of fishermen.
(c)   Percent of fishermen who are women.
(d)   Total fish caught.

CHAPTER 13

Coverage, Canvassing, and Measurement

◇   ◇   ◇

13.1. Introduction

For many statistical surveys once the sample has been selected, the rest of the work is fairly routine and simple—observations are obtained on the sampling units, the data processed, estimates made, and inferences drawn. The survey problem is hardly more than a simple problem in applied sampling theory. But many surveys seem to have rather serious problems even after the sample has been selected. These are the problems of carrying out a survey, given the set of sampling units to which it is to be confined. They are essentially the same problems that would be faced if the survey were a complete census.

There are three rather distinct operations in the data gathering portion of a survey: (i) *coverage*, seeking out and obtaining data on every OU (observation unit) in the sample, (ii) *canvassing*, the task of examining each SU (survey unit) to detect the identity of each OU appropriate to that SU and its exact boundaries or extent, and (iii) *measurement*, the classifying, counting, or measuring needed on each OU to obtain the basic data required of the survey. Hence coverage here will mean those activities concerned with getting a Y_i from each OU; canvassing, those activities dealing with scrutinizing each potential OU to see that it meets the appropriate test of fitness; and measurement, the activities of obtaining the final observations. Each topic will be dealt with separately below.

Up to now it has generally been assumed that when a sample has been designated, the required Y_is will be forthcoming and without error on each and every OU in that sample. In other words, the survey conducted on the sample will be both complete and accurate. The theory presented is valid only when the data are complete and accurate. However, in practice this requirement is frequently not met and, in some cases, cannot possibly be met. What is to be done in these cases depends on a number of considerations, some of which will be presented in this chapter.

The problems of coverage will be considered first. What happens to the validity of our inferences if observations are not obtained from all OUs in the sample? Does incomplete coverage merely reduce sample size, or are there more serious consequences? If so, what, if anything, can be done? These and related matters will be dealt with in Part A.

Part B will discuss the problems of canvassing, the errors that can be expected, and their consequences. The problems of measurement, particularly those of interviewing, will be considered in Parts C and D. Part E will give some suggestions on questionnaire construction.

13.2. Missing (Nonresponse) and Hard-to-Get Data

In the actual conduct of statistical surveys it is frequently found that observations on some elements are difficult to obtain and in some cases cannot be obtained at all. In common interview surveys, for example, where homes are visited without appointment, a number of them will be found with no one at home. Even after repeated "call-backs" there may remain a group of homes in which no one is at home during the period of the survey, or of those found at home, some may refuse to provide the requested data. The resulting *missing data* cause difficulties in continuing with the survey. Can they be ignored with impunity?

This and the following section consider the likely consequences of ignoring missing data and the merits of some techniques for dealing with the problem. Obviously one method of avoiding "missing data" is simply to spend more and more time, money, and other resources until the data are obtained. For example, "not-at-homes" can be eliminated from the missing-data category by repeated call-backs. Some problems of the economics of collecting "hard-to-get" data will also be considered in this chapter.

The principle causes of missing data vary by type of survey and the nature of the elements on which observations are to be taken. Listed below are examples in each of three classes of surveys.

Mail	Interview	Direct observation or measurement
Wrong address	Overlooks the household	Element (observation unit) not available
Won't respond	Not at home	Miscellaneous reasons
Doesn't complete the questionnaire	Refuses to give interview	
	Doesn't answer question; either doesn't know or refuses	

The missing-data problem is certainly not confined to surveys that depend on people to produce the desired information. In "crop-cutting" surveys, where small plots of the crop are harvested and weighed in order to determine yield, and where it is necessary to wait until the crop is nearly ripe before its yield can be properly assessed, some fields will be harvested by the farmer before the survey investigator arrives—thus producing a missing datum.

Unless a sample design specifically requires certain elements to be observed, the amount of missing data may not be easily ascertainable. Many so-called quota samples fail to show the number of missed respondents simply because only those persons are approached who are at least visibly present. Even samples requiring a fixed set of elements will not have a missing-data problem if substitutions are allowed. In these cases the interviewer goes to the designated households first and obtains interviews for those at home and who don't refuse. For those not-at-home, or for the refusals, he then goes to substitute households, according to whatever procedure was laid down for him, until he gets the required number of interviews. Actually in this case all not-at-homes and refusals on designated homes are technically missing data for which substitutions were supplied by field procedures.

The true extent of missing data is probably greater than we may be aware, owing partly to its insidious nature in some sampling designs and procedures and partly to a failure to properly report it in many cases where it is known or could be made known. The sections to follow will deal with the extent of the problem in more detail, and its possible effects on the accuracy of results will be presented for a few cases.

Kinds and Extent.   Although only a few examples will be given here, the reader may gain from them some idea of how serious or how trivial the problem might be.

Missing data will be regarded as the data not obtained for any reason on valid elements designated for observation in the sample design, either explicitly or implicitly. Since the amount of missing data in a survey depends on a number of factors, such as type of survey (mail, interview, objective measurement, etc.), type of respondent (whether any person in a home is a specified one, whether a cross-section of all families is sought or some occupational or income group, etc.), skill and diligence of the interviewer (in case of interview surveys), appeal of the questionnaire (in case of mail surveys), the nature of the inquiry (whether appealing or not), time of day or year the survey is made, number and nature of repeat attempts made (such as a series of mail-outs or several "call-backs" for an interview rather than one), it will suffice here to give only some rough ideas of the effects these factors may have on the extent of missing data. Table 13.1 gives some general experiences with missing-data rates by type for surveys involving the four commonly used methods of eliciting observations. Table 13.2 gives the completion rates by type for a number of surveys using personal interviewing. It is common for interview surveys to have from 10% to 30% of the data "missing."

Effects of Missed Data on Survey Accuracy.   Since the properties of the samples discussed in this book presume that all observations have been taken as prescribed by the sampling procedure, any incompleteness or pseudo-

Table 13.1. Rough non-completion rates by type commonly experienced by surveys using different methods of observation

1. Personal interview surveys	5 to 50%
NCV—Not covered (cannot locate, overlook)	0–3%
NAH—Not at home (away at work, vacation, etc.)	5–50%
RNA—Respondent not available (too busy, sick)	1–20%
REF—Refusal	1–10%
2. Telephone interview surveys	20 to 40%
NCV—Not covered (no telephone, not listed, etc.)	30%
NAH—Not at home (no answer)	18%
RNA—Respondent not available (line busy)	11%
REF—Refusal	0%
3. Mail surveys	5 to 90%
NCV—Not covered (wrong address, returned undelivered)	5–20%
NAH—Not at home ⎫	
RNA—Respondent not available ⎬ (No reply)	5–90%
REF—Refusal (return unusable, etc.)	1–5%
4. "Objective" surveys (direct measurement)	0 to 10%
NCV—Not covered (roads out, overlooked, etc.)	0–3%
NAH—Not at home (too early or too late for crop cutting)	5–10%
RNA—Respondent not available (crop or trees under spray)	4%
REF—Refusal (farmers refuse trespassing)	0–3%

Table 13.2. Completion rates experienced in various interview surveys[a]

	Survey, time and place	Type of respondent required	Completeness and other experience	Rules and remarks
1.	Iowa Farm Survey (Dec. 1938–Jan. 1939)	Operator of a farm	71% completed 13% NAH, 16% "refusals" on single call	One call only (substitutes were taken from nearest farms)
2.	Greek II Survey (July–Aug, 1946)	Responsible adult	99.7% completed, 0.3% "refusals"	Repeated calls
3.	National Sample of HHs, MRCA (Winter 1947)	Responsible adult in HH	Urban: 93.0% completed, 72.0–19.4–8.6 by call Rural: 97.0% completed, 83.5–10.3–6.2 by call	3 calls

4.	Iowa Food Habit and Preference Survey (Jan.–June 1947)	Individual persons: (1) Age 17–19 (2) Age 46–58	(1) 73.0% completed (2) 75.1% completed	? calls
5.	North Central Housing Survey Spring 1948)	Homemaker	84.5% completed	? calls
6.	Elmira Study, Elmira, N. Y. (Summer 1948)	Randomly selected adult from HH	81.9% completed, 38.0–32.9–29.1 by call; 6.3% NAHs, 12.6% refusals	3 calls
7.	Sources of Information Survey, Iowa (Oct. 1948)	(1) Operator of a farm (2) Homemaker of a farm	(1) 84.6% completed (2) 90.8% completed	? calls
8.	Survey of Consumer Finances, USA (Winter 1950	Head of "spending unit"	95.5% completed	Repeared call-backs
9.	Agriculture Survey in Ecuador Sept. 1952)	Operator of farm	100% completed	Repeated calls; operators traced to town residences when necessary
10.	TV Ratings (See Report, p. 77) (1961)	Anyone competent	Personal interview: 52% Telephone recall: 59% Telephone coincidental: 52% Meter: 62% Diary: 54%	
11.	Calif. Immunization Survey (Spring and Summer 1964)	Mother of children under 5 years old	92.0% completed, 4.0% refusals, 4.0% NAH and other	3 calls, 1 hour questionnaire
12.	Mexican-American Survey, L.A. County (Summer 1965)	Either spouse ("balanced")	85% completed, 4.0% refusals, 8.5% NAH's, 2.5% other	3 calls
13.	Mental retardation Study, Riverside, Calif. (Summer 1964)	Responsible adult?	90.7% completed, 6.9% refusals, 2.4% NAH	$1\frac{1}{2}$ hour questionnaire, 10.2% refusals on first attempt

[a] HH = household; NAH = not at home.

457

completeness (acceptance of substitute observations, etc.) makes the theory inapplicable. The degree to which the theory leads to incorrect results depends on two things: (1) the differences, if any, between the characteristics of observed and unobserved elements and (2) the fraction of sample elements observed—that is, the *completeness* fraction. The latter can be obtained by keeping proper records during the course of the survey; the former can best be determined by special studies undertaken for that purpose.

Finkner (1950) studied the response of 3116 North Carolina fruit growers with known orchard sizes in three successive mailings of a questionnaire. He found that the growers responding to each successive mailing request, and those not responding to any of the three, differed in average number of trees (one of a number of questions asked) as shown in Table 13.3. The larger

Table 13.3   Finkner's fruit orchard data

	First mailing	Second mailing	Third mailing	Non-respondents after three mailings	All farmers
No. in sample	300	543	434	1839	3116
Percent returned (or remaining)	10	17	14	(59)	100
Item: "Number of fruit trees per farm"	456	386	340	290	329

growers responded to the first mailing, and the average size of orchard became smaller with each successive mailing. The respondents after three mailings consisted of 41% of all growers and had an average of 387 trees per farm as contrasted with the nonresponders with an average of 290. In this instance the missed observations are different enough to cause rather serious consequences if inferences on the average orchard size of all growers is based on the respondents as a representative sample of all growers. And the results would be even worse if one confined his sample to those easiest to reach—that is, those responding to the first mailing!

Clausen and Ford (1947) came to similar conclusions in studying the results of three mailings to recently discharged military personnel from World War II, where the item of particular interest was the fraction of veterans planning to enroll in school (and hence utilize the GI Education Bill). The results are shown in Table 13.4. Here again the differences in the proportion of veterans

Table 13.4. Clausen and Ford data

	First mailing	Second mailing	Third mailing	Non-respondents After three mailings	All veterans
No. in sample	7900	3300	1600	1800	14,600
Percent returned (or remaining)	54	23	11	(12)	100
Item: Percent "Considering enrolling in school"	10	15	21	(25)	(14)

considering enrolling in school runs from 10 % for respondents on first mailing to 21 % for the third and an estimated 25 % for nonrespondents. Clausen and Ford estimate from the trend appearing here that the true proportion is 14 %. If only the results of the first mailing were used, the estimate would be 10 %. Note that after three mailings 88 % of the observations were received, and their simple average of 12.7 % is not far from the estimated true proportion of 14 %. However, this is in part due to the rather high response rate—88 %, whereas the grower survey was only 41 % after three attempts.

An example of respondent characteristics by call in interview surveys is given by Stephan and McCarthy (1958) reporting results from the study of a survey at Elmira, N.Y. The data given in Table 13.5 show a characteristic trend on calls. Women with children under two years old are likely to be at home when survey interviewers appear.

Mercer and Butler (1967), in a study of the characteristics of refusals in a general household survey of Riverside, Calif., found that the refusal rate was

Table 13.5. Elmira, N.Y., data

	Interview first call	Interview second call	Interview third or later call	Non-respondents after 3+ calls	All HHs interviewed
No. in sample	2072	726	467	0	3265
Percent interviewed	63.5	22.2	14.3	–	100.0
Item: "HHs having children under 2 years of age"	17.2	9.5	6.2	–	13.9

about 2% for households with eligible respondents aged 18–20, gradually rising to 16% for those of ages 60–70. Apparently older people were less cooperative in giving information to survey takers.

These are only a small sample of the investigations carried out on response-nonresponse differences. Differences seem to be present in almost all cases examined. However, the magnitude is not easy to predict without some experience, and even there it is hazardous.

Since in general observed and nonobserved elements in a universe are found to differ, it is convenient to regard them as belonging to two strata, the observed (response) and unobserved (nonresponse). The sample provides information on the mean of $\bar{Y}_R$, the mean of the observed stratum, and on N_R/N, the fraction of elements in the observed stratum, but nothing on the mean, say $\bar{Y}_{NR}$, of the unobserved stratum. Since

(13.1)
$$Y = \left(\frac{N_R}{N}\right)\bar{Y}_R + \left(\frac{N_{NR}}{N}\right)\bar{Y}_{NR}$$

and since the sample mean $\bar{y}_R$ is estimating $\bar{Y}_R$, then the bias in $\bar{y}_R$ in estimating $\bar{Y}$ is

(13.2)
$$\text{bias}(\bar{y}_R) = \frac{N_{NR}}{N}(\bar{Y}_R - \bar{Y}_{NR}),$$

and this becomes zero only if $\bar{Y}_{NR} = \bar{Y}_R$ or $N_{NR}/N = 0$.

In the case where the Y_is are 0 or 1, hence $\bar{Y}_R$ and $\bar{Y}_{NR} = P_R$ and P_{NR}, limits of 0 and 1 can be put on P_{NR} and calculations can be made on conservative confidence limits for p_R. Hence the conservative 67% confidence limits are

(13.3)
$$\hat{P}_L = \frac{N_R}{N}\left(p_R - \sqrt{\frac{p_R q_R}{n_R}}\right) + \frac{N_{NR}}{N}(0),$$

$$P_U = \frac{N_R}{N}\left(p_R + \sqrt{\frac{p_R q_R}{n_R}}\right) + \frac{N_{NR}}{N}(1),$$

where nonrespondents are regarded as either all 0s or all 1s. By taking a larger sample, one can reduce the term $\sqrt{p_R q_R/n_R}$ in Eq. 12.40. But of course the terms involving N_{NR}/N will remain constant, and unless N_{NR}/N, the nonresponse rate, is small, the confidence limits will not reduce very much. Birnbaum and Sirken (1950) give a method for determining appropriate sample size in this case. Cochran (1963) gives some worked-out examples under a useful range of possible conditions.

Reduction and Prevention of Missed Data. Although in most cases it is impossible, or nearly so, to reduce N_{NR} to zero (which implies that n_{NR} in the sample is reduced to zero), it is possible in many cases to minimize it, and sometimes with little cost. The NCV (not covered) type can usually be minimized in area-list survey units by good instructions and aids such as maps and photos, in mailed and phone surveys by accurate mailing and phone lists, and in all cases by trained investigators and good detection schemes so that errors can be detected and resolved. NAH (not-at-homes) can be reduced by call-backs, and REF (refusals) can be minimized by appealing inquiries, skillful interviews, and sensible questionnaires. Economical call-backs can be made by skillful detection of the likelihoods of respondents' being at home and proper scheduling of interview assignments. Although some procedures will be given in the next section to minimize the undesirable effects of missed data, the best policy still seems to be to use a lot of care and cleverness in the prevention of incompleteness rather than in the treatment of its crippled data.

13.3.　Schemes for Dealing with Missing Data

There are two things one can do with missing data: (1) nothing or (2) make certain adjustments that appear appropriate to reduce the undesired effects. Actually, "doing nothing" is an implicit adjustment scheme, so in all cases something is done. A variety of schemes have been proposed and/or used. They may be put into four categories: (1) direct estimation of missing values, (2) estimation of the probabilities of response, (3) nonresponse as a double-sampling problem, and (4) miscellaneous schemes. All will be dealt with briefly here.

Direct Estimation of Missed Values.　　The unobserved value is regarded as a function of some characteristic of the elements that have been observed or otherwise known. For example, suppose income of a household has been missed but its location (block, area, etc.) and its housing expenditure (rent paid or assessed home value, etc.) are known. A simple model to supply the missing datum might be

$$(13.4) \qquad\qquad Y_{ij} = \mu + \alpha_i + \beta X_{ij} + \varepsilon_{ij},$$

where μ = the overall population mean of y,
　　　α_i = the component of y common to all elements in category i (say
　　　　　　block or area),
　X_{ij} = the observed X on the ij^{th} element,
　　β = some constant,
　　ε_{ij} = the residual deviation　of y after removing the μ, α_i, and βX_{ij}
　　　　　　"effects."

In practice it seems to be common that only crude estimates of $\mu + \alpha_i$ and β need be obtained in order reasonably to satisfy our needs—which appear to be the restoration of a balance that the original sample was designed to possess. If nothing is done for missed data, one is implicitly estimating $\alpha_i = 0$ and $\beta = 0$.

Another common practice is to duplicate the OU by a neighbor or a randomly drawn household from the same block (or area, etc.) to replace the missing household. This method, sometimes called the "dubbin" technique, is a crude way of estimating $\mu + \alpha_i$ and letting $\beta = 0$. If care is taken to choose a neighbor with a similar X_{ij}, then both α_i and β are being estimated. This method provides some semblance of balance without the need of complicated estimates of α_i and β or the use of weights (rather than duplicates) in the processing.

It may be noted that if X_{ij} is ignored, or β is regarded as zero, the model 13.4 becomes a poststratified type estimator, where weights are employed to estimate the overall mean, μ, say

$$(13.5) \qquad\qquad \hat{\mu} = \sum_i \frac{n_i}{n} \frac{\sum_j^{n_i'} Y_{ij}}{n_i'} ,$$

where it is assumed that $n_i/n = N_i/N$ in the original design and where n_i' elements of the n_i designated in category i have actually been observed and $n_i - n_i'$ have been missed. The advantages and disadvantages of duplication ("dubbin") of cards or entries versus direct estimation by some method versus the use of weights will not be discussed here. This area appears still to be only partially explored.

The "Resistance Function" Estimator.   Another model for estimating missed values is based on the assumption that the observed characteristic is functionally related to the difficulty of observing it. Indeed, this seems to be correct when one observes the means of $\bar{y}_k$ obtained for various calls (in interview surveys) and mailings (for mailed surveys). (See the data of Tables 13.3–13.5.) By fitting simple functions to data where, say, three calls or mailings have been used, one can then attempt an estimate of the values expected on further calls or mailings. Using these estimates for the missing values, an overall estimate can be obtained. The techniques are beguiling [see Clausen and Ford (1947) and Hendricks (1949)], but unless the missing fraction is small, the dangers are great at the present stage of development.

The "Probability of Miss-Out" Estimator.   Suppose a survey of HHs is taken at a fixed time (say 6–7 p.m.) during a week. A random sample of HHs is selected and contacted in a random order. When a household is contacted and a respondent is found at home, the interview is completed in the normal manner. However, an additional question is asked: "During the past week, how many evenings were you at home during this hour?" If everyone was at home during one or more of the K evenings through the week, the probability of obtaining an interview from each HH, given the sample, is k/K, where k is the number of evenings at home. To estimate the total of y for the sample, we have

(13.6)
$$\hat{Y} = \sum_{k=1}^{K} n'_k \bar{y}'_k \left(\frac{K}{k}\right),$$

where $\bar{y}'_k$ is the sample mean of y for "at homeness," class k, and $k = 1, 2, \ldots, 6$, say.

An estimate of $\bar{Y}$ is obtained by the ratio, denoted $\bar{y}_{PS}$, after Politz and Simmons (1949 and 1950),

(13.7)
$$\bar{y}_{PS} = \frac{\sum n'_k \bar{y}'_k / k}{\sum n'_k / k},$$

where n'_k is the number of sample elements found at home and classified in class k. Since the estimator is a ratio of two random variables, it is not unbiased, but the bias for reasonably large samples should not be worrisome. However, the variance may be. The method, suggested by Hartley (1946) and developed by Politz and Simmons (1949 and 1950), is intriguing since it is free from call-backs. However, the cost of this freedom is not trivial. Studies indicate that considering costs, the method is competitive but not generally superior to call-back methods. However, where survey time is short, it should offer distinct advantages over call-back methods.

Subsampling Missed Data. In this approach the position is taken that a survey should be carried out in the usual manner and as a consequence will have a large enough fraction of missing data that cannot be ignored. However, it is presumed that by using an alternative measuring scheme, the missing data can be ascertained—but at a somewhat higher cost. The problem then is how best to design the survey so as to optimally allocate the available resources properly between the conventional portion (the initial survey) and the followup.

This scheme is due to Hansen and Hurwitz (1946); only the general idea will be presented here. Essentially it is a type of double sampling (see Chapter 10). As a simple example, suppose one wishes to conduct a survey of HHs or business establishments by mail, but missing data appear to be a problem—as they should be! To reduce those worries, a portion of the sampling resources are set aside for an interview followup on a sample of the nonrespondents in order to determine their characteristics. The overall estimator of $\bar{Y}$, then, is the weighted value of the two portions:

(13.8)
$$\bar{y}_{HH} = \frac{1}{n} (n_R \bar{y}_R + n_{NR} \bar{y}_{NR}).$$

where n = the sample size of phase I—that is, the number of questionnaires
mailed out,

n_R = the number of responses, the number of mailed questionnaires
mailed back.

n_{NR} = the number of nonresponses: $n = n_R + n_{NR}$,

$\bar{y}_{NR}$ = the mean of y in the sample of r taken from the n_{NR} nonresponses:

$\bar{y}_{NR} = \sum_{i=1}^{r} Y_i / r$ $(i = 1, 2, \ldots, r)$.

If it is assumed that the personal interview results in no further nonresponses,
then 13.8 is an unbiased estimator of $\bar{Y}$. The variance of 13.8 is given by

$$(13.9) \qquad \overline{\text{var}}\,(\bar{y}_{HH}) = \left(\frac{N-n}{N}\right)\frac{S^2}{n} + \left(\frac{k-1}{n}\right) W_{NR}S^2_{NR},$$

where $k = n_{NR}/r$, the inverse of the sampling fraction used on the
nonrespondents,

W_{NR} = true fraction of nonrespondents,

S^2_{NR} = variance of y among the nonrespondents.

Note that if there are no nonrespondents, the second term drops out and 13.9
reduces to the conventional random sampling variance. If $k = 1$—that is, all
nonrespondents are sampled (by interview)—then the variance also reduces to
the conventional case, since $n_R + n_{NR} = n$. Hence when $k > 1$, the second term
on the right gives the additional variance due to sampling the nonrespondents.
Hansen and Hurwitz (1946) present optimal choices of n and k, where costs
and total budget are known. If $S^2_{NR} = S^2$ or approximately so, optimal k turns
out to be

$$(13.10) \qquad k_{\text{opt}} = \sqrt{\frac{c_{NR}W_R}{c_0 + c_1 W_R}},$$

where c_{NR} = the unit cost of obtaining and processing the observations taken
on the "nonrespondents,"

c_0, c_1 = the unit costs of mailing and of processing observations taken in
step 1 (that is, of the mailed survey) and the unit costs of
interviewing and processing in step 2, respectively,

W_R = the fraction of the universe that are respondents.

Since $c_{NR}W_R$ is the initial expected cost of obtaining and processing a
nonrespondent (since $W_R = 1 - W_{NR}$) and $c_0 + c_1 W_R$ is the initial expected cost
of obtaining and processing a respondent, k_{opt} is simply the square root of the
cost ratio of the two portions of the sample. A large saving—that is, a large
k_{opt}—will only be possible when the differential costs are rather large.
However, to avoid unknown and hence worrisome biases, it may be wise to do
some subsampling—that is, choose some $k > 1$.

Miscellaneous Techniques.   Kish and Hess (1959) suggest that where surveys are taken repeatedly, "nonrespondents" can be identified and information from them stockpiled and used in current surveys. In such repeated surveys Madow et al. (1961, p. 78) suggest that the previously identified nonrespondents be used as a regular portion of the current survey, and thereby all currently surveyed elements appear as receiving only one contact. Hence no call-backs are made on any current survey.

13.4.   Canvassing Problems

Nature and Types of Canvassing Operations.   Canvassing problems become the OU, and SU, if different (e.g., clusters, or a frame including duds), are not clearly recognizable either visibly or conceptually. If a store plans to survey its customers, then a question can be raised, "What is a customer of the store?" Likewise, what is a TV-watcher, a poor person, an employee, a household, a city block, a farm, a plot of 1/10,000 acres? Each of these OUs or SUs has been used in surveys; each has been, or perhaps should have been, operationally defined for the purpose. When an investigator is sent to a city block designated on a map with instructions to make a list of all HHs living there and to obtain some information on each, he may perform his task without error, or, more likely, he will bungle things because the "block" on the map is somehow different from what he finds, and when he starts making his list of households it is not clear to him what one really is. So he decides to go ahead and "do the job." In doing so he may miss a portion of the block because he stops at an alley instead of the street. Also he may fail to list four households living in trailers in the back and may "overlook" three students living in a basement apartment. He may be unable to reckon with the living arrangements of seven persons so he lists them as a single household, rather than the correct two.

    The list of examples can be easily extended. About how extensive and how damaging are such errors in practice?

Examples of Canvassing Errors.   In crop-cutting surveys for estimating yields of non-row crops using area plots of 1/10,000 acre and larger, estimates have been found to be from 5 % to 45 % too high, depending on plot size and type of crop. Homeyer and Black (1946) report a $2' \times 2'$ plot gave 8 % more "yield" than a $3' \times 3'$ plot on oats. Sukhatme (1946) likewise found yields overestimated by 4 % to 43 % on wheat where plot sizes ranged from 422 to 13 square feet, respectively. Hence the bias was related inversely to plot size. When studied, it appeared that the overmeasurements were attributable to two sources: (i) failure to properly randomize the placement of plots in the field, and (ii) failure to determine the exact boundary of the plot to be measured (in this case the grain within the plot was cut and subsequently threshed). The rules given the investigators for locating the plots were not rigorous enough to prevent the average investigator from favoring higher-yielding portions of the field for his sample plots and also gathering an excess of those plant stalks located along the boundary into the plot. The plot was incorrectly located and incorrectly defined operationally; hence serious errors in the Y_is were obtained.

In surveys where the area SU is used as a grid (e.g., to obtain clusters), there is a tendency to undercount the correct number of farms (using a headquarters rule of inclusion) and households contained therein. Kish and Hess (1958) report that a 10% undercount of households in block SUs can be reduced to 3% if proper measures are taken. Likewise farm undercounts up to 15% have been reported, although in a survey in Argentina 3% too many farms were "accounted for" using the headquarters rule of inclusion. (The overcount was due primarily to large farms, over 1000 hectares, many of which had apparently more than one "headquarters"—hence investigators tended to include such farms too frequently.) [See Jessen and Thompson (1958).]

In cases where the SU and OU are co-extensive, there may also be a problem of identification and delimitation of boundaries. A farm has been selected from a list of some sort—say, avocado growers. The person presumed to be the grower, when reading a questionnaire sent him, or if interviewed by the investigator, presents a canvassing problem. Is he a proper "farm"? If so, how many and which of his various activities belong to his "farm"? With households there are similar problems of determining their proper composition.

Detection and Control of Canvassing Errors.   The detection of canvassing errors is usually not very difficult except where concepts are "soft," as they may well be in dealing with economic and social problems. However, softness should not be an excuse for ignoring the problem. Fairly simple operational definitions have proved to be useful—for example, the "household" of the Bureau of the Census. "Business establishment" is also useful. But farm and firm still seem to be less useful operational concepts than they need be.

Investigators will perform their tasks better when concepts are operationally simple and they are provided with useful aids and equipment, such as good maps and aerial photos, and trained in their use.

In an attempt to control canvassing errors in an agricultural survey in Ecuador, where some farming was carried out by Indian organizations in a part-communal and part-private arrangement, and in non-Indian areas where the land was held and worked under various arrangements, it appeared feasible to design a field form that would require an accounting of the total land area within the SU along with the relationship of each plot or holding to the ownership. This permitted a rigorous two-way accounting of "farms" and "subfarms" on one hand and land ownership on the other. It also permitted an accounting of all land use, whether in "farms" or not. [See Jessen (1954).]

13.5. Basic Observation (Response) Error Problem

Errors in the observed values of the OUs under survey can arise for many reasons, depending on the nature of OUs, the characteristic being observed, and the manner in which it is observed and recorded. Generally the errors are greatest when the OU is a person and he is being asked by an interviewer to recall some events over some long period of time, or perhaps his opinion on some matter to which he has given little consideration. Moreover, errors are usually smallest when the OU is clearly recognizable, the characteristic being measured is unequivocal, and the means to measure it are accurate and at hand. An example might be a survey of TV sets to determine the proportion of them that are turned on, where mechanical devices can be used to detect and record the data. Most cases are somewhere in between. Since the case where a person is supplying information either by mailed questionnaire or by interview is most abundant with errors, we shall use it for illustrations and examples.

Nature of Interview Response Errors.   Whether a person is interviewed by "himself" (mailed questionnaire) or by someone else, whether the question pertains to a quantitative fact (e.g., year of birth) or to an opinion, errors can be produced by such factors as whether he understands the concept being dealt with or not, whether he is ignorant of the answer, whether he wishes to conceal the answer, and whether he wishes to present a certain image (e.g., being poor or rich, smart or dumb, good or bad). Moreover, memory is not always error-free, and it is frequently called upon to supply critical answers to apparently simple questions.

Examples of Response Errors.   In an attempt to obtain some measure of the nature and extent of error in farmers' reports of various business transactions during the year, when reported at the end of the year, a sample of 11 farmers, whose business records in January were being processed at Iowa State University, were called upon to answer the same questionnaire put to a sample of about 800 farmers being surveyed over the state of Iowa [Jessen (1942)]. Many of the replies to the questions (provided by memory) could be checked against entries in the farmer's business records, which were made available at the University. It may be argued that the results are not representative, since these farmers depended on their books and therefore were worse than average in their ability to recall business events. The results for hogs and cattle sold are shown in Table 13.6. In the case of hogs the interview values were exact only for those having no sales at all; on the average, 17% of the hogs sold were not reported. In the case of cattle, interview values were exact for all zero sales and for two nonzero cases. Moreover, on the average, cattle sales show no systematic error or bias in this small sample.

It may be argued that hog sales, being more frequent and somewhat

Table 13.6. Comparison of interview response with record check, 11 Iowa farms

Farm no. *i*	Number of hogs, all ages, sold during 1938		Number of cattle, all ages, sold during 1938	
	Interview X_i	Record Y_i	Interview X_i	Record Y_i
1	0	0	5	7
2	85	126	0	0
3	0	0	20	40
4	137	165	67	54
5	70	80	69	69
6	0	0	137	130
7	0	0	57	54
8	160	153	30	30
9	31	37	6	9
10	37	115	20	21
11	270	276	107	103
Σ	790	952	518	517
	790/952 = 0.830		518/517 = 1.002	

sporadic during the year, are more difficult to recall than those of cattle, which occur less frequently and perhaps in a somewhat more memorable manner.

An illustration of response errors where financial accounts are involved is given by Ferber (1965). A sample of persons having savings accounts in a number of savings institutions in a metropolitan area were selected and interviewed on a number of financial matters, of which savings deposits in specific banks was one. Access to the exact status of the savings accounts in the banks permitted a comparison of interview responses versus record values. The results for the 196 persons interviewed are as follows:

	Interviewed	Bank record	Percent discrepancy
Average size of account	1476	2663	44.6

Hence stated deposits were about one-half of the true!

For health data obtained by interviewing a cross section of HHs in the US, Koons (1973) reports the results of a reinterview of a subsample of these HHs using supervisory staff, who are presumably more skillful in obtaining accurate responses from respondents than the ordinary interviewers employed in the survey. Since both the main survey and the reinterview followup have been going on for at least $7\frac{1}{2}$ years, the results are fairly reliable. Some examples are given in Table 13.7. Note that the reinterviews altered some results rather markedly—particularly the proportion of people in the US "having one or more chronic conditions."

Table 13.7.   *Estimated proportions and net difference rates*

Survey item, persons with	Percent in class on original interview	Percent in class on reinterview	Net difference rate
One or more chronic conditions	42.3	49.2	−7.0
One or more hospital episodes in past 12 months	9.3	10.0	−0.6
One or more restricted-activity days in past two weeks	10.6	12.1	−1.5
One or more bed days in past two weeks	5.6	6.4	−0.8

Methods of Measuring. The two common methods of detecting and measuring response errors are: (i) checks against records (which are presumed to be the true values) and (ii) reinterviewing the same respondents with more expert interviewers and using the earlier responses to resolve or conciliate apparent discrepancies. The record check is rather limited, and it is not always particularly relevant even if authoritative. The reinterview method really produces two responses for each interviewee, where the later response is presumed to be the "authoritative" one. This can be seriously questioned in many cases. Hence it may be useful even though somewhat short of some unavailable ideal.

13.6.   Schemes for Dealing with Response Error Problems

Errors in people's responses to survey questions can be quite vexing to detect, to assess the seriousness of, and to resolve into information useful for the purposes of the survey. But evasion or avoidance of the problem, however carefully carried out, has yet to be proved desirable, even if a more forthright course may be less than satisfying. It is hoped that research-minded persons in the academic and professional fields who are heavily dependent on people's responses to questions will continue the search for a fuller understanding of the problem. Presumably a lot is already known by lawyers, sociologists, psychiatrists, anthropologists, social psychologists, and others, and perhaps a lot of this knowledge is in the literature—even in a form useful for statistical survey work. But here we shall merely present some of the more conspicuous practices and views that seem most relevant.

General Approach to Interviewing.   To obtain good-quality interview data it is helpful to adopt an approach appropriate to the occasion. The respondent does a better job answering questions if he feels that he is getting something out of the imposition on his time. Perhaps all he needs is an opportunity to sound off—on anything you ask. Many people like to talk if they are regarded as an "authority" on something, or as a "typical citizen" or someone sought out for "advice." Making the respondent feel at ease and willing to cooperate is commonly called *establishing rapport*. Without rapport the interviewer may get bad data or an outright refusal.

In addition to rapport, quality of responses can be improved by choosing the proper line of questioning, with appropriate probing questions, gestures, reactions, and the like to clarify concepts and to facilitate memory recall. Carefully trained interviewers, using carefully planned lines of questioning, can greatly improve quality of data. [See Hyman (1954) and Marquis and Cannell (1971).]

Questionneering.   Some errors in response are directly chargeable to poor or careless phrasing of questions. A common fault of many academic-based survey directors is the use of pedantic and vague wording. Moreover, the sequencing and arranging of questions, if not carefully considered, can easily confuse the respondent to the point where he has lost direction and sense of purpose of the interview and his responses will be poor if relevant at all. (More on this in Section 13.11.)

Studies show, as one would expect, that the length of the recall period has an inverse effect on the quality of replies. Reducing our greed for too much data by substituting short periods of recall appropriately sampled over the period of interest can effectively reduce *memory bias*.

A simple but effevtive method for detecting possible errors in response is that of placing in the questionnaire one or more alternative questions dealing with the same concept. (This should be done in a way that doesn't appear to be naive or entrapping.) Comparisons may turn up discrepancies that can't be ascribed to obvious phrasing or location effects.

Double Sampling.   Sometimes it is more efficient not to devote a lot of resources to improving the quality of responses but rather just to accept the lower quality and remove the effects of low quality statistically. For example, in counting oranges on randomly located branches (where fruits were counted in place) it was found that counting errors were being made and that from 2% to 4% were being missed. Rather than spend more resources on training and developing special counting aids, it was decided to use a form of double sampling instead. In phase II a *very* careful count was made on a small sample of the original phase I sample of branches, using a few well-trained and disciplined investigators. The correlation between the two counts being high ($r = 0.98 +$), a ratio estimator was very effective in removing the bias due to counting errors. [See Jessen (1955 and 1972).]

13.7. Investigator (Interviewer Bias) Problem

The investigator, particularly but not only if he is an interviewer, can be a source of errors in the data, which, like response errors, can be far more harmful to final results than those due to sampling. The nature and extent of these errors has received much attention by researchers from the 1950s on, but some illuminating work goes back at least to 1929 [Rice (1929)]. Here we shall briefly consider the nature and extent of such errors and some schemes for their detection and control.

Nature of Investigator's Influence.   In interview surveys the interviewer can influence the responses obtained by his mere presence, by his (or her) sex, skin color, age, or dress. His behavior and manner of asking (or not asking) questions and his "clarifying" or probing statements can affect the results. The influence is particularly great if the questions are of the "open" rather than "closed" type (see Section 13.10) and the nature of the matter dealt with is not obvious or concise or is not easily recalled on the spur of the moment.

A number of studies on the matter have shown that interviewer's prejudices can seriously affect the respondent and thereby the responses obtained. Rice (1929), studying 2000 destitute men for the reasons they gave for their plight, found that responses were related to their interviewer. Antidrink interviewers tended to come up with a preponderance of responses blaming drink; socialist-oriented interviewers came up with a high number of their respondents blaming industrial causes. Apparently the men were willing to give interviewers who took an interest in them the kinds of answers that would please them.

Since the interviewer's influence is so great when concepts are "soft" or where memory is depended upon to supply the desired information, the usefulness of the data is quite dependent on him. His influence may ruin the data, or it can be put to good use. A good interviewer can "harden" a soft concept into something more meaningful for each of his respondents, keeping the basic concept but varying the language and presentation to accommodate each respondent's needs for understanding. On the memory problem, some interviewer techniques can be quite helpful in enhancing recall ability. Those techniques should be considered for large-scale surveys, but to be used only by those who know their power and can control them.

Examples of Interviewer Errors.   Hanson and Marks (1958), in a study of the quality of the 1950 Census of Population, used a design wherein 1000 interviewers had randomized assignment areas to determine the effects of interviewer differences on some 97 items. Those with the largest differences relative to sampling variance are presented in Table 13.8. On four of the six items nonresponse (or possibly nonasking) rates varied most among interviewers, and on the two others some ambiguity seemed to exist on determining education status.

Table 13.8. Items having largest interviewer differences[a]
(All expressed as ratios as indicated.)

Item no.	Characteristic	Ratio	F-value[b]
59	Persons, age 5–24, now attending school/total population	0.029	4.14
53	Persons, age 25 and older, reporting highest grade not completed/population 25 and older	0.162	3.79
78	Persons, age 14 and older, reporting "not answered" to whether self-employed or not/population 14 and older	0.052	3.77
74	Persons, age 14 and older, reporting "not answered" to whether any wage and salary income/population 14 and older	0.050	3.75
82	Persons, age 14 and older, reporting "not answered" to amount of unearned income received/population 14 and older	0.054	3.66
93	Females, age 14 and older, and not working, "not answered" to "did any work last week (other than housework)"/females 14 and older	0.036	3.27

[a] Data from Hanson and Marks (1958, p. 643).
[b] The ratio of between-interviewer mean square to within-interviewer mean square. A conservative average number of degrees of freedom is 125 for both.

Examples of Investigator's Canvassing Errors.   Some examples of the kinds of errors occurring in the canvassing operation, when the SU or OU is scrutinized to determine the correct OUs in it, or its correct nature and extent, have been dealt with in Section 13.4.

Other Investigator Errors and Effects.   Besides the errors of measurement and canvassing, investigators can cause other disturbances in the accuracy of results and in the general environment of survey taking. Cheating, the outright falsification of survey data, can be a serious problem in poorly managed surveys. Some interviewers may obtain data and otherwise seem to do complete and accurate work, but leave the respondent feeling he has been put through a third-degree type of grilling. Such public relations are particularly serious where surveys are repeated on panels.

At best the investigator has a "treatment effect" on the respondent. This effect may be either good or bad so far as obtaining valid data is concerned. If the issues are important, then it is important that the "investigator effect" be kept under control. That control should be used to achieve valid results with good respondent relations.

13.8. Schemes for Dealing with Investigator Errors

In those surveys in which a person is interviewed, or where a person is required to carry out certain canvassing operations, measurements, or counts, errors are almost sure to happen. In some surveys the errors committed by investigators can be ruinous if nothing is done to prevent or minimize them. These measures are the obvious ones: select good people, train them well, direct them properly, and check their work to see if all is in order. In addition to these platitudes, here are some more specific suggestions.

Recruitment. Generally, professional interviewers are better than nonprofessional. However, some professionals may need retraining for inquiries requiring special techniques, and it may be that retraining is more difficult than training a new staff. Also, most, but not all people, appear to be more comfortable and franker when interviewed by people like themselves in terms of ethnicity, color, age, language, sex, same or slightly higher social status and competence, and so on. In recruiting nonprofessionals, look for persons who like to talk with people, but who, while being good listeners, are able to evaluate the reasonableness of what they hear and ask nondisturbing questions where the validity of what they hear seems doubtful. Simple procedures can be used to detect those who would have problems (e.g., can't read maps, have speech difficulties), but actual field trials are hard to beat.

Training. Eliciting accurate information by interviewing requires skill. To acquire this skill quickly and effectively, a training program is helpful. Aids such as training manuals are commonly used by organizations such as the U.S. Bureau of the Census and the larger survey and polling organizations. Copies of these can usually be obtained by asking, and they may be very useful either directly or as a basis for one's own tailored version. Such training aids as taped interviews, movies, or video tapes are also useful in the training sessions. Simulated interviews and actual field trials are very important.

In addition to training on interviewing techniques in general, it is well to have special training on the particular survey being undertaken. This survey may have a number of unusual features that require special handling, such as field sampling procedure, call-back technique, and concepts in the questionnaire. Again, as much as possible should be put into writing, so that the interviewer can consult or review the guidelines when problems arise.

Equipment. Special forms may be prepared to help the investigator do his job efficiently and accurately—forms for listing HHs within sample blocks or street segments, farms within sample areas, and so on. In the case of agricultural surveys involving the use of an area as an SU, it may be helpful to apply both the grid and plot concepts to the area and, by using a two-way bookkeeping scheme, account for all land in the area, whether in farms or not, using proprietorship in the basic listing of the land. By mapping these properties and then following up on the families having holding arrangements on them—whether they live there or not—one can obtain a complete determination of the land-people relationships on each property. Forms to aid in the work, together with paper for sketching, can help reduce canvassing errors substantially. [See Jessen (1954).]

Management. It is usually wise to plan for the supervision of the interviewers and other field staff during the course of the survey. As carefully as the training may have been carried out, problems are still likely to develop. Unusual situations arise, investigators must make many on-the-spot decisions, which may or may not be correct. To minimize difficulties that such decisions may bring, prudent use of telephones, rapid delivery of completed work, and frequent inquiries on goings-on can uncover faults and get the proper procedures into operation quickly. Many of these cannot be blamed on the investigator—they just might not have been anticipated by the planners!

In surveys where a considerable amount of work is required to carry out the canvassing of area survey units, such as making lists of the farms or households as well as interviewing them, it may be helpful to split the operation into its interviewing and canvassing (e.g., listing) portions and assign investigators specially trained for each. Skilled interviewers usually do not like to do listing and other more technical operations, and skilled listers may not like to interview. The two specialists can be organized into teams—especially in regions or counties where travel is difficult—so both can work together. [See Jessen and Strand (1953).]

If the nature of the survey is such that errors due to interviewers can be worrisome, it may be feasible to consider the problem of optimal number of investigators. The problem has been looked into by Hansen et al. (1951).

Evaluation Methods. There are two general methods for determining the errors committed by investigators and respondents and separating them out from sampling errors. The *interpenetrating-samples* method involves the assignment of the investigators to the SUs to be random (in the simple case). The *reinterviewing* method requires two sets of measurements on the survey sample or some portion of it.

The essential idea in the interpenetrating-samples scheme is to regard the differences in the results between the investigators, exceeding those ascribable to sampling variations, to be due to interviewers. Hence an ANOVA performed on the data will provide two estimates of the between-interviewer mean square, that from calculating the mean square of the various interviewer means and that on the average within-interviewer variance. The comparison can be evaluated by the simple F-test.

Although formal calculations were not made, the interpenetrating-samples method was used to determine whether nationality of investigators had any effect on results in the Greek Election Survey referred to previously. Here there were 60 British, 60 American, and 40 French investigators. Assignments of these national groups to voting places over Greece were randomized separately for each of the three groups, despite objections from the military organization that this would unduly burden the logistics problems in supplying the investigator teams (consisting of a military officer, his driver, and his interpreter, together with a jeep and trailer stocked to provide the team the means of support wherever it was assigned in Greece). The random assignments assured us, after examining the results, that nationality differences in this case were insignificant. [See Barnes (1946) and Jessen et al. (1947).]

The reinterview method is a more precise means to measure nonsampling errors, provided that the double interviewing does not in itself cause disturbances. The effects of these disturbances depend to a certain extent on the nature of the inquiry dealt with and the care with which the whole thing is carried out. The Bureau of the Census and the National Center for Health Statistics have carried out a number of evaluations of their main survey operations using the reinterview technique. The results have been very revealing and very helpful in providing guidance in the use of their published information and in suggesting whether certain data being collected are really helpful. [See Tepping and Boland (1972), U.S. Bureau of the Census (1972), and Koons (1973).]

Removal of Errors by Survey Design.   If it is known that errors are present and that they can be measured, such as by reinterviewing (or in general by remeasuring), and the error is predominantly a bias rather than random, then it may be feasible to remove the effects of the error by survey design. See the discussion of double sampling in Section 13.6 for an illustration on orange counts.

13.9. Objectives of Good Questionnaire Design

The questionnaire, whether it be mailed to the informant or read to him by the investigator, will, if well designed, accomplish the following:

1.   Obtain accurate data. This is accomplished through (i) careful phrasing of the questions, (ii) careful choice of words, (iii) planned type, size, and position of phrases for answers to minimize posting errors, (iv) planned built-in instructions for particular aspects of the questions, so that these are made known to every informant as part of the questions.

2.   Obtain information in less time, by (i) careful choice of questions, so that only those essential to the investigator are included, (ii) clear statement of concepts dealt with, (iii) careful planning of order and arrangement of questions and places for posting replies.

3.   Obtain a complete set of replies by (i) avoiding questions that make people mad, (ii) making the inquiry as attractive to the informant as possible, so he feels it is worth his while to contribute his time and effort, (iii) making questions simple so that they are easy to comprehend and to reply to.

4.   Simplify and speed the processing of the data by (i) planning for the coding in the questionnaire, and (ii) planning the physical structure of the question and other forms required for checking, editing, and so on for convenience in handling.

13.10. General Considerations for Questionnaire Design

Kinds of Questions.   Questions can be classified into several kinds or types. The "open question" does not include designated alternative answers. The "closed question" does include them; the respondent must choose among the alternatives offered and exclude all others. Between these "pure" types there are many mixtures, because clearly an "open question" must ordinarily be limited to some more or less specific area of interest.

Examples of "open question":

What do you think of the government's farm program?
Why haven't you terraced your hilly land?

Examples of "closed question":

Do you think the government's farm program is generally good, bad, or indifferent?

Have you not terraced your hilly land because:
     a–You don't think it will be a good investment?
     b–You don't know how?
     c–You can't get a loan?
     d–You don't need the land for crops?
     e–Any other reasons?_____ What are they?_____

Notice in the last example that subquestion "e" is an open type.

Another useful classification is that of "direct" questions as contrasted with "indirect." For example, suppose we want to obtain the questionee's age. We can ask directly:

How old are you?

or we can ask for presumably the same information indirectly:

In what year were you born? What month?

Some tests indicate that in this case the indirect question elicits more accurate answers.

Choosing the Kind of Questions to Use.   There are at least five criteria to consider in dealing with this problem: (i) the characteristics of the informants, (ii) how the information is to be elicited (personal interview, telephone, through the mail, etc.), (iii) the kind of people to be used as interviewers, (iv) the objectives of the investigation, and (v) the length of the questionnaire (in terms of time required to complete).

Ordinarily the open question is more satisfactory for the better educated and more intelligent informant. Also, if the inquiry is directed toward some specific area of interest rather than things in general, the open type is more appropriate—for example, if we are asking farmers about farming, or grocers about the grocery business.

The open question seems to be more suitable to the personal interview than to the mailed questionnaire. This may be because the answers are usually longer, requiring so much effort that the informant does not care to write them down but does not mind dictating his reply to interviewer.

The open question generally requires a more skilled interviewer in order to prevent evasive and tangential answers.

Often the objectives of the investigation determine the type of questions to be used. Studies dealing with *why* people do or do not believe or expect certain things, such as opinion and attitudes studies, generally are quite suitable for the open type of question. Where so-called "facts" are sought, such as whether or not a household has a radio or how many acres are in a farm, the open type of question is very common. However, sometimes such factual questions can be made easier to answer by being closed. Instead of asking "How many acres are there in this farm?" ask, "Does this farm have (a) less than 80 acres, (b) between 80 and 200 acres, or (c) more than 200 acres?" This may greatly reduce the difficulty of determining the answer but it also may greatly reduce amount of information.

Last, the total length of the questioning period may be an important consideration in determining the choice of questions. In order to keep the informant from feeling that it was an unfair demand on his time, it should be kept within some reasonable length. Usually an hour is about as long as one should go, but this will vary considerably according to circumstances. For farmers during their busy seasons this is far too long, but during their slack seasons $1\frac{1}{2}$ or 2 hours may not be too long. The time requirements of a questionnaire may be cut by converting open questions to closed ones, though at some sacrifice of information. Another technique is to employ a "split questionnaire."

Covering the Objectives.   It is surprising how frequently this rather obvious point is lost sight of in the preparation of questionnaires. The writer may be so involved in details of his work, such as searching for the right word or the best phrasing for a particular concept, that he forgets just what he is after in the first place. This is particularly likely where indirect techniques of questioning are used. A good way to detect a poor job on covering objectives is to make out a set of dummy tables for the analysis to see if one is getting the right kind of data to go into them.

Making the Questions Understandable.   This is a very common failing of questionnaires. It is not enough that a question be worded so that it can be understood; rather, it should be such that it cannot possibly be misunderstood.

Building Rapport.   Rapport is an intimate or harmonious relationship. To build and to maintain rapport during the interview is to just use good sense in public relations. Here are some hints for accomplishing this:

1.   Reveal the seriousness of your study in the way you choose and phrase your questions.
2.   Avoid questions that to the respondent may appear to be silly.
3.   Avoid questions that indicate a lack of understanding of people or that show lack of consideration or good taste.
4.   It is surprising what people are willing to talk about—if they are questioned in the right manner.

Sequence of Questions.   If several persons examine a questionnaire, each will probably arrive at a different opinion on the "best" sequence for the questions. This lack of agreement arises because it is difficult to predict or even to measure the effects of possible sequences on the results. For example, one may contend that the answer to a given question will be "conditioned" by the presence of a previous question, or by a set of previous questions. Some of these problems can be solved by an empirical test: try them on a small sample and observe the results. Others are more difficult. Since the sequence of questions may be important to the reader of your results, it is advisable to include as part of your report a complete replica of the questionnaire, so that he can make his own conjectures on possible sequence effects.

Ordinarily it appears advisable to adopt a sequence that would be logical for the respondent, even though this may be illogical from the research point of view. It is usually a simple matter to regroup the data later for your analysis. Forcing your analytical scheme on the respondent may lengthen the time of the interview (because you are forcing the respondent to think and recall in a unnatural order) and also make the answers less accurate.

Transitions.   A large number of the studies using interview data deal with more than one topic. Hence if a study is made on farmers' use of fertilizers, it may be of interest to get information on the following topics.

1. The description of the farm, its area, use of land, the kinds and general nature of its enterprises, etc.
2. The description of the operator or operators, education, age, experience, activities other than farming, etc.
3. Knowledge of kinds of fertilizer and their probable effects on yields, use in regard to specific crops and on specific fields, etc.

In order to avoid possibly awkward gaps in the interview, it may be helpful to insert questions between widely different topics that serve to bring about a smooth transition from one to the other, even though these questions provide no information of interest. Sometimes smoothness of transition can be most conveniently accomplished by a statement explaining that now you are going to take up a new topic and why. The interviewer might read this statement aloud, or he may wish to provide his own words extemporaneously.

"Touchy" Questions.   Ordinarily questions that annoy or rile should be omitted, but sometimes answers to these questions are essential to the study. In this case it appears to be common practice to put such questions at the end of the interview, so that other questions will be safeguarded in case the touchy one either terminates the interview or seriously affects rapport. "Touchy" questions are those that, although they are in good taste, tend to arouse emotions and might subject the interviewer to a long harangue. Touchy questions in bad taste, like any question in bad taste, should not be tolerated.

From General to Specific.   Usually it is advisable to have "general" concepts precede "specific" ones. For example, in obtaining the membership of a household it may be preferable first to ask, "How many people live here?" and then proceed to get information on each to see if in fact they are members of the household as you define it. Also it is advisable to ask questions at the end that may uncover other persons who are validly members of the household (infants, persons now on vacation, etc.).

13.11. Specific Considerations for Questions

Questionnaire design appears to be essentially an art, but that does not mean some principles cannot be formulated. Many studies are being made on how to put more logic into the subject, and the reader is advised at least to consult the references at the end of this chapter. This section will discuss some examples of bad practice, some limits, and some aspects of effective technique.

Frame of Reference.   In filling out a form for the records of the Brooklyn Dodgers, one outfielder, on the line asking "length of residence in home town," wrote: "about 40 feet." An applicant at the University of Pennsylvania filling out an entrance form came to the question, "Are you a natural-born citizen of the U.S.?" He puzzled a while, then wrote: "No—Caesarean" (*Readers Digest,* June 1952). A survey was taken using the question, "Do you smoke Winston cigarettes?" Nearly 95% replied, "Yes—what else do you do with them?"

Lazarsfeld (1935), writing in a more serious vein, discussed "The Art of Asking Why." If we ask a person "Why did you buy Spearmint toothpaste?" we may get answers on three different levels, or in three different *frames of reference.* The person may say, "Well, I saw the box in the window of the drugstore and it was more attractive—that nice blue with pink around it—so I went in and bought it." In other words, he mentioned a characteristic of the product itself. Another person may say, "Well, I was walking to work this morning and my friend said, 'Why don't you try Spearmint toothpaste, it's just wonderful,' and so I went in and bought it." He is now mentioning stimulation from some other person. Or a person may say, "Well, I ran out of toothpaste this morning and so I went in and bought some." Now he is citing his own needs. Each of these persons has used a different frame of reference to explain his behavior. These frames are not necessarily mutually exclusive, and therefore we cannot obtain meaningful answers when the results are summarized. This is because the person who told us that he bought some toothpaste because he ran out may also have noticed the package of Spearmint in a window, but since he didn't happen to think of that frame, he answered us in another.

In August 1943, Gallup (American Institute of Public Opinion) asked the following question: "After the war, would you like to see many changes or reforms made in the United States or would you rather have the country remain pretty much the same as it was before the war?" Gallup reported that 58% of the people wanted things to "remain the same." It occurred to the Division of Program Surveys of the United States Bureau of Agricultural Economics [Crutchfield and Gordon (1947)] that people might take various views of what the words "changes and reforms" mean, so it made a checkup on a cross-section sample of 114 New Yorkers asking them the same question and getting about the same results. However, the interviewers followed up with these questions: "Well, what do you have in mind?" and "Why do you want to see these changes?" They covered seven frames of references, as follows:

72 were thinking of domestic reforms,

 3 were thinking of foreign affairs,

12 were thinking of the basic structure of the country,

 2 were thinking of technological change,

 4 were thinking of comparison with wartime conditions,

 8 were thinking of their personal affairs,

 4 were thinking of a desirable state of affairs in general,

 9 interpretations were unascertainable.

What do Gallup's results mean?

Loaded Questions.   A simple illustration of a "loaded" question is as follows:

(a)   "Do you approve of having Thanksgiving a week earlier this year?"

(b)   "Do you approve of President Roosevelt's idea of having Thanksgiving a week earlier this year?"

The second version is said to be loaded because it includes not only the issue under consideration but also something else, which may have a considerable influence on the answer obtained.

It is said that if a question is to conjure up any ideas at all, it must have "loading" of some sort, so that the problem is not to formulate load-free questions but rather to use various loadings properly. Campbell gives an illustration: A study was made of the attitudes of the American public on the feeding of "starving" Europeans after the war. The first question was, "What do you think of our feeding the starving people of Europe after the war?" The use of the word "starving" brought the expected results: 90% said "Yes." ("Why sure, you've got to feed starving people.") Campbell points out that use of a word like "needy" would have given fewer "yes" responses, and perhaps omitting the adjective altogether would have given even fewer yeses. But now the concept is no longer sharp—one is being asked if he approves of feeding people, regardless of the circumstances of their need. But suppose we use the "starving" question and then follow up on the yeses with "Do you think they ought to pay for it?" (This is reverse loading—Americans think Europe should pay for everything.) The "yes" respondents to this then were asked, "Well, supposed they can't pay for it; should we send it anyhow?" From a set of questions of this sort Campbell believes we can get a more complete notion of how a person stands on an issue than through some "unloaded" question.

Choice of Words.   Only the most general remarks will be made here. Try to avoid "touchy" and "technical" words when good simple words exist. Avoid words with many meanings. For example, the English word "you" means you, singular, or you and your family. When a question is to be used over a wide geographical area or over different groups, there is a possibility that the same word may have different meanings. For example, in some peanut-growing areas peanuts are unknown by the name "peanut" but are called "ground peas" or "goobers." Avoid words with disturbing connotations, such as "just," "only," or "merely," unless they are serving a planned function in the question.

Double-Barreled Questions. "What do you think of the economic and political situation" is a double-barreled question. Such questions can confuse the respondent, who does not know whether you want some kind of an "average" of the two or whether you are asking him two questions at the same time.

13.12. Field Testing

After a draft of the questionnaire is completed, a simple office test may turn up important weaknesses. This test might consist of simply trying it out on some hypothetical cases. Another useful test is to set up the various steps planned for the analysis and then see if the various concepts required are in fact dealt with in the questionnaire. It is amazing how many things, particularly in a long and involved inquiry, can be overlooked in an early draft of a questionnaire.

Unless the survey is a repetition of one done before, the first real test of one's newborn questionnaire is the field test. This need not be done on an expensive sample—as a matter of fact it may be better to purposely choose cases where trouble is expected. Where the questionnaire is to be mailed, the questionee may be visited personally after he completes it, in order to determine what went on. Ordinarily, if a test questionnaire is completed and returned, one can examine it to see if everything seemed to be understood; if it was not returned, usually an interview is required to find out why it was not. Maybe only a part of ghe questionnaire was at fault.

Another function of the field test is to obtain information and experience useful in teaching interviewers. This information is invaluable to illustrate the problems that your field workers will have with real-life examples.

13.13. Special Topics

A number of devices and techniques developed by survey workers to deal with problems peculiar to their survey may have wider applicability. A few of the more common ones will be mentioned here.

Check Questions. On numerous occasions, especially when numerical data are being dealt with, a question might be asked merely to see if certain things actually are in agreement. This sort of question may disturb rapport unless it is used with discretion and tact. Sometimes it is advisable to include as trouble indicators the sample concepts already measured by some statistical agency. On household surveys, for example, by obtaining information on such items as skin color and whether the household has a telephone, one can compare his results with that of the last census to see if anything unexplainable has arisen. Sometimes serious flaws in the field work can be detected this way, or perhaps there is a previously unnoticed difference in concepts between the census and the survey.

Incomplete Questionnaire.   When the questions desired on a questionnaire become so numerous that difficulty is expected with rapport, yet the arbitrary omission of any will seriously reduce the effectiveness of the study, it may be possible to arrange a sampling of questions so that only so many will be presented to any one respondent. The appropriate scheme for doing this depends on the nature of the study and other considerations; only the idea will be presented here. Notice that if certain questions are put to all respondents and if there are relationships between these and the questions put only to a subsample, conditions exist for a possible double-sampling design. The partial-questionnaire technique appears particularly appropriate for certain mail inquiries because of the severe limitations on the number of questions that can be put to a single respondent.

Split Questionnaire.   In this case a possible difference exists between two or more ways of asking a question on a single concept. This may be resolved, in the case of questions A and B, by making half the questionnaires contain A and half contain B. If a difference is detected, it may be difficult to decide what to do with the results.

Precoding.   Devices for designing in advance some of the steps required for coding the data for possible transfer to punched cards may be referred to as precoding. This provision in a questionnaire is an excellent way to reduce costs of this rather expensive operation, particularly where large numbers of units are being dealt with. A newer device, known as "document-sensing," is a further step in adapting the questionnaire in a machine age.

Instructions on the Questionnaire.   Opinions vary whether special instructions necessary to explain a concept to the interviewer should appear on the questionnaire or whether the interviewer should commit them to memory. If the instructions go on the questionnaire, they should be set in small type and suitably boxed so that they do not clutter it up. On the other hand, if they are entrusted to the interviewer, some precautions ought to be taken that he does not follow the old adage, "Out of sight, out of mind."

Special Aids.   A technique frequently used to assist the respondent in recalling, say, the names of magazines that he receives, is to present him a "flash card" listing the names of a number of them; he glances at it and names the ones he recognizes as receiving. Some designers of mailed or other self-administered questionnaires have made liberal use of drawings to illustrate their questions. These aids may increase the number returned as well as clarify and simplify the questions.

Position Bias.   It is widely suspected, and much evidence has been provided to support the suspicion, that in checklist type questions the first entry is more likely to be noticed and therefore responded to than one far down the list. For short lists the effect of position may be negligible, but it should be seriously considered when lists are long and the entries are not vivid or concise in concept. Some question may be raised on the usefulness of any result under these conditions, but at least some scheme of rotation should be adopted to balance out position effects. A simple randomization of the ordering is not very helpful.

◇   ◇   ◇

13.14.   Discussion and Summary

This chapter has considered some of the nonsampling problems of statistical surveys. The treatment has been brief, considering the extent and complexity of the problems in this area and their importance in obtaining satisfactory survey results, yet hopefully the nature and extent of the problems have been illuminated. Too often research workers dealing with statistical survey data regard the sampling portion of the inquiry to be the sole determinant of quality results: all we need is a "representative sample" and somehow perfect data will come from every element in it. Many of us seem to have a reasonable appreciation of data problems when dealing with individuals or even with censuses, yet we feel that the right kind of sampling will somehow, by statistical magic perhaps, eliminate them. One purpose of this chapter is to remind ourselves again of the problem of quality data.

13.15.   Review Illustrations

1.   Suppose we have a city in which half the homes are listed in the telephone directory and half are not. It so happens that 70% of the nontelephone homes watch a particular program on TV, whereas 20% of the telephone homes do. A random sample of 100 homes is to be selected from the telephone directory and asked if they watch this program. The results are to be used to estimate the proportion of all homes in the city watching the program.

(a) Will the estimate be unbiased? Why or why not?

(b) Determine the mean square error for p, the estimate, its variance, and its (bias)^2, if any.

Solution

(a) Yes, since sample is not random over all homes and the two groups differ in TV-watching characteristics.

(b) $P = (.7 + .2)/2 = .45$. $E[p] = .20$ (for the telephone homes only).

$$\text{bias} = E[p] - P = .20 - .45 = .25,$$

$$(\text{bias})^2 = .0625,$$

$$\text{var } (p) = \frac{PQ}{n} = \frac{(.2)(.8)}{100} = .0016,$$

$$\text{MSE } (p) = \text{var } (p) + (\text{bias})^2 = .0016 + .0625 = .0641.$$

2. We have 40,000 stores on which we wish to determine the value of $\bar{Y}$, the mean of y per store. A random sample of 1000 is drawn and sent a questionnaire by mail; 500 (or 50%) respond. A sample of 50 (or 10%) of the nonrespondents is drawn and visited to obtain the desired information by personal interview.

(a) What is an appropriate unbiased estimator of $\bar{Y}$ for this case?

(b) If unit costs of mailing are $5 and of processing a returned questionnaire are $6, and the cost of obtaining and processing an interviewed questionnaire is $140, what should be the optimal rate at which to subsample nonrespondents here?

(c) If the variability of both respondents and nonrespondents (by mail) is the same, say a coefficient of variation of 50%, and an accuracy of 2.5% is desired (with confidence of 67%), how large should the mail and interview samples be?

(d) Suppose the survey planner was convinced that mean of y for nonrespondents was identical to that of respondents and designed the sample accordingly. Using the accuracy specifications of (c), how large would his sample be? What would the costs be for (c) and (d)?

(e) What if all stores are responders—how large a sample and what are the costs? Comment.

Solution

(a) In this case we can regard the universe of stores as comprising, at the time of the survey, about 20,000 responders and 20,000 nonresponders to the survey undertaken. An unbiased estimator is given by

$$\bar{y} = \left(\frac{N_R}{N}\right)\bar{y}_R + \left(1 - \frac{N_R}{N}\right)\bar{y}_{NR}$$

or (Eq. 13.8)

$$\bar{y} = \frac{1}{n}(n_R\bar{y}_R + n_{NR}\bar{y}_{NR}).$$

(b) Given: $c_0 = 5$, $c_1 = 6$, $C_{NR} = 140$. Optimal k is given by

$$k = \sqrt{\frac{C_{NR}W_R}{c_0 + c_1 W_R}} = \sqrt{\frac{(140)\,(0.5)}{5 + 6\,(0.5)}} \doteq 3.0.$$

In the survey described, k was 10, hence the interview sample was too small.

(c) If $S^2 = S_{NR}^2 = (0.50\bar{Y})^2$, and $e = 0.025\bar{y}$ with $\theta = 0.67$, then, using Eq. 13.9,

$$\overline{\text{var}}\,(\bar{y}) = \frac{S^2}{n}[1 + (k-1)W_{NR}]$$

and

$$n = \frac{Z^2\sigma^2}{e^2} = \frac{(1)^2(0.50\bar{y})^2[1 + (3-1)(0.5)]}{(0.025\bar{y})^2} = 800;$$

n_{NR} will be approximately 400 and r, the size of the interview sample, will be about 133.

(d) If $\bar{Y}_R = \bar{Y}_{NR}$, there would be no need for an interview followup. Now

$$n = \frac{Z^2\sigma^2}{e^2} = \frac{(1)^2[s^2 = (0.50\bar{Y})^2]}{(0.025\bar{Y})^2} = 400 \text{ responses.}$$

To obtain these, 800 must be mailed out. The cost will be

$$C = n(c_0 + c_1 W_r) = 800[5 + 6(0.5)] = \$6400.$$

Previously the cost was

$$C = n[(c_0 + c_1 W_R) + (C_{NR}W_{NR}/k)]$$

$$= 800[5 + 6(0.5) + 140(0.5)/3]$$

$$= \$25{,}000, \quad \text{or about eight times as much.}$$

(e) If all respond on a mail survey, $n = 400$ and its cost is $n(c_0 + c_1 W_R) = 400$ $[5 + 6(1.0)] = \$4400$. Therefore for the same accuracy, \$4400 will suffice if everybody is a responder, \$6400 will suffice if only one-half respond but they are on the average identical to responders, and $25,000 if only half respond to mail but everyone does to an interview and responders and nonresponders are different and an unbiased estimate is required.

3. Critically evaluate the following statement:

In the Elmira Study, where a carefully designed probability sample was employed, the number of people not reached after three call-backs was still quite large, and there was strong evidence that the unreachables were quite different sorts of people from those who were reached. This experience at Elmira raised a very fundamental question about probability sampling in actual practice. If no substitutions are permitted, there is a danger that the final sample will contain a bias due to the people who refuse or were unavailable.

Solution

Substitution will not solve the problem of getting information from not-at-homes and refusals. It simply provides more information on those who are at home and who cooperate. The scheme does provide a means to balance out better on geographic and other proximity factors, but in most inquiries these seem minor relative to the "not-at-home" and "refusal" factors.

4. Some years ago a manufacturer of home rugs carried out an investigation to determine the colors, textures, material, and other characteristics that consumers really desire in rugs. Investigators were sent to the homes of a "representative sample" of housewives and, by showing them various samples of rugs, elicited a lot of information on the housewives' preferences. At the close of the interview the investigator told the housewife that a rug would be sent free to a number of interviewees as a sort of "door" prize if their names happened to be drawn: "If you should be one of the lucky ones, which of the rugs I showed you would you want to have?" When the data were being analyzed, it was discovered that the housewives' replies to the door-prize question were quite inconsistent with those to the questions in the survey proper.

(a) What is your explanation of the different results?
(b) Which results might be more useful to the manufacturer? Discuss.

Solution

(a) Apparently the housewife thought the inquiry was a game in which she was to state the preferences of the "modern intelligent woman" who knows what the current styles and vogues are, or should be. But when it came to the matter of what sort of rug would she have on *her* floor, she became quite practical and less fanciful. This is an entirely different matter and she responded accordingly.

(*b*)   The prize selection.

5.   "Sampling, for all the claims, is still not so precise as to warrant much less than 3 % margin of error (and this is on the low side)" [Louis Harris, "Election Polling and Research," *Public Opinion Quarterly* **21** (1957): 115]. The author of this statement was commenting on problems of election polling.

(*a*)   What do you think the author of the statement meant by sampling?
(*b*)   Do you agree with his presumed definition of sampling? Why or why not?

Solution

(*a*)   The author probably includes errors in population shifts and of measurement and processing, of nonrandomness, of missing data, and so on in addition to *sampling* error. Actually, with proper sampling, the sampling error decreases to zero with increase in sample size.

(*b*)   No. It is unfortunate that the author implies these errors are due to *sampling*. It would be more precise to say, "*Social surveys*, for all the claims,. . ."

13.16.   References

Adams, J. S.
1958
Interviewing Procedures. Chapel Hill, N.C.: The University of North Carolina Press. (A manual for interviewers.)

Barnes, W.
1946
Allied Observation of Elections in Greece. *The American Foreign Service Journal,* **23** (6).

Bartholomew, D. J.
1961
A Method of Allowing for "Not-at-Home" Bias in Sample Surveys. *Applied Statistics.* **10**: 52–59.

Birnbaum, Z. W.
Sirken, M. G.
1950
Bias Due to Nonavailability in sampling Surveys. *Journal of the American Statistical Association,* **45**: 98–111 (March).

Blankenship, A. B.
1961
Creativity in Consumer Research. *Journal of Marketing,* **25**: 34–38 (October). (Good example of how to ask questions to avoid social stigma.)

Campbell, A.
Unpublished mimeographed material.

Cash, W. S.
Moss, A. J.
1972
Optimum Recall Period for Reporting Persons Injured in Motor Vehicle Accidents. DHEW Publication No. (HSM) 72–1050. Washington, D.C.: U.S. Government Printing Office.

Clausen, J. A.
Ford, R. N.
1947
Controlling Bias in Mail Questionnaires. *Journal of the American Statistical Association,* **42**: 497–511.

Cochran, W. G.   *Sampling Techniques.* (2nd ed.) New York: John Wiley & Sons, Inc.
1953, 1963   (Chapter 13.)

Crutchfield, R. S.   Variations in Respondents' Interpretation of an Opinion Poll Question.
Gordon, D. A.   *International Journal of Opinion and Attitude Research,* 1: 1–12.
1947

Dalenius, T.   *Bibliography on Nonsampling Errors.* Institute of Statistics, University of
1971   Stockholm. (Mimeographed. An excellent bibliography.)

Deming, W. E.   On Errors in Surveys. *American Sociological Review,* 9: 359–369.
1944

Durbin, J.   Callbacks and Clustering in Sample Surveys: An Experimental Study.
Stuart, A.   *Journal of the Royal Statistical Society,* 117: 388–428. (A comparison of
1954   the H-P and call-back scheme.)

Fenlason, Anne   *Essentials in Interviewing: For the Interviewer Offering Professional
1962   Services.* (372 pp.) New York: Harper & Row, Publishers.

Ferber, R.   The Reliability of Consumer Surveys of Financial Holdings: Time-
1965   Deposits. *Journal of the American Statistical Association,* 60: 148–163.

Finkner, A. L.   Methods of Sampling for Estimating Commercial Peach Production in
1950   North Carolina. *North Carolina Agricultural Experiment Station
                        Technical Bulletin* 91.

Gleeson, G. A.   *Interviewing Methods in the Health Interview Survey.* Public Health Service
1972   Publication No. (HSM) 72–1048, Series 2, No. 48. Washington, D.C.: U.S.
                        Government Printing Office.

Hansen, M. H.   The Problem of Non-response. *Journal of the American Statistical
Hurwitz, W. N.   Association,* 41: 517–529.
1946

Hansen, M. H.   Response Errors in Surveys. *Journal of the American Statistical
Hurwitz, W. N.   Association,* 46: 147–190. (Considers optimal number of interviewers.)
Marks, E. S.
Mauldin, W. P.
1951

Hanson. R. H.   Influence of the Interviewer on the Accuracy of Survey Results. *Journal of
Marks, E. S.   the American Statistical Association,* 53: 635–655 (September).
1958

Hartley, H. O.   Discussion quoted in F. Yates, A Review of Recent Statistical
1946   Developments in Sampling and Sampling Surveys. *Journal of the Royal
                        Statistical Society,* 109: 12–43.

Hendricks, W. A.   Adjustment for Bias by Nonresponse in Mailed Surveys. *Agricultural
1949   Economic Research,* 1: 52–56.

Homeyer, P. G.   Sampling Replicated Field Experiments on Oats for Yield
Black, C. A.   Determinations. *Proceedings of Soil Science Society of America,* 11:
1946   341–344. (Example of investigator bias in cropcutting.)

Houseman, E. E.   Statistical Treatment of Nonresponse Problem. *Agricultural Economic
1953   Research,* 5: 12–18.

Hyman, H. H. (ed.)   *Interviewing in Social Research.* Chicago: University of Chicago Press.
1954

Ingram, J. J.   Measuring Completeness of Coverage in the 1969 Census of Agriculture.
Prochaska, D. D.   1972 *Proceedings of the Business and Economics Section, American
1972   Statistical Association.*

Jessen, R. J.
1942

Statistical Investigation of a Sample Survey for Obtaining Farm Facts. *Iowa Agricultural Experiment Station Research Bulletin* 304. Ames, Iowa: Iowa State University.

Jessen, R. J.
1954

Agricultural Sample Survey of the Province of Pichincha, Ecuador, 1952. *Estadistica: Journal of the Inter-American Statistical Institute*, **12**(44): 401–421. (Example of use of field form to account for farms and nonfarms.)

Jessen, R. J.
1955

Determining the Fruit Count on a Tree by Randomized Branch Sampling. *Biometrics*, **11**: 99–109 (March).

Jessen, R. J.
1972

On an Experiment in Forecasting a Tree Crop by Counts and Measurements. Article (pp. 145–165) in T. A. Bancroft (ed.), *Statistical Papers in Honor of George W. Snedecor*. Ames, Iowa: Iowa State University Press.

Jessen, R. J.
Blythe, R. H.
Kempthorne, O.
Deming, W. E.
1947

On a Population Sample for Greece. *Journal of the American Statistical Association*, **42**: 357–384 (September).

Jessen, R. J.
Kempthorne, O.
Daly, J. F.
Deming, W. E.
1949

Observations on the 1946 Elections in Greece. *American Sociological Review*, **14**: 11–16 (February).

Jessen, R, J,
Strand, N. V.
1953

Statistical Methodology in Multipurpose Surveys of Crete, Greece. Appendix (pp. 235–566) of L. G. Allbaugh, *Crete: A Case Study of an Underdeveloped Area*. Princeton, N. J.: Princeton University Press. (Example of use of investigator teams, consumption versus production checks.)

Jessen, R. J.
Thompson, D. J.
1958

Encuesta por Muestreo de las Fincas en la Provincia de Buenos Aires, Argentina. *Estadistica: Journal of the Inter-American Statistical Institute*, **16**(61): 464–504.

Kahn, R. L.
Cannell, C. F.
1957

The Dynamics of Interviewing: Theory, Technique, and Cases. New York: John Wiley & Sons, Inc.

Kish, L.
1962

Studies of Interviewer Variance for Attitudinal Variables. *Journal of the American Statistical Association*, **57**: 92–115 (March).

Kish, L.
1965

Survey Sampling. New York: John Wiley & Sons, Inc.

Kish, L.
Hess, I.
1958

On Noncoverage of Sample Dwellings. *Journal of the American Statistical Association*, **53**: 509–524.

Kish, L.
Hess, I.
1959

A "Replacement" Procedure for Reducing the Bias of Nonresponse. *The American Statistician*, **13**: 17–19 (October).

Koons, D. A.
1973

Quality Control and Measurement of Nonsampling Error in the Health Interview Survey, DHEW Pub. No. (HSM) 73–1328. (53 pp.) (Example of assessment of interviewer variability and total error.)

Lageman, J. K.
1965

The Delicate Art of Asking Questions. *The Reader's Digest*, June 1965, pp. 87–91. (A popular and perceptive article on question asking in general.)

Laurent, A.
Cannell, C. F.
Marquis, K. H.
1972

Reporting Health Events in Household Interviews: Effects of an Extensive Questionnaire and a Diary Procedure. DHEW Publication No. (HSM) 72–1049. Washington, D.C.: U.S. Government Printing Office.

Lazersfeld, P. F.
1935

The Art of Asking Why. *National Marketing Review* (National Association of Marketing Teachers). Chicago: Business Publications, Inc.

Maccoby, E. E.
Maccoby, N.
1954

The Interview: A Tool of Social Science, in Gardner Lindzey (ed.), *Handbook of Social Psychology.* Cambridge, Mass.: Addison–Wesley Publishing Co.

Madow, W. G.
Hyman, H. H.
Jessen, R. J.
1961

Evaluation of Statistical Methods Used in Obtaining Broadcast Ratings. House Report No. 193, 87th Congress, 1st Session. (163 pp.) Washington, D.C.: U.S. Government Printing Office.

Mahalanobis, P. C.
1946

Recent Experiments in Statistical Sampling in the Indian Statistical Statistical Institute. *Journal of the Royal Statistical Society,* **109**: 325–378. (Presentation of the interpenetrating-samples idea.)

Marquis, K. H.
Cannell, C. F.
1971

Effect of Some Experimental Interviewing Techniques on Reporting in the Health Interview Survey. Public Health Service Publication No. 1000 Series 2 No. 41. Washington, D.C.: U.S. Government Printing Office.

Marquis, K. H.
Cannell, C. L.
Laurent, A.
1972

Reporting Health Events in Household Interviews: Effects of Reinforcement, Question Length, and Reinterviews. DHEW Publication No. (HSM) 72–1028. Washington, D.C.: U.S. Government Printing Office.

Mercer, J. A.
Butler, E. W.
1967

Disengagement of Aged Population and Response Differentials in Survey Research. *Social Forces,* **46**: 89–96 (September).

Morgenstern, O.
1950, 1963

On the Accuracy of Economic Observations. Princeton, N.J.: Princeton University Press. (Excellent presentation of errors and their effects in economics.)

Moser, C. A.
1951

Interviewer Bias. *Review of International Statistical Institute,* **19**: 1–13. (Good review of literature to 1950.)

Payne, S. L.
1951

The Art of Asking Questions. Princeton, N.J.: Princeton University Press.

Politz, A.
Simmons, W. R.
1949, 1950

An Attempt to Get the Not-at-Homes into the Sample without Call-backs. *Journal of the American Statistical Association,* **44**: 9–31, and **45**: 136–137.

Rice, S.
1929

Contagious Bias in the Interview. *American Journal of Sociology.* **35**: 420–423.

Stephen, F. F.
McCarthy, P. J.
1958

Sampling Opinions. (451 pp.) New York: John Wiley & Sons, Inc. (An analysis of survey procedures.)

Sukhatme, P. V.
1946

Bias in the Use of Small-Size Plots in Sample Surveys for Yield. *Nature,* **157**: 630. (Early presentation of investigator errors in crop-cutting surveys.)

Tepping, B. J.
Boland, K. L.
1972

Response Variance in the Current Population Survey. Working Paper No. 36. (20 pp.) Bureau of the Census, U.S. Dept. of Commerce. (Estimates of investigator variances where they are experienced.)

U.S. Bureau of the Census 1947 — *How to Read Aerial Photographs for Census Work.* (44 pp.) Washington, D.C.: Government Printing Office.

U.S. Bureau of the Census 1972 — *Evaluation and Research Program of the U.S. Censuses of Population and Housing, 1960: Effects of Different Reinterview Techniques on Estimates of Simple Response Variance.* Series ER60, No. 11. Washington, D.C.: U.S. Government Printing Office.

Yates, F. 1949, 1953 1960, 1971 — *Sampling Methods for Censuses and Surveys.* (1st ed., 1949; 2nd ed., 1953; 3rd ed., 1960.) London: Charles Griffin & Sons.

13.17. Exercises

1. By means of a search of the literature, examine the evidence available on the relationship of various factors with refusal rates in interview surveys—for example (say HH head is respondent): (*a*) sex, (*b*) age, (*c*) economic level, (*d*) religion, (*e*) years of schooling, (*f*) marital status, (*g*) telephone presence, (*h*) race, (*i*) size of HH, (*j*) occupation, (*k*) rent or own home, (*l*) length of residence.

2. (*a*) Distinguish between a "missed observation" and a "don't know." Illustrate with examples.
    (*b*) How is each dealt with in estimation? Explain fully.

3. What are the common sources of interviewer errors or bias? How might they be avoided?

4. The following questionnaire was handed to each patron as he arrived at a theater. Critically evaluate its possible use to management (either the producer or the theater owner).

We earnestly request *everyone* to check the items on the form after tonight's performance and leave it in the lobby. The usefulness of the form depends upon a complete audience survey. Thank you for your help.

Would you recommend TAKE A GIANT STEP to your friends if it appears again in a regular theater?	____Yes ____No
Would you ever go to see it again?	____Yes ____No
Would you recommend PULL MY DAISY to your friends?	____Yes ____No
Would you see it again?	____Yes ____No

5. A survey of students at UCLA was taken during January 1967 to examine how they were getting along with the newly instituted quarter system. A questionnaire was sent to all students enrolled at the time. Following is a full quote of a section of the report on findings under the heading *Statistical Validity*. Critically evaluate the statement.

Approximately 4.4% of the graduate and undergraduate student body returned the questionnaire. We received 1090 completed forms. Although this may not seem like a large enough sample, the data received did indicate, for almost every question, significant trends. These are especially apparent in the graphs accompanying the questions in the analysis section. It is possible that the sample may not be random. The survey, because of its voluntary nature, may have attracted more of the discontented students than a random sample would have found. This is not substantiated from personal experience. The responses seem to be what would be expected from talking to students. Finally, this report is an attempt to find the problems of the quarter system. In this respect the sample taken is a valid one.

CHAPTER 14

Analysis and Presentation

14.1. Introduction

After the survey is designed and the observations taken, the investigator examines his data, organizes and analyzes them to see what they show. He then may try to determine the reliability of his statistical summaries and perhaps develop some simple models to hang his results on. This may well be the most satisfying phase of the whole undertaking—yet it can bring forth surprises, some even unpleasant.

This chapter will take note of some of the steps along the way, commenting on what might be expected and describing some experiences. In general, most topics will be dealt with only briefly, but further readings will be pointed out for those interested.

14.2. Data Processing

When data start coming in, attention is often attracted to the rate of response—particularly with mailed inquiries but scarcely less so with interview surveys. If the response rate is good, no special measures to improve it are necessary, so attention is then directed to quality of response.

Canvass Check. In the case where cluster units are used, and canvassing is required, an immediate monitoring of this operation should be undertaken. To facilitate this check, it is helpful to prepare for it. Each cluster SU is designed to have some actual or expected number of OUs reporting from it. By calculating this M_i for each SU and comparing the responses, say the M_i' with them, one can watch for unexpected occurrences—somebody erred or some unusual but real event occurred. This requires an SU by SU check-in and perhaps some material (maps, census block data, worksheets, etc.) to provide the means for checking reasonability of disturbing departures from the expected. In the case of farm or HH surveys using area SUs, investigators' use of sketch maps to explain their findings at each SU can be very helpful to detect and resolve errors.

Scanning and Cleaning the Data. In small-scale surveys a quick visual look at the data can detect many of the grosser errors. A pattern forms, and wide departures from that pattern are usually easily detected. These should be examined to see if they are genuine or goofs, and dealt with accordingly. For large-scale surveys this operation can be done by electronic computer.

Coding. The need for coding is considerably reduced when electronic computers are used. Since there is usually some loss in information when numerical data are categorized, this is a welcome gain—particularly to those who like to thoroughly analyze their data and hence like maximum freedom in their activities. Moreover, coding is a source of errors and therefore requires additional monitoring and checking. [See the references for further information on this section—particularly Yates (1971).]

14.3. Quality Check

Though it is not always possible, a set of questions and procedures built into the survey can be helpful in assuring oneself that all is well, at least on some feature of the investigation. These checks are generally of two types: (i) internal consistency and (ii) external compatibility. Some examples will be given of each type.

Internal Consistency.   Two surveys of Greece to observe and evaluate the fairness of the 1946 parliamentary elections dealt with determining the number of Greeks validly registered to vote. One survey of a sample of names taken off the official registration lists had the objective of determining the number of names on the list belonging to persons alive and registered to vote. The other survey was of households, where the objective was to determine the number of family members validly registered to vote and the number who were not and for what reason. Taken together, the two surveys provided two independent estimates of total registered voters. When results were examined, it appeared the two estimates were within reasonable allowable sampling error of each other. This result was reassuring, because so many things could have gone wrong due to circumstances.

In a farm survey of Argentina, a new method of linking farms to area sampling units was adopted. This method required all farms, any part of which extended into the area SU (a grid). to be included in the sample. In analysis it was to be weighted by the fraction contained in the SU. Just to see how the more traditional method of inclusion compared, it was adopted too. Hence, all farms whose headquarters were located within the area SU were so designated. If both methods are valid, they should provide unbiased estimates of farm characteristics in the universe. A test showed that they did not—the headquarters method of inclusion was found to be biased.

External Compatibility.   In a survey of California households having children under five years of age, a general household was first chosen and screened to obtain its qualification for the survey proper. However, it was thought to be a useful check to include a small subsample of the excluded households to obtain their characteristics. When combined, we could check with the Census (adjusted to the current time) to see if we were on the right course in accounting for all households and all persons (in HHs). We were.

In most surveys of households using area-probability methods, independent estimates of total households can be made from the sample and checked with some extrasample source such as the Census, with appropriate adjustments to make it current.

In agricultural surveys similar "benchmarks" can be found to check for compatibility. In the Iowa Farm Survey of 1939 the sample results were compared with the State Farm Census (carried out by the Assessor) and the U.S. Department of Agriculture. Here are some of the results:

Item	Survey	State Census	USDA
Land in farms (acres)	34,080,000	34,403,000	—
Land in corn (acres)	10,149,000	10,270,000	10,417,000
Land in oats (acres)	5,980,000	5,923,000	5,972,000
Corn harvested (bus.)	455,550,000	449,509,000	452,824,000
Cattle, all ages (no.)	4,295,000	4,001,000	4,465,000
Sows for farowing (no.)	1,765,000	1,707,000	1,643,000

It appeared that the data were in good order and that nothing seriously wrong had happened in the design of the sample and the field operations.

Sometimes a slight modification of the universe of inquiry, such as making it coextensive with some established region for which sound statistics are available, can provide a useful check. Sometimes a slight change in definition of the OU can help—such as making the definition of a household or farm or business establishment agree with that used by the Bureau of the Census—and can also be easily dealt with if anticipated and planned.

Opportunities to provide consistency and compatibility checks are nearly always available. Sometimes some extra expense is involved, but it is usually a good investment—for assurance.

14.4. Analysis of Descriptive Surveys

When we have the data in hand and, in the case of highly structured sample designs, properly identified within the structure, the procedures for obtaining the desired estimates for the population as a whole, or for its domains, can proceed fairly straightforwardly, provided the estimators are simple. This is particularly so where means and ratios are being calculated.

When totals are being estimated rather than means, proportions, or ratios, the problem becomes a little more complicated, because alternative estimating methods become available. Moreover, extrasample information may be available, and it may be necessary to incorporate it into the procedure or, if serious discrepancies appear, to attempt reconciliation. Furthermore, when both domain and overall estimates are made, inconsistencies may develop between domain totals and the overall collective total. This is particularly likely when domains are identified after the sample is drawn and where an effort is made to use the best available estimator for each estimating operation.

Adjustment for Consistency.   In the case where extrasample information is available to establish marginal totals in a two-way table and the sample is used to estimate cell proportions or totals, an adjustment procedure such as that suggested by El-Badry and Stephan (1955) can be used to arrive at results that are consistent both internally and externally. A procedure such as this can be theoretically justified only if discrepancies are due to sampling variations. If discrepancies are due to nonresponse, response errors, and the like, then adjustment for consistency not only may be cosmetic but also can diminish whatever accuracy the unadjusted estimates contained.

Special Estimates for Domains.   Where domains are of known sizes and are not already strata, it is possible sometimes to set them up as another dimension of stratification, thus assuring stability to the number of SUs assigned to them (see Chapter 11). Where the number of SUs allocated to each is small, two-way stratification will improve the accuracy of domain estimates over that obtained by poststratification methods.

In some cases national surveys are required to supply usable estimates for small areas (e.g., states, counties), where sample representation may be very sparse or not at all. Some methods have been proposed to supply such estimates; see Abbey (1972), National Center for Health Statistics (1968) and Woodruff, (1966).

14.5.   Estimation of Sampling Variances; Shortcut Methods

For the simpler sampling designs, the calculation of estimated sampling variance is usually fairly straightforward. However, in surveys where the design is sophisticated, the estimators are computationally complex, the number of characteristics covered is large, and the analysis elaborate, the problem of how to carry out the estimation of sampling variances can become very troublesome. As a consequence, they are often not computed at all, or some very poor approximations are used.

There are two general objectives for obtaining sampling variances: (i) to determine the reliability (confidence interval) of an estimate and (ii) to determine the nature and sources of the overall variance—that is, to determine the magnitude of the components of variance. The latter is of help primarily to the sample designer, telling him where the variance is coming from and what might be done in future designs to deal with it most economically.

Probably the most commonly used approximate method for estimating sampling variance is that of assuming the sample is random and calculating the variance accordingly. Hence, ignoring FPC,

$$\widehat{\text{var}}\ (\bar{y}) = \frac{s^2}{n}\quad \text{and}\quad \widehat{\text{var}}\ (p) = \frac{pq}{n}.$$

Design Effect.   The accuracy of the above approximations depends on the efficiency of the sample design used relative to that of a random one. Efficiency is expressed (in Chapter 4) as

$$(14.1) \qquad \text{RE (design/random)} = \frac{\text{var (random)}}{\text{var (design)}},$$

Where the sampling fraction is held constant. Then

$$(14.2) \qquad \text{var (design)} = \frac{\text{var (random)}}{\text{RE (design/random)}}$$

$$(14.3) \qquad\qquad\qquad\qquad\qquad = (D) \text{ var (random)}.$$

The factor D is called "the design effect" by Kish (1965). Note that the assumption of a random sample variance is correct only when $D = 1$. In practice, D can be either less than or greater than 1 according to the structure of the population dealt with and the sample design used. Since many large-scale surveys involve several stages of sampling and rather a lot of clustering, the value of D can be considerably greater than 1, hence the random sample assumption can seriously underestimate the true variance.

Shortcut Methods; Replicate Sampling. If computations become burdensome, a convenient method of estimating sampling variance is by simply splitting the sample into two parts—preferably halves. The method is sometimes called "replicate sampling." If the sample is simple random, and the split is random, the difference between the two means will be the exact 50% confidence interval for the population mean, provided the underlying population is symmetrical. If samples are reasonably large, the approximation will be quite good for many statistics as well as the mean. Likewise, a stratified random sample with two SUs in each of L strata can be split into two samples of L SUs each. The differences in means, ratios, and so on between the two provide useful 50% confidence intervals with a minimum of computation. For greater accuracy, the number of mutually exclusive splits can be increased, or, as in the case of the stratified designs with two SUs per stratum, a number of overlapping samples can be generated.

Approximations of the method can be made for complex designs. Even in lattice sampling, where the restrictions seem to be quite confining, one can allow for at least a split sample. See the references for more refined techniques.

Other Methods.   Producers of results from large-scale or continuing surveys, such as the Bureau of the Census, the National Center for Health Statistics, and the University of Michigan Survey Research Center, have found that tables and charts can be useful for estimating the sampling errors where many estimates and their combinations are made. The principle is based on the observation that many items seem to follow certain empirical relationships quite well and, once established, the sampling errors for individual estimates can be approximately determined by interpolation or extrapolation on this function. Provisions can be made for making comparisons between groups as well as for the usual individual estimates.

14.6. Analysis of Analytic Surveys

Under this heading we shall consider the problems of analyzing differences among groups where groups may be simply different categories, different levels of a single factor, or the result of categorizing by two or more factors in a cross classification. The case where factors are continuous, and hence appropriate for regression, will be acknowledged but not considered. Since these analyses utilize an expertise that is covered well in standard books on statistics for research, only highlights and references will be presented here.

Simple One-Way Case—Randomized Survey.   In this case the sample is one-stage, unclustered, and the domains of study are either pre- or poststrata. The group differences can be analyzed as a completely randomized experiment, using ANOVA to determine the statistical significance of group differences. If data are binomial and samples are small, a chi-square analysis would be appropriate. If the SUs are clusters of OUs and the analysis is on an OU basis, an ANOVA can be used where the pooled between-cluster mean square within groups becomes an appropriate error variance for making F-tests.

Simple Two-Way Case.   If strata can be regarded as blocks in the experimental design sense and each group is represented in each stratum with the same proportion of unclustered SUs, then a simple two-way ANOVA may be appropriate for analysis. By choosing the appropriate variance for clustered SUs, as indicated above for the one-way case, we do not add any complications.

Other Cases.   When disproportionality enters and when clusters cross domain boundaries, when several stages are involved and when strata and domains partially overlap, the appropriate analysis becomes more complicated. To some extent this complication can be avoided or minimized by designing an analytic survey appropriately, but too often it cannot. Other methods of analysis are required. See the references for some attempts along these lines: Hartley (1959), Walsh (1947), Morgan and Sonquist (1963), and Nathan (1973).

"Treatment" Effects.   The problems of inferring that the presence or absence of certain "treatments," such as smoking or not smoking cigarettes, has brought about the observed "effects," such as higher or lower incidence of lung cancer, can be dealt with in properly designed *experiments*. In surveys the "treatments," not having been randomly assigned, may very well be

confounded with other factors and therefore present additional difficulties in drawing causal inferences, hence such conclusions should be tentative.

14.7. Presentation of Results

Effective presentation of statistical survey results depends on the nature of the investigation, the kind of reader, and the personality of the report writer. Certain features of reporting seem worth mentioning, even though they may appear obvious.

1.   Consult a book or manual on the art and science of presenting statistical information so that it is both clear and informative. Much can be presented in simple charts or graphs rather than detailed tables.

2.   If a complete explanation of some statistical material seems to be complex and lengthy, separate the presentation into at least two parts: one for the report proper, one for an appendix for elaboration, and if further detail may be of possible interest, add a note that the author will supply such detail on request.

3.   In the case of attitude and opinion data, the exact wording of the question put to the respondent should be presented with the statistical results.

4.   Present sampling errors along with the estimate with some explanation somewhere as to what confidence level is being used. Simple standard errors are very convenient unless samples are very small, in which case the exact confidence level should be given.

5.   Where many statistical results are given and the presentation of sampling errors on each becomes cumbersome, provide tables or charts so that the reader can calculate at least an approximate estimate for the statistic of interest to him.

6.   If the investigation is regarded as a possible contribution to the scientific world, then a very careful description of procedures followed should be an integral part of the report. Measurements without a clear explanation of what was measured, how it was measured, and when, can be just another batch of meaningless numbers.

7.   When questionnaires are used, a copy should be attached to the report. The critical reader of the report may be interested in the sequencing of questions and in those questions or behavior that may have a conditioning effect on certain responses.

◇   ◇   ◇

14.8.   Remarks and Summary

At various places in the book a topic may be brought up and briefly discussed, and then perhaps references made to related sections. One such topic has been the descriptive versus the analytic survey. Another has been the survey having one or more purposes. Still another has been surveys that are of quite different types. Table 14.1 presents an arrangement of these topics, their interrelationships, and some of the names used in that context. The numbers refer to sections in which the topic is dealt with.

Table 14.1. Reference Table

Survey type	Purpose or objective	Control of sample allocation	
		Control (fixed n_h)	No control (variable n_h)
Descriptive	1. Collective mean	Strata "stratification" 7.3	Strata "poststratification" 7.8
	2. Individual means	Subuniverses "regionalization" 7.11	Subuniverses "postregionalization" 7.11
	3. Both collective and individual	Domains "multiestimation" 12.8	Domains "multiestimation" 12.8
Analytic	4. Collective mean	Strata (blocks) (not dealt with)	Strata (blocks) (not dealt with)
	5. Individual means	Groups "group comparisons" 12.9	Groups "group comparisons" 12.9
Hybrid	6. Various	12.10	12.10

14.9. Review Illustrations

1. A sample of registered voters in a city is desired to poll attitudes and opinions on a proposed referendum. Following is the procedure used:

A sample of 500 registered voters with maximum geographic scatter was desired. Working with census Block Statistics, 500 blocks (of the total of about 1100) were selected with PPS (homes). Within-block procedures were established to obtain the random selection of a single home in each sample block (except where a block was so large it required a "quota" of two). Interviewers were instructed to go to the designated home and inquire if it contained a registered voter or not. If so, an interview was obtained from him. If there were more than one, the most convenient one was chosen as the respondent (with effort made to obtain a "balance" of both sexes). If none, the interviewer went to the neighboring one and, if necessary, other homes, until one containing a registered voter was found. An interview was taken in his home in the manner described above.

Critically evaluate the scheme.

Solution

(1) Appropriate sampling rate for each block should be $1/M_i$, hence variations in "registration density" would result in variable numbers of interviews from each block. The current scheme underrepresents high "registration density" blocks and vice versa.
(2) Fails to account for changes in population since last census.
(3) "Within-home" selection bias.
(4) Costly if only one HH is taken per block.

2. In May–June 1964 an investigation of financial needs was made of students attending four-year colleges and universities in California. A sample was drawn in the following manner:

(1) A list of all four-year colleges and universities in the state was obtained, together with enrollment figures for autumn of 1962.
(2) A sample of n schools was selected with PPS (replacement), where enrollment, M_i, was the measure of size.
(3) A sampling fraction for each school was calculated as $f_i = k/M_i$.
(4) A random systematic sample of 1 in $1/f_i$ students was taken from the ith school in the sample.

Assuming that correct information is obtained from each and every student selected for the sample:
    (*a*) Is the simple mean, $\bar{y}$, an unbiased estimator of $\bar{Y}$, the true value of y for all students in the state enrolled in 4-year colleges during Spring 1964? Explain.
    (*b*) It is found that some students attend more than one school. What do you suggest be done about this matter in the course of the survey? Give logical support.
    (*c*) What about the effects of changes in enrollments since Fall 1962. Do these changes introduce possible biases? Discuss.

Solution

(*a*)   Yes, $\bar{y}$ should be unbiased, assuming all students are enrolled in one and only one school. $P_i \propto (M_i/M)(k/M_i) = k/M$ for all schools.

(*b*)   If multiple enrollments are discovered, such students will be overrepresented unless properly dealt with, either by (i) weighting the data from these students by $1/r$ (where r = number of different schools attended), (ii) retaining them only if selected in the school, say, first in alphabetical order, or (iii) retaining each with a $1/r$ probability.

(*c*)   Enrollment changes will not affect the validity of the design. The simple sample mean remains unbiased even with enrollment changes, since these are automatically accommodated by the design.

3.   Prepare a reply for the following letter:

Gentlemen:

Our organization is in the process of beginning public opinion surveys for political aspirants on a contract basis.

Prior to this time our work did not require such great emphasis upon economy and we simply worked from a sample which we felt to be sufficient. But now we must enter a competitive area and give advance estimates. It is because of this we write you, hoping for assistance.

Most simply stated, we have an immediate need to know what method or percentage should be employed in a universe of, say, ninety thousand (90,000), to assure the sample would be sufficient to give us reasonable confidence in our basis for projection. Any formulas, percentages, or even any reference sources giving this information would be highly appreciated.

In the absence of any proven method, even estimates from your experience would assist us greatly now. Any other thoughts and suggestions you might care to make would be equally welcome.

Thank you for your kindness and consideration.

Very truly yours,

Lyndon B. Goldwater
Director

LBG:jj

14.10. References

Abbey, D.
1972
Some Estimators of Subuniverse Means for Use with Lattice Sampling. Ph.D. Dissertation, University of California, Los Angeles. (196 pp.)

Bean, J. A.
1970
Estimation and Sampling Variance in the Health Interview Survey. Public Health Service Publication No. 1000, Series 2, No. 38. Washington, D.C.: U.S. Government Printing Office.

El-Badry, M. A.
Stephan, F. F.
1955
On Adjusting Sample Tabulations to Census Counts. *Journal of the American Statistical Association*, **50**: 738–762 (September).

Hartley, H. O.
1959
Analytic Studies of Survey Data. Rome: Instituto di Statistica. Volume in *Onora di Corrado Gini*, 1959.

Hyman, H. H.
1955
Survey Design and Analysis. New York: The Free Press.

Jessen, R. J.
1954
Agricultural Sample Survey of the Province of Pichincha, Ecuador, 1952. *Estadistica: Journal of the Inter-American Statistical Institute*, **12**: 401–421 (September).

Jessen, R. J.
Kempthorne, O.
Daly, J. F.
Deming, W. E.
1949
Observations on the 1946 Election in Greece. *American Sociological Review*, **14**: 11–16 (February).

Jessen, R. J., et al.
1951
Accessibility and Availability of Farm Machinery in Iowa. (128 pp.) J. I. Case Company.

Jessen, R. J.
Strand, N. V.
1953
Statistical Methodology in Multipurpose Surveys of Crete, Greece. In L. G. Allbaugh, *Crete: A Case Study of an Underdeveloped Area.* Princeton, N.J.: Princeton University Press, 1953.

Jessen, R. J.
Thompson, D. J.
1958
Encuesta por Muestreo de las Fincas en la Provincia de Buenos Aires, Argentina. *Estadistica: Journal of the Inter-American Statistical Institute*, **16**: 464–504 (December).

Kish, L.
1962
Variances for Indexes from Complex Surveys. *Proceedings of the Social Statistics Section, American Statistical Association*, 190–199.

Kish, L.
1965
Survey Sampling. New York: John Wiley & Sons, Inc.

Klein, L. R. (ed.)
1954
Contributions of Survey Methods to Economics. New York: Columbia University Press.

McCarthy, P. C.
1966
Replication: An Approach to the Analysis of Data from Complex Surveys. U.S. Public Health Service Publication No. 1000, Series 2, No. 14. (38 pp.)

Miller, H. W.
1973
Plan and Operation of the Health and Nutrition Examination Survey. DHEW Publication No. (HSM) 73–1310. Washington, D.C.: U.S. Government Printing Office.

Morgan, J. N.
Sonquist, J. A.
1963
Problems in the Analysis of Survey Data and a Proposal. *Journal of the American Statistical Association*, **58**: 415–434 (June).

Nathan, G.
1973
Approximate Tests of Independence in Contingency Tables from Stratified Cluster Samples. DHEW Publication No. (HSM) 73–1327, Series 2, No. 53. Washington, D.C.: U.S. Government Printing Office.

National Center for Health Statistics 1968	*Synthetic State Estimates of Disability.* U.S. Public Health Service Publication No. 1759. (16 pp.) Washington, D.C.: U.S. Government Printing Office.
United Nations Statistical Office 1964	*Recommendations for the Preparation of Sample Survey Reports.* United Nations Publication ST/STAT/SER.C/1/Rev. 2.
U.S. Bureau of the Census 1949	*Bureau of the Census Manual of Tabular Presentation.* (266 pp.) Washington, D.C.: U.S. Government Printing Office.
Walsh, J. E. 1947	Concerning the Effect of Intraclass Correlation on Certain Significance Tests. *Annals of Mathematical Statistics,* **18**: 88–96.
Williams, W. H. 1964	Sample Selection and the Choice of Estimator in Two-Way Stratified Populations. *Journal of the American Statistical Association,* **59**: 1054–1062 (December).
Woodruff, R. S. 1966	Use of a Regression Technique to Produce Area Breakdowns of the Monthly National Estimates of Retail Trade. *Journal of the American Statistical Association,* **61**: 496–504 (June).
Yates, F. 1949, 1971	*Sampling Methods for Censuses and Surveys.* London: Charles Griffin & Sons.

14.11. Exercises

1. "Jasper County, Iowa, which has been a 'political barometer' for the entire nation for 56 years, again reveals political opinion of national importance in a recent [1952] survey conducted by the Iowa Poll for *Look* Magazine. Voters in Jasper County are considered a mirror of all U.S. voters because they have voted for the winning candidate in every presidential election since 1896. The margin of victory for the president has been about the same in Jasper County as in the nation." Critically evaluate this statement.

2. An important social study is under consideration. One group has proposed that with the resources available, the sampling of families for the study should be selected using the quota method in order that important variables such as size of family, religion, socioeconomic class, and ethnic group are properly balanced in accordance with those of the universe under study. Another group opposes this proposal and advocates a strictly probability method of selection. You are called in to referee the dispute. Write a brief, concise, and germane report on the matter. If you feel assumptions are needed, state them.

3. Explain your understanding of the meaning of "case study." Can you distinguish this "method of study" from a "noncase" study? Discuss.

4.   Many studies are carried out by examining the students of a university class or members of a clinical group. Variability is estimated by calculating the variance of the characteristic under study for those observed. In testing hypotheses or estimating population parameters it is generally assumed that one has obtained a random sample from some more general population. In general, do these "hypothetical" samples provide "unbiased" estimates? If not, are they likely to be under- or overestimating the "true" variance? Discuss.

Appendix

Table A: Random Digits

	1	2	3	4	5	6	7	8	9	10
1	03991	10461	93716	16894	98953	73231	39528	72484	82474	25593
	38555	95554	32886	59780	09958	18065	81616	18711	53342	44276
	17546	73704	92052	46215	15917	06253	07586	16120	82641	22820
	32643	52861	95819	06831	19640	99413	90767	04235	13574	17200
	69572	68777	39510	35905	85244	35159	40188	28193	29593	88627
2	24122	66591	27699	06494	03152	19121	34414	82157	86887	55087
	61196	30231	92962	61773	22109	78508	63439	75363	44989	16822
	30532	21704	10274	12202	94205	20380	67049	09070	93399	45547
	03788	97599	75867	20717	82037	10268	79495	04146	52162	90286
	48228	63379	85783	47619	87481	37220	91704	30552	04737	21031
3	88618	19161	41290	67312	71857	15957	48545	35247	18619	13674
	71299	23853	05870	01119	92784	26340	75122	11724	74627	73707
	27954	58909	82444	99005	04921	73701	92904	13141	32392	19763
	80863	00514	20247	81759	45197	25332	69902	63742	78464	22501
	33564	60780	48460	85558	15191	18782	94972	11598	62095	36787
4	90899	75754	60833	25983	01291	41349	19152	00023	12302	80783
	78038	70267	43529	06318	38384	74761	36024	00867	76378	41605
	55986	66485	88722	56736	66164	49431	94458	74284	05041	49807
	87539	08823	94813	31900	54155	83436	54158	34243	46978	35482
	16818	60311	74457	90561	72848	11834	75051	93029	47665	64382
5	34677	58300	74910	64345	19325	81549	60365	94653	35075	33949
	45305	07521	61318	31855	14413	70951	83799	42402	56623	34442
	59747	67277	76503	34513	39663	77544	32960	07405	36409	83232
	16520	69676	11654	99893	02181	68161	19322	53845	57620	52606
	68652	27376	92852	55866	88448	03584	11220	94747	07399	37408
6	79375	95220	01159	63267	10622	48391	31751	57260	68980	05339
	33521	26665	55823	47641	86225	31704	88492	99382	14454	04504
	59589	49067	66821	41575	49767	04037	30934	47744	07481	83828
	20554	91409	96277	48257	50816	97616	22888	48893	27499	98748
	59404	72059	43947	51680	43852	59693	78212	16993	35902	91386
7	42614	29297	01918	28316	25163	01889	70014	15021	68971	11403
	34994	41374	70071	14736	65251	07629	37239	33295	18477	65622
	99385	41600	11133	07586	36815	43625	18637	37509	14707	93997
	66497	68646	78138	66559	64397	11692	05327	82162	83745	22567
	48509	23929	27482	45476	04515	25624	95096	67946	16930	33361
8	15470	48355	88651	22596	83761	60873	43253	84145	20368	07126
	20094	98977	74843	93413	14387	06345	80854	09279	41196	37480
	73788	06533	28597	20405	51321	92246	80088	77074	66919	31678
	60530	45128	74022	84617	72472	00008	80890	18002	35352	54131
	44372	15486	65741	14014	05466	55306	93128	18464	79982	68416
9	18611	19241	66083	24653	84609	58232	41849	84547	46850	52326
	58319	15997	08355	60860	29735	47762	46352	33049	69248	93460
	61199	67940	55121	29281	59076	07936	11087	96294	14013	31792
	18627	90872	00911	98936	76355	93779	52701	08337	56303	87315
	00441	58997	14060	40619	29549	69616	57275	36898	81304	48585
10	32624	68691	14845	46672	61958	77100	20857	73156	70284	24326
	65961	73488	41839	55382	17267	70943	15633	84924	90415	93614
	20288	34060	39685	23309	10061	68829	92694	48297	39904	02115
	59362	95938	74416	53166	35208	33374	77613	19019	88152	00080
	99782	93478	53152	67433	35663	52972	38688	32486	45134	63545

TABLE B Areas of a Standard Normal Distribution

An entry in the table is the proportion under the entire curve which is between $z = 0$ and a positive value of z. Areas for negative values of z are obtained by symmetry.

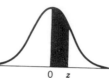

z	.00	.01	.02	.03	.04	.05	.06	.07	.08	.09
0.0	.0000	.0040	.0080	.0120	.0160	.0199	.0239	.0279	.0319	.0359
0.1	.0398	.0438	.0478	.0517	.0557	.0596	.0636	.0675	.0714	.0753
0.2	.0793	.0832	.0871	.0910	.0948	.0987	.1026	.1064	.1103	.1141
0.3	.1179	.1217	.1255	.1293	.1331	.1368	.1406	.1443	.1480	.1517
0.4	.1554	.1591	.1628	.1664	.1700	.1736	.1772	.1808	.1844	.1879
0.5	.1915	.1950	.1985	.2019	.2054	.2088	.2123	.2157	.2190	.2224
0.6	.2257	.2291	.2324	.2357	.2389	.2422	.2454	.2486	.2517	.2549
0.7	.2580	.2611	.2642	.2673	.2703	.2734	.2764	.2794	.2823	.2852
0.8	.2881	.2910	.2939	.2967	.2995	.3023	.3051	.3078	.3106	.3133
0.9	.3159	.3186	.3212	.3238	.3264	.3289	.3315	.3340	.3365	.3389
1.0	.3413	.3438	.3461	.3485	.3508	.3531	.3554	.3577	.3599	.3621
1.1	.3643	.3665	.3686	.3708	.3729	.3749	.3770	.3790	.3810	.3830
1.2	.3849	.3869	.3888	.3907	.3925	.3944	.3962	.3980	.3997	.4015
1.3	.4032	.4049	.4066	.4082	.4099	.4115	.4131	.4147	.4162	.4177
1.4	.4192	.4207	.4222	.4236	.4251	.4265	.4279	.4292	.4306	.4319
1.5	.4332	.4345	.4357	.4370	.4382	.4394	.4406	.4418	.4429	.4441
1.6	.4452	.4463	.4474	.4484	.4495	.4505	.4515	.4525	.4535	.4545
1.7	.4554	.4564	.4573	.4582	.4591	.4599	.4608	.4616	.4625	.4633
1.8	.4641	.4649	.4656	.4664	.4671	.4678	.4686	.4693	.4699	.4706
1.9	.4713	.4719	.4726	.4732	.4738	.4744	.4750	.4756	.4761	.4767
2.0	.4772	.4778	.4783	.4788	.4793	.4798	.4803	.4808	.4812	.4817
2.1	.4821	.4826	.4830	.4834	.4838	.4842	.4846	.4850	.4854	.4857
2.2	.4861	.4864	.4868	.4871	.4875	.4878	.4881	.4884	.4887	.4890
2.3	.4893	.4896	.4898	.4901	.4904	.4906	.4909	.4911	.4913	.4916
2.4	.4918	.4920	.4922	.4925	.4927	.4929	.4931	.4932	.4934	.4936
2.5	.4938	.4940	.4941	.4943	.4945	.4946	.4948	.4949	.4951	.4952
2.6	.4953	.4955	.4956	.4957	.4959	.4960	.4961	.4962	.4963	.4964
2.7	.4965	.4966	.4967	.4968	.4969	.4970	.4971	.4972	.4973	.4974
2.8	.4974	.4975	.4976	.4977	.4977	.4978	.4979	.4979	.4980	.4981
2.9	.4981	.4982	.4982	.4983	.4984	.4984	.4985	.4985	.4986	.4986
3.0	.4987	.4987	.4987	.4988	.4988	.4989	.4989	.4989	.4990	.4990
3.1	.4990	.4991	.4991	.4991	.4992	.4992	.4992	.4992	.4993	.4993
3.2	.4993	.4993	.4994	.4994	.4994	.4994	.4994	.4995	.4995	.4995
3.3	.4995	.4995	.4995	.4996	.4996	.4996	.4996	.4996	.4996	.4997

Table C: Confidence Interval for Binomial Distribution

95% Confidence Interval (Per Cent) for Binomial Distribution (1)*

Number Observed f	Size of Sample, n 10		15		20		30		50		100		Fraction Observed f/n	Size of Sample 250		1000	
0	0	27	0	20	0	15	0	10	0	07	0	4	0.00	0	1	0	0
1	0	40	0	31	0	23	0	17	0	11	0	5	.01	0	4	0	2
2	3	61	2	37	1	30	1	21	0	14	0	7	.02	1	5	1	3
3	8	62	5	45	4	36	2	25	1	17	1	8	.03	1	6	2	4
4	15	74	9	56	7	42	4	30	2	19	1	10	.04	2	7	3	5
5	22	78	14	64	10	47	6	33	3	22	2	11	.05	3	9	4	7
6	26	85	19	67	14	54	9	37	5	24	2	12	.06	3	10	5	8
7	38	92	19	71	14	59	10	41	6	27	3	14	.07	4	11	6	9
8	39	97	29	81	20	65	13	44	7	29	4	15	.08	5	12	6	10
9	60	100	33	81	22	71	16	48	9	31	4	16	.09	6	13	7	11
10	73	100	36	86	29	71	17	53	10	34	5	18	.10	7	14	8	12
11			44	91	29	78	20	56	12	36	5	19	.11	7	16	9	13
12			55	95	35	80	23	60	13	38	6	20	.12	8	17	10	14
13			63	98	41	86	24	64	15	41	7	21	.13	9	18	11	15
14			69	100	47	86	29	68	16	43	8	22	.14	10	19	12	16
15			80	100	53	90	32	68	18	44	9	24	.15	10	20	13	17
16					58	93	32	71	20	46	9	25	.16	11	21	14	18
17					64	96	36	76	21	48	10	26	.17	12	22	15	19
18					70	99	40	77	23	50	11	27	.18	13	23	16	21
19					77	100	44	80	25	53	12	28	.19	14	24	17	22
20					85	100	47	83	27	55	13	29	.20	15	26	18	23
21							52	84	28	57	14	30	.21	16	27	19	24
22							56	87	30	59	14	31	.22	17	28	19	25
23							59	90	32	61	15	32	.23	18	29	20	26
24							63	91	34	63	16	33	.24	19	30	21	27
25							67	94	36	64	17	35	.25	20	31	22	28
26							70	96	37	66	18	36	.26	20	32	23	29
27							75	98	39	68	19	37	.27	21	33	24	30
28							79	99	41	70	19	38	.28	22	34	25	31
29							83	100	43	72	20	39	.29	23	35	26	32
30							90	100	45	73	21	40	.30	24	36	27	33
31									47	75	22	41	.31	25	37	28	34
32									50	77	23	42	.32	26	38	29	35
33									52	79	24	43	.33	27	39	30	36
34									54	80	25	44	.34	28	40	31	37
35									56	82	26	45	.35	29	41	32	38
36									57	84	27	46	.36	30	42	33	39
37									59	85	28	47	.37	31	43	34	40
38									62	87	28	48	.38	32	44	35	41
39									64	88	29	49	.39	33	45	36	42
40									66	90	30	50	.40	34	46	37	43
41									69	91	31	51	.41	35	47	38	44
42									71	93	32	52	.42	36	48	39	45
43									73	94	33	53	.43	37	49	40	46
44									76	95	34	54	.44	38	50	41	47
45									78	97	35	55	.45	39	51	42	48
46									81	98	36	56	.46	40	52	43	49
47									83	93	37	57	.47	41	53	44	50
48									86	100	38	58	.48	42	54	45	51
49									89	100	39	59	.49	43	55	46	52
50									93	100	40	60	.50	44	56	47	53
										†				++		++	

* Reference (1) at end of chapter.
† If f exceeds 50, read 100 − f = number observed and subtract each confidence limit from 100.
†† If f/n exceeds 0.50, read 1.00 − f/n = fraction observed and subtract each confidence limit from 100.

Table D: Values of *t* for Given Levels of Probability

				Probability level				
df	.5	.6	.7	.8	.9	.95	.99	.999
1	1.000	1.376	1.963	3.078	6.314	12.706	63.657	636.619
2	.816	1.061	1.386	1.886	2.920	4.303	9.925	31.598
3	.765	.978	1.250	1.638	2.353	3.182	5.841	12.924
4	.741	.941	1.190	1.533	2.132	2.776	4.604	8.610
5	.727	.920	1.156	1.476	2.015	2.571	4.032	6.869
6	.718	.906	1.134	1.440	1.943	2.447	3.707	5.959
7	.711	.896	1.119	1.415	1.895	2.365	3.499	5.408
8	.706	.889	1.108	1.397	1.860	2.306	3.355	5.041
9	.703	.883	1.100	1.383	1.833	2.262	3.250	4.781
10	.700	.879	1.093	1.372	1.812	2.228	3.169	4.587
11	.697	.876	1.088	1.363	1.796	2.201	3.106	4.437
12	.695	.873	1.083	1.356	1.782	2.179	3.055	4.318
13	.694	.870	1.079	1.350	1.771	2.160	3.012	4.221
14	.692	.868	1.076	1.345	1.761	2.145	2.977	4.140
15	.691	.866	1.074	1.341	1.753	2.131	2.947	4.073
16	.690	.866	1.071	1.337	1.746	2.120	2.921	4.015
17	.689	.863	1.069	1.333	1.740	2.110	2.898	3.965
18	.688	.862	1.067	1.330	1.734	2.101	2.878	3.922
19	.688	.861	1.066	1.328	1.729	2.093	2.861	3.883
20	.687	.860	1.064	1.325	1.725	2.086	2.845	3.850
21	.686	.859	1.063	1.323	1.721	2.080	2.831	3.819
22	.686	.858	1.061	1.321	1.717	2.074	2.819	3.792
23	.685	.858	1.060	1.319	1.714	2.069	2.807	3.768
24	.685	.857	1.059	1.318	1.711	2.064	2.797	3.745
25	.684	.856	1.058	1.316	1.708	2.060	2.787	3.725
26	.684	.856	1.058	1.315	1.706	2.056	2.779	3.707
27	.684	.855	1.057	1.314	1.703	2.052	2.771	3.690
28	.683	.855	1.056	1.313	1.701	2.048	2.763	3.674
29	.683	.854	1.055	1.311	1.699	2.045	2.756	3.659
30	.683	.854	1.055	1.310	1.697	2.042	2.750	3.646
35	.682	.852	1.052	1.306	1.690	2.030	2.724	3.591
40	.681	.851	1.050	1.303	1.684	2.021	2.704	3.551
45	.680	.850	1.048	1.301	1.679	2.014	2.690	3.520
50	.679	.849	1.047	1.299	1.676	2.009	2.678	3.496
55	.679	.849	1.047	1.297	1.673	2.004	2.668	3.476

60	.679	.848	1.046	1.296	1.671	2.000	2.660	3.460
70	.678	.847	1.045	1.294	1.667	1.994	2.648	3.435
80	.678	.847	1.044	1.292	1.664	1.990	2.639	3.416
90	.677	.846	1.043	1.291	1.662	1.987	2.632	3.402
100	.677	.846	1.042	1.290	1.660	1.984	2.626	3.390
200	.676	.844	1.039	1.286	1.652	1.972	2.601	3.340
300	.676	.843	1.038	1.285	1.650	1.968	2.592	3.323
400	.676	.843	1.038	1.284	1.649	1.966	2.588	3.315
500	.675	.843	1.037	1.283	1.648	1.965	2.586	3.310
1000	.675	.842	1.037	1.282	1.646	1.962	2.581	3.300
∞	.67449	.84162	1.03643	1.28155	1.64485	1.95996	2.57582	3.29053

Source: "Extended Tables of the Percentage Points of Student's t-Distribution," Enrico T. Federighi, *Journal of the American Statistical Association*, September, 1959, p. 684, and on additional computations.

Acknowledgments of Tables

A: From Hoel and Jessen's *Basic Statistics for Business and Economics*. New York: John Wiley & Sons, Inc., 1971.

B: From Hoel's *Introduction to Mathematical Statistics*. (3rd ed.) New York: John Wiley & Sons, Inc., 1962.

C: From Snedecor and Cochran's *Statistical Method*. (6th ed.) Ames: Iowa State University Press, 1967.

D: Partly from Federighi's "Extended Tables of the Percentage Points of Student's t-Distribution" *Journal of the American Statistical Association*, **54**: 683–688 (September), 1959, supplemented by the author.

Index